John Spears
6910 Shelby St.
Indeanapolis, In. 46227

317-787-0183

EXPLORING ELECTRICITY
TECHNIQUES AND TROUBLESHOOTING

Michael Merchant
Houston Community College

Prentice Hall

Englewood Cliffs, New Jersey   Columbus, Ohio

Library of Congress Cataloging-in-Publication Data

Merchant, Michael.
    Exploring electricity : techniques and troubleshooting / Michael
Merchant.
      p. cm.
    Includes index.
    ISBN 0-02-380555-2
    1. Electric engineering.   I. Title.
TK146.M493   1996
821.319—dc20

95-30312
CIP

Cover photo: © Al Francekevich/The Stock Market
Editor: Dave Garza
Production Coordination: Lisa Garboski
Photo Editor: Monica Ohlinger
Design Coordinator: Jill E. Bonar
Text Designer: Susan E. Frankenberry
Cover Designer: Thomas Mack
Production Manager: Laura Messerly
Marketing Manager: Debbie Yarnell
Illustrations: Academy ArtWorks

This book was set in Times and Frutiger by The Clarinda Company and was
printed and bound by Von Hauffman Press. The cover was printed by Von Hauffman, Inc.

Photo credits: Figure 3.8 B & K Precision; Figure 3.11 a Allen-Bradley, a Rockwell automation business
b Ohmite Manufacturing; Figure 3.12 Allen-Bradley, a Rockwell automation business; Figure 3.16 ClaroStat;
Figure 3.19 a, b, d Eaton Corp; c, e Grayhill, Inc.; Figure 4.10 B & K Precision; Figure 4.11 B & K Precision; Figure 4.20 amprobe B & K Precision; megger Weschler Instruments; wattmeter Weschler Instruments;
kWh meter ABB; Figure 9.3 Exide Corp.; Figure 9.8 Energizer Power Systems; Figure 9.12 Belden Wire and
Cable; Figure 10.4 Goldstar U.S.A., Inc.; Figure 12.4 b Delevan Division of American Precision Industries,
Inc.; Figure 13.39 Leeson Electric Corp.; Figure 17.4 Magnetek, Inc.; Figure 17.23 a Superior Electric; Figure
18.7 Leeson Electric Corp.

Line art credits: Figure 9.29 Used with permission of Leviton Manufacturing Company; Figures 9.13, 9.15,
9.16, 13.16 Use with permission of Premier Industrial Corporation and Newark Electronics; Figures 1-7, 3-4a,
3-6, 3-10a, 3-13, 3-14, 3-22, 3-23, 4-8, 4-10, 4-13, 4-21, 4-24, 9-1a, 9-1b, 9-9, 10-18, 10-19, 10-20, 10-21,
10-22, 10-27, 10-29, 11-1, 11-7, 11-14, 12-3, 12-4, 12-5, 12-6, 12-7, 12-9, 13-8, 13-9, 13-20, 13-21, 13-22,
13-28, 13-29, 13-30, 17-1, 17-3, 17-14, 17-17, 17-20, 17-23 from Floyd, Thomas, Electric Circuit Fundamen-
tals, 2/E, © 1990, pp. 31, 33, 35, 40, 42, 47, 55, 58, 64, 255, 256, 258, 253, 263, 262, 267, 268, 315, 316,
281, 385, 389, 390, 391, 341, 344, 345, 355, 357, 360, 420, 422, 443, 442, 425. Reprinted with permission of
Prentice Hall, Upper Saddle River, NJ.

Printed in the United States of America

10 9 8 7 6 5 4 3 2 1

ISBN: 0-02-380555-2

Prentice-Hall International (UK) Limited, *London*
Prentice-Hall of Australia Pty. Limited, *Sydney*
Prentice-Hall of Canada, Inc., *Toronto*
Prentice-Hall Hispanoamericana, S. A., *Mexico*
Prentice-Hall of India Private Limited, *New Delhi*
Prentice-Hall of Japan, Inc., *Tokyo*
Simon & Schuster Asia Pte. Ltd., *Singapore*
Editora Prentice-Hall do Brasil, Ltda., *Rio de Janeiro*

**To Dr. Kuppaswami,
my greatest teacher.**

PREFACE

Exploring Electricity: Techniques and Troubleshooting is for vocational technical students with no previous technical background who intend to become electronic technicians. The primary goals of this text are to view the introductory material from the student's perspective, and to present the material clearly, simply, and completely. This book is meant to make a career in electronic technology accessible to all who are willing to learn.

The text supports both the student's need to retain the material and the school's need to retain the student by developing the student's self-confidence, sense of progress, and commitment to further training. To serve the needs of students with diversified backgrounds, the text incorporates a number of unique features.

Exploring Electricity: Techniques and Troubleshooting is written in a friendly and informal manner, avoiding academic style, figurative language, and idiomatic expressions. The text is designed to be immediately relevant and useful. Thus, in the first few weeks of the course, the student will each day understand more about electrical circuits and devices in his or her home and automobile, so both students and family members know that the student is gaining solid information about electricity.

Math and formulas are introduced in small doses, often separated by chapters with no new math, to give the student time to absorb and master the math skills. The first math example of any new topic uses small integer values so the student may focus on the electrical relationships without being confused by the arithmetic. The text requires no knowledge of algebra. There are no manipulations of equations with the expectation that the student should be able to follow the steps.

Instead of the usual examples of math problems, the text offers the Skill-Builder, a unique feature that walks the student through the solution of a math problem with margin notes identifying formulas, explaining key steps, and showing calculator keystroke sequences. After the problem is solved, the Skill-Builder offers a set of practice data that allows the student to repeat the Skill-Builder up to five times for immediate reinforcement and retention of the material. The text also incorporates several helpful pedagogical features such as the Math Tip, a sidebox that tutors the student on specific math topics at the point in the text where these topics first appear.

The text enables the student to perform lab experiments as soon as possible. Topics such as meter readings, meter loading effects, polarity, electrical paths and points, and so on are treated in detail to develop the student's comprehension of laboratory situations. In addition, the text offers a unique and thorough presentation on using an oscilloscope with numerous exercises in measuring amplitude, period, and phase.

Organization of the Chapters

Each chapter begins with a troubleshooting situation, which the student will be able to solve after studying the chapter. These examples are plausible, interesting, and amusing, to enliven the material.

An overview of the chapter provides the context and need for the material. The overview leads to the chapter learning objectives, consisting of a small number of essential theoretical and practical skills.

Each chapter section begins with a group of key statements and definitions, highlighted in color, which summarize the section. These statements and definitions are the subject of the Self-Tests at the end of each section. Thus, the answers to the Self-Tests at the end of the chapter constitute the chapter summary.

The end-of-chapter material includes a summary of schematic symbols and formulas, followed by a set of questions to test each of the chapter learning objectives. Next comes the solution to the troubleshooting situation, followed by the chapter summary and glossary. The chapter review questions and problems progress through four levels: true-false, multiple choice, basic math (repetitions of the chapter Skill-Builders), and challenging math (requiring creative application of the material learned in the chapter).

For futher practice, students should consult the accompanying manual to this text, *Experiments for Exploring Electricity*, by Howard B. Brown.

Organization of the Text

The chapters are organized to present the material in a familiar sequence without overwhelming the student in the early chapters. Chapter 1 covers atomic theory and electromagnetic energy. Chapters 2 through 4 (electrical quantities, units, components, calculations, and measurements) focus on preparing the student for basic dc lab experiments. The math in these early chapters is intentionally kept to a minimum to ease the student back into the classroom environment.

Chapters 5 through 7 (series, parallel, and combination circuits) are built around a unique feature known as the *RIPE* table. This is a highly successful classroom-tested method for supporting the student in solving dc circuit problems. Each step in using the *RIPE* table reinforces the student's comprehension of the four basic laws of electrical theory: Ohm's Law, Watt's Law, Kirchhoff's Voltage Law in simplified form, and Kirchhoff's Current Law.

Chapter 8 (circuit theorems) presents Thevenin's Theorem as a basis for understanding voltage regulation and impedance matching. The Superposition Theorem is presented as a basis for understanding how a single current can be a combination of many currents. Networks are presented only to show the limits of series-parallel analysis. The text does not contain any network analysis other than simple Wheatstone bridges.

Chapter 9 offers the student a math-free change of pace before studying alternating current. This chapter on components, tools, and residential wiring will excite students eager for hands-on experience who by now may be exploring electronics stores and magazines for projects and experiments, or wanting to fix minor electrical problems around the house.

Chapter 10 (magnetism and electromagnetism) provides a widely accepted transition into ac theory. Chapter 11 (alternating current) thoroughly defines the key quantities of amplitude, time, and phase. The chapter strongly emphasizes the concept that the sine wave, like all waveforms, is a graph that the student must learn to read.

Chapters 12 (inductance) and 13 (capacitance) present the physical characteristics and ratings of coils and capacitors, and show how these devices produce dc transient response and ac reactance. The chapters also discuss the practical applications of these devices. Chapter 13 (inductive and capacitive circuits) begins with a presentation of trigonometry, including simple, practical methods of applying trigonometry to problem solving. The chapter then presents *RL, RC,* and *RCL* circuits in both series and parallel configurations. The emphasis is on understanding how these circuits affect ac current and phase.

Chapter 15 (ac frequency response) begins by developing the concept of waveforms and harmonics based on a nonmathematical presentation of the Fourier Series. This material, missing from most introductory texts, is the key to understanding a wide range of frequency-selective circuits. The chapter then addresses each of the standard pass and band filters. This is done in a two-level approach: a lower-level analysis of voltage and current values and a higher-level analysis of signal processing. Chapter 16 (dc frequency response) develops the topic of dc pulse waveforms as digital signals, then presents the filters of the previous chapter as pulse modifiers: integrators and differentiators.

Chapters 17 (transformers) and 18 (motors and ac power) end the text with simpler material and a minimum of math. Chapter 18 is optional in a purely electronic program, though most students will find it interesting for its coverage of torque, rpm, horsepower, and the relationships between electrical quantities and well-known heating/cooling quantities such as Btu and ton ratings.

Acknowledgments

I thank the reviewers of this text: William M. Adams, Wentworth Technical School; John Littlefield, Eastern Short Community College; Gordon W. Martin, Isothermal Community College; Michael Lanouette, Southeast College of Technology; Rick Hoover, Owens Technical College; Thomas Gibson, Mohave Community College; Kurt Osadchuk, National Institute of Technology; Edward Waller; Terri L. Valenti, Kent State University; Ernest A. Joerg, SUNY–West Chester Community College. I am deeply grateful for the support, encouragement, and guidance of many people, including my loving wife, Lauretta; my editors and production associates Dave Garza, Steve Helba, Monica Ohlinger, Patty Skidmore, and Lisa Garboski; and my friends and colleagues Lynn and Robert Cale, Bob and Betty Obenhaus, Richard Bemis, Gil Gray, and John Six. Most of all I am grateful to all my students, who taught me how to teach and how to learn.

Mike Merchant

CONTENTS

PREFACE   v

1   BASICS OF ELECTRICITY   xvi

      Section Outline   1
      Troubleshooting: Applying Your Knowledge   1
      Overview and Learning Objectives   1
1–1   Atoms   2
1–2   Electrical Charge   3
1–3   Conductors and Insulators   5
1–4   Charged Objects   8
1–5   The Electric Field   13
1–6   Electromagnetic Energy   16
      Troubleshooting: Applying Your Knowledge   20
      Chapter Summary: Answers to Self-Tests   20
      Chapter Review Questions   21
      Glossary   22

2   ELECTRICAL QUANTITIES AND UNITS   24

      Section Outline   25
      Troubleshooting: Applying Your Knowledge   25
      Overview and Learning Objectives   25
2–1   Work and Energy   26
2–2   Voltage   28
2–3   Current   29
2–4   Resistance   33
2–5   Power   38
2–6   International Numerical Prefixes   44
      Troubleshooting: Applying Your Knowledge   50
      Chapter Summary: Answers to Self-Tests   50
      Chapter Review Questions   51
      Glossary   52

3   CIRCUIT COMPONENTS   54

      Section Outline   55
      Troubleshooting: Applying Your Knowledge   55
      Overview and Learning Objectives   55

3–1   The Electric Circuit   56
3–2   dc Power Supplies   60
3–3   Resistors   61
3–4   Switches   69
3–5   Fuses   72
3–6   Electrical Symbols and Diagrams   74
        Troubleshooting: Applying Your Knowledge   77
        Chapter Summary: Answers to Self-Tests   77
        Chapter Review Questions   77
        Glossary   79

4   ELECTRICAL CALCULATIONS AND MEASUREMENTS   80

        Section Outline   81
        Troubleshooting: Applying Your Knowledge   81
        Overview and Learning Objectives   81
4–1   Ohm's Law   82
4–2   Watt's Law   84
4–3   The Formula Wheel   88
4–4   Multimeters   95
4–5   Measuring Voltage   100
4–6   Measuring Current   104
4–7   Measuring Resistance   105
4–8   Other Electrical Meters and Measurements   107
        Troubleshooting: Applying Your Knowledge   110
        Chapter Summary: Answers to Self-Tests   111
        Chapter Review Questions   111
        Glossary   113

5   SERIES CIRCUITS   114

        Section Outline   115
        Troubleshooting: Applying Your Knowledge   115
        Overview and Learning Objectives   115
5–1   Definition of Series Circuits   116
5–2   Resistance in Series Circuits   116
5–3   Current in Series Circuits   117
5–4   Voltage in Series Circuits   119
5–5   Power in Series Circuits   123
5–6   Polarity in Series Circuits   125
5–7   Multiple Voltage Sources in Series Circuits   126
5–8   The *RIPE* Table in Series Circuits   128
5–9   Ground-Referenced Voltage Readings   131
5–10   Shorts and Opens in Series Circuits   136
5–11   Practical Examples of Series Circuits   142
        Troubleshooting: Applying Your Knowledge   143
        Chapter Summary: Answers to Self-Tests   144
        Chapter Review Questions   144
        Glossary   146

6   PARALLEL CIRCUITS   148

   Section Outline   149
   Troubleshooting: Applying Your Knowledge   149
   Overview and Learning Objectives   149
6–1   Definition of Parallel Circuits   150
6–2   Current in Parallel Circuits   150
6–3   Resistance in Parallel Circuits   154
6–4   Voltage in Parallel Circuits   158
6–5   Power in Parallel Circuits   160
6–6   Multiple Voltage Sources in Parallel Circuits   162
6–7   The *RIPE* Table in Parallel Circuits   163
6–8   Shorts and Opens in Parallel Circuits   166
6–9   Practical Examples of Parallel Circuits   169
   Troubleshooting: Applying Your Knowledge   171
   Chapter Summary: Answers to Self-Tests   172
   Chapter Review Questions   172
   Glossary   173

7   SERIES-PARALLEL CIRCUITS   174

   Section Outline   175
   Troubleshooting: Applying Your Knowledge   175
   Overview and Learning Objectives   175
7–1   Definition of Series-Parallel Circuits   176
7–2   The *RIPE* Table in Series-Parallel Circuits   177
7–3   Opens and Shorts in Series-Parallel Circuits   182
7–4   Loading Effects of Meters   187
   Troubleshooting: Applying Your Knowledge   191
   Chapter Summary: Answers to Self-Tests   191
   Chapter Review Questions   191
   Glossary   000

8   CIRCUIT THEOREMS   194

   Section Outline   195
   Troubleshooting: Applying Your Knowledge   195
   Overview and Learning Objectives   195
8–1   Viewing a Circuit from a Terminal Pair   196
8–2   Internal Resistance   197
8–3   Thevenin's Theorem   201
8–4   Networks   208
8–5   Bridges   210
8–6   Superposition Theorem   213
   Troubleshooting: Applying Your Knowledge   218
   Chapter Summary: Answers to Self-Tests   218
   Chapter Review Questions   218
   Glossary   221

9   CIRCUIT COMPONENTS II   222

        Section Outline   223
        Troubleshooting: Applying Your Knowledge   223
        Overview and Learning Objectives   223
9–1   Cells and Batteries   224
9–2   Wire and Cable   230
9–3   Insulators   242
9–4   Circuit Breakers and Residential Wiring   244
        Troubleshooting: Applying Your Knowledge   252
        Chapter Summary: Answers to Self-Tests   252
        Chapter Review Questions   252
        Glossary   253

10   MAGNETISM AND ELECTROMAGNETISM   254

        Section Outline   255
        Troubleshooting: Applying Your Knowledge   255
        Overview and Learning Objectives   255
10–1   The Magnetic Field   256
10–2   Magnets   259
10–3   Electromagnetism   266
10–4   Magnetic Devices   268
10–5   Magnetic Quantities and Units   271
10–6   Induced Voltage   277
        Troubleshooting: Applying Your Knowledge   282
        Chapter Summary: Answers to Self-Tests   282
        Chapter Review Questions   283
        Glossary   283

11   ALTERNATING CURRENT   284

        Section Outline   285
        Troubleshooting: Applying Your Knowledge   285
        Overview and Learning Objectives   285
11–1   Alternators   286
11–2   The ac Sine Wave   288
11–3   ac Voltage Values   292
11–4   Frequency and Period   300
11–5   Phase   306
        Troubleshooting: Applying Your Knowledge   314
        Chapter Summary: Answers to Self-Tests   315
        Chapter Review Questions   315
        Glossary   317

12   INDUCTANCE   318

        Section Outline   319
        Troubleshooting: Applying Your Knowledge   319

Overview and Learning Objectives   319
12–1   Inductance   320
12–2   Inductors   322
12–3   Inductive Opposition to dc Current: Transient Time   327
12–4   Inductive Opposition to ac Current: Reactance   334
Troubleshooting: Applying Your Knowledge   341
Chapter Summary: Answers to Self-Tests   341
Chapter Review Questions   341
Glossary   345

13   CAPACITANCE   344

Section Outline   345
Troubleshooting: Applying Your Knowledge   345
Overview and Learning Objectives   345
13–1   Capacitance   346
13–2   Capacitors   349
13–3   Capacitors in Series and Parallel   364
13–4   Effect of Capacitance on Current   369
13–5   Capacitor Applications   377
13–6   Testing and Troubleshooting Capacitors   380
Troubleshooting: Applying Your Knowledge   384
Chapter Summary: Answers to Self-Tests   384
Chapter Review Questions   385
Glossary   386

14   ac INDUCTIVE AND CAPACITIVE CIRCUITS   388

Section Outline   389
Troubleshooting: Applying Your Knowledge   389
Overview and Learning Objectives   389
14–1   Trigonometry   390
14–2   Series *RL* Circuits   399
14–3   Series *RC* Circuits   405
14–4   Series *RLC* Circuits   410
14–5   Parallel *RL* Circuits   417
14–6   Parallel *RC* Circuits   423
14–7   Series *RLC* Circuits   427
Troubleshooting: Applying Your Knowledge   434
Chapter Summary: Answers to Self-Tests   434
Chapter Review Questions   435
Glossary   437

15   ac FREQUENCY RESPONSE   438

Section Outline   439
Overview and Learning Objectives   439
15–1   Waveforms   440

15–2   Frequency Bands   441
15–3   Filters   445
15–4   ac Frequency Response: *RL* and *RC* Pass Filters   452
15–5   ac Frequency Response: Resonant Band Filters   462
        Chapter Summary: Answers to Self-Tests   479
        Chapter Review Questions   479
        Glossary   481

16   dc FREQUENCY RESPONSE   482

        Section Outline   483
        Troubleshooting: Applying Your Knowledge   483
        Overview and Learning Objectives   483
16–1   Pulses   484
16–2   Integration   488
16–3   Differentiation   493
16–4   Using Pulses to Test Circuits   500
        Troubleshooting: Applying Your Knowledge   502
        Chapter Summary: Answers to Self-Tests   503
        Chapter Review Questions   503
        Glossary   503

17   TRANSFORMERS   504

        Section Outline   505
        Troubleshooting: Applying Your Knowledge   505
        Overview and Learning Objectives   505
17–1   Mutual Inductance   506
17–2   Basic Transformer Characteristics   508
17–3   Transformer Applications   518
17–4   Transformer Variations   521
17–5   Transformer Transformers   529
        Troubleshooting: Applying Your Knowledge   531
        Chapter Summary: Answers to Self-Tests   531
        Chapter Review Questions   532
        Glossary   533

18   ELECTRIC MOTORS AND ac POWER   534

        Section Outline   534
        Troubleshooting: Applying Your Knowledge   535
        Overview and Learning Objectives   535
18–1   Principles of Electric Motors   536
18–2   Types of Electric Motors   544
18–3   ac Power   548
18–4   Miscellaneous Power Ratings   559
        Troubleshooting: Applying Your Knowledge   561
        Chapter Summary: Answers to Self-Tests   561

Chapter Review Questions   562
Glossary   563

ANSWERS TO CHAPTER REVIEW QUESTIONS   564

INDEX   569

1

BASICS OF ELECTRICITY

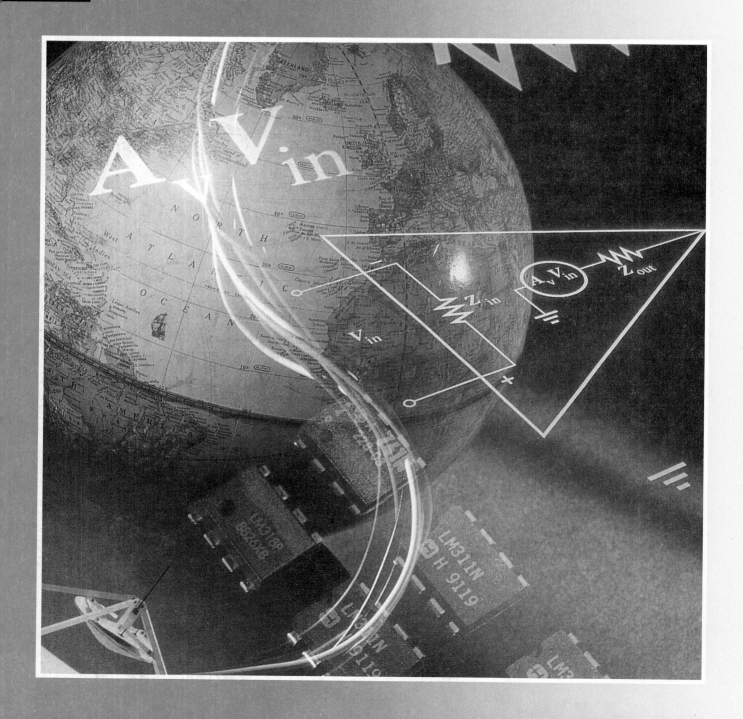

● SECTION OUTLINE

1–1 Atoms

1–2 Electrical Charge

1–3 Conductors and Insulators

1–4 Charged Objects

1–5 The Electric Field

1–6 Electromagnetic Energy

● TROUBLESHOOTING: Applying Your Knowledge

Your friend uses a microcomputer at his office. Most of the time the computer works fine, but occasionally when he returns to his desk and puts in a diskette that was good when he left, he finds the computer cannot read the information on the diskette. He has taken the computer into the repair shop several times, and they cannot find anything wrong with it. He is losing a lot of time and work, and he doesn't know what to do. Can you think of anything that might be causing his problem?

● OVERVIEW AND LEARNING OBJECTIVES

The electrical force is one of the basic forces of nature. Without the electrical force, the entire universe would be just a cloud of high-energy particles without shape or form. The electrical force is one of the forces that pulls these particles together to form atoms.

The electrical force is involved in all chemical reactions. It pulls atoms together into molecules, or allows molecules to break apart and form other molecules.

The electrical force produces heat and light which can travel through space. The heat and light from a star, a campfire, or a candle are caused by the electrical force.

In other words, the electrical force maintains a balance of energy among all atoms. The electrical force is extremely strong, and if we disturb this force even slightly, we release a large amount of energy. This is what we call electricity.

After completing this chapter, you will be able to:

1. Describe basic atomic structure.
2. Define electrical charge and state the law of charges.
3. Identify common electrical conductors and insulators, and explain the reason for their electrical characteristics.
4. Explain how static charge is created.
5. Describe the characteristics of an electric field.
6. Name several common forms of electromagnetic energy.

SECTION 1–1: Atoms

In this section you will learn the following:

→ An atom is an arrangement of protons, neutrons, and electrons.
→ Protons and neutrons are located in the center of the atom, an area called the nucleus.
→ Electrons are located in zones at a distance from the nucleus. These zones are called shells.
→ Atoms have different physical and chemical characteristics according to the number of protons, neutrons, and electrons each atom contains.
→ Atoms join together in groups called molecules.

Atomic Structure

The world around us is made of matter that has mass and occupies space in the form of solids, liquids, and gases. The smallest units of matter are called particles. There are many types of particles. The three most important particles are called *proton, neutron,* and *electron.*

An atom is an arrangement of protons, neutrons, and electrons. Natural forces hold the particles in certain positions. The protons and neutrons form a central cluster called the *nucleus* of the atom. Protons and neutrons are heavy compared to electrons and are held together in the nucleus by strong forces.

The electrons are found around the nucleus in zones or regions called *shells.* Atoms may have up to seven shells. Figure 1–1 illustrates the structure of a typical atom.

The shells are fairly close together compared to the distance between the shells and the nucleus. Therefore, all atoms are approximately the same size, regardless of the number of shells. If we could imagine single atoms lined up in a row one inch long, the row would contain approximately 127 million atoms.

The nucleus is extremely small compared to the overall size of the atom, as shown in Figure 1–2. In other words, the interior of the atom is mostly empty space. If we could enlarge an atom until the shells were half a mile across, the nucleus would be somewhere between the size of a pea and a baseball.

Elements and Compounds

Atoms have different physical and chemical characteristics according to the number of protons, neutrons, and electrons each atom contains. The simplest atom is made of one proton, one electron, and no neutron. This basic atom is called hydrogen, the most com-

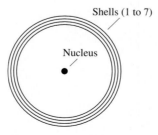

FIGURE 1–1
Nucleus and shells of an atom.

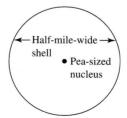

FIGURE 1–2
The nucleus is small compared to the shells.

mon substance in the universe. Table 1–1 lists several familiar substances and gives the number of protons, electrons, and neutrons in these atoms. A normal atom has an equal number of protons and electrons. The number of neutrons given in the table is an average value. The actual number of neutrons in a particular atom may vary.

Atoms join together in groups called molecules. An *element* is a molecule whose atoms are all of the same type. A *compound* is a molecule whose atoms are a mixture of different types.

→ *Self-Test*

1. What is an atom?
2. Where are protons and neutrons located?
3. Where are electrons located?
4. What causes atoms to have different physical and chemical characteristics?
5. What is the name for a group of atoms joined together?

SECTION 1–2: Electrical Charge

In this section you will learn the following:

→ A charged particle produces the electrical force.
→ Protons and electrons are charged particles.
→ An uncharged particle does not produce the electrical force.

TABLE 1–1
Partial list of elements.

ELEMENT	PROTONS	ELECTRONS	NEUTRONS
Hydrogen	1	1	0
Helium	2	2	2
Carbon	6	6	6
Nitrogen	7	7	7
Oxygen	8	8	8
Sodium	11	11	12
Aluminum	13	13	14
Silicon	14	14	14
Calcium	20	20	20
Iron	26	26	30
Cobalt	27	27	32
Nickel	28	28	30
Copper	29	29	34
Zinc	30	30	35
Silver	47	47	61
Gold	79	79	118
Mercury	80	80	120
Uranium	92	92	146

→ Neutrons are uncharged particles.
→ There are two types of electrical charge, called positive and negative.
→ Protons have positive electrical charge.
→ Electrons have negative electrical charge.
→ The law of charges states that like charges repel, and unlike charges attract.
→ Normal atoms contain equal numbers of protons and electrons. Therefore, normal atoms have a neutral charge.

Charged and Uncharged Particles

The electrical force is one of the basic forces of nature. The electrical force is produced by protons and electrons. *Charge* is another word for the electrical force. Thus, because protons and electrons produce the electrical force, we say that protons and electrons are *charged particles,* or that protons and electrons have charge. The neutron does not produce the electrical force; thus, we say the neutron is an uncharged particle.

Protons and electrons each have their own kind of charge. The charge of a proton is called *positive charge*. The charge of an electron is called *negative charge*. The words *positive* and *negative* are just names for the two types of charge. We say that an uncharged particle like the neutron has *neutral charge*. Neutral charge means zero charge. (See Figure 1–3.)

The Law of Charges

The electrical force acts at a distance, and the action is either a push or a pull. This means that charged particles repel or attract each other across an open space without touching. If two charged particles are like each other (the same type), they will repel each other (push apart). If two charged particles are unlike each other (opposite types), they will attract each other (pull together). These two actions are called the *law of charges* and are illustrated in Figure 1–4.

Thus, protons repel other protons, and electrons repel other electrons; but protons and electrons are attracted toward each other. The electrical attraction between protons and electrons is the force that holds the electrons in their shells around the nucleus.

A normal atom contains an equal number of protons and electrons. This produces a balance in the electrical forces, and the atom has an overall charge of zero. Thus, we say an atom normally has neutral charge.

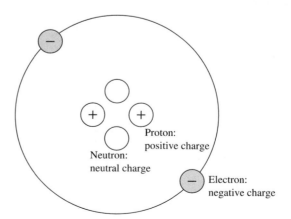

FIGURE 1–3
Charged and uncharged particles.

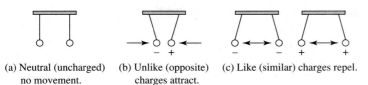

(a) Neutral (uncharged)   (b) Unlike (opposite)   (c) Like (similar) charges repel.
    no movement.   charges attract.

FIGURE 1–4
Attraction and repulsion of electric charges.

→ *Self-Test*

6. What type of particle produces the electrical force?
7. Which particles are charged?
8. What type of particle does not produce the electrical force?
9. Which particles are uncharged?
10. How many types of electrical charge are there, and what are they called?
11. Which particles have positive electrical charge?
12. Which particles have negative electrical charge?
13. What is the law of charges?
14. What is the relationship between protons and electrons in normal atoms, and what effect does this have on the atom's electrical charge?

SECTION 1–3: Conductors and Insulators

In this section you will learn the following:

→ The outermost shell of an atom is called the valence shell.
→ Atoms are most stable with eight electrons in the valence shell and combine with other atoms to achieve this. This tendency is called the Rule of Eight.
→ A free electron is one that has gained enough energy to leave the valence shell. Like all electrons, free electrons have a negative charge.
→ A hole is a vacancy in the valence shell created by the departure of a free electron. A hole acts like a positive charge.
→ An ion is an atom with an unequal number of protons and electrons.
→ A negative ion has extra electrons, and thus has a negative charge.
→ A positive ion has missing electrons, and thus has unfilled holes, which give the ion a positive charge.
→ Free electrons, holes, and ions are called charge carriers.
→ A conductor is a material with a large number of charge carriers, which makes it easy for electrical charges to move through the material.
→ Metals and ionized liquids and gases are conductors.
→ An insulator is a material with a small number of charge carriers, which makes it difficult for electrical charges to move through the material.
→ Nonmetallic solids and nonionized liquids and gases are insulators.

The Valence Shell

The outside shell of an atom is called the *valence shell*. Ordinary chemical reactions involve electrons in the valence shell.

Atoms that contain eight electrons in the valence shell are chemically stable, which means they do not combine with other atoms. Atoms with fewer than eight valence electrons become more stable by attaching themselves to other atoms and sharing valence electrons. This tendency accounts for a large number of chemical reactions. We call this tendency the *Rule of Eight,* which means that atoms prefer to have eight valence electrons. In Figure 1–5, an oxygen atom with six valence electrons has combined with two hydrogen atoms, each with one valence electron. Thus, the oxygen atom "feels" eight electrons in its valence shell. The three atoms have formed a molecule of water.

Free Electrons and Holes

A *free electron* is a valence electron that gains enough energy to leave the shell. Free electrons may be pulled off the valence shell by outside electrical forces, or they may be knocked off by waves of heat and light.

When a free electron leaves the valence shell, it creates a *hole* in the shell. A hole is just a vacant spot in the valence shell, yet it acts like a positive charge as shown in Figure 1–6. When another free electron comes along, the hole may pull the free electron onto the valence shell.

According to the Rule of Eight, an atom that has only one or two valence electrons will easily let go of them. Thus, atoms with only one or two valence electrons easily produce lots of free electrons and holes.

Ions

Normally, an atom has an equal number of protons and electrons. However, because of chemical reactions or external electrical forces, an atom may have an unequal number of electrons and protons. Such an atom is called an *ion.*

Since protons stay in the nucleus, ions are atoms that have gained or lost electrons. An ion with extra valence electrons has an overall negative charge and is called a *negative ion.* An ion with missing valence electrons has an overall positive charge and is called a *positive ion.*

Figure 1–7 illustrates the formation of ions. Salt molecules are made of sodium atoms and chlorine atoms. When salt dissolves in water, the molecule breaks apart. The chlorine atom keeps an electron that really belongs to the sodium atom. Thus, the chlorine atom has an extra electron and becomes a negative ion. The sodium atom has lost an electron and becomes a positive ion.

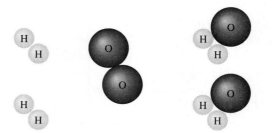

(a) Two molecules of hydrogen and one molecule of oxygen.   (b) Two molecules of water.

FIGURE 1–5
Oxygen and hydrogen molecules recombine to form molecules of water.

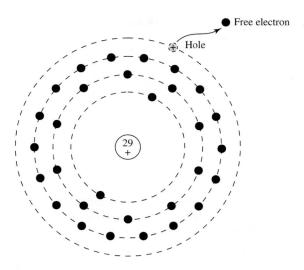

FIGURE 1–6
A free electron leaves the valence shell of a
copper atom, creating a hole in the valence
shell. The holes acts like a positive charge.

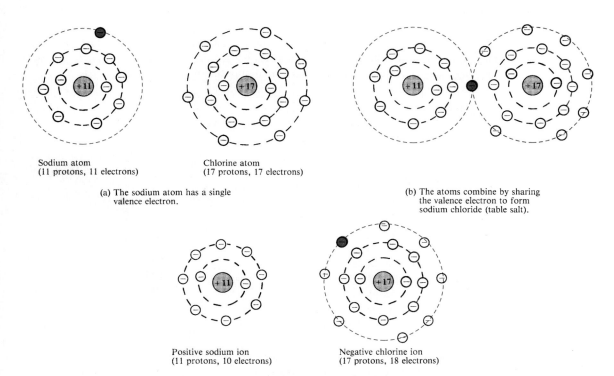

Sodium atom
(11 protons, 11 electrons)

Chlorine atom
(17 protons, 17 electrons)

(a) The sodium atom has a single
valence electron.

(b) The atoms combine by sharing
the valence electron to form
sodium chloride (table salt).

Positive sodium ion
(11 protons, 10 electrons)

Negative chlorine ion
(17 protons, 18 electrons)

(c) When dissolved, the sodium atom gives up the valence electron to become a positive
ion, and the chlorine atom retains the extra valence electron to become a negative ion.

FIGURE 1–7
(a) The sodium atom has a single valence electron. (b) The atoms combine by sharing the valence
electron to form sodium chloride (table salt). (c) When dissolved, the sodium atom gives up the valence
electron to become a positive ion, and the chlorine atom retains the extra electron to become a
negative ion.

Electrical Conductors

Metals such as copper, aluminum, gold, and silver have atoms that contain one or two valence electrons. At ordinary temperatures, there is enough heat to produce a large number of free electrons. The abundance of free electrons allows electrical charges to pass easily through the material. Thus, we use these metals to make electrical wires and connectors.

Electrical charges will also pass easily through ionized liquids (such as salt water) and ionized gases, since the ions have charge and can move around easily. A material that allows electrical charge to pass easily through it is classified as an electrical *conductor.* Therefore, metals and ionized liquids or gases are conductors.

When metals are heated in a vacuum, free electrons jump completely off the metal and float in space near the surface. This is called thermionic emission. A positively charged object in another location in the vacuum will attract these electrons, and they will move across the empty space in a stream called an electron beam. The electron beam acts as a conductor and is used in vacuum tubes and television picture tubes.

Electrical Insulators

Atoms with six or seven valence electrons are likely to combine with other atoms to complete their valence shells and form stable molecules, which produce only a small number of free electrons. Therefore, electrical charges have great difficulty in passing through nonmetallic solids, such as glass, rubber, plastic, stone, and wood, and nonionized liquids and gases, such as distilled water and air. Thus, all these materials are classified as electrical *insulators.*

Free electrons, holes, and ions are often called *charge carriers* because they allow charge to pass easily through materials. Thus, we say that a conductor has a large number of charge carriers, and an insulator has a small number of charge carriers.

→ *Self-Test*

15. What is the valence shell?
16. What condition causes atoms to be most stable, and what is the name for this tendency?
17. What is a free electron and what is its electrical charge?
18. What is a hole and what is its electrical charge?
19. What is an ion?
20. Describe a negative ion and explain its electrical charge.
21. Describe a positive ion and explain its electrical charge.
22. What is a charge carrier?
23. What is a conductor?
24. Name some examples of conductors.
25. What is an insulator?
26. Name some examples of insulators.

SECTION 1–4: Charged Objects

In this section you will learn the following:

→ When an object gains electrons, it becomes negatively charged. When an object loses electrons, it becomes positively charged.

→ Electrical polarity describes the direction charge carriers would move from one charged point to another.

→ Charge (Q) is measured in units of coulombs (C).
→ One coulomb is the charge of approximately 6.25 billion billion electrons.
→ An insulator may become charged by rubbing.
→ Both insulators and conductors may become charged by applying an external voltage.
→ A charged object will induce a charge into a nearby uncharged object.

If we ask ourselves whether an object has an electrical charge, there are three possible answers:

1. If the object contains equal numbers of protons and electrons, then the object is electrically neutral or uncharged.
2. If the object contains extra electrons, then the object is negatively charged.
3. If the object contains extra holes because of missing electrons, then the object is positively charged.

Electrical charge is created by moving electrons from one object to another, as illustrated in Figure 1–8. The object that loses electrons becomes positively charged, and the object that gains electrons becomes negatively charged.

For example, the battery in your car has two terminals. The acid in the battery has taken electrons away from the positive terminal and forced the electrons onto the negative terminal. (See Figure 1–9.) This difference in charge between the two terminals gives the battery its voltage.

Polarity

If two points have different amounts of charge, charge carriers will move from one point to the other. To understand which way the charge carriers would move, we speak of the electrical *polarity* of the two points.

If the two points have opposite charges, the polarity is obvious. In Figure 1–10, point A is missing four valence electrons and point B is missing two valence electrons. Therefore, points A and B both have positive charge. These electrons have been transferred, two to point C and four to point D. Therefore, points C and D both have negative charge.

If we compare point A to point D, we see that a free electron (negative charge) would move away from point D toward point A. Therefore, we say that point A has positive polarity and point D has negative polarity. In a similar way, if we compare point B to point C, then point B has positive polarity and point C has negative polarity.

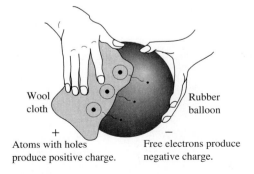

Wool cloth

Rubber balloon

+

−

Atoms with holes produce positive charge.

Free electrons produce negative charge.

FIGURE 1–8
When the rubber balloon and wool cloth are rubbed together, electrons pass from the wool to the rubber, giving both objects an electrical charge.

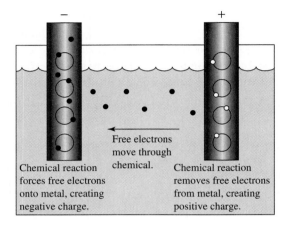

FIGURE 1–9
Chemical reactions in an auto battery
produce charge on the metal plates.

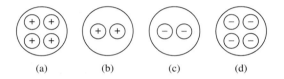

(a)   (b)   (c)   (d)

FIGURE 1–10
Relative polarity.

However, if a free electron were placed between point *A* and point *B,* it would move away from *B* toward *A* because *A* is more positive than *B.* Therefore, if we ignore *C* and *D* and look only at *A* and *B,* we would have to say that although point *B* actually has a positive charge, its polarity is negative compared to point *A* because a negative charge carrier would move away from *B* to *A.*

By similar reasoning, a free electron placed between point *C* and point *D* would move away from *D* toward *C* because *D* is more negative than *C.* Therefore, although point *C* actually has negative charge, its polarity is positive compared to point *D.*

In other words, charge is an absolute condition; but polarity is relative, one point compared to another. There are many situations in electrical and electronic theory where you must understand this important distinction between charge and polarity. Charge describes an accumulation of positive or negative charge carriers at a point. Polarity describes the direction charge carriers would move between two points.

 PRACTICAL TIP

When working with any electrical device or machine, you must always ask yourself whether the device has polarity. In other words, you want to know whether it matters which way you connect the device to the circuit. Any device that has polarity, such as a battery, diode, or transistor, will not work if you connect it the wrong way, and may cause damage to the circuit when you turn on the power.

The Coulomb

The symbol for charge is the letter Q. The unit of charge is the *coulomb (C)*.

The coulomb is a unit of quantity. A coulomb is a specific amount of charge, just as a gallon is a specific amount of liquid. One coulomb is the amount of charge produced by approximately 6.25 billion billion electrons. (See Figure 1–11.)

Static Charge

Static charge, often called static electricity, is charge that builds up on the surface of a material. When two materials rub together, friction produces heat. The heat gives a few electrons enough energy to leave their valence shells and become free electrons.

If either material is a conductor, the free electrons will move freely through the material. No charge will build up on the surfaces.

However, if the two materials are insulators, free electrons will move from one surface to the other and remain trapped on the surface, since charge cannot easily pass through insulators. Thus, a difference in charge builds up on the two surfaces.

If the two insulators are different materials, the buildup of charge is greater because of differences in the molecular structures of the materials. In the eighteenth century, scientists learned how to list materials according to their ability to develop static charge when rubbed together. If two materials were near each other on the list, the static charge between them would be small. On the other hand, if the materials were far apart on the list, the static charge would be greater.

Static charge is more likely to build up when the air is dry because moisture in the air allows the charge to escape from the surfaces. Thus, on a dry winter's day, when you walk across a rug and rub your shoes against the rug, a static charge may build up in your body. When you touch a doorknob, the charge flows between your hand and the doorknob, giving you an electrical shock.

When you comb long hair with a rubber or plastic comb, charge builds up both on the comb and on the strands of hair. Since like charges repel each other, the strands of hair stand up and repel each other, producing flyaway hair.

When you put wool and polyester fabrics in a clothes dryer, the rubbing and dry air produce static charges in the fabrics. Since unlike charges attract each other, the wool and polyester fabrics will stick to each other, and you will hear the crackle of tiny electrical sparks when you pull the fabrics apart.

Charge Transfer

If a charged object touches an uncharged object, some of the charge will transfer, giving both objects the same type of charge. Since a voltage is the result of an accumulation of

1 coulomb =
the charge of
6,250,000,000,000,000,000
(6.25 billion billion)
electrons

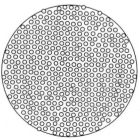

FIGURE 1–11
The coulomb.

charge, applying a voltage to an object allows some of the accumulated charge to transfer into the object.

For example, a battery produces a charge in its metal terminals from the chemical action inside the battery. When the chemical action becomes weak, we say the battery is dead. Depending on the type of battery, we may be able to recharge the battery. A battery charger forces electrons into the battery and restores the chemical to its original condition, allowing the chemical to create charge and produce voltage once again.

Induced Charge

If a charged object is brought near an uncharged object but does not touch it, the electric force will pull any charge carriers of the opposite type to the surface of the uncharged object and repel into the interior any charge carriers of a similar type. Thus, the uncharged object will act as if it were charged, as shown in Figure 1–12. Since the surface charges are of opposite types, the two objects will attract each other. When the charged object is removed, the uncharged object seems to lose its charge. We describe all this by saying that the uncharged object receives an *induced charge* from the charged object.

For example, the glass screen of a color television is charged while the television is on but loses the charge when the television is turned off. The screen is charged by a high voltage inside the television. If you hold your arm near the screen while the television is on, the hair on your arm will stand up and be drawn toward the screen. In other words, the charge in the screen has induced a charge into the hair. When you step back from the screen, the hair loses its charge.

Beneficial Uses of Static Charges

Static charges can be put to good use. For example, in spray-painting automobiles, we can give opposite charges to the automobile body and the tiny particles of paint. When the paint particles reach the metal, the opposite electrical charges cause them to stick tightly, producing a much better coat of paint.

Electrostatic precipitators in central air-conditioning systems use static charges to remove dust from the air. Fans blow the air across charged rods, which transfer charge to the dust particles. The air then moves past plates of opposite charge, which pull the dust particles out of the air.

If a charged object has a tip or sharp point, the charge is more intense at the tip. In making coated abrasives such as sandpaper, the abrasive particles are given a static charge to make them stand up on their sharp points when they are glued to the paper.

In copiers and laser printers, a beam of light creates a pattern of static charges on a drum made of selenium, a light-sensitive metal. The static charges on the drum are transferred to the blank paper, attracting toner particles to the paper. The paper then passes through heated rollers, which melt the particles into the paper to produce a permanent image.

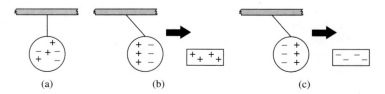

(a)   (b)   (c)

FIGURE 1–12
(a) Uncharged ball with charge carriers randomly distributed.
(b) Positively charged rod induces charge in ball and attracts it.
(c) Negatively charged rod has same effect.

Hazards of Static Charge

When flammable vapors are in the air, sparks from static discharge can cause fires or explosions. For example, static charge builds up in gasoline tanker trucks from the rubber tires rubbing against the road. To prevent sparks while the truck is pumping gasoline into a storage tank, the driver first discharges the truck by connecting it through a wire to a metal rod or pipe that goes into the ground.

Computer diskettes and delicate electronic circuits can be ruined by static discharges from a person's body. To prevent this, electronic technicians wear a special strap on the wrist to drain away static charge while working on delicate circuits. For the same reason, computers often are placed on a special mat which drains away static charge. Both the wrist strap and the mat have a wire which attaches to a grounded point in the building's wiring.

→ *Self-Test*

27. What causes an object to become negatively or positively charged?
28. What is electrical polarity?
29. What is the symbol of charge? What is the measurement unit of charge? What is the unit symbol?
30. What is the definition of one coulomb?
31. What is the effect of rubbing an insulator?
32. What is the effect of applying an external voltage to either an insulator or conductor?
33. What is the effect of bringing a charged object near an uncharged object?

SECTION 1–5: The Electric Field

In this section you will learn the following:

→ An electric field is a region in space where an electrical force exists. At every point in the field, the force has a certain strength and direction.

→ An electric line of force is a method of visualizing the strength and direction of an electric field.

→ If a charged object enters an electric field, the object will move along the lines of force within the field.

→ If the surface of a charged object has a sharp point, the electric field is most intense at that point, and sparks and discharges are more likely to occur.

→ Between parallel plates of opposite polarity, an electric field has a constant intensity, represented by lines of force that are straight, parallel, and evenly spaced.

An *electric field* is a region of space around a charged object. The electrical force can be felt within this space. If another charged object enters the field, the object is affected by the electrical forces in the field. The field will cause the charged object to move. For example, in television picture tubes, strong electric fields are used to focus beams of electrons onto the screen, producing a sharp picture.

Lines of Force

We imagine the electric field around a charged object as a space filled with invisible *electric lines of force.* By thinking that an electric field contains lines of force, we are able to visualize the way a charged object would move at that location in the field.

Electric fields begin or end in charged objects. Figure 1–13 illustrates an electric field around two charged objects.

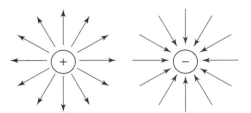

FIGURE 1–13
Invisible lines of force create an electric
field around isolated charges.

The polarity of the field and the object determines the direction of the object's movement. Positive charges would move along the lines in the direction of the arrowheads. Negative charges would move along the lines against the arrowheads.

The electric field is more intense at the surface of a charged object than in the space around the object. The lines of force are always perpendicular to the surface.

Notice how the lines of force spread out away from the object. The field is more intense where the lines of force are shown spaced more closely together. This means the electrical force feels stronger at these locations. The spacing of the lines of force shows us that the electrical force is more intense near the object and less intense at a distance.

Field Intensity

Field intensity refers to how concentrated the electrical force is at a specific location within an electric field. We have just learned that the electric field has its greatest intensity at the surface of a charged object. If the object has a smooth shape, the electric field will spread out evenly along the surface. However, if the surface has a sharp point, the electric field will become intense at that point. Sparks and discharges are more likely to occur at a sharp point than anywhere else on the surface of a charged object, as shown in Figure 1–14. Devices that produce strong electric fields, such as television picture tubes and automobile ignition wires, must be smooth and free of sharp points to avoid sparks and discharges.

Occasionally, the electric field near a sharp point is intense enough to ionize air molecules and cause them to glow. This faint light is called a corona, and thus the electrical discharges that may occur around a sharp point are called corona discharges.

The concept of electric field intensity helps us understand how a thunderstorm produces lightning bolts. The turbulent air in the clouds of a thunderstorm ionizes the air by

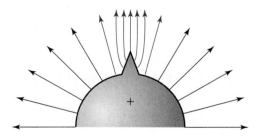

FIGURE 1–14
The electric field is more intense around
a sharp point projecting above the
surface of a charged object.

friction. Charges accumulate in the clouds and create a complex electric field. When the field becomes intense enough, the ionized air molecules allow the charges to move, discharging the field. The energy of the moving charges heats the air, producing the visible light we see as a lightning bolt.

In this process, lightning often strikes the ground. Buildings, trees, and people standing in an open field all create sharp points on the earth's surface. The electric field between earth and cloud is more intense at these points, and the lightning is more likely to strike there. This is why you are advised to stay away from open spaces and trees if you are caught outside during a thunderstorm.

Lightning rods on houses use the principle of the electric field being most intense at a sharp point. If a house has a lightning rod, the lightning is more likely to strike the rod than the house, as shown in Figure 1–15. A wire from the rod carries the discharge safely into the ground, protecting the house from being set on fire from the heat of the lightning.

The electric field between two parallel plates of opposite polarity is particularly interesting and important, as shown in Figure 1–16. A plate is just a flat piece of metal. The plates are charged by connecting them to a battery. The electric field between the plates consists of parallel lines of force.

In this field, the lines of force are evenly spaced, showing that the field has equal intensity at all locations. The arrows show that a positive charge placed between the plates would move away from the positive plate and toward the negative plate. An electric field between parallel plates is found in many electrical devices such as capacitors and picture tubes.

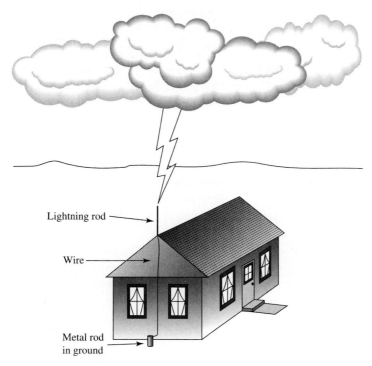

FIGURE 1–15
The lightning strikes the rod because the electric field is more intense at that point. The heavy wire carries the discharge safely into the ground.

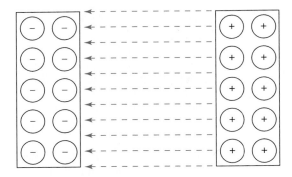

FIGURE 1–16
Electric field between parallel plates.

→ *Self-Test*

34. What is an electric field, and what are its two principal characteristics?
35. What is an electric line of force?
36. What happens to a charged object that enters an electric field?
37. What are the consequences of a sharp point on the surface of a charged object?
38. Describe the electric field between parallel plates of opposite polarity.

SECTION 1–6: Electromagnetic Energy

In this section you will learn the following:

→ Electromagnetic energy, such as magnetism, heat, and light, is produced by the movement of charged particles, particularly electrons.

→ Electromagnetic energy that travels in waves or rays through space is called electromagnetic radiation.

→ The wavelength of an electromagnetic beam is the distance between each wave in the beam.

→ The frequency of an electromagnetic beam is the number of waves per second arriving in the beam.

→ The speed of an electromagnetic wave is the speed of light, approximately 186,000 miles per second.

→ When atoms absorb electromagnetic energy, electrons jump to higher shells, or leave the atom entirely as free electrons.

→ When electrons fall back onto the atom, or fall to lower shells, atoms release electromagnetic energy as heat or light.

→ Infrared waves are commonly called heat waves. Infrared waves are light waves with lower frequencies (longer wavelengths) than visible light.

→ Ultraviolet light is an ionizing radiation capable of producing chemical changes in materials, plants, and animals. Ultraviolet light has higher frequencies (shorter wavelengths) than visible light.

The Electromagnetic Force

Electricity and magnetism are closely related and are considered to be the same force, called the electromagnetic force. Electromagnetic forces are produced by the movements of charged particles, especially electrons. Energy released by electromagnetic forces is called *electromagnetic energy*.

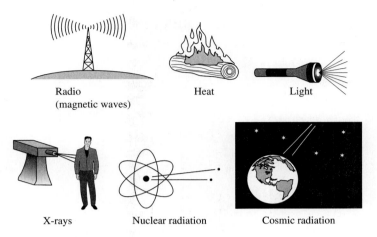

Radio
(magnetic waves)   Heat   Light

X-rays   Nuclear radiation   Cosmic radiation

FIGURE 1–17
Examples of electromagnetic radiation.

Electromagnetic Radiation

When electromagnetic energy travels in waves or rays through space, it is called *electromagnetic radiation*. The most common forms of electromagnetic radiation, shown in Figure 1–17, include radio waves (which are actually waves of magnetism), heat, light, X-rays, nuclear radiation, and cosmic rays from outer space.

Wavelength and Frequency

The *wavelength* is the distance between each wave in a beam of electromagnetic radiation. Typical wavelengths vary from a few meters for magnetic radio waves to a small fraction of a millimeter for heat and light waves. Differences in wavelength give visible light its different colors.

The *frequency* is the number of waves per second arriving in the beam. We are able to tune in particular radio stations because they are sending out waves of different frequencies. Radio waves have frequencies of millions of waves per second, and heat and light waves have frequencies of trillions of waves per second.

The relationship between wavelength and frequency is illustrated in Figure 1–18. A lower frequency has a longer wavelength. The energy is vibrating at a slower rate and, thus, the individual waves are farther apart. Similarly, a higher frequency has a shorter wavelength. The energy is vibrating at a faster rate, and the individual waves are closer together.

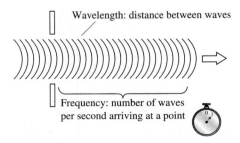

Wavelength: distance between waves

Frequency: number of waves
per second arriving at a point

FIGURE 1–18
Wavelength and frequency.

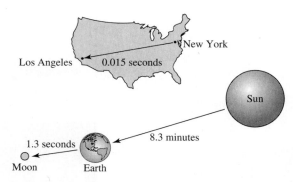

FIGURE 1–19
The speed of light.

The Speed of Light

Electromagnetic waves travel at the *speed of light,* approximately 186,000 miles per second. Thus, in a telephone call between New York and Los Angeles, it takes about 0.015 seconds for the electrical signal to travel across the United States. A radio signal from the earth takes about 1.3 seconds to reach the moon, and the light from the sun takes about 8.3 minutes to reach the earth. (See Figure 1–19.) According to present scientific theory, nothing can travel faster than the speed of light.

Electromagnetic Waves and Atoms

Electromagnetic energy is closely related to movements of charged particles, particularly electrons. For example, when a wave of heat or light hits an atom, the atom absorbs the energy. Some of the electrons jump into shells farther from the nucleus. We say that the electrons are *excited* by the radiation. When these electrons fall back into their normal shells, the electrons release energy by sending out waves of heat and light, which may be of a different wavelength than the original waves the atom absorbed.

When electrical charge moves through materials and collides with atoms, the atoms become excited, releasing waves of heat and light. Therefore, electrical circuits, machines, and devices such as electric heaters and ordinary light bulbs produce a mixture of heat and light, as shown in Figure 1–20.

There are a number of electrical devices which produce light without much heat, such as LEDs (light-emitting diodes), used as indicator lights in electronic equipment; phosphors, which are chemical coatings used to produce light in television screens and fluorescent lamps; and gas-filled lamps such as neon, sodium vapor, and mercury vapor. These devices use materials whose atoms release electromagnetic energy as light waves of a specific color rather than as heat waves.

When certain crystals absorb electromagnetic energy, the excited atoms release intense waves of light of a pure color, which allow the light to be focused into a narrow beam. This is the basic principle of a laser.

Infrared and Ultraviolet Waves

The main difference between heat waves and visible light waves is the frequency at which they vibrate. Heat waves have lower frequencies (longer wavelengths) than visible light and are called *infrared* waves. We have cameras and films which are sensitive to infrared waves, allowing scientists and military personnel to see heat-producing objects in the dark.

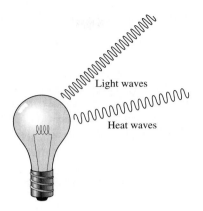

FIGURE 1–20
Filament of an incandescent
lamp produces a mixture of
heat and light.

Many electronic devices are sensitive to infrared energy rather than visible light. For example, infrared light is the invisible beam used in hand-held remote controls and automatic door openers.

At the other end of the scale, electromagnetic waves with frequencies higher than visible light contain enough energy to ionize atoms, causing chemical changes in materials. In the case of living plants and animals, the chemical changes produced by this ionizing radiation can be harmful or deadly. Thus, we must protect our bodies and our planet from ionizing radiation, such as X-rays, nuclear radiation, and cosmic rays.

Ultraviolet light is a type of ionizing radiation. Ultraviolet light has higher frequencies (shorter wavelengths) than visible light, and extends into the X-ray range. Sunlight contains a large amount of ultraviolet light. The ultraviolet (UV) component of sunlight produces chemical changes that cause suntans and sunburns in our skin. UV light also causes colors to fade in paint, ink, and fabrics and causes rubber, plastic, and synthetic fabrics to crack and crumble.

UV light has many beneficial industrial applications, including curing (drying and hardening) ink and rubber, and sterilizing (killing harmful bacteria).

Some electronic devices are sensitive to ultraviolet light. For example, there is a certain type of computer chip called an EPROM, which means erasable programmable read-only memory. The chip stores a large amount of information and holds the information even when the power is turned off. The chip has a small window on top. Shining UV light through this window affects the material inside the chip and erases the information so the chip can be reprogrammed.

→ *Self-Test*

39. How is electromagnetic energy produced? What are some common examples of electromagnetic energy?
40. What is electromagnetic radiation?
41. What is the wavelength of an electromagnetic beam?
42. What is the frequency of an electromagnetic beam?
43. What is the speed of an electromagnetic wave?
44. What happens when atoms absorb electromagnetic energy?
45. What happens when atoms release electromagnetic energy?
46. What are infrared waves?
47. What is ultraviolet light?

● TROUBLESHOOTING: Applying Your Knowledge

Ask your friend whether he has a thick carpet in his office, and what kind of shoes he wears. It is possible that when he returns to his desk, static charge has built up in his body from walking across the carpet. When he picks up the diskette, a static discharge may occur that damages the diskette, making it impossible for the computer to read it.

If static discharge is the problem, the solution would be to place the computer and keyboard on a conductive mat grounded to the electrical outlet, and touch the mat to discharge the body before handling anything else.

CHAPTER SUMMARY: ANSWERS TO SELF-TESTS

1. An atom is an arrangement of protons, neutrons, and electrons.
2. Protons and neutrons are located in the center of the atom, an area called the nucleus.
3. Electrons are located in zones at a distance from the nucleus. These zones are called shells.
4. Atoms have different physical and chemical characteristics according to the number of protons, neutrons, and electrons each atom contains.
5. Atoms join together in groups called molecules.
6. A charged particle produces the electrical force.
7. Protons and electrons are charged particles.
8. An uncharged particle does not produce the electrical force.
9. Neutrons are uncharged particles.
10. There are two types of electrical charge, called positive and negative.
11. Protons have positive electrical charge.
12. Electrons have negative electrical charge.
13. The law of charges states that like charges repel, and unlike charges attract.
14. Normal atoms contain equal numbers of protons and electrons, and therefore have a neutral charge.
15. The outermost shell of an atom is called the valence shell.
16. Atoms are most stable with eight electrons in the valence shell and combine with other atoms to achieve this. This tendency is called the Rule of Eight.
17. A free electron is one that has gained enough energy to leave the valence shell. Like all electrons, free electrons have a negative charge.
18. A hole is a vacancy in the valence shell created by the departure of a free electron. A hole acts like a positive charge.
19. An ion is an atom with an unequal number of protons and electrons.
20. A negative ion has extra electrons, and thus has a negative charge.
21. A positive ion has missing electrons, and thus has unfilled holes, which give the ion a positive charge.
22. Free electrons, holes, and ions are called charge carriers.
23. A conductor is a material with a large number of charge carriers, which makes it easy for electrical charges to move through the material.
24. Metals and ionized liquids and gases are conductors.
25. An insulator is a material with a small number of charge carriers, which makes it difficult for electrical charges to move through the material.
26. Nonmetallic solids and nonionized liquids and gases are insulators.
27. When an object gains electrons, it becomes negatively charged. When an object loses electrons, it becomes positively charged.
28. Electrical polarity describes the direction charge carriers would move from one charged point to another.
29. Charge *(Q)* is measured in units of coulombs *(C)*.
30. One coulomb is the charge of approximately 6.25 billion billion electrons.
31. An insulator may become charged by rubbing.
32. Both insulators and conductors may become charged by applying an external voltage.
33. A charged object will induce a charge into a nearby uncharged object.
34. An electric field is a region in space where an electrical force exists. At every point in the field, the force has a certain strength and direction.
35. An electric line of force is a method of visualizing the strength and direction of an electric field.
36. If a charged object enters an electric field, the object will move along the lines of force within the field.
37. If the surface of a charged object has a sharp point, the electric field is most intense at that point, and sparks and discharges are more likely to occur.
38. Between parallel plates of opposite polarity, an electric field has a constant intensity, represented by lines of force that are straight, parallel, and evenly spaced.
39. Electromagnetic energy, such as magnetism, heat, and light, is produced by the movement of charged particles, particularly electrons.
40. Electromagnetic energy that travels in waves or rays through space is called electromagnetic radiation.
41. The wavelength of an electromagnetic beam is the distance between each wave in the beam.
42. The frequency of an electromagnetic beam is the number of waves per second arriving in the beam.
43. The speed of an electromagnetic wave is the speed of light, approximately 186,000 miles per second.
44. When atoms absorb electromagnetic energy, electrons jump to higher shells, or leave the atom entirely as free electrons.
45. When electrons fall back onto the atom or fall to lower shells, atoms release electromagnetic energy as heat or light.
46. Infrared waves are commonly called heat waves. Infrared

waves are light waves with lower frequencies (longer wavelengths) than visible light.

47. Ultraviolet light is an ionizing radiation capable of producing chemical changes in materials, plants, and animals. Ultraviolet light has higher frequencies (shorter wavelengths) than visible light.

CHAPTER REVIEW QUESTIONS

Determine whether the following questions are true or false.

1–1. The protons and neutrons form a central cluster called the shell of the atom.

1–2. Atoms may have only one shell.

1–3. All atoms are approximately the same size.

1–4. The interior of the atom is mostly empty space.

1–5. An element is a molecule whose atoms are all of the same type.

1–6. Protons repel neutrons.

1–7. An atom normally has positive charge.

1–8. Atoms prefer to have a total of eight electrons.

1–9. Ions have an unequal number of protons and electrons.

1–10. Free electrons, holes, and ions are charge carriers.

1–11. An electric field is more intense where the lines of force are spaced more closely together.

1–12. Heat and light are forms of electromagnetic radiation.

Select the response that best answers the question or completes the statement.

1–13. The electrons are held in the shells by
  a. electromagnetic radiation.
  b. the nucleus.
  c. electrical forces.
  d. nuclear forces.

1–14. The charged particles in an atom are
  a. only the protons.
  b. only the electrons.
  c. protons, electrons, and neutrons.
  d. protons and electrons.

1–15. Atoms join together in groups called
  a. ions.
  b. molecules.
  c. fields.
  d. charges.

1–16. The type of electrical force produced by the proton is called
  a. positive charge.
  b. negative charge.
  c. nuclear charge.
  d. neutral charge.

1–17. The type of electrical force produced by the electron is called
  a. positive charge.
  b. negative charge.
  c. valence charge.
  d. neutral charge.

1–18. The law of charges states that
  a. like charges attract, and unlike charges repel.
  b. like charges repel, and unlike charges attract.
  c. like charges are negative, and unlike charges are positive.
  d. like charges are positive, and unlike charges are negative.

1–19. Normal atoms contain
  a. more electrons than protons.
  b. more protons than electrons.
  c. equal numbers of protons and electrons.
  d. equal numbers of protons, electrons, and neutrons.

1–20. The outermost shell of any atom
  a. contains equal numbers of electrons and protons.
  b. cannot lose or gain electrons.
  c. is called the valence shell.
  d. contains equal numbers of free electrons and holes.

1–21. If valence electrons gain enough energy,
  a. they may fall into the nucleus.
  b. they may leave the shell as free electrons.
  c. they may disappear into holes.
  d. they may release nuclear forces.

1–22. Atoms with only one or two valence electrons easily produce lots of
  a. molecules.
  b. free electrons and holes.
  c. elements.
  d. compounds.

1–23. A hole acts like
  a. a positive charge.
  b. a negative charge.
  c. a neutral charge.
  d. a zero charge.

1–24. Conductors are materials with
  a. a large number of shells.
  b. a small number of shells.
  c. a small number of charge carriers.
  d. a large number of charge carriers.

1–25. Electrical charge is created by moving _____ from one object to another.
  a. protons
  b. electrons
  c. neutrons
  d. holes

1–26. Charge that builds up on the surface of a material is called
  a. free charge.
  b. valence charge.
  c. electromagnetic charge.
  d. static charge.

1–27. Charge transfer may be produced by
  a. rubbing two insulators together.
  b. applying an external voltage.
  c. touching a charged object to an uncharged object.
  d. all of the above.

1–28. The unit of charge is the
   a. volt.
   b. amp.
   c. meter.
   d. coulomb.

1–29. An electric field is
   a. the space around a nucleus.
   b. the space around a charged object where the electrical force is present.
   c. the space inside a shell.
   d. the space around a ray of light.

1–30. In an electric field, positive charges move
   a. along the lines of force in the direction of the arrowheads.
   b. along the lines of force against the arrowheads.
   c. across the lines of force.
   d. between the lines of force.

1–31. The question of which way charges will move is called
   a. valence.
   b. charge induction.
   c. radiation.
   d. polarity.

1–32. If a charged surface has a sharp point,
   a. no electric field can form on the surface.
   b. the electric field will become intense at that point.
   c. the electric field will become weak at that point.
   d. lightning will strike the point.

1–33. Radiation is
   a. energy at a dangerous level.
   b. electromagnetic energy that moves through space.
   c. nuclear energy not related to electromagnetic energy.
   d. unable to produce ions.

1–34. Electromagnetic forces are produced by the movements of
   a. magnetic particles.
   b. polarized particles.
   c. charged particles.
   d. uncharged particles.

1–35. The frequency of an electromagnetic beam is
   a. the distance between the waves in the beam.
   b. the number of waves per second arriving in the beam.
   c. the speed of the beam.
   d. the ionizing intensity of the beam.

1–36. Electromagnetic energy travels at the speed of
   a. electrons.
   b. atoms.
   c. heat.
   d. light.

1–37. Infrared energy is the same as
   a. radio waves.
   b. heat.
   c. visible light.
   d. X-rays.

1–38. Ultraviolet energy
   a. is found in radio and television waves.
   b. is an ionizing radiation capable of producing chemical changes in materials.
   c. is another name for infrared.
   d. is another name for X-rays.

GLOSSARY

charge   Another word for the electrical force.

charge carriers   A general term for free electrons, holes, and ions, so called because they allow charge to pass easily through materials.

charge transfer   Creating charge in a material by applying an external voltage.

charged particle   A particle that produces the electrical force, usually a proton or an electron.

compound   A molecule whose atoms are a mixture of different types.

conductor   A material with a large number of charge carriers, thus allowing charge to pass through the material easily.

coulomb   The unit of charge; the charge of 6.25 billion billion electrons.

electric field   A region of space surrounding a charged object, wherein the electrical force is felt.

electric line of force   A way of visualizing the movement of a charged object within an electric field.

electromagnetic energy   Energy released by electromagnetic forces, caused by the movements of charged particles.

electromagnetic radiation   Electromagnetic energy that travels through space as waves or beams, such as magnetic radio waves, heat, light, X-rays, nuclear radiation, and cosmic rays.

electron   An atomic particle with negative charge.

element   A molecule whose atoms are all of the same type.

excited   Referring to an atom or electron that has gained extra energy.

field intensity   The degree of concentration of electrical force at a specific location within an electric field.

free electron   An electron that has gained energy and jumped off the valence shell to wander in the space between atoms.

frequency   The number of waves arriving per second in a beam of electromagnetic energy.

hole   A vacancy in a valence shell created by the departure of a free electron.

induced charge   Charge created in an object by a nearby charged object.

infrared   Heat waves at frequencies just below visible light.

insulator   A material with a small number of charge carriers, thus making it difficult for charge to pass through the material.

ion   An atom with an unequal number of protons and electrons; thus, a charged atom.

law of charges   The statement that like charges repel and unlike charges attract.

negative charge   The type of charge produced by an electron.

negative ion   An atom with negative charge due to extra electrons in the valence shell.

neutral charge   An overall charge of zero due to an equal number of positive and negative charges.

neutron   An atomic particle with neutral charge.

nucleus   The central region of the atom, consisting of a cluster of protons and neutrons.

polarity   A way of speaking about the two types of electrical charge.

positive charge   The type of charge produced by a proton.

positive ion   An atom with positive charge due to electrons missing in the valence shell.

proton   An atomic particle with positive charge.

Rule of Eight   The tendency of atoms to seek chemical stability by having eight valence electrons.

shells   Outer zones of the atom where the electrons are located.

speed of light   The speed at which electromagnetic energy travels through space, approximately 186,000 miles per second.

static charge   Charge that builds up on the surface of an insulator.

ultraviolet   Ionizing radiation at frequencies just above visible light.

valence shell   The outside shell of an atom.

wavelength   The distance between each wave in a beam of electromagnetic energy.

2

ELECTRICAL QUANTITIES AND UNITS

● SECTION OUTLINE

2–1 Work and Energy

2–2 Voltage

2–3 Current

2–4 Resistance

2–5 Power

2–6 International Numerical Prefixes

● TROUBLESHOOTING: Applying Your Knowledge

A rock hits your car's headlight, and you buy a new one at the auto parts store. You read the box and notice the headlight is rated at 60 watts. This makes no sense to you because you have a table lamp at home that uses a 60-watt bulb and is plugged into a 120-volt wall outlet. How can the headlight be the same wattage as the table lamp, when the car battery produces only 12 volts?

● OVERVIEW AND LEARNING OBJECTIVES

In this chapter, you will learn about the four basic electrical quantities: voltage, current, resistance, and power. You must understand these quantities and their units of measurement in order to begin your study of electrical circuits. To appreciate how these electrical quantities fit into the larger picture of general science, you will first learn about the basic concepts of energy and work.

In this chapter, you will also learn how to express large and small numbers using an international system called engineering notation. You will often find such numbers in working with voltage, current, resistance, and power.

After completing this chapter, you should be able to:

1. Calculate mechanical work in both British and mks units.
2. Solve basic problems relating units of resistance, voltage, and current.
3. State the four factors that determine resistance, and the effect that a change in these factors has on the resistance of an object.
4. Solve basic problems relating units of wattage, voltage, and current.
5. Solve basic problems relating units of wattage, kilowatt-hours, and horsepower.
6. Convert electrical quantities from standard to international numeric prefix notation, and vice versa.

SECTION 2–1: Work and Energy

In this section you will learn the following:

→ Work equals force multiplied by distance.
→ In the British system of measurement, one foot-pound of work is a force of 1 pound acting through a distance of 1 foot.
→ In the metric system of measurement, 1 newton-meter of work is a force of 1 newton acting through a distance of 1 meter.
→ One joule is the energy necessary to do 1 newton-meter of work. Therefore, joules of energy are equivalent to newton-meters of work.

Work

Work is the result of a force acting through a distance, causing an object to move. We calculate work by multiplying force and distance. Formula Diagram 2–1 expresses this definition.

Formula Diagram 2–1

$$\frac{\text{Work}}{\text{Force} \mid \text{Distance}}$$

The Foot-Pound

In the British system of measurement, the unit of distance is the foot (ft), and the unit of force is the pound (lb). When we multiply these quantities, we get a unit of work called the *foot-pound* (ft-lb).

● *SKILL-BUILDER 2–1*

How much work is done in lifting a 50-pound box off the floor and setting it on a table that is 3 feet high? (See Figure 2–1.)

GIVEN:
  Force = 50 lb
  Distance = 3 ft

FIND:
  Work

SOLUTION:
  Work = Force × Distance   Formula Diagram 2–1
       = 50 lb × 3 ft   substitute values and calculate
       = 150 ft-lb   result

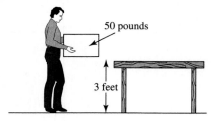

FIGURE 2–1
Work is done when a force acts through a distance.

Repeat the Skill-Builder using the following values. You should obtain the results that follow.

Given:

Force	25 lb	60 lb	75 lb	15 lb	35 lb
Distance	16 ft	4 ft	8 ft	20 ft	15 ft

Find:

Work	400 ft-lb	240 ft-lb	600 ft-lb	300 ft-lb	525 ft-lb

The Newton-Meter

The mks (meter-kilogram-second) system of measurement is used internationally. The unit of force is the newton (N), and the unit of distance is the meter (m). Thus, the mks unit of work is the *newton-meter* (Nm). One newton equals 0.225 pounds of force, and 1 meter equals 3.281 feet.

Formula Diagram 2–2 converts work between newton-meters and foot-pounds. One foot-pound of work equals 1.355 newton-meters of work.

Formula Diagram 2–2

$$\frac{\text{Newton-meters}}{1.355 \mid \text{Foot-pounds}}$$

Energy

Energy is the ability to do work. Energy is measured by the work it can do. The amount of energy in a gallon of gasoline, a pound of coal, or a battery is measured by the amount of work the energy can do when it is released. The letter W is used as the symbol for both work and energy.

The Joule

The mks unit of energy is the *joule* (J).

In Skill-Builder 2–1 we calculated that we did 150 foot-pounds of work. Using Formula Diagram 2–2, we calculate that 150 foot-pounds equals 203.25 newton-meters. Since 1 joule of energy equals 1 newton-meter of work, we spent 203.25 joules of energy lifting the box from the floor to the table.

Energy Conversion

Energy can do many kinds of work by turning into different forms, such as electricity, magnetism, heat, light, and mechanical force. This process is called energy conversion. According to a basic scientific law, no energy is lost in the process. For example, a lamp converts electrical energy into both light and heat. The total joules of heat and light will equal the joules of electrical energy.

→ *Self-Test*

1. What is work?
2. What is a foot-pound?
3. What is a newton-meter?
4. What is a joule?

SECTION 2–2: Voltage

In this section you will learn the following:

→ Voltage (V) is a difference in electrical energy (potential) between two points due to a difference in charge.
→ Electromotive force (emf) is the ability of a voltage source to maintain its electrical potential and produce a steady movement of charge.
→ One volt equals 1 joule of energy per coulomb of charge.
→ Low-voltage electrons have low energy. High-voltage electrons have high energy.

Electrical Potential and Voltage

Voltage is created when there is a difference in charge between two points. When charge moves from one point to another, work is done. Thus, when a battery or generator creates voltage, the electrons gain energy. The difference in energy between the two electrically charged terminals is called an *electrical potential.* Electrical potential is another name for voltage.

Voltage acts like an electrical pressure because of the difference in energy between the two points. This energy is released when the electrons move from the point of higher energy to the point of lower energy. The release of potential energy is what gives electricity its ability to do work for us.

Electromotive Force

All voltage sources produce an electrical potential. However, useful voltage sources such as batteries and generators can maintain the electrical potential while the electrons are moving. In other words, a useful voltage source is able to restore the potential energy in the electrons as fast as the energy is being released.

To do this, the voltage source must contain some sort of force inside itself. The force may be chemical as in batteries, or magnetic and mechanical as in generators. The force inside the voltage source takes the returning electrons off the positive terminal and restores their potential energy by putting them back on the negative terminal, ready to deliver more energy to the circuit. Thus, the force inside the voltage source causes the electrons to move continuously. Since the voltage source produces a force that maintains potential energy while electrons are moving, we say the voltage source produces an *electromotive force* (emf).

The Volt

The unit of voltage is the *volt* (V). One volt equals 1 joule of energy per coulomb of charge.

Remember that potential energy is equal to the work done in moving something from one place to another, and that a coulomb is simply a certain number of electrons. If only a small amount of work has been done in moving each coulomb of electrons to the negative terminal, then the electrons will have a small amount of potential energy, and the voltage will be low. This is like pulling a bowstring back only a little. The bowstring has a low potential energy, and so the arrow has a low kinetic energy when it is released.

However, if a large amount of work has been done in moving each coulomb of electrons, then the electrons will have a large amount of potential energy and the voltage will be high. This is like pulling the bowstring all the way back. When the arrow is released, it will have a high kinetic energy.

The relationship between voltage, energy, and charge is given in Formula Diagram 2–3.

Formula Diagram 2–3

$$\frac{\text{Energy } (W) \text{ joules}}{\text{Voltage volts } (V) \mid \text{Charge } (Q) \text{ coulombs}}$$

● *SKILL-BUILDER 2–2*
- -

An automobile battery contains 108,000 coulombs of charge and has a potential of 12 volts across its terminals. How much energy does the battery contain?

GIVEN:
   Voltage = 12 V
   Charge = 108,000 C

FIND:
   Energy

SOLUTION:
   Energy = Voltage × Charge   Formula Diagram 2–3
         = 12 V × 108,000 C   substitute values and calculate
         = 1,296,000 J   result

   Repeat the Skill-Builder using the following values. You should obtain the results that follow.

Given:

Voltage	15 V	6 V	24 V	18 V	9 V
Charge	124,000 C	53,000 C	256,000 C	165,000 C	87,000 C

Find:

Energy	1,860,000 J	318,000 J	6,144,000 J	2,970,000 J	783,000 J

- -

→ *Self-Test*
- - - - - - - -

5. What is voltage?
6. What is electromotive force?
7. What is a volt?
8. What is the meaning of low and high voltage?
- -

SECTION 2–3: Current

In this section you will learn the following:

   → Current (I) is the movement or flow of charge carriers.
   → One ampere (A) is a rate of flow of charge equal to 1 coulomb per second.

   In Chapter 1, we learned that free electrons, holes, and ions act as charge carriers. If a material contains charge carriers, and if there is a difference in electrical potential between two points in the material, the charge carriers will begin to move through the material, away from one point and toward the other. This movement of charge carriers is an electric *current*. The symbol for current is *I*, which means intensity.

Current in Solids

In solid conductors such as copper and aluminum, free electrons are the charge carriers. If a voltage source such as a battery is applied to the two ends of a copper wire, an electric field is established from one end of the wire to the other. Under the influence of the electric field, free electrons begin to move away from the negative end and toward the positive end, as shown in Figure 2–2.

Notice that the movement of the electrons is irregular. In fact, some are actually moving against the electric field due to heat energy. However, there is an overall trend or tendency for the electrons to move from negative to positive.

As they move through the wire, the free electrons collide with atoms and bounce off in all directions, losing energy in the collisions, then regaining energy as the influence of the electric field gets the electrons moving again in the general direction of the current.

To better visualize electric current, imagine a hall with balloons floating in the air and fans at each end of the hall, blowing the air from one end to the other, as shown in Figure 2–3. Under the influence of the breeze created by the fans, the balloons drift down the hall. The hall contains obstacles, symbolized by the boxes in the picture. The drifting balloons collide with the boxes, bouncing off, losing energy, then regaining energy as they begin to drift again in the breeze.

In this illustration, the hall is like a copper wire, the balloons are like the free electrons in the wire, the fans are like the voltage source applied to the wire, the breeze is like the electric field in the wire created by the voltage source, and the boxes are like the copper atoms held in place in the wire.

When the voltage is low, the intensity of the electric field is weak. The free electrons drift slowly through the wire. This is like the balloons drifting slowly when the fans

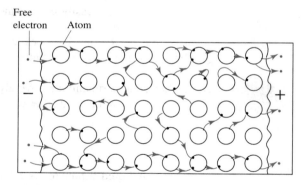

FIGURE 2–2
Current in solid conductors. The electric field causes free electrons to drift toward positive terminal. Most current flows on the surface of the conductor.

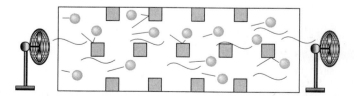

FIGURE 2–3
Balloons drifting through a room with obstacles resemble electrons moving through a conductor, colliding with atoms.

are on a low speed. However, when the voltage is high, the electrons are pushed swiftly through the material in large numbers, like the balloons flying across the room when the fans are on a high speed.

Even in a strong current, the electrons have a top speed of only a few hundred feet per second. However, the electrons begin to move as soon as the electric field is established in the wire, which happens at the speed of light. For example, suppose that a wire extends from New York City to Los Angeles (approximately 2,850 miles), with a voltage source applied to the wire. When the switch is turned on in New York, it takes about 0.015 seconds for the electric field to become established in the wire. This is the time it takes for the electromagnetic energy to move through the wire at the speed of light. Once the field is established, the electrons begin to move through the wire at a speed of a few hundred feet per second, far below the speed of light. An electron that leaves New York will probably never travel all the way to Los Angeles because of collisions with atoms. This doesn't matter because all the free electrons in the wire are moving together under the influence of the electric field, like the balloons all moving together in the breeze. The wire has current throughout its length.

Semiconductors

Materials such as carbon, silicon, and germanium have four valence electrons. The atoms have an orderly arrangement called a crystalline structure. The crystalline structure contains few charge carriers, and charges move through these materials with considerable difficulty. Therefore, these materials are classified as *semiconductors*.

By adding additional materials through chemical processes, we can increase either the number of free electrons or the number of holes in the crystalline structure, making it easier for charges to move. Thus, we can control how easily current moves through the material by controlling the number of charge carriers.

If we increase the number of free electrons, the material is called an N-type semiconductor. If we increase the number of holes, the material is called a P-type semiconductor.

Semiconductor materials such as silicon and germanium are widely used in modern electronics to make diodes, transistors, integrated circuits, and other components.

Current in Liquids

In liquids, current is the movement of ions. A liquid conductor is called an *electrolyte*. The chemicals in batteries are electrolytes. The complex chemicals inside your brain and nerve cells are also electrolytes. Your brain acts like a battery, producing tiny electrical currents which flow through the nerves to the muscles. When you exercise vigorously, some of your body's electrolytes are lost as you sweat. As a result, nerve and muscle function begins to deteriorate. Many of the special drinks for athletes contain electrolytes to restore nerve and muscle performance.

Current in Gases

In gases, molecules become ionized by the electric field, creating a mixture of free electrons and positive ions that move in opposite directions to become the current. A voltage of several hundred volts is necessary to ionize the gas molecules. Many types of lamps use gas discharges to produce light, including neon, fluorescent, mercury vapor, and sodium lamps. These lamps all require special transformers to produce the necessary voltages. Lightning in a thunderstorm is also an example of current in a gas.

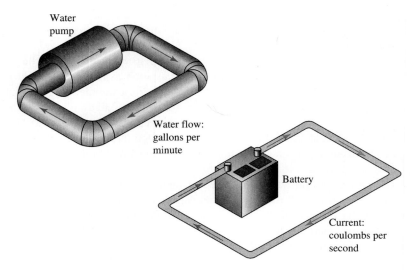

Water pump

Water flow: gallons per minute

Battery

Current: coulombs per second

FIGURE 2–4
Flow is measured in unit volume per unit time.

The Ampere

The unit of current is the *ampere* (A) and is commonly called the amp. One ampere is a rate of flow of charge equal to 1 coulomb per second.

Measuring the flow of electrical charge in coulombs per second is similar to measuring the flow of water in gallons per minute, as shown in Figure 2–4. A current of 1 ampere means that 1 coulomb of electrons per second moves through the wire.

Formula Diagram 2–4 expresses the relationship between amperes, coulombs, and seconds.

Formula Diagram 2–4

$$\frac{\text{Charge } (Q)}{\text{coulombs}}$$

Current (I)	Time (t)
amps	seconds

● *SKILL-BUILDER 2–3*

A circuit has a current of 3 A. How many coulombs of charge flow through the circuit in 5 seconds?

GIVEN:
    Current = 3 A
    Time = 5 s

FIND:
    Charge

SOLUTION:
    Coulombs = Amperes × Seconds   Formula Diagram 2–4
            = 3 A × 5 s   substitute values and calculate
            = 15 C   result

Repeat the Skill-Builder using the following values. You should obtain the results that follow.

Given:

Current	6 A	4 A	9 A	7 A	5 A
Time	3 s	5 s	6 s	2 s	4 s

Find:

Charge	18 C	20 C	54 C	14 C	20 C

→ *Self-Test*

9. What is current?
10. What is an ampere?

SECTION 2–4: Resistance

In this section you will learn the following:

→ Resistance (R) is a measure of how difficult it is for current to flow through an object.

→ Ohm's Law states that the resistance of an object is the ratio of the applied voltage and the resulting current.

→ The unit of resistance is the ohm (Ω). One ohm is the resistance of an object with an applied voltage of 1 volt and a current of 1 ampere.

→ Objects with more cross-sectional area have less resistance; objects with less cross-sectional area have more resistance.

→ Objects with more length have more resistance; objects with less length have less resistance.

→ Metals at higher temperatures have more resistance. Carbon and silicon at higher temperatures have less resistance.

Resistance (R) is a measure of the difficulty current has in flowing through various materials. Resistance is caused by the atomic structure of materials.

Resistance explains why the same amount of voltage may produce different amounts of current. For example, a hair dryer and a night light may be plugged into the same outlet, which gives them the same amount of voltage. The hair dryer may draw a current of 10 A, while the night light may draw a current of only 0.06 A. The difference in current occurs because the night light has more resistance than the hair dryer.

Resistance is related to the number of collisions between electrons and atoms as the electrons move through the material, and the amount of energy lost in each collision. The formulas that describe electrical resistance resemble the formulas that describe mechanical friction. Therefore, we think of resistance as a kind of electrical friction.

When electrons move through a material and collide with atoms, the kinetic energy in the current is converted into heat and released. Since all electrical devices contain resistance, all electrical equipment produces heat. In electronic devices, this heat can be harmful to the equipment. Proper ventilation is necessary to prolong the life of the equipment. The unwanted heat wastes electrical energy and thus increases the cost of operating the equipment.

Superconductors

The detailed study of atomic behavior is called quantum mechanics, which can be quite complicated and unusual. In certain materials, at extremely cold temperatures, quantum

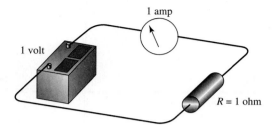

FIGURE 2–5
An object has a resistance of 1 ohm if a
voltage of 1 volt produces a current of
1 amp.

effects allow free electrons to pass through the material without any collisions with atoms. Under these conditions, the material has zero resistance and is called a *superconductor.*

Superconductors can carry extremely large currents over long distances without any energy loss. If we could replace all our copper and aluminum wires with superconductor materials, our world would look completely different. All the huge high-voltage power lines, heavy electrical cables, and transformer substations would be gone. Instead, ordinary wires could carry enough power for an entire city.

Unfortunately, superconductors must be kept extremely cold and are made of a brittle material similar to ceramic. Therefore, superconductors are not practical except in special scientific and military applications.

The Ohm

The unit of resistance is the *ohm,* symbolized Ω (the Greek letter omega). Since resistance is the ratio of voltage and current, 1 ohm is the resistance of an object with an applied voltage of 1 volt and a current of 1 ampere.

The definition of the ohm is illustrated in Figure 2–5 and shown in Formula Diagram 2–5. This important relationship between voltage and current is called Ohm's Law. We will learn more about Ohm's Law in a later chapter.

Formula Diagram 2–5: Ohm's Law

$$\frac{\text{Voltage } (V)}{\text{volts}}$$

| Resistance (R) | Current (I) |
| ohms Ω | amps |

● *SKILL-BUILDER 2–4*

A stove element has a resistance of 16 Ω. The applied voltage is 240 V. What is the current?

GIVEN:
  Resistance = 16 Ω
  Voltage = 240 V

FIND:
  Current

SOLUTION:
  Current = Voltage ÷ Resistance   Ohm's Law
       = 240 V ÷ 16 Ω substitute values and calculate
       = 15 A   result

Repeat the Skill-Builder using the following values. You should obtain the results that follow.

Given:

Resistance	21 Ω	12 Ω	30 Ω	8 Ω	17 Ω
Voltage	105 V	180 V	240 V	160 V	85 V

Find:

Current	5 A	15 A	8 A	20 A	5 A

Electrical Shock

The electrical power in our homes is typically 110 volts. Many people have been shocked by this voltage at some time. Most people survive ordinary electrical shocks, yet a large number die. What makes the difference? Part of the answer is the amount of resistance in the human body.

The resistance in your body depends on several factors, including your size, age, and body chemistry. However, the most important factors are how firmly you are in contact with the voltage source and whether your skin is dry or wet.

For example, suppose you accidentally touch a "live wire" while doing electrical work. If you are lightly touching the wire and your skin is dry, your body resistance may be high. The voltage may produce enough current to give you an unpleasant shock, but no harm will come to you. On the other hand, if you have a tight grip on the wire and your skin is wet or sweaty, your body resistance may be low. A current of 0.1 A passing through your chest for 1 second will almost certainly kill you. Thirty volts or more can produce this much current when your body resistance is low.

The Four Factors of Resistance

The resistance of an object is determined by four factors: cross-sectional area, length, temperature, and resistivity.

Cross-sectional Area

Objects with more cross-sectional area have less resistance, and objects with less cross-sectional area have more resistance. Just as water flows with less resistance in a big pipe, current flows with less resistance in a big wire, as shown in Figure 2–6.

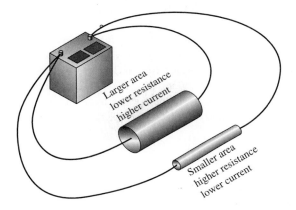

FIGURE 2–6
Effects of cross-sectional area on resistance.

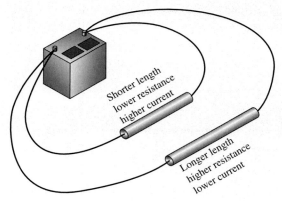

FIGURE 2–7
Effects of length on resistance.

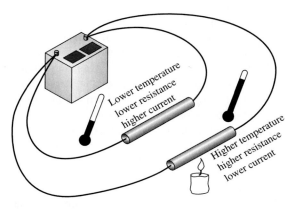

FIGURE 2–8
Effects of temperature on resistance.

Length

Objects with more length have more resistance, and objects with less length have less resistance.

If electrons must travel farther in an object, more collisions with atoms will occur, which produces more resistance. Therefore, if the only difference between two wires is length, a long wire will have more resistance than a short wire, as shown in Figure 2–7.

Electrical extension cords are a good example of the relationship of cross-sectional area, length, and resistance. You need a larger wire size to carry current over a longer distance. The larger wire size compensates for the longer distance. If the wire size is too small, the combination of high current and high resistance produces too much heat, which wastes energy and reduces the performance of the load. In extreme cases, the excessive heat could melt the insulation and create a fire or electrocution hazard.

Temperature

The effect of temperature on resistance depends on the atomic structure of the material. Metals generally have more resistance at higher temperatures. Thus, the tungsten metal filament of an ordinary incandescent light bulb has low resistance when the bulb is off

TABLE 2–1
Resistivity of several metals.

ELEMENT	RESISTIVITY $\mu \Omega/CM^3$
Aluminum	2.733
Copper	1.725
Gold	2.271
Iron	9.98
Silver	1.629
Tungsten	5.4

and the filament is cool. However, when the bulb is on and the filament is hot, the resistance is hundreds of times greater. Figure 2–8 illustrates the effects of temperature on the resistance of metals.

On the other hand, materials such as carbon and silicon have less resistance at higher temperatures. This fact is important in the design of electronic circuits, since resistors are often made of carbon, and transistors and integrated circuits (IC chips) are often made of silicon. When these components overheat, their lower resistance allows more current, which produces more heat, which lowers the resistance even more. This process is called thermal runaway and can damage or destroy the components. Thus, adequate cooling and ventilation are essential for electronic equipment.

Temperature Coefficient

A *temperature coefficient* is a value used to calculate an object's change in resistance for a given change in temperature. A positive temperature coefficient means the resistance increases when the object becomes hotter. On the other hand, a negative temperature coefficient means the resistance increases when the object becomes cooler.

Resistivity

Resistivity *(K)* is an ohmic value that is characteristic of a particular material. For example, copper, gold, carbon, and silicon all have different values of resistivity because they are different materials.

The symbol for resistivity is *K*. The unit of resistivity is micro-ohms per cubic centimeter ($\mu\Omega/cm^3$). Table 2–1 lists the resistivity of several materials at room temperature (27°C).

→ Self-Test

11. What is resistance?
12. How is resistance calculated?
13. What is an ohm?
14. What is the relationship between resistance and cross-sectional area?
15. What is the relationship between resistance and length?
16. What is the relationship between resistance and temperature?

SECTION 2–5: POWER

In this section you will learn the following:

→ Power *(P)* is the time rate at which a source produces energy, or a load dissipates energy.

→ One watt (W) is a power of 1 joule per second.

→ Watt's Law states that electrical power is the product of voltage and current.

→ A kilowatt-hour (kWh) is the energy necessary to operate a 1,000-watt load for 1 hour.

→ One horsepower (hp) equals 550 foot-pounds per second, or 33,000 foot-pounds per minute.

→ One horsepower equals 746 watts.

→ Efficiency is the ratio of useful power to total power.

Power (P) is the time rate at which a source produces energy, or a load dissipates (uses up) energy.

To understand the difference between energy and power, imagine two people (Jim and Joe) working in a warehouse unloading boxes, as shown in Figure 2–9. Suppose that all boxes are exactly the same weight, and both Jim and Joe move equal numbers of boxes through equal distances. Thus, the work done is the same for both people. Since the work is the same, the energy is the same. This means that Jim and Joe use up an equal number of joules of energy to do the work.

Now suppose that Jim does his job in 1 hour, but Joe does his job in 2 hours. Jim uses up his energy more quickly than Joe and gets the job done in less time. Therefore, we say that Jim has more power than Joe.

The Watt

The mks unit of power is the *watt* (W). One watt is a power of 1 joule per second. Formula Diagram 2–6 expresses this relationship.

Remember that *W* is also the symbol for work or energy. When *W* appears by itself on the left side of an equation, it stands for work or energy. When W appears after a number on the right side of an equation, it stands for watts.

Formula Diagram 2–6

$$\frac{\text{Energy } (W)}{\text{joules}}$$
$$\frac{\quad}{\text{Power } (P) \mid \text{Time } (t)}$$
$$\text{watts} \mid \text{seconds}$$

All forms of power can be measured in watts, but we are most familiar with watts as units of power for electrical circuits and devices. For example, a generator may be rated at 1,000 watts, which means the generator can produce 1,000 joules per second. Similarly, a heater may be rated at 1,000 watts, which means the heater dissipates 1,000 joules per second.

In all electrical machines, more power means more performance, as illustrated in Figure 2–10. For example, a 100-watt lamp produces more light than a 60-watt lamp; a 1,200-watt heater produces more heat than a 900-watt heater; and a 100-watt stereo produces louder music than a 30-watt stereo. Whatever the machine is supposed to do, more power means more performance, since more energy per second is released.

In heaters, we can easily associate time with power. We can warm up a room, dry our hair, and cook food in less time when we use heaters at higher power levels

FIGURE 2–9
(a) Jim and Joe have the same amount of work to do and begin at the same time. (b) Because Jim has more power than Joe, he does more work per unit of time.

(more watts). However, in many other electrical devices, we do not ordinarily associate time with power. For example, we know that a 100-watt lamp produces more light than a 60-watt lamp, but we have no sense of the 100-watt lamp producing light in less time.

Nonetheless, time is always involved in power, as shown in Figure 2–11. Batteries run down quicker and generators burn up fuel faster when the power level increases. Thus, at the power source, the energy in the fuel is being released more quickly when the load has a higher power rating.

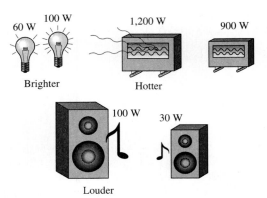

FIGURE 2–10
More power means more performance.

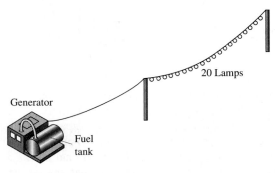

FIGURE 2–11
The power level of the lamps determines how
long the fuel will last.

Power Is the Product of Voltage and Current

The voltage represents the joules of energy in each coulomb of electrons. The current represents the coulombs of electrons per second arriving at the load. Therefore, the combination of voltage and current represents the joules of energy per second arriving at the load, which is power.

Mathematically, power (watts) equals voltage multiplied by current (volts times amps). This important relationship is called Watt's Law and is given in Formula Diagram 2–7. We will learn more about Watt's Law in a later chapter.

Formula Diagram 2–7: Watt's Law

$$\frac{\text{Power } (P)}{\text{watts}}$$
$$\frac{\text{Voltage } (V)}{\text{volts}} \bigg| \frac{\text{Current } (I)}{\text{amps}}$$

The Kilowatt-hour

A *kilowatt-hour* (kWh) is the energy necessary to operate a 1,000-watt load for 1 hour. Electric utility companies bill their customers according to the energy that each customer has used, measured in kilowatt-hours.

Kilo is the international word for thousand. Thus, a kilowatt is 1,000 watts. Formula Diagram 2–8 expresses the relationship between watts and kilowatts.

Formula Diagram 2–8

$$
\begin{array}{c}
\text{Power } (P) \\
\text{watts} \\
\hline
\begin{array}{c|c}
\text{Power } (P) & \\
\text{kilowatts} & 1{,}000
\end{array}
\end{array}
$$

We have learned that a 1-watt load uses up 1 joule of energy per second. Therefore, a 1-kilowatt load uses up 1,000 joules per second. There are 60 seconds in a minute, and 60 minutes in an hour. Therefore, a 1-kilowatt load that operates for 1 hour uses $1{,}000 \times 60 \times 60 = 3.6$ million joules of energy. Thus, a kilowatt-hour is an amount of energy equal to 3.6 million joules.

Formula Diagram 2–9 expresses the kilowatt-hour relationship.

Formula Diagram 2–9

$$
\begin{array}{c}
\text{Energy } (W) \\
\text{kilowatt-hours} \\
\hline
\begin{array}{c|c}
\text{Power } (P) & \text{Time } (t) \\
\text{kilowatts} & \text{hours}
\end{array}
\end{array}
$$

The utility company installs a kilowatt-hour meter for each customer, reads the meter once a month, and subtracts the present reading from the previous reading to determine the kilowatt-hours used during the month. The company charges the customer a certain price per kilowatt-hour to pay for the fuel burned to produce electrical energy for that customer. Of course, the total bill also includes additional costs for service and equipment; but the kilowatt-hour is the primary value that determines the amount of the electric bill.

● SKILL-BUILDER 2–5

A homeowner operates a 1,500-W heater for 6 hours. How many kilowatt-hours of energy are consumed?

GIVEN:
  Power = 1,500 W
  Time = 6 hr

FIND:
  Kilowatt-hours

SOLUTION:

  1. Kilowatts = Watts ÷ 1,000   convert watts to kilowatts
            = 1,500 W ÷ 1,000   substitute values and calculate
            = 1.5 kW   result

  2. Kilowatt-hours = Kilowatts × Hours   determine correct formula
            = 1.5 kW × 6 hr   substitute values and calculate
            = 9 kWh   result

Repeat the Skill-Builder using the following values. You should obtain the results that follow.

Given:

Power	1,200 W	800 W	650 W	200 W	1,500 W
Time	5 h	3.5 h	4 h	19 h	7 h

Find:

| Kilowatt-hours | 6 kWh | 2.8 kWh | 2.6 kWh | 3.8 kWh | 10.5 kWh |

- -

Horsepower

The British unit for power is the *horsepower* (hp). Mechanical power is often measured in horsepower. One horsepower equals 550 foot-pounds per second, or 33,000 foot-pounds per minute. Figure 2–12 illustrates the definition of 1 horsepower.

One horsepower equals 746 watts. Formula Diagram 2–10 and Formula Diagram 2–11 show the relationships concerning horsepower and watts.

Formula Diagram 2–10

Work or Energy (*W*) foot-pounds	
Power (*P*) horsepower	Time (*t*) seconds I 550

Formula Diagram 2–11

Power (*P*) watts
Power (*P*) horsepower I 746

The relationship between watts and horsepower is particularly useful in dealing with generators and motors, which are rated electrically in watts and mechanically in horsepower.

We can calculate the amount of current an electric motor will draw by first converting the horsepower rating to watts with Formula Diagram 2–11, then finding current with Formula Diagram 2–7.

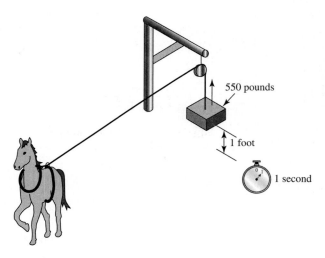

FIGURE 2–12
One horsepower can lift 550 pounds a distance of 1 foot in a time of 1 second.

● *SKILL-BUILDER 2–6*

A table saw has a 3.5-hp motor and operates on 110 V. How much current does it draw?

GIVEN:
   Power = 3.5 hp
   Voltage = 110 V

FIND:
   Current

SOLUTION:

1. Watts = 746 × hp   from Formula Diagram 2–11
         = 746 × 3.5 hp   substitute values and calculate
         = 2,611 W   result
2. Current = Power ÷ Voltage   from Formula Diagram 2–7
             = 2,611 W ÷ 110 V   substitute values and calculate
             = 23.74 A   result

Repeat the Skill-Builder using the following values. You should obtain the results that follow.

Given:

Hp	2.4 hp	4.8 hp	15.3 Hp	1.7 hp	34.2 hp
Voltage	115 V	125 V	240 V	110 V	480 V

Find:

Watts	1,790 W	3,581 W	11,414 W	1,268 W	25,513 W
Current	15.57 A	28.65 A	47.56 A	11.53 A	53.15 A

Efficiency

Efficiency is the ratio of useful power to total power. Most electrical machines waste some power. In other words, the useful power they produce is less than the total power they dissipate. Naturally, we want the useful power to be as close to the total power as possible. To evaluate the performance of a machine, we measure the useful and total power, and calculate the machine's efficiency.

Formula Diagram 2–12

$$\frac{\text{Useful Power}}{\text{Efficiency} \mid \text{Total Power}}$$

Efficiency may be expressed either as a decimal fraction or as a percentage.

→ *Self-Test*

17. What is power?
18. What is a watt?
19. What is Watt's law?
20. What is a kilowatt-hour?
21. What is a horsepower?
22. What is the relationship between horsepower and watts?

SECTION 2–6: International Numerical Prefixes

In this section you will learn the following:

→ An international numerical prefix is a symbol for a large number or a small fraction.

→ Scientific notation is a way of writing large numbers and small fractions using exponents called powers of ten.

→ Scientific notation expresses numbers as a coefficient between 1.00 and 9.99, followed by a power of ten. The exponent indicates the number of positions the decimal point has been moved from its true position.

→ A positive exponent indicates a number larger than one. Thus, the true location of the decimal point is farther to the right.

→ A negative exponent indicates a number smaller than one (a fraction). Thus, the true location of the decimal point is farther to the left.

→ Engineering notation is a version of scientific notation that expresses numbers as a coefficient between 1 and 999, followed by a power of ten whose exponent is a multiple of three.

→ Engineering notation matches the system of international numerical prefixes.

The word *kilo* is an international word for thousand. There are several other words that are used internationally for large numbers and small fractions. These words and their symbols are called international numeric prefixes. The international numeric prefixes are part of a shorthand system of writing numbers called *engineering notation,* which is a version of *scientific notation.*

Scientific Notation

Scientists and engineers often have to work with large numbers such as 3,240,000,000,000,000 or small decimal fractions like 0.00000000000237. The zeros in these numbers are not significant, since they do not affect the precision of the number; yet the zeros are necessary to show the correct position of the decimal point. Scientific notation is a way of writing large numbers and small fractions without writing all the zeros.

Scientific notation uses powers of ten to indicate the true position of the decimal point. Numbers written in this way are shown in Figure 2–13.

In scientific notation, the number ten is called the base. The small number written as a superscript (above and to the right) of the base is called the exponent, or power of the base. The mixed number before the base is called the coefficient. The decimal point in the coefficient is not in its true position. The exponent (power of ten) shows how many positions the decimal point has been moved.

When the Exponent Is Zero

When the exponent is zero, the decimal point of the coefficient is in its true location, as shown in Figure 2–14(a).

When the Exponent Is Positive

A positive exponent means that the true position of the coefficient's decimal is farther to the right, as shown in Figure 2–14(b). Thus, if the exponent is positive, when the number is rewritten in standard notation, it will be larger than one.

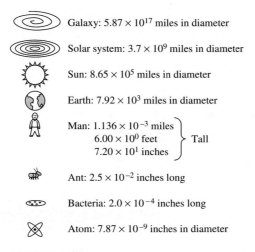

Galaxy: 5.87×10^{17} miles in diameter

Solar system: 3.7×10^9 miles in diameter

Sun: 8.65×10^5 miles in diameter

Earth: 7.92×10^3 miles in diameter

Man: $\left.\begin{array}{l} 1.136 \times 10^{-3} \text{ miles} \\ 6.00 \times 10^0 \text{ feet} \\ 7.20 \times 10^1 \text{ inches} \end{array}\right\}$ Tall

Ant: 2.5×10^{-2} inches long

Bacteria: 2.0×10^{-4} inches long

Atom: 7.87×10^{-9} inches in diameter

FIGURE 2–13
Measurements written in scientific notation.

(a) $2.54 \times 10^0 = 2.54$
(b) $2.54 \times 10^2 = 254$
(c) $2.54 \times 10^{-2} = 0.0254$

FIGURE 2–14

In the value 5.03×10^4, the exponent 4 tells us that the true position of the decimal point in the coefficient is four positions to the right. Therefore, 5.03×10^4 is actually 50,300. When we move the decimal to convert from scientific to standard notation, we fill all empty positions with zeros.

When the Exponent Is Negative

A negative exponent means that the true position of the coefficient's decimal is farther to the left, as shown in Figure 2–14(c). Thus, if the exponent is negative, when the number is rewritten in standard notation, it will be smaller than one (a fraction).

In the value 6.27×10^{-3}, the exponent -3 tells us that the true position of the decimal point in the coefficient is three positions to the left. Therefore, 6.27×10^{-3} is actually 0.00627. When we move the decimal to convert from scientific to standard notation, we fill all empty positions with zeros.

Converting from Standard to Scientific Notation

To convert a number from standard notation to scientific notation, follow these steps:

1. Move the decimal point until the coefficient is a mixed number with an integer portion between one and nine. Round off the coefficient at the third significant digit. In other words, the value of the coefficient will be between 1.00 and 9.99.
2. After the coefficient, write the term "$\times$ 10" to show that the decimal point has been moved.

3. Write an exponent to the base ten to show the number of positions the decimal point has been moved. If the original number was larger than one, the exponent will be positive. If the original number was smaller than one (a fraction), the exponent will be negative.

● SKILL-BUILDER 2–7

Write the number of electrons in a coulomb of charge in both standard and scientific notation.

6,240,000,000,000,000,000   standard notation
$= 6.24 \times 10^{18}$ move decimal 18 places to the left
                       discard all the zeros
                       write "$\times 10$"
                       write a positive exponent of 18

Repeat the Skill-Builder using the following values. You should obtain the results that follow.

3,150,000,000,000,000	$= 3.15 \times 10^{15}$
26,400,000,000,000,000,000	$= 2.64 \times 10^{19}$
483,000,000,000,000,000	$= 4.83 \times 10^{17}$
61,300,000,000,000,000,000,000	$= 6.13 \times 10^{22}$
5,170,000,000,000,000,000,000,000	$= 5.17 \times 10^{24}$

● SKILL-BUILDER 2–8

The following fraction is the charge of a single electron in coulombs. Convert the fraction from standard to scientific notation.

0.000000000000000000160 C   standard notation
$= 1.60 \times 10^{-19}$ move decimal 19 places to the right
                       discard all the zeros
                       write "$\times 10$"
                       write a negative exponent of 19

Repeat the Skill-Builder using the following values. You should obtain the results that follow.

0.000000000000000327	$= 3.27 \times 10^{-16}$
0.00000000000000000000000468	$= 4.68 \times 10^{-24}$
0.000000000000539	$= 5.39 \times 10^{-13}$
0.00000000000000000762	$= 7.62 \times 10^{-18}$
0.0000000000000000000933	$= 9.33 \times 10^{-20}$

Converting from Scientific to Standard Notation

To convert a number in scientific notation back into standard notation, follow these steps:

1. Move the decimal point the number of positions shown by the exponent of the base. If the exponent is positive, move the decimal point to the right; if the exponent is negative, move the decimal point to the left.
2. While moving the decimal point, fill all empty positions with zeros.

● SKILL-BUILDER 2–9

	Move the decimal point:
$2.57 \times 10^{-9} = 0.00000000257$	9 places to the left
$3.54 \times 10^{10} = 35,400,000,000$	10 places to the right
$23.468 \times 10^{-4} = 0.0023468$	4 places to the left
$193.6 \times 10^{2} = 19,360$	2 places to the right

The Power of Ten

The mathematical meaning of the power of ten is the number of times the coefficient should be multiplied by ten (for positive exponents) or divided by ten (for negative exponents).

● SKILL-BUILDER 2–10

5.03×10^{2}	exponent $= +2$
$= 5.03 \times 10 \times 10$	multiply by ten 2 times
$= 503$	result
6.27×10^{-3}	exponent $= -3$
$= 6.27 / 10 / 10 / 10$	divide by ten 3 times
$= 0.00627$	result

Thus, we can move the decimal point manually counting the positions, or we can move the decimal point mathematically by multiplying and dividing. Either way, the result is the same.

When the coefficient is the number one, we get the set of numbers shown in Table 2–2.

TABLE 2–2
Powers of ten act as multipliers.

1×10^{6}	$= 1,000,000$
1×10^{5}	$= 100,000$
1×10^{4}	$= 10,000$
1×10^{3}	$= 1,000$
1×10^{2}	$= 100$
1×10^{1}	$= 10$
1×10^{0}	$= 1$
1×10^{-1}	$= 0.1$
1×10^{-2}	$= 0.01$
1×10^{-3}	$= 0.001$
1×10^{-4}	$= 0.0001$
1×10^{-5}	$= 0.00001$
1×10^{-6}	$= 0.000001$

TABLE 2–3
The same number written in standard, scientific, and engineering notation.

Standard	123,000,000
Scientific	1.23×10^{8}
Engineering	123×10^{6}

Engineering Notation

Engineering notation is similar to scientific notation. There are two major differences in engineering notation:

1. The exponents are always multiples of three.
2. The coefficients range in value from 1 to 999.

Since the exponents must be multiples of three, engineering notation multiplies and divides numbers by thousands instead of by tens. Table 2–3 compares several numbers written in standard, scientific, and engineering notation.

Figure 2–15 shows an exponent line for engineering notation. Notice that each exponent is a multiple of three.

Associated with each exponent value is one of the *international numerical prefixes*. The English equivalents are also shown. Notice that each prefix has a symbol. For large numbers the exponents are positive and the symbols are upper-case letters (capitals) except for kilo. For fractions, the exponents are negative and the prefix symbols are lower-case letters. The symbol for micro is a Greek letter, μ (mu). When mu is used as a prefix, it is always pronounced "micro."

Tera	Giga	Mega	Kilo	Unit value	milli	micro	nano	pico
T	G	M	k		m	μ	n	p
+12	+9	+6	+3	0	−3	−6	−9	−12

FIGURE 2–15
The system of international numeric prefixes and corresponding powers of ten.

● SKILL-BUILDER 2–11

$345{,}000 \text{ V} = 345 \times 10^3 \text{ volts}$
$= 345 \text{ thousand volts}$
$= 345 \text{ kV}$
  k, or kilo, is the prefix for thousand

$0.020 \text{ A} = 20 \times 10^{-3} \text{ amps}$
$= 20 \text{ thousandths of an amp}$
$= 20 \text{ mA}$
  m, or milli, is the prefix for thousandth

$1{,}250{,}000{,}000 \text{ W} = 1.25 \times 10^9 \text{ watts}$
$= 1.25 \text{ billion watts}$
$= 1.25 \text{ GW}$
  G, or Giga, is the prefix for billion

$0.000015 \text{ s} = 15 \times 10^{-6} \text{ seconds}$
$= 15 \text{ millionths of a second}$
$= 15 \text{ μs}$
  μ, or micro, is the prefix for millionth

→ Self-Test

23. What is an international numerical prefix?
24. What is scientific notation?
25. How does scientific notation express numbers?

26. What does a positive exponent indicate?
27. What does a negative exponent indicate?
28. What is engineering notation?
29. What is the relationship between engineering notation and the system of international numeric prefixes?

- -

Formula List

Formula Diagram 2–1

$$\frac{\text{Work}}{\text{Force} \mid \text{Distance}}$$

Formula Diagram 2–2

$$\frac{\text{Newton-meters}}{1.355 \mid \text{Foot-pounds}}$$

Formula Diagram 2–3

$$\frac{\begin{array}{c}\text{Energy }(W)\\ \text{joules}\end{array}}{\begin{array}{c|c}\text{Voltage }(V) & \text{Charge }(Q)\\ \text{volts} & \text{coulombs}\end{array}}$$

Formula Diagram 2–4

$$\frac{\begin{array}{c}\text{Charge }(Q)\\ \text{coulombs}\end{array}}{\begin{array}{c|c}\text{Current }(I) & \text{Time }(t)\\ \text{amps} & \text{seconds}\end{array}}$$

Formula Diagram 2–5: Ohm's Law

$$\frac{\begin{array}{c}\text{Voltage }(V)\\ \text{volts}\end{array}}{\begin{array}{c|c}\text{Resistance }(R) & \text{Current }(I)\\ \text{ohms }\Omega & \text{amps}\end{array}}$$

Formula Diagram 2–6

$$\frac{\begin{array}{c}\text{Energy }(W)\\ \text{joules}\end{array}}{\begin{array}{c|c}\text{Power }(P) & \text{Time }(T)\\ \text{watts} & \text{seconds}\end{array}}$$

Formula Diagram 2–7: Watt's Law

$$\frac{\begin{array}{c}\text{Power }(P)\\ \text{watts}\end{array}}{\begin{array}{c|c}\text{Voltage }(V) & \text{Current }(I)\\ \text{volts} & \text{amps}\end{array}}$$

Formula Diagram 2–8

$$\frac{\begin{array}{c}\text{Power }(P)\\ \text{watts}\end{array}}{\begin{array}{c|c}\text{Power }(P) & \\ \text{kilowatts} & 1,000\end{array}}$$

Formula Diagram 2–9

$$\frac{\text{Energy } (W)}{\text{kilowatt-hours}}$$

Power (P)	Time (t)
kilowatts	hours

Formula Diagram 2–10

$$\frac{\text{Work or Energy } (W)}{\text{foot-pounds}}$$

Power (P)	Time (t)	
horsepower	seconds	550

Formula Diagram 2–11

$$\frac{\text{Power } (P)}{\text{watts}}$$

Power (P)	
horsepower	746

Formula Diagram 2–12

$$\frac{\text{Useful Power}}{\text{Efficiency} \mid \text{Total Power}}$$

Formula Diagram 2–13

$$\frac{\text{Useful Power}}{\text{Efficiency} \mid \text{Total Power}}$$

● TROUBLESHOOTING: Applying Your Knowledge

After learning that power is the combination of voltage and current, you realize that the headlight and table lamp can both be 60 W even though one operates at 12 V and the other at 120 V. The headlight makes up for the low voltage by drawing more current, 5 A instead of the 0.5 A that the table lamp draws.

CHAPTER SUMMARY:
ANSWERS TO SELF-TESTS

1. Work equals force multiplied by distance.
2. In the British system of measurement, one foot-pound of work is a force of 1 pound acting through a distance of 1 foot.
3. In the metric system of measurement, 1 newton-meter of work is a force of 1 newton acting through a distance of 1 meter.
4. One joule is the energy necessary to do 1 newton-meter of work. Therefore, joules of energy are equivalent to newton-meters of work.
5. Voltage (V) is a difference in electrical energy (potential) between two points due to a difference in charge.
6. Electromotive force (emf) is the ability of a voltage source to maintain its electrical potential and produce a steady movement of charge.

7. One volt equals 1 joule of potential energy per coulomb of charge.
8. Low voltage means electrons have low energy. High voltage means electrons have high energy.
9. Current (I) is the movement or flow of charge carriers.
10. One ampere (A) is a rate of flow of charge equal to 1 coulomb per second.
11. Resistance (R) is a measure of how difficult it is for current to flow through an object.
12. The resistance of an object is the ratio of the applied voltage and the resulting current.
13. One ohm (Ω) is the resistance of an object with an applied voltage of 1 volt and a current of 1 ampere.
14. Objects with more cross-sectional area have less resistance; objects with less cross-sectional area have more resistance.
15. Objects with more length have more resistance; objects with less length have less resistance.

16. Metals at higher temperatures have more resistance. Carbon and silicon at higher temperatures have less resistance.
17. Power *(P)* is the time rate at which a source produces energy, or a load dissipates energy.
18. One watt (W) is a power of 1 joule per second.
19. Watt's Law states that electrical power is the product of voltage and current.
20. A kilowatt-hour (kWh) is the energy necessary to operate a 1,000-watt load for 1 hour.
21. One horsepower (hp) equals 550 foot-pounds per second, or 33,000 foot-pounds per minute.
22. One horsepower equals 746 watts.
23. An international numerical prefix is a symbol for a large number or a small fraction.
24. Scientific notation is a way of writing large numbers and small fractions using exponents called powers of ten.
25. Scientific notation expresses numbers as a coefficient between 1.00 and 9.99, followed by a power of ten. The exponent indicates the number of positions the decimal point has been moved from its true position.
26. A positive exponent indicates a number larger than one. Thus, the true location of the decimal point is farther to the right.
27. A negative exponent indicates a number smaller than one (a fraction). Thus, the true location of the decimal point is farther to the left.
28. Engineering notation is a version of scientific notation that expresses numbers as a coefficient between 1 and 999, followed by a power of ten whose exponent is a multiple of three.
29. Engineering notation matches the system of international numerical prefixes.

CHAPTER REVIEW QUESTIONS

Determine whether the following questions are true or false.

2–1. Work is done when a force acts through a distance.
2–2. Work equals force divided by distance.
2–3. Resistance acts like an electrical pressure.
2–4. One volt equals 1 joule of energy per coulomb of charge.
2–5. One ampere is a rate of flow of charge equal to 1 joule per second.
2–6. Resistance is a measure of the difficulty voltage has in flowing through various materials.
2–7. Resistance is a kind of electrical friction.
2–8. One ohm is the resistance of an object with an applied voltage of 1 volt and a current of 1 ampere.
2–9. Objects with more cross-sectional area have less current.
2–10. Objects with more length have more resistance.
2–11. Metals at higher temperatures have more resistance.
2–12. Power is the time rate at which a source produces energy, or a load dissipates energy.
2–13. One watt is a power of 1 joule per second.
2–14. A kilowatt-hour is the energy necessary to operate a 1,000-watt load for 1 hour.

2–15. One horsepower equals 550 foot-pounds per second.
2–16. One watt equals 746 horsepower.
2–17. When the exponent is zero, the decimal point of the coefficient is in its true location.

Select the response that best answers the question or completes the statement.

2–18. The foot-pound is a unit of
　　a. power
　　b. force
　　c. weight
　　d. work
2–19. The mks unit of work is the
　　a. volt-ampere　　c. joule-second
　　b. newton-meter　　d. foot-pound
2–20. Which of the following statements is false?
　　a. Energy is the ability to do work.
　　b. Energy is measured by the work it can do.
　　c. The mks unit of energy is the joule.
　　d. Energy plus work equals power.
2–21. _____ is created when there is a difference in charge between two points.
　　a. Voltage　　c. Resistance
　　b. Current　　d. Power
2–22. Electrical potential is another name for
　　a. power　　c. current
　　b. voltage　　d. resistance
2–23. The ability of a voltage source to maintain its electrical potential is called
　　a. power　　c. energy
　　b. emf　　d. charge
2–24. _____ is the movement or flow of charge carriers.
　　a. Voltage　　c. Resistance
　　b. Current　　d. Power
2–25. In solids, the charge carriers are
　　a. ions
　　b. free electrons
　　c. ions and free electrons
　　d. electrolytes
2–26. In liquids, the charge carriers are
　　a. ions
　　b. free electrons
　　c. ions and free electrons
　　d. electrolytes
2–27. In gases, the charge carriers are
　　a. ions
　　b. free electrons
　　c. ions and free electrons
　　d. electrolytes
2–28. An insulator is sometimes called
　　a. a charge carrier
　　b. a semiconductor
　　c. a dielectric
　　d. an electrolyte
2–29. The unit of current is the
　　a. volt　　c. ohm
　　b. ampere　　d. watt

2–30. _____ have zero resistance.
   a. Superconductors   **c.** Electrolytes
   b. Semiconductors   **d.** Dielectrics
2–31. The ratio of the applied voltage and the resulting current is _____.
   a. power   **c.** potential
   b. emf   **d.** resistance
2–32. The unit of resistance is the
   a. volt   **c.** ohm
   b. ampere   **d.** watt
2–33. The unit of power is the
   a. volt   **c.** ohm
   b. ampere   **d.** watt
2–34. Electrical power is the product of
   a. current and resistance
   b. resistance and voltage
   c. voltage and current
   d. energy and work
2–35. Which of the following is *not* an international numeric prefix?
   a. kilo   **d.** milli
   b. micro   **e.** nilo
   c. mega   **f.** giga
2–36. Which of the following numbers is written in scientific notation?
   a. 34.6378×10^{-4} **d.** 0.00037
   b. 285×10^{6} **e.** 3,450,000
   c. 9.01×10^{-7} **f.** 22.5 μ
2–37. Which of the following numbers is written in engineering notation?
   a. 34.6378×10^{-4} **d.** 0.00037
   b. 285×10^{6} **e.** 3,450,000
   c. 9.01×10^{-7} **f.** 22.5 μ
2–38. In the number 98.07×10^{-3}, the exponent is
   a. 98   **c.** −3
   b. .07   **d.** × 10

Solve the following problems.

2–39. How much work is done in lifting a 70-pound box off the floor and setting it on a table that is 2 feet high?
2–40. A boat is pulled across a distance of 700 feet. We know that 14,000 foot-pounds of work is done. How much force does it take to pull the boat?
2–41. How many joules of energy are spent in doing 500 ft-lb of work?
2–42. How many joules of energy are spent in doing 500 Nm of work?
2–43. A radio battery contains 12,000 coulombs of charge and has a potential of 9 volts across its terminals. How much energy does the battery contain?
2–44. A circuit has a current of 4 A. How many coulombs of charge flow through the circuit in 7 seconds?
2–45. A voltage of 9 V applied to an object produces a current of 2 A. What is the resistance?
2–46. A hair dryer has a resistance of 12 Ω. The applied voltage is 120 V. What is the current?

2–47. A 52–gallon hot water heater operates on 240 V at a power level of 6,000 W. How many amps does the water heater draw?
2–48. A homeowner operates a 3,000-W heater for 8 hours. How many kilowatt-hours of energy are consumed?
2–49. An electric motor operates a hoist which lifts 495 pounds through 50 feet in 15 seconds. What is the power of the motor both in horsepower and in watts?
2–50. A circular saw has a 2.8-hp motor and operates on 120 V. How much current does it draw?
2–51. Write 34,600,000,000 in scientific notation.
2–52. Write 0.0000635 in engineering notation.
2–53. Write 0.0000000734 in scientific notation.
2–54. Write 4,630,000 in engineering notation.
2–55. Write 3.76×10^{7} in standard notation.
2–56. Write 2.58×10^{-4} in standard notation.
2–57. Write 23.9×10^{9} in standard notation.
2–58. Write 112.83×10^{-6} in standard notation.
2–59. Write each of the following values in standard notation.
   a. 37.5 kV   **c.** 22.3 mA
   b. 7.5 GW   **d.** 15 μV
2–60. Write each of the following values in international numeric prefix notation.
   a. 510,000 Ω   **c.** 0.000035 V
   b. 3,600,000 W   **d.** 0.05 A

GLOSSARY

ampere   The unit of current; a rate of flow of charge equal to 1 coulomb per second.
current   The movement or flow of charge carriers.
efficiency   The ratio of useful power to total power.
electrical potential   The difference in potential energy between two electrically charged terminals.
electrolyte   A liquid conductor.
electromotive force   The ability of a voltage source to maintain its electrical potential and produce a steady movement of charge.
energy   The ability to do work.
engineering notation   A method of expressing numbers as a coefficient between 1 and 999, followed by a power of ten whose exponent is a multiple of three.
foot-pound   The British unit of work; a force of 1 pound acting through a distance of 1 foot.
horsepower   A power of 550 foot-pounds per second, or 33,000 foot-pounds per minute; equal to 746 watts.
joule   The mks unit of energy; the energy necessary to do 1 newton-meter of work.
kilowatt-hour   The energy necessary to operate a 1,000-watt load for 1 hour.
international numeric prefix   A word or its symbol representing an engineering notation exponent.
newton-meter   The mks unit of work; a force of 1 newton acting through a distance of 1 meter.

ohm   The unit of resistance; the resistance of an object with an applied voltage of 1 volt and a current of 1 ampere.

power   The time rate at which a source produces energy, or a load dissipates energy.

resistance   A measure of how difficult it is for current to flow through an object; the ratio of the applied voltage and the resulting current.

scientific notation   A method of expressing numbers as a coefficient between 1.00 and 9.99, followed by a power of ten.

semiconductor   A material with a crystalline structure that contains few charge carriers. Thus, charges move through semiconductor materials with considerable difficulty.

superconductor   A material which has zero resistance at extremely low temperatures due to quantum atomic effects.

temperature coefficient   A value used to calculate an object's change in resistance for a given change in temperature.

volt   The unit of voltage; 1 volt equals 1 joule of potential energy per coulomb of charge.

voltage   A difference in electrical energy (potential) between two points due to a difference in charge.

watt   The mks unit of power; a power of 1 joule per second.

work   The result of a force acting through a distance.

● SECTION OUTLINE

3–1 The Electric Circuit

3–2 dc Power Supplies

3–3 Resistors

3–4 Switches

3–5 Fuses

3–6 Electrical Symbols and Diagrams

● TROUBLESHOOTING: Applying Your Knowledge

Your neighbor recently replaced his doorbell button, and tells you that now the doorbell rings all the time, except when he pushes the button. Can you help him find the problem?

● OVERVIEW AND LEARNING OBJECTIVES

All electrical circuits, simple and complex, can be analyzed in terms of five basic circuit elements. You will be able to troubleshoot circuits more effectively by understanding these five elements.

In this chapter, you will learn the basic circuit components necessary to build your first circuits. These components are represented by symbols in electrical schematic drawings.

After completing this chapter, you should be able to:

1. Name the five elements of an electric circuit.
2. Describe the basic differences between dc and ac voltage sources.
3. Describe basic styles and types of switches.
4. Read resistor color bands to determine ohmic value and tolerance.
5. Read a simple schematic drawing.

SECTION 3–1: The Electric Circuit

In this section you will learn the following:

→ A circuit requires a voltage source, a current path, a load, a current control device, and an overcurrent protection device.

→ A practical voltage source can provide only a limited amount of current. As the output current increases, output voltage decreases.

→ A voltage regulator is a special circuit or device used to maintain constant output voltage in a practical voltage source.

→ The current path provides a low-resistance path for current to travel from one terminal of the voltage source to the other terminal.

→ The energy of the current is transferred to the load, producing useful work.

→ The current control device may be simple or complex, mechanical or electronic. The most common current control device is a switch.

→ The overcurrent protection device stops all current if the current becomes abnormally high, protecting the circuit from damage due to heat and fire.

An electric circuit has five standard elements: voltage source, current path, load, current control device, and overcurrent protection device. These five elements are shown in Figure 3–1.

When troubleshooting a circuit, you must consider each of the five elements as the possible location of the trouble. In deciding which of the five circuit elements to examine and test, you must ask yourself which element is the most likely cause of the trouble, and also which element is the quickest and easiest to test.

When working on any electrical problem, always use pencil and paper. Write down everything you do and what you find. Draw sketches of wires and devices before you move them. Note carefully any information printed on a device. If you become confused and need the help of another person, your written notes will be valuable in explaining what you have done.

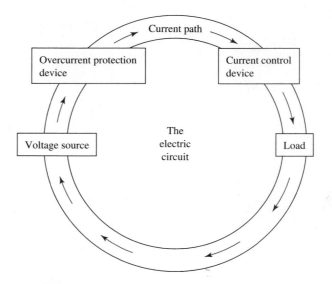

FIGURE 3–1
The five elements of an electric circuit.

The Voltage Source

The voltage source creates voltage to produce current and deliver energy to the circuit. We must understand the difference between an ideal voltage source and a practical voltage source.

An ideal voltage source can maintain a constant voltage while delivering an unlimited amount of current to any kind of load. A practical voltage source can deliver only a limited amount of current, and its output voltage decreases as the load current increases. A practical voltage source often requires an extra device called a *voltage regulator* to keep the voltage at a steady value when the load current changes.

You will see the terms *ideal* and *practical* many times in your study of electrical and electronic theory. Remember that any ideal device is an imaginary model with perfect characteristics, and any practical device is a real-world example with imperfect characteristics.

Direct Current (dc) and Alternating Current (ac)

There are two types of voltage sources: direct current *(dc)* and alternating current *(ac)*.

A dc voltage source has terminals labeled positive and negative. The polarity never changes. One terminal is always positive and the other always negative. Therefore, direct current always flows in the same direction.

A direct current source produces a steady voltage. Figure 3–2 is the symbol for a dc voltage source.

An ac voltage source does not have terminals labeled positive and negative because the polarity of the terminals alternates (changes) quickly. In other words, one terminal will be positive and the other negative for a short time. Then the positive terminal becomes negative, and the negative terminal becomes positive, for an equal time. Because the terminals change polarity, alternating current flows back and forth, changing directions when the polarity reverses. Figure 3–3 is the symbol for an ac voltage source.

An ac voltage source produces a voltage whose value is always changing, increasing from zero volts to a peak value, then decreasing back to zero. The polarity of the terminals reverses each time the voltage reaches zero volts. Commercial electric utility companies provide ac voltage to our homes, offices, factories, and so on. In the United States, commercial ac voltage changes polarity 120 times each second. This standard pattern is called 60 cycles per second or 60 Hz (hertz). In Europe, the standard is 50 Hz (100 changes in polarity per second).

Cells and Batteries

Cells and batteries are familiar dc voltage sources. A *cell* consists of two electrodes of different metals placed in a liquid or paste electrolyte, as shown in Figure 3–4. Chemical

FIGURE 3–2
The symbol for a
dc voltage source.

FIGURE 3–3
The symbol for an
ac voltage source.

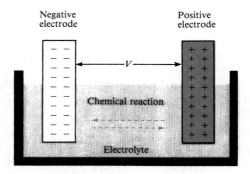

FIGURE 3–4
A cell produces voltage by chemical reactions between the electrodes and the electrolyte.

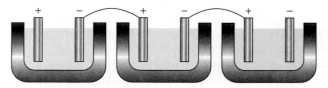

FIGURE 3–5
Several cells are connected to increase voltage and make a battery.

reactions between the electrolyte and the electrodes remove electrons from one electrode and transfer them to the other, producing a voltage.

The voltage of a cell is a natural value, determined by the metals used for the electrodes and the chemical used for the electrolyte. Cell voltages are low, between 0.6 V and 3.8 V. To create higher voltages, several cells are connected together to make a *battery,* as shown in Figure 3–5. We will study cells and batteries in detail in a later chapter.

Alternators

An *alternator* is a machine that produces ac voltage. An automobile alternator is shown in Figure 3–6. The operation of alternators will be explained in detail in a later chapter. The basic principle involves mechanical rotation between a coil of wire and a magnetic field. Thus, alternators need to be driven by an external mechanical power source. The automobile alternator is driven by fan belts from the automobile engine.

The Current Path

The current path allows current to move through the circuit between the terminals of the voltage source, delivering energy to the load. An ideal current path would not use up any of the energy in the current, allowing all the energy to be released in the load. However, all practical current paths use up energy due to resistance and other electrical characteristics. These energy losses are sometimes called line losses. Therefore, the current path is carefully designed to keep the line losses acceptably low.

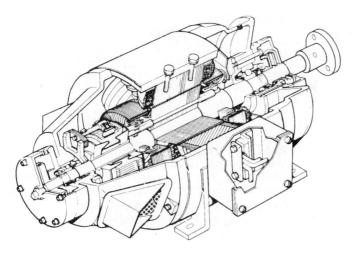

FIGURE 3–6
An automobile alternator.

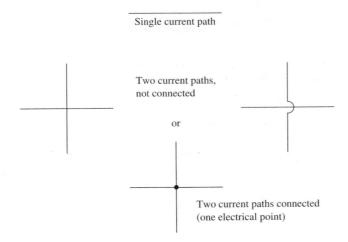

Single current path

Two current paths,
not connected

or

Two current paths connected
(one electrical point)

FIGURE 3–7
Current path symbols.

The symbol for a current path is a straight line. Figure 3–7 shows several symbols used to represent current paths and connections.

The Load

The load limits current and converts the energy of the current into useful work. The current is really seeking the other terminal of the voltage source. The load is simply something we put in the way of the current to absorb and release its energy.

If the current path goes only to the load and does not return to the voltage source, no current will flow. Therefore, the current path consists of two parts: a supply path from the voltage source to the load, and a return path from the load back to the voltage source.

In a complete circuit, current flows through the voltage source, current path, and load. Current cannot flow unless all three circuit elements are working and connected properly.

The Current Control Device

We need some way to control the current in the circuit. A *switch* controls current by opening or closing the current path. The switch may be a simple mechanical device (ordinary switch), a more complicated electromechanical device (relay or contactor), an electronic device (transistor, integrated circuit, and so on), or complex electronic equipment (computer, programmable controller, and so on).

The Overcurrent Protection Device

All circuits contain resistance and when current flows through resistance, heat is produced. If the current becomes abnormally high, the heat could damage or destroy the circuit. In extreme cases, the heat could start a fire. Therefore, all circuits need an overcurrent protection device to quickly stop all current if the current becomes too high. Fuses and circuit breakers are overcurrent protection devices.

→ *Self-Test*

1. What five elements does a circuit require?
2. What are the characteristics of a practical voltage source?
3. What is a voltage regulator?
4. What is the function of the current path?
5. What is the function of the load?
6. Describe the current control device.
7. What is the function of the overcurrent protection device?

SECTION 3–2: dc Power Supplies

In this section you will learn the following:

→ Electronic devices such as vacuum tubes, transistors, and integrated circuits require a dc voltage source.
→ A power supply is a circuit that converts ac voltage to regulated dc voltage.

Electronic circuits use vacuum tubes, transistors, or integrated circuits to control current. These devices all operate on direct current. Therefore, all electronic circuits require a dc voltage source. The source may be a battery or a special circuit called a *power supply,* which converts high-voltage ac into low-voltage dc. The power supply may be built into the equipment or may be mounted in its own capsule outside the equipment. You are probably familiar with converters that plug into wall outlets to operate portable radios, tape recorders, and so on.

There are dc power supplies available as separate units. These power supplies provide well-regulated dc voltage and current. A typical power supply, shown in Figure 3–8, produces a variable voltage of 0 to 20 V and several fixed voltages, such as +12 V, +5 V, and −5 V. Current capacity is typically 2 A to 5 A, and voltage and current are extremely well regulated by the electronic circuits.

A power supply often includes overcurrent protection circuits to protect both the power supply and the circuit from damage if the supply is connected improperly. In some cases, the user can adjust the maximum output current. If the current exceeds this amount, the power supply will shut off all current until the problem is corrected and the power supply is reset. The overcurrent protection circuit acts like an adjustable circuit breaker.

FIGURE 3–8
Regulated power supplies.

Other power supplies have a maximum value of output current. Since the supply will not exceed this value of current even with the terminals shorted (directly connected without a load), no harm can come to either the power supply or the circuit.

→ *Self-Test*

8. What devices require a dc voltage source?
9. What is a power supply?

SECTION 3-3: Resistors

In this section you will learn the following:

→ A resistor is a component used in electrical and electronic circuits to limit current and to produce specific voltage drops.

→ Resistors are rated by ohmic value, precision, and power rating.

→ The colored bands on a resistor indicate the ohmic value (significant digits and multiplier) and tolerance.

→ The tolerance of a resistor is the allowable difference between the nominal value and the actual value of a resistor, expressed as a percent of the nominal value.

→ The physical size of the resistor, or a value printed on the resistor, indicates the power rating.

→ A rheostat is a two-terminal variable resistor used to adjust current.

→ A potentiometer is a three-terminal variable resistor used to adjust voltage as a means of controlling vacuum tubes, transistors, and integrated circuits.

Resistors are used in electrical and electronic circuits to limit current and to produce specific voltages. Resistors are rated by ohmic value, precision, and power rating. Ohmic value and precision are often indicated by a series of colored bands on the resistor. Power rating is indicated by the physical size of the resistor. Some resistors are large enough to allow all this information to be printed on the body of the resistor. Figure 3–9 is the symbol for a resistor.

—\/\/\—

FIGURE 3–9
Symbol for a resistor.

TABLE 3–1
The electronic color code.

Black	0	Black hole (zero)
Brown	1	Brown one
Red	2	Two red lips
Orange	3	Orange tree (three)
Yellow	4	Yell for help (four)
Green	5	Five dollar bill (green)
Blue	6	Sick (six) and blue
Violet	7	Violet heaven (seven)
Grey	8	Great! (grey-eight)
White	9	White wine (nine)

The Electronic Color Code

The electronic color code is given in Table 3–1. Each color represents a digit, 0 through 9. You should memorize the color code because you will use it frequently in your career. Table 3–1 is a way to remember the color code.

Ohmic Value

Figure 3–10 shows a resistor with four colored bands, the most common style. The first and second bands are called the first and second significant digit bands. The third band is called the multiplier band. The fourth band is called the tolerance band.

Notice that one of the bands is on the end of the resistor. This is the first band. Bands one, two, and three are used to indicate the ohmic value of the resistor.

To find the ohmic value from the first three bands, follow these two steps:

1. Write down the digits indicated by the colors of the first two bands.
2. Write down the number of zeros indicated by the color of the third band.

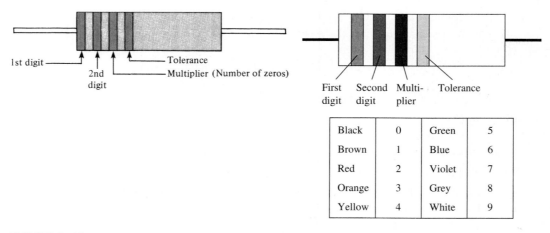

FIGURE 3–10
Resistor color code bands.

● *SKILL-BUILDER 3–1*

Suppose the first three bands of a resistor are colored brown, black, and red. What is the ohmic value?

STEP 1:
Write down the digits indicated
by the first two bands: brown represents 1
  1 0   black represents 0

STEP 2:
Write down a number of zeros
as indicated by the color of
the third band: red represents 2
  1 0 <u>0 0</u>   write down 2 zeros

ANSWER:
The ohmic value is 1,000 Ω, or 1 kΩ.

● *SKILL-BUILDER 3–2*

Suppose the first three bands of a resistor are colored green, brown, and black. What is the ohmic value?

STEP 1:
Write down the digits indicated
by the first two bands: green represents 5
  5 1   brown represents 1

STEP 2:
Write down a number of zeros as
indicated by the color of the
third band: black represents 0
  5 1   write down 0 zeros

ANSWER:
The ohmic value is 51 Ω.

When the third band is black, write down no zeros (0 zeros). A resistor whose third band is black will have an ohmic value between 10 Ω and 99 Ω.

Repeat the Skill-Builder using the following values. You should obtain the results that follow.

Band 1	Band 2	Band 3	Ohmic Value
red	violet	red	2,700 Ω
blue	grey	orange	68,000 Ω
orange	white	black	39 Ω
violet	green	yellow	750,000 Ω
blue	red	brown	620 Ω
brown	green	green	1,500,000 Ω

TABLE 3–2
Resistor tolerance.

NUMBER OF STRIPES	TOLERANCE COLOR	TOLERANCE VALUE
3	none	20%
4	silver	10%
4	gold	5%
5	silver	2%
5	gold	1%

Precision and Tolerance

There is usually a difference between the actual value of an individual resistor and the *nominal value* indicated by the first three bands. In other words, the nominal value of the bands is an approximate value rather than a precise value.

The *tolerance* is the allowable difference between the nominal value indicated by the bands and the actual value of a particular resistor. Tolerance is expressed as a percent. The color of the fourth band indicates the tolerance of the resistor. Table 3–2 indicates the standard tolerances and corresponding colors.

If a resistor has a small tolerance, its actual value will be close to its nominal value, and the resistor is a high-precision component. For example, a resistor with 1% tolerance is a high-precision resistor, and a resistor with 20% tolerance is a low-precision resistor. Most resistors in electronic circuits have a tolerance of 5% or less.

The nominal value and the tolerance are used to calculate two other ohmic values, the maximum and minimum values, in three steps:

1. Convert the tolerance from a percentage into a decimal fraction and multiply by the nominal value. The result is the tolerance in ohms.
2. The maximum value equals the nominal value plus the tolerance.
3. The minimum value equals the nominal value minus the tolerance.

● *SKILL-BUILDER 3–3*
- -

A resistor has four bands colored red-black-red-gold. Find the nominal value, tolerance, maximum value, and minimum value.

NOMINAL VALUE:

First significant digit:	2	red represents 2
Second significant digit:	0	black represents 0
Multiplier:	0 0	red represents 2
Nominal value:	2,000 Ω	

TOLERANCE: 5%   gold represents 5%

MAXIMUM AND MINIMUM VALUES:

1. $5\% = 0.05$ convert percentage to decimal fraction

    $0.05 \times 2{,}000\ \Omega = 100\ \Omega$ multiply decimal fraction by nominal value to find tolerance in ohms

2. $2,000\ \Omega + 100\ \Omega = 2,100\ \Omega$ maximum value equals nominal value plus tolerance

3. $2,000\ \Omega - 100\ \Omega = 1,900\ \Omega$ minimum value equals nominal value minus tolerance

Repeat the Skill-Builder using the following values. You should obtain the results that follow.

Given:

Band 1	blue	brown	yellow	orange
Band 2	grey	green	violet	blue
Band 3	orange	green	red	brown
Band 4	silver	none	silver	gold

Find:

Value	$68,000\ \Omega$	$1,500,000\ \Omega$	$4,700\ \Omega$	$360\ \Omega$
Tolerance %	10%	20%	10%	5%
Tolerance Ω	$6,800\ \Omega$	$300,000\ \Omega$	$470\ \Omega$	$18\ \Omega$
Maximum value	$74,800\ \Omega$	$1,800,000\ \Omega$	$5,170\ \Omega$	$378\ \Omega$
Minimum value	$61,200\ \Omega$	$1,200,000\ \Omega$	$4,230\ \Omega$	$342\ \Omega$

If the actual value of a resistor is between the maximum and minimum values, the resistor is within tolerance and acceptable for use in the circuit. On the other hand, if the actual value is greater than the maximum or less than the minimum, the resistor is out of tolerance and cannot be used.

Resistors Less Than 10 Ω

If the resistor has an ohmic value between 1 Ω and 9.9 Ω, the multiplier band is gold. For example, a blue-grey-gold-gold resistor has an ohmic value of 6.1 Ω with a tolerance of 5%.

If the resistor has an ohmic value less than 1 Ω, the multiplier band is silver. For example, a yellow-violet-silver-silver resistor has an ohmic value of 0.47 Ω with a tolerance of 10%.

High-Precision Resistors

A high-precision resistor has five bands and a tolerance of either 1% or 2%. The first three bands are significant digits, the fourth band is the multiplier, and the fifth band is the tolerance: Gold is 1% and silver is 2%.

Standard Values

Resistors have standard values. The standard values depend on the tolerance range and are chosen so that the maximum value of one resistor overlaps the minimum value of the next resistor.

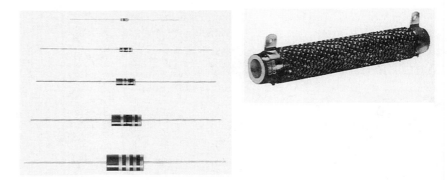

FIGURE 3–11
A resistor's power rating depends on the physical size of the resistor.

Power Rating

Heat and temperature are different. Heat is the amount of energy in an object. Temperature is the concentration of heat and is related to the size of an object. If resistors are bigger, they can contain more heat because the heat can spread out and the temperature will stay within a normal range.

The heat produced in a resistor depends on the ohmic value of the resistor and the amount of current. Heat is directly related to power, and so the power rating of a resistor is given in watts. The power rating is the maximum amount of power that the resistor can have and still stay at a safe temperature.

Resistors have standard power ratings of 1/8 W, 1/4 W, 1/2 W, 1 W, and 2 W. Larger power resistors may have much higher power ratings. Figure 3–11(a) shows the actual body size of standard power rating resistors.

Types of Resistors

The carbon-composition resistor shown in Figure 3–12 is common and inexpensive. The resistor is a mixture of carbon and clay, with a plastic coating and two wire leads embedded in the ends. When the resistor becomes too hot, the coating produces an odor which is easily recognized.

Wire-Wound Resistor

A wire-wound resistor is made of fine, high-resistance wire wrapped around an insulated body. Wire-wound resistors may be large with high power ratings of 50 W or more, as shown in Figure 3–11(b).

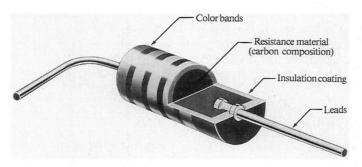

FIGURE 3–12
Cutaway view of a carbon-composition resistor.

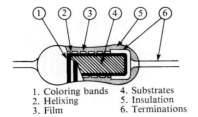

1. Coloring bands   4. Substrates
2. Helixing   5. Insulation
3. Film   6. Terminations

FIGURE 3–13
A film-type resistor.

Film-Type Resistors

Film-type resistors have a film strip of either carbon or metal wrapped around an insulated core. These resistors are easily recognized by the ends being slightly larger than the body, as shown in Figure 3–13.

Adjustable Resistors: Rheostats and Potentiometers

The resistors described previously are fixed resistors, which means that the ohmic value is not adjustable.

There are two types of adjustable resistors: the *rheostat* and the *potentiometer.* Rheostats are used to control current, and potentiometers are used to control voltage. The symbols for a rheostat and potentiometer are shown in Figure 3–14.

A rheostat is a two-terminal device. Inside the rheostat, the resistive element is exposed. One contact is fixed at one end of the resistive element. The other contact is a metal arm that slides along the resistive element when a shaft or screw is turned. The resistance between the contacts changes according to the amount of the resistive element between the fixed contact and the sliding contact.

Potentiometers are similar in construction to rheostats. The potentiometer is a three-terminal device with fixed contacts at both ends. The middle contact is attached to a moveable arm called a wiper, which touches the resistive element. A potentiometer may be converted into a rheostat by connecting the wiper to an end contact.

Figure 3–15 shows how potentiometers are used to control voltage. When the sliding arm touches terminal 3, the voltage between terminals 1 and 2 is 12 V. When the arm touches terminal 1, the voltage between terminals 1 and 2 is zero volts. Therefore, when the arm touches the resistive element somewhere between terminals 1 and 3, the voltage between terminals 1 and 2 is somewhere between 12 V and zero volts. In other words, the voltage between terminals 1 and 2 may be adjusted by moving the sliding arm.

Potentiometers are widely used as controls in electronic equipment, such as a volume, tone, or balance control on a stereo, or a brightness, contrast, or color control on a video monitor. A potentiometer usually controls the voltage applied to a transistor or integrated circuit. Figure 3–16 shows a variety of typical potentiometers. A potentiometer is often called a pot.

Potentiometers are available in two styles: linear and tapered. In a linear pot, each section of the resistive element has an equal amount of resistance. In a tapered pot, the sections of the element have unequal amounts of resistance. Tapered pots are often used for volume controls because the human ear responds unevenly to changes in loudness.

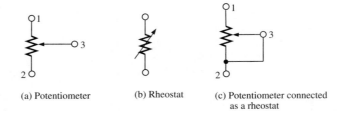

(a) Potentiometer   (b) Rheostat   (c) Potentiometer connected as a rheostat

Sliding contactor (wiper)

1

Rotating shaft

2

3

Resistive carbon arc

(d) Basic potentiometer construction

FIGURE 3–14
(a) Potentiometer. (b) Rheostat. (c) Potentiometer connected as a rheostat. (d) Potentiometer construction.

→ *Self-Test*

10. What is a resistor?
11. How are resistors rated?
12. What is indicated by the colored bands on a resistor?
13. What is the tolerance of a resistor?
14. What indicates the power rating of a resistor?
15. What is a rheostat?
16. What is a potentiometer?

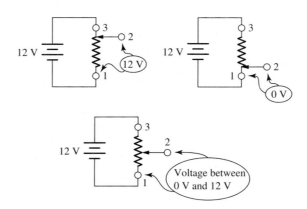

FIGURE 3–15
A potentiometer is often used to adjust voltage.

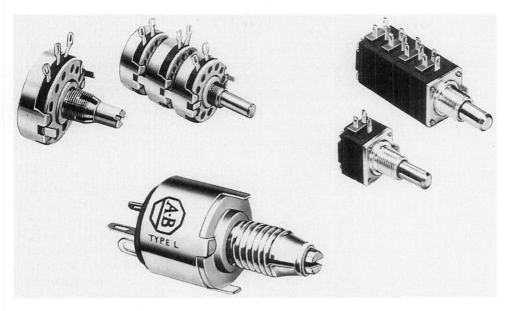

FIGURE 3–16
Potentiometers.

SECTION 3–4: Switches

In this section you will learn the following:

- ➜ A switch pole is a moveable contact inside a switch, controlling a current path.
- ➜ A closed position completes a current path. An open position interrupts a current path.
- ➜ A switch throw is a closed position of a switch.

Mechanical switches consist of moveable metal contacts that are part of the current path. When the switch is in its on position, the metal contacts touch, and current can flow. When the switch is in its off position, the contacts separate, and current cannot flow. The simplest style of switch is the knife switch, illustrated in Figure 3–17, where the action of the metal contact is easy to see.

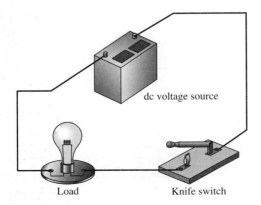

FIGURE 3–17
The knife switch is the simplest type of current control device.

Strictly speaking, the words *on* and *off* apply to the load, not to the switch. We describe the switch as being open or closed. For example, think of an ordinary wall switch controlling a ceiling light. When the switch is closed (contacts touching), the light is on. When the switch is open (contacts separated), the light is off.

Switches are classified by the number of *poles* and *throws* they have. The word *pole* has a number of different meanings in electricity and electronics. In the case of switches, the pole is the moveable contact inside the switch. In practical terms, the number of switch poles is the number of current paths that the switch can control. Thus, a single-pole switch can control only one current path, and a double-pole switch can control two current paths at the same time.

The word *throw* refers to the number of contacts controlled by each pole. In practical terms, the throw is the number of closed or on positions. For example, an ordinary wall switch has two positions, but only one of them is a closed position. Therefore, the wall switch is a single-throw switch. However, there is another type of wall switch, commonly called a three-way switch, which is used in halls and stairs to control the lights from either end. Both positions of the switch are closed positions, and thus the switch is a double-throw switch. Figure 3–18 illustrates single-throw and double-throw (three-way) switches used to control lights. In the case of the pair of three-way switches, no matter which position one switch is in, the other switch can open or close the current path.

The terms *pole* and *throw* are generally abbreviated, as follows:

SPST	single-pole, single-throw
SPDT	single-pole, double-throw
DPST	double-pole, single-throw
DPDT	double-pole, double-throw

Toggle switches (Figure 3–19(a)) and rocker switches (Figure 3–19(b)) are widely used in electronic equipment. They are available in a variety of sizes.

A slide switch is often small in size. The handle does not protrude much, and the switch is quite rugged.

A DIP (double in-line package) switch (Figure 3–19(c)) is a group of miniature rocker or slide switches packaged in a standard size to fit on a printed circuit board. DIP switches are widely used in computers and printers to allow the user to customize the machine's operation.

Push-button (Figure 3–19(d)) switches may be either momentary-contact or push-on-push-off types. A momentary-contact switch has a spring that holds the contacts in a

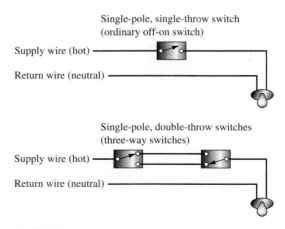

FIGURE 3–18
Single-pole and double pole switches.

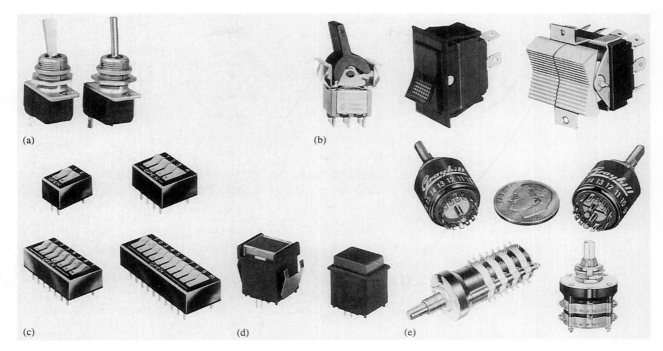

FIGURE 3–19
Switches: (a) toggle, (b) rocker, (c) DIP, (d) push-button, (e) rotary.

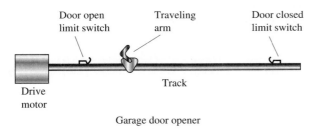

Garage door opener

FIGURE 3–20
Limit switches on a garage door opener.

normal position. Pressing the button places the contacts momentarily (temporarily) in the other position. Thus, a switch may be a NOPB (normally open push-button) type, or an NCPB (normally closed push-button) type. An ordinary doorbell switch is an example of an NOPB.

Rotary switches (Figure 3–19(e)) are frequently used as multifunction selector switches. They are usually custom-designed for a particular machine.

Micro switches are miniature momentary-contact rocker switches used to detect the physical position of an object. Thus, they are frequently called limit switches. For example, a garage door opener has two limit switches mounted on the track, as shown in Figure 3–20. The switches are located so the traveling arm will press the switch rockers when the door is fully opened or fully closed. The switches then turn off the drive motor to stop the door from moving further.

Industrial machines used in automated manufacturing may contain dozens of limit switches interconnected to allow safe and automatic operation of the machine. The switches may be operated by mechanical movements of the machine or by other forces such as temperature, pressures of liquids and gases, or the level of solid or liquid contents

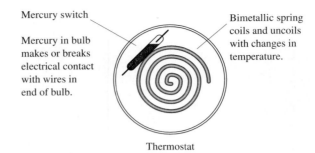

FIGURE 3–21
Mercury switch in a thermostat

in a tank. Many of the warning lights on an automobile instrument panel are operated by limit switches that respond to low fuel level, low oil pressure, or high coolant temperature.

Mercury switches (Figure 3–21) use a small quantity of mercury in a bulb to open or close the current path between two electrodes mounted in the bulb. Thus, mercury switches respond to gravity. Tilting the bulb one way causes the mercury to flow over the contacts, closing the switch. Tilting the other way causes the mercury to flow away from the contacts, opening the switch.

Mercury switches are used in wall-mounted thermostats that control central heating and air-conditioning units. The mercury bulb is mounted on a spring that coils and uncoils with changes in temperature. The small mercury switch controls a larger magnetically operated switch called a contactor, mounted near the central air unit. In a later chapter, we will learn more about contactors and relays.

Figure 3–22 shows the symbols for several switch types.

→ *Self-Test*

17. What is a switch pole?
18. Describe the closed and open positions of a switch pole.
19. What is a switch throw?

SECTION 3–5: Fuses

In this section you will learn the following:

→ A fuse is an overcurrent protection device placed in the current path and designed to melt from the heat produced by a rated amount of current, thus opening the current path.

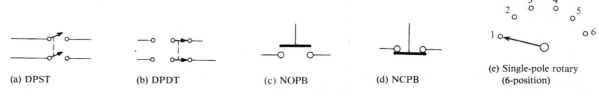

FIGURE 3–22
Switch symbols: (a) SPST, (b) SPDT, (c) DPST, (d) DPDT, (e) NOPB, (f) NCPB, (g) rotary.

FIGURE 3–23
Fuses and fuse holders.

→ An overload is an excessive current caused by placing too many loads in the circuit.

→ A surge is a sudden extreme increase in source voltage.

→ A short is a decrease in load resistance to nearly zero ohms.

→ A time-delay fuse tolerates a brief overload of two to three times rated current.

As shown in Figure 3–23, a *fuse* is a short piece of metal that is thin and narrow. The metal is an alloy with a lower melting point than the metal in the wires. The fuse is mounted inside a hollow glass or cardboard container called the fuse body and attached to metal contacts on the outside of the fuse body. The fuse is mounted in a fuse holder which is part of the current path.

Because the fuse is thin and narrow, its resistance is higher than other parts of the current path. The fuse is the weak point in the current path. As current increases, the fuse becomes hotter than the wires. Eventually, the fuse becomes so hot that it melts, opening the current path and stopping all current. The current rating of a fuse, such as 0.5 A or 30 A, is the value of current that will produce enough heat to melt or blow the fuse. A fuse is an inexpensive, replaceable component that is sacrificed to protect the rest of the circuit in case of excessive current.

There are basically three circuit problems that can blow a fuse: an overload, a surge, and a short.

An overload is an excessive current caused by placing too many loads in the circuit. The time it takes for a moderate overload to melt a fuse may vary between a few seconds and several minutes. Because the fuse melts slowly, the glass fuse body will remain clear.

A surge is a sudden extreme increase in voltage, such as lightning striking a power line. Voltage surges may damage delicate electronic circuits before the fuse has time to blow, and often the surge is so brief that the fuse never blows. Therefore, it is a good idea to use a surge suppressor to protect electronic equipment.

A short is a decrease in load resistance to nearly zero ohms, usually caused by wires touching each other so current flows around the load instead of through it. Either condition, a surge or a short, will produce a current much higher than the rating of the fuse. The fuse melts violently, sputtering tiny beads of metal on the glass and turning it dark.

A fuse rarely goes bad by itself. A blown fuse is almost always a sign of trouble somewhere else in the circuit. By examining the glass body of a blown fuse, you may get a clue to the problem. Clear glass suggests an overload, and darkened glass covered with tiny beads of metal suggests a surge or a short. Table 3–3 lists typical fuse conditions and possible circuit troubles that could cause these conditions.

TABLE 3–3
Blown fuse causes and symptoms.

CAUSE	APPEARANCE OF BLOWN FUSE	RESULT OF REPLACING BLOWN FUSE
Overload	Glass is clear. No sign of melting or splattering.	Circuit operates normally for some time, then blows fuse again.
Short	Glass is dark or smudged. Tiny bits of melted metal splattered on glass.	Circuit blows new fuse immediately when power is applied.
Surge	Same as short.	Circuit operates normally.

Many electric motors draw two or three times normal current while starting up and gaining speed. A normal fuse may blow under these conditions. Therefore, a *time-delay fuse* (nicknamed a slow-blow fuse) is often used for motors. This fuse take a little longer to melt than a normal fuse, giving the motor time to get up to speed and operate at a normal value of current. In other words, a time-delay fuse will tolerate a temporary overload, but not a continuous overload. When replacing a fuse, use the correct type: normal or slow-blow.

→ *Self-Test*

20. What is a fuse?
21. What is an overload?
22. What is a surge?
23. What is a short?
24. Describe a time-delay fuse.

SECTION 3–6: Electrical Symbols and Diagrams

In this section you will learn the following:

→ In any electrical circuit, current is used either as a source of power or as a means of carrying a signal.
→ A schematic diagram is a drawing that shows how current or signals move through a circuit.
→ A schematic diagram uses symbols to represent the components in the circuit.
→ The location of symbols in the schematic may be different from the location of components in the actual circuit.

In any electrical circuit, current is used either as a source of power or as a means of carrying a signal. For example, in a heater, lamp, or motor, the current is a source of power. However, in a television, stereo, or computer, the current is used as a signal and represents information: pictures, sounds, words, numbers, and commands.

A *schematic diagram* is a drawing that shows how current or signals move through a circuit. When current is used solely for power, the schematic generally provides enough detail to allow you to trace the complete flow of current throughout the circuit. Figure 3–24, a schematic for a flashlight, shows the complete details of the current path.

However, when current is used as a signal or control, the schematic may not show all the details of the current path. This is especially true of integrated circuits (ICs or chips) mounted on printed circuit boards. These schematics show the flow of signals and

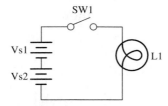

FIGURE 3–24
Schematic of a flashlight.

commands but do not show all the details of how the current moves through the circuit. Because the signals and commands may be peculiar to the particular circuit, a service manual that explains the circuit is often necessary to understand the schematic.

For now, we will limit our study to basic schematics that show all details of the current path. The circuit components are represented in the schematics by standard symbols. Figure 3–25 shows the symbols for the components we have studied thus far. These basic symbols should be memorized.

The location of symbols in the schematic may be quite different from the location of components in the actual circuit. For example, Figure 3–26 is a schematic diagram for

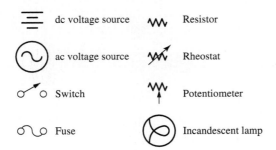

FIGURE 3–25
Basic schematic symbols.

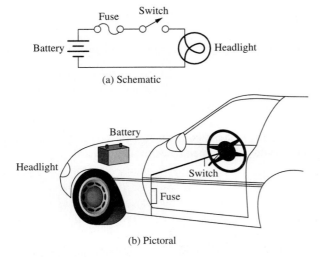

FIGURE 3–26
(a) Automobile headlight schematic. (b) Placement
of actual components.

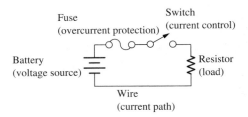

FIGURE 3–27
Schematic of the most basic five-element electric circuit.

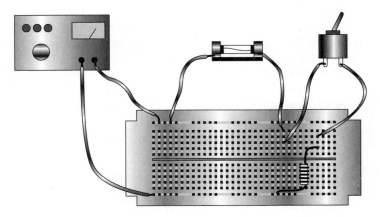

FIGURE 3–28
The circuit of Figure 3–27 mounted on an experiment board.

a headlight in an automobile. In the schematic, the battery, the switch, and the headlight appear to be near each other and connected in a simple straight line. However, in the actual automobile, the components are in widely separated locations. On the other hand, a schematic of a printed circuit board may show two components widely separated, yet the components may be next to each other on the board.

A Basic Circuit Schematic

Figure 3–27 illustrates the most basic circuit: a dc voltage source, fuse, SPST switch, and resistive load. The lines connecting the components indicate a current path. The details of the current path are not given, so we do not know whether the path is made of wire, a printed circuit board, an experiment board, or something else. The schematic diagram simply tells us that somehow these components are connected as shown. Figure 3–28 shows how the circuit could be set up on an experiment board.

→ *Self-Test*

25. What are the two uses of current?
26. What is a schematic diagram?
27. How does a schematic diagram represent components in a circuit?
28. What is the relationship between the location of symbols in the schematic diagram and the location of components in the actual circuit?

● TROUBLESHOOTING: Applying Your Knowledge

You discover that your neighbor has installed a normally closed push-button (NCPB) switch. You inform him that a doorbell switch should be a normally open push-button (NOPB) switch.

CHAPTER SUMMARY: ANSWERS TO SELF-TESTS

1. A circuit requires a voltage source, a current path, a load, a current control device, and an overcurrent protection device.
2. A practical voltage source can provide only a limited amount of current. As the output current increases, output voltage decreases.
3. A voltage regulator is a special circuit or device used to maintain constant output voltage in a practical voltage source.
4. The current path provides a low-resistance path for current to travel from one terminal of the voltage source to the other terminal.
5. The energy of the current is transferred to the load, producing useful work.
6. The current control device may be simple or complex, mechanical or electronic. The most common current control device is a switch.
7. The overcurrent protection device stops all current if the current becomes abnormally high, protecting the circuit from damage due to heat and fire.
8. Electronic devices such as vacuum tubes, transistors, and integrated circuits require a dc voltage source.
9. A power supply is a circuit that converts ac voltage to regulated dc voltage.
10. A resistor is a component used in electrical and electronic circuits to limit current and to produce specific voltage drops.
11. Resistors are rated by ohmic value, precision, and power rating.
12. The colored bands on a resistor indicate the ohmic value (significant digits and multiplier) and tolerance.
13. The tolerance of a resistor is the allowable difference between the nominal value and the actual value of a resistor, expressed as a percent of the nominal value.
14. The physical size of the resistor, or a value printed on the resistor, indicates the power rating.
15. A rheostat is a two-terminal variable resistor used to adjust current.
16. A potentiometer is a three-terminal variable resistor used to adjust voltage as a means of controlling vacuum tubes, transistors, and integrated circuits.
17. A switch pole is a moveable contact inside a switch, controlling a current path.
18. A closed position completes a current path. An open position interrupts a current path.
19. A switch throw is a closed position of a switch.
20. A fuse is an overcurrent protection device placed in the current path and designed to melt from the heat produced by a rated amount of current, thus opening the current path.
21. An overload is an excessive current caused by placing too many loads in the circuit.
22. A surge is a sudden extreme increase in source voltage.
23. A short is a decrease in load resistance to nearly zero ohms.
24. A time-delay fuse tolerates a brief overload of two to three times rated current.
25. In any electrical circuit, current is used either as a source of power or as a means of carrying a signal.
26. A schematic diagram is a drawing that shows how current or signals move through a circuit.
27. A schematic diagram uses symbols to represent the components in the circuit.
28. The location of symbols in the schematic may be different from the location of components in the actual circuit.

CHAPTER REVIEW QUESTIONS

Determine whether the following questions are true or false.

3–1. Because the current reverses polarity, an ac circuit does not require overcurrent protection.

3–2. A voltage regulator protects a circuit against surges by opening the current path.

3–3. A dc voltage source has terminals that never change polarity.

3–4. A cell consists of two electrodes of identical metals placed in a liquid or paste electrolyte.

3–5. An alternator is an ideal voltage source.

3–6. All electronic equipment requires a dc voltage source.

3–7. Resistor power rating is indicated by a series of colored bands on the resistor.

3–8. Rheostats are used to adjust voltage.

3–9. A fuse is made of a metal with a lower melting point than the metal in the wires.

Select the response that best answers the question or completes the statement.

3–10. The ability to maintain a constant value of voltage under all load conditions is called
   a. overload protection
   b. voltage regulation
   c. precision
   d. polarity

3–11. In the United States, commercial ac voltage changes polarity 120 times each second. This standard pattern is called _____ hertz ac.
   a. 50   c. 100
   b. 60   d. 120

3–12. Cell voltages vary between
   a. 1.5 V and 12 V   c. 0.6 V and 3.8 V
   b. 50 V and 60 V   d. 120 V and 240 V

3–13. In Figure 3–29, which is the correct symbol for connected current paths?

3–14. A power supply is a circuit that converts
  a. dc voltage to regulated ac voltage.
  b. unregulated dc voltage to regulated dc voltage.
  c. unregulated ac voltage to regulated ac voltage.
  d. ac voltage to regulated dc voltage.

3–15. Which of the following is *not* a resistor rating?
  a. ohmic value   **c.** precision
  b. voltage rating   **d.** power rating

3–16. Which of the following is *not* a standard value for 5% tolerance resistors?
  a. 2.7 kΩ   **d.** 78 kΩ
  b. 560 Ω   **e.** 68 Ω
  c. 4.7 MΩ   **f.** 910 Ω

3–17. The maximum amount of power that a resistor can dissipate at a safe temperature is called the
  a. tolerance   **c.** maximum value
  b. power rating   **d.** nominal value

3–18. Potentiometers are used to adjust
  a. current   **c.** power
  b. tolerance   **d.** voltage

3–19. Potentiometers are available in two styles:
  a. ac and dc   **c.** 5% and 10%
  b. linear and tapered   **d.** high and low

3–20. Which of the following is *not* a standard type of switch?
  a. NCPB   **d.** SPST
  b. DPDT   **e.** DPST
  c. NPST   **f.** NOPB

3–21. What type of switch would most likely be found on the main circuit board of a printer?
  a. DIP   **c.** rotary
  b. toggle   **d.** push-button

3–22. What type of switch is usually custom-designed for a particular machine?
  a. toggle   **c.** rocker
  b. rotary   **d.** slide

3–23. What type of switch would most likely be found on a machine used in automated manufacturing?
  a. DIP   **c.** limit
  b. three-way   **d.** fused

3–24. What type of switch would most likely be found on a thermostat?
  a. relay   **c.** rotary
  b. knife   **d.** mercury

3–25. An excessive current caused by placing too many loads in the circuit is called
  a. a short   **c.** an overload
  b. a surge   **d.** a line loss

3–26. A sudden extreme increase in source voltage is called
  a. a short   **c.** an overload
  b. a surge   **d.** a line loss

3–27. A decrease in load resistance to nearly zero ohms is called
  a. a short   **c.** an overload
  b. a surge   **d.** a line loss

3–28. Which type of fuse is often used with motors?
  a. regular   **c.** tapered
  b. linear   **d.** time-delay

3–29. Refer to Figure 3–30 and identify the schematic symbols for each of the following components
  _____ resistor   _____ fuse
  _____ dc voltage source   _____ rheostat
  _____ potentiometer   _____ switch
  _____ ac voltage source   _____ lamp

Solve the following problems.

3–30. You examine a group of resistors and find the colors of the first three bands are as follows. What is the nominal ohmic value of each resistor?
  a. yellow-violet-green
  b. brown-black-black
  c. blue-grey-red
  d. white-brown-orange
  e. green-brown-gold
  f. red-violet-silver

3–31. What are the maximum and minimum values for each of the following resistors?
  a. red-black-red-gold
  b. green-blue-yellow-silver
  c. brown-green-orange

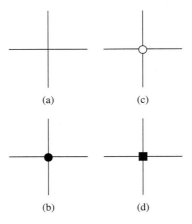

(a)   (c)

(b)   (d)

FIGURE 3–29

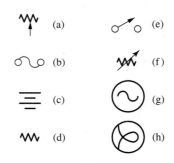

(a)   (e)

(b)   (f)

(c)   (g)

(d)   (h)

FIGURE 3–30

3–32. What sequence of colored bands represents each of the following resistors?

    a. 1.5 kΩ ± 10%   **c.** 360 Ω ± 20%

    b. 5.6 MΩ ± 5%   **d.** 4.7 Ω ± 5%

The following problems are more challenging.

3–33. The following list contains the nominal value, tolerance, and actual value of several resistors. For each one, determine whether the resistor is within tolerance, and show calculations to support your answer.

Nominal	Actual
a. 2.7 kΩ ± 10%	2,850 Ω
b. 33 kΩ ± 5%	31,800 Ω
c. 470 kΩ ± 2%	453,000 Ω
d. 12 Ω ± 5%	11.28 Ω
e. 5.1 MΩ ± 20%	6,250,000 Ω
f. 3.6 kΩ ± 2%	3,660 Ω

3–34. A dc voltage source has an output of 12.1 V with a 20-kΩ load, and an output of 10.8 V with a 50-Ω load. Is the source ideal or practical? Explain your answer.

3–35. You believe a circuit has a short in it because replacement fuses blow instantly and violently when power is applied. How can you cause a small, safe amount of current to flow in the circuit long enough for you to find the short?

GLOSSARY

ac Alternating current: current whose polarity reverses periodically and whose value constantly varies between zero and a maximum value.

alternator A machine that produces ac voltage.

battery A dc voltage source consisting of two or more cells connected together.

cell A dc voltage source consisting of two electrodes of different metals placed in a liquid or paste electrolyte.

dc Direct current: current that maintains constant polarity and value.

fuse An overcurrent protection device designed to melt from the heat produced by a rated amount of current.

nominal value The value of a component indicated by its markings.

overload An excessive current caused by placing too many loads in the circuit.

pole A moveable contact arm inside a switch.

potentiometer A three-terminal variable resistor used to adjust voltage.

power supply The section of an electronic circuit that converts ac voltage to dc voltage.

resistor Electronic component used to limit current or to produce specific voltage drops.

rheostat A two-terminal variable resistor used to adjust current.

schematic diagram An electrical diagram that shows the scheme or plan by which current or signals move through a circuit.

short A decrease in load resistance to nearly zero ohms.

surge A sudden extreme increase in source voltage.

switch A device that controls current by completing or interrupting the current path.

throw A closed position of a switch.

time-delay fuse A fuse that tolerates a brief overload of two to three times rated current.

tolerance the amount by which the actual value of a component is allowed to differ from the nominal value, expressed as a percent of nominal value.

voltage regulator A device that maintains a constant value of voltage under all load conditions.

4 ELECTRICAL CALCULATIONS AND MEASUREMENTS

● SECTION OUTLINE

4–1 Ohm's Law

4–2 Watt's Law

4–3 The Formula Wheel

4–4 Multimeters

4–5 Measuring Voltage

4–6 Measuring Current

4–7 Measuring Resistance

4–8 Other Electrical Meters and Measurements

● TROUBLESHOOTING: Applying Your Knowledge

Your mother is worried about her new microwave oven. The 120-volt electrical outlets in her kitchen are controlled by a 20-amp circuit breaker. She already has a 1,200-W toaster and a 1,000-W coffee maker plugged in, and they work normally. Her new microwave oven is rated at 2.6 kW, and every time she turns it on, the circuit breaker trips open. She wants your advice. Do you think something is wrong with the new microwave oven? Should she take it back to the store? What should she do? You'll find the answer at the end of this chapter.

● OVERVIEW AND LEARNING OBJECTIVES

The relationship between current, voltage, and resistance is known as Ohm's Law. The relationship between power, voltage, and current is known as Watt's Law. These two laws are expressed as simple formulas, which are widely used in electrical calculations.

We can rewrite the two basic formulas into 12 formulas, which are traditionally arranged in a format known as the Formula Wheel. If any two basic quantities (voltage, current, resistance, or power) are known, the other two can be easily found using the Formula Wheel.

Electrical meters are necessary to measure values of voltage, current, resistance, and power in circuits. The measured values indicate whether the circuit is performing properly or not, and where the trouble may be.

Student technicians have an additional need for electrical meters. As you begin to construct circuits, you will use meter readings to verify the electrical laws and rules you learn. Also when you make a mistake in constructing a circuit, the meter readings may help you identify your error quickly.

After completing this chapter, you should be able to:

1. Predict how current will change when voltage or resistance changes, according to Ohm's Law.

2. Predict how power will change when voltage or current changes, according to Watt's Law.

3. Use the Formula Wheel to perform basic calculations involving Ohm's Law and Watt's Law.

4. Name the parts of a VOM and DMM, and explain the function of each part.

5. Explain how to set up the meter to measure voltage, current, or resistance.

6. Explain how to apply the probes to the circuit or device in order to measure voltage, current, or resistance.

7. Describe electrical meters other than the VOM and DMM.

SECTION 4–1: Ohm's Law

In this section you will learn the following:

→ Ohm's Law expresses the relationship between voltage, current, and resistance.
→ Ohm's Law states that resistance is the ratio of voltage and current: $R = V \div I$.
→ If resistance is constant, current is directly proportional to voltage: An increase in voltage produces an increase in current.
→ If voltage is constant, current is inversely proportional to resistance: An increase in resistance produces a decrease in current.

Ohm's Law defines the mathematical relationships between current, voltage, and resistance. In its most basic form, Ohm's Law states that resistance is the ratio of voltage and current: $R = V \div I$.

Throughout your study of electronics, always remember that any value of ohms is simply the result of this formula: voltage divided by current. For example, if you have a 12-V battery producing 2 A of current, you can say that you have a resistance of 6 Ω because 12 divided by 2 equals 6. This is an important point. Voltage is real, and current is real; but ohms are just a ratio, the answer you get when you divide voltage by current. Formula Diagram 4–1 expresses Ohm's Law. Ohm's Law is one of the most basic and important of all electrical formulas, and you should memorize it.

Formula Diagram 4–1: Ohm's Law

$$\begin{array}{c|c} \multicolumn{2}{c}{\text{Voltage } (V)} \\ \multicolumn{2}{c}{\text{volts V}} \\ \hline \text{Resistance } (R) & \text{Current } (I) \\ \text{ohms } \Omega & \text{amps A} \end{array}$$

From Formula Diagram 4–1, we can see that current equals voltage divided by resistance. Because we are often interested in current, we can express Ohm's Law as follows:

1. If resistance is constant, current is directly proportional to voltage. This means that an increase in voltage produces an increase in current.
2. If voltage is constant, current is inversely proportional to resistance. This means that an increase in resistance produces a decrease in current.

● SKILL-BUILDER 4–1

- -

What is the current in the circuit in Figure 4–1?

GIVEN:
  $V = 36$ V
  $R_L = 18$ Ω

FIND:
  I

SOLUTION:
  $I = V \div R_L$ select formula
    $= 36$ V $\div 18$ Ω   substitute values
    $= 2$ A   divide

     Repeat the Skill-Builder using the following values. You should obtain the results that follow.

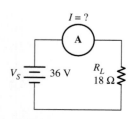

FIGURE 4–1

Given:

V	48 V	35 V	27 V	91 V	36 V
R_L	12 Ω	7 Ω	9 Ω	14 Ω	5 Ω

Find:

I	4 A	5 A	3 A	6.5 A	7.2 A

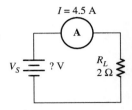

FIGURE 4–2

● *SKILL-BUILDER 4–2*

What is the source voltage in the circuit in Figure 4–2?

GIVEN:
   $I = 4.5$ A
   $R_L = 2$ Ω

FIND:
   V_S

SOLUTION:
   $V_S = I \times R_L$ select formula
      $= 4.5$ A $\times 2$ Ω   substitute values
      $= 9$ V   multiply

   Repeat the Skill-Builder using the following values. You should obtain the results that follow.

Given:

I	16 A	12 A	14.2 A	11.6 A	3.7 A
R_L	4 Ω	6 Ω	5 Ω	3 Ω	6.3 Ω

Find:

V	64 V	72 V	71 V	34.8 V	23.31 V

● *SKILL-BUILDER 4–3*

What is the load resistance in the circuit in Figure 4–3?

GIVEN:
   $V_S = 6$ V
   $I = 1.5$ A

FIND:
   R_L

SOLUTION:
   $R_L = V_S \div I$ select formula
      $= 6$ V $\div 1.5$ A   substitute values
      $= 4$ Ω   divide

   Repeat the Skill-Builder using the following values. You should obtain the results that follow.

Given:

V_S	68 V	42 V	135.6 V	54.12 V	97.44 V
I	17 A	6 A	12 A	3.3 A	11.2 A

Find:

R_L	4 Ω	7 Ω	11.3 Ω	16.4 Ω	8.7 Ω

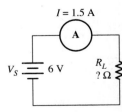

FIGURE 4–3

→ *Self-Test*

1. What relationship does Ohm's Law express?
2. What does Ohm's Law state?
3. According to Ohm's Law, if resistance is constant, what is the relationship between current and voltage?
4. According to Ohm's Law, if voltage is constant, what is the relationship between current and resistance?

SECTION 4–2: Watt's Law

In this section you will learn the following:

→ Watt's Law expresses the relationship between voltage, current, and power.
→ Watt's Law states that power is the product (combination) of voltage and current: $P = V \times I$.
→ A low-voltage circuit needs high current to produce large amounts of power. A high-voltage circuit can produce the same amount of power with a low current.

Watt's Law defines the mathematical relationships between power, voltage, and current. As stated earlier, the power of a circuit is the product of voltage and current: Watts equals volts multiplied by amps.

Formula Diagram 4–2: Watt's Law

Power (P)	
watts W	
Voltage (V)	Current (I)
volts V	amps A

● *SKILL-BUILDER 4–4*

What is the power in the circuit in Figure 4–4?

GIVEN:
  $V_S = 12$ V
  $I = 3$ A

FIND:
  P_L

SOLUTION:

$P_L = V_S \times I$	select formula
$= 12$ V $\times 3$ A	substitute values
$= 36$ W	multiply

Repeat the Skill-Builder using the following values. You should obtain the results that follow.

Given:

V_S	16 V	28 V	41 V	23.5 V	34.8 V
I	5 A	4 A	3.6 A	12 A	13.5 A

Find:

P_L	80 W	112 W	147.6 W	282 W	469.8 W

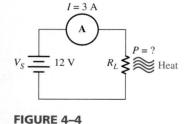

FIGURE 4–4

● *SKILL-BUILDER 4–5*

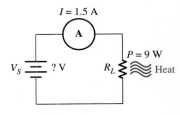

FIGURE 4–5

What is the source voltage in the circuit in Figure 4–5?

GIVEN:
  $P_L = 9$ W
  $I = 1.5$ A

FIND:
  V_S

SOLUTION:
  $V_S = P_L \div I$ select formula
    $= 9$ W $\div 1.5$ A   substitute values
    $= 6$ V   divide

  Repeat the Skill-Builder using the following values. You should obtain the results that follow.

Given:

P_L	437 W	136 W	185 W	2,160 W	2,760 W
I	3.8 A	8.5 A	14.8 A	9.6 A	11.5 A

Find:

V_S	115 V	16 V	12.5 V	225 V	240 V

● *SKILL-BUILDER 4–6*

What is the current in the circuit in Figure 4–6?

GIVEN:
  $V_S = 6$ V
  $P_L = 10$ W

FIND:
  I

SOLUTION:
  $I = P_L \div V_S$ select formula
    $= 10$ W $\div 6$ V   substitute values
    $= 1.67$ A   divide

  Repeat the Skill-Builder using the following values. You should obtain the results that follow.

Given:

V_S	16 V	76 V	50 V	120 V	24 V
P_L	140 W	247 W	1,175 W	2,100 W	750 W

Find:

I	8.75 A	3.25 A	23.5 A	17.5 A	31.25 A

Power Is the Combination of Voltage and Current

Power is the combination of voltage and current. As Figure 4–7 illustrates, in your home you may have a 60-W lamp operating on a 120-V outlet. The current in the lamp equals

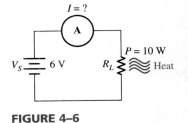

FIGURE 4–6

60 W ÷ 120 V, or 0.5 A. However, in your car you may have a 60-W headlight operating on a 12-V battery. The current in the headlight equals 60 W ÷ 12 V, or 5 A. Notice that both the lamp and the headlight have equal power and produce an equal amount of light. The lamp in the house uses a combination of high voltage and low current: 120 V and 0.5 A. The headlight in the car produces the same power with a different combination of low voltage and high current: 12 V and 5 A.

To understand Watt's Law, let's think of an ordinary situation involving work. Suppose we have a large job that requires physical labor, like building a road. Now suppose we ask a question: How much work can be done in one day?

The answer depends on two things: how many people work on the road, and how hard each person works. On one road you might have a small number of workers, but each worker has a lot of energy and works hard. On another road, you might have a large number of workers, but they are all tired and lazy and don't work as hard. Still, it is possible that the same amount of work gets done on both roads each day. If you have high-energy workers, you don't need as many of them. If you have low-energy workers, you just put more of them on the job.

In this example, the workers are like electrons. A higher current means more electrons per second pass through the load. This is like having more workers on the job. The energy that each worker has is like voltage. A higher voltage means each electron has more energy and can do more work.

Thus, the 120-V lamp that draws 0.5 A of current is like having a small number of high-energy workers. Because of the higher voltage, each electron in the current can do more work. This makes up for the fact that the number of electrons in the current is fairly low.

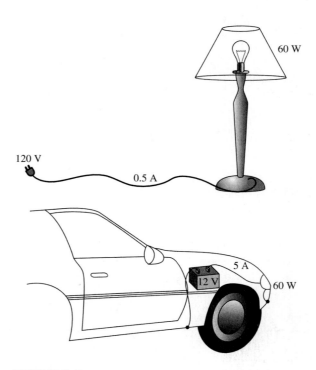

60 W

120 V

0.5 A

5 A

12 V

60 W

FIGURE 4–7
(a) Table lamp: high voltage, low current.
(b) Automobile headlight: low voltage, high current.

The 12-V automobile headlight that draws 5 A is like having a large number of low-energy workers: you can still get the same job done. The low voltage means that each electron has less energy, but because of the higher current, a large number of electrons are passing through the load and a lot of work is still getting done.

This is why, according to Watt's Law, electrical power is the combination of voltage and current. The voltage represents the energy each electron has, and the current represents the number of electrons per second delivering energy to the load. The combination of voltage and current is power, the total amount of energy per second being turned into work.

● SKILL-BUILDER 4–7

- -

A subdivision of homes consumes power at a rate of 1 megawatt. The utility company generates the power at 18 kV and steps up the voltage to 138 kV for transmission to the substation. The substation drops the voltage down to 69 kV and again to 12 kV before feeding it to the subdivision. The transformers in the subdivision drop the voltage down to 240 V for delivery to the homes. What is the current at each of the voltage levels along the line?

SOLUTION:

$I = P \div V$ select formula

Generator:

$I = 1 \text{ MW} \div 18 \text{ kV}$ substitute values
$= 55.55 \text{ A}$ divide

Long lines:

$I = 1 \text{ MW} \div 138 \text{ kV}$ substitute values
$= 7.25 \text{ A}$ divide

Substation:

$I = 1 \text{ MW} \div 69 \text{ kV}$ substitute values
$= 14.49 \text{ A}$ divide

Trunk lines:

$I = 1 \text{ MW} \div 12 \text{ kV}$ substitute values
$= 83.33 \text{ A}$ divide

Homes:

$I = 1 \text{ MW} \div 240 \text{ V}$ substitute values
$= 4,167 \text{ A}$ divide

- -

Skill-Builder 4–7 shows us why the utility company uses high voltages in its power lines: to keep the current as low as possible until it reaches the customer.

→ Self-Test

- - - - - - -

5. What relationship does Watt's Law express?
6. What does Watt's Law state?
7. What effect does voltage have on the amount of current necessary to produce a large amount of power?

- -

MATH TIP

The Exponent Key
Scientific calculators allow you to enter numbers directly in scientific or engineering notation. This is a great timesaver, since you do not have to convert the numbers back to standard notation. It is also safer, since you are more likely to make a keystroke error in standard notation when putting in a long series of zeros. Finally, it is the only way to enter large or small numbers with more zeros than the calculator can display.

The calculator has a special key, usually labeled either EXP or EE, used to enter powers of ten. Follow these steps:
1. Enter the coefficient.
2. Press the EXP or EE key.
3. Enter the exponent. For negative exponents, enter the value of the exponent; then change the sign with the +/– key.

Warning: Do not multiply by ten! Students often make the mistake of entering a number into a calculator exactly the way the number is written. A number in scientific or engineering notation will have "×10" written as part of the number; but if you enter "× 10" into the calculator, your result will be wrong. Pressing the EXP or EE key is the equivalent of entering "× 10." Follow the steps exactly as given in the preceding steps.

All scientific calculators give you a choice of normal or scientific display modes. Consult the instructions for your calculator to learn how to select these modes.

Many scientific calculators include ENG keys,

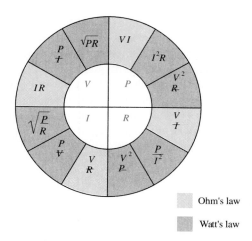

Ohm's law

Watt's law

FIGURE 4–8
The formula wheel.

SECTION 4–3: The Formula Wheel

In this section you will learn the following:

→ Ohm's Law and Watt's Law may be rearranged and combined to produce 12 formulas known as the Formula Wheel.

→ If any two basic electrical quantities are known, the other two can be found, using the Formula Wheel.

Figure 4–8 is called the *Formula Wheel.* Using the rules of algebra, we have rearranged Ohm's Law and Watt's Law to create 12 formulas. The shading indicates the origin of each formula.

The inner circle shows the four basic electrical quantities: power, voltage, current, and resistance. The outer circle gives three formulas for each of the quantities.

We will use the circuit in Figure 4–9 to illustrate the use of the 12 formulas in the wheel. In this circuit, all four quantities are known. We will see that each of the quantities can be found by three formulas from the wheel.

POWER:

1. $P = VI$ select formula
   $= 6\,V \times 2\,A$ substitute values
   $= 12\,W$ multiply

2. $P = I^2 R$ select formula
   $= 2^2\,A \times 3\,\Omega$ substitute values
   $= 4 \times 3$ square
   $= 12\,W$ multiply

3. $P = V^2 \div R$ select formula
   $= 6^2\,V \div 3\,\Omega$ substitute values
   $= 36 \div 3$ square
   $= 12\,W$ divide

All three power formulas give an answer of 12 W.

RESISTANCE:

1. $R = V \div I$ — select formula
   $= 6\,V \div 2\,A$ — substitute values
   $= 3\,\Omega$ — divide

2. $R = P \div I^2$ — select formula
   $= 12\,W \div 2^2\,A$ — substitute values
   $= 12 \div 4$ — square
   $= 3\,\Omega$ — divide

3. $R = V^2 \div P$ — select formula
   $= 6^2\,V \div 12\,W$ — substitute values
   $= 36 \div 12$ — square
   $= 3\,\Omega$ — divide

All three formulas give an answer of 3 Ω.

CURRENT:

1. $I = V \div R$ — select formula
   $= 6\,V \div 3\,\Omega$ — substitute values
   $= 2\,A$ — divide

2. $I = P \div V$ — select formula
   $= 12\,W \div 6\,V$ — substitute values
   $= 2\,A$ — divide

3. $I = \sqrt{P \div R}$ — select formula
   $= \sqrt{12\,W \div 3\,\Omega}$ — substitute values
   $= \sqrt{4}$ — divide
   $= 2\,A$ — square root

All three formulas give an answer of 2 A.

VOLTAGE:

1. $V = IR$ — select formula
   $= 2\,A \times 3\,\Omega$ — substitute values
   $= 6\,V$ — multiply

2. $V = P \div I$ — select formula
   $= 12\,W \div 2\,A$ — substitute values
   $= 6\,V$ — divide

3. $V = \sqrt{PR}$ — select formula
   $= \sqrt{12\,W \times 3\,\Omega}$ — substitute values
   $= \sqrt{36}$ — multiply
   $= 6\,V$ — square root

All three formulas give an answer of 6 V.

Using the Formula Wheel

1. On the inner circle of the wheel, locate the quantity you want to find.
2. For each quantity, the wheel offers three formulas. Choose the formula that contains the two given quantities you know.

 MATH TIP

which change the display into correct engineering notation. This is a helpful feature to look for when choosing a calculator.

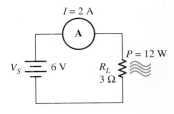

FIGURE 4–9

 MATH TIP

Square and Square Root
Some formulas require the square of a number. In these formulas, the number should be squared before any other arithmetic. For example, $4^2 \div 8 = 16 \div 8 = 2$.

Other formulas require the *square root* of a number. The square root of a number is a smaller number that squares to give the larger number. For example, the square root of 64 is 8 because $8^2 = 64$.

The symbol of the square root is $\sqrt{\ }$, called the *radical*. Scientific calculators have a special key to automatically extract the square root, and the key has the $\sqrt{\ }$ symbol on it. *Note:* All calculations inside the radical must be completed before taking the square root. With calculators, this means doing the arithmetic inside the radical and *pressing the =* key before pressing the $\sqrt{\ }$ *key.* For example, $\sqrt{15} \times \sqrt{8} = \sqrt{120} = 10.95$.

● *SKILL-BUILDER 4–8*

- -

A 115-V toaster has a resistance of 12.6 Ω. What is the power rating?

GIVEN:
   $V = 115$ V
   $R = 12.6$ Ω

FIND:
   P

SOLUTION:

$P = V^2 \div R$	select formula
$= 115^2$ V $\div$ 12.6 Ω	substitute values
$= 13{,}225 \div 12.6$	square
$= 1{,}050$ W	divide
$= 1.05$ kW	apply metric prefix

   Repeat the Skill-Builder using the following values. You should obtain the results that follow.

Given:

V	38 V	125 V	14 V	235 V	600 V
R	6.02 Ω	3.62 Ω	4.9 Ω	8 Ω	27.97 Ω

Find:

P	240 W	4.32 kW	40 W	6.9 kW	12.87 kW

- -

● *SKILL-BUILDER 4–9*

- -

A clothes dryer operates on 240 V. Its heater draws 23 A. What is the resistance?

GIVEN:
   $V = 240$ V
   $I = 23$ A

FIND:
   R

SOLUTION:

$R = V \div I$	select formula
$= 240$ V $\div$ 23 A	substitute values
$= 10.43$ Ω	divide

   Repeat the Skill-Builder using the following values. You should obtain the results that follow.

Given:

V	115 V	12 V	6.8 V	235 V	48 V
I	13.7 A	6.3 A	730 mA	23.4 A	16.8 A

Find:

R	8.39 Ω	1.9 Ω	9.32 Ω	10.04 Ω	2.86 Ω

- -

● *SKILL-BUILDER 4–10*

- -

An electric blanket operates on 117 V and has a resistance of 76 Ω. What is the current?

GIVEN:
  $V = 117$ V
  $R = 76\ \Omega$

FIND:
  I

SOLUTION:
  $I = V \div R$ select formula
    $= 117$ V $\div 76\ \Omega$ substitute values
    $= 1.54$ A   divide

    Repeat the Skill-Builder using the following values. You should obtain the results that follow.

Given:

V	480 V	277 V	16.5 V	169 V	12.6 V
R	38.5 Ω	15.9 Ω	22 Ω	13.7 Ω	2.48 Ω

Find:

I	12.47 A	17.42 A	0.75 A	12.34 A	5.08 A

● SKILL-BUILDER 4–11

An electric chain saw draws 12 A and has a resistance of 9.58 Ω. What is the voltage?

GIVEN:
  $I = 12$ A
  $R = 9.58\ \Omega$

FIND:
  V

SOLUTION:
  $V = IR$ select formula
    $= 12$ A $\times 9.58\ \Omega$ substitute values
    $= 115$ V   multiply

    Repeat the Skill-Builder using the following values. You should obtain the results that follow.

Given:

I	16.5 A	28.3 A	19.4 A	63.2 A	12.7 A
R	14.24 Ω	4.13 Ω	3.87 Ω	7.59 Ω	21.81 Ω

Find:

V	235 V	117 V	75 V	480 V	277 V

● SKILL-BUILDER 4–12

A table saw operates on 115 V and draws 14 A. What is the power rating?

GIVEN:
  $V = 115$ V
  $I = 14$ A

FIND:
  P

SOLUTION:

$P = VI$	select formula
$= 115\,V \times 14\,A$	substitute values
$= 1{,}610\,W$	multiply
$= 1.61\,kW$	apply metric prefix

Repeat the Skill-Builder using the following values. You should obtain the results that follow.

Given:

V	115 V	230 V	277 V	208 V	480 V
I	22.96 A	14.83 A	21.91 A	6.49 A	8.73 A

Find:

P	2.62 kW	3.41 kW	6.07 kW	1.35 kW	4.19 kW

● *SKILL-BUILDER 4–13*

A steam iron has a power rating of 1,200 W and draws 10.43 A. What is the resistance?

GIVEN:
$P = 1{,}200\,W$
$I = 10.43\,A$

FIND:
R

SOLUTION:

$R = P \div I^2$	select formula
$= 1{,}200\,W \div 10.43^2\,A$	substitute values
$= 1{,}200 \div 108.785$	square
$= 11.03\,\Omega$	divide

Repeat the Skill-Builder using the following values. You should obtain the results that follow.

Given:

P	550 W	1,670 W	4,350 W	230 W	675 W
I	14.32 A	15.67 A	19.63 A	6.84 A	5.41 A

Find:

R	2.68 Ω	6.8 Ω	11.29 Ω	4.92 Ω	23.06 Ω

● *SKILL-BUILDER 4–14*

A large color television has a power rating of 116 W and operates on 120 V. What is the current?

GIVEN:
$P = 116\,W$
$V = 120\,V$

FIND:
I

SOLUTION:

$I = P \div V$ select formula
  $= 116\text{ W} \div 120\text{ V}$ substitute values
  $= 0.97\ \Omega$ divide

Repeat the Skill-Builder using the following values. You should obtain the results that follow.

Given:

P	648 W	1,735 W	2,312 W	125 W	7.5 W
V	220 V	117 V	235 V	12.6 V	115 V

Find:

I	2.95 A	14.83 A	9.84 A	9.92 A	65 mA

- -

● *SKILL-BUILDER 4–15*

A 60-gallon air compressor draws 20 A and has a power rating of 4.8 kW. What is the voltage?

GIVEN:
  $P = 4.8\text{ kW}$
  $I = 20\text{ A}$

FIND:
  V

SOLUTION:

$V = P \div I$ select formula
  $= 4.8\text{ kW} \div 20\text{ A}$ substitute values
  $= 4.8\text{ EXP }3 \div 20$ enter metric prefix
  $= 240\text{ V}$ divide

Repeat the Skill-Builder using the following values. You should obtain the results that follow.

Given:

P	6.37 kW	240 W	1.12 kW	2.06 kW	9.3 kW
I	30.63 A	2.09 A	9.33 A	9.36 A	33.57 A

Find:

V	208 V	115 V	120 V	220 V	277 V

- -

● *SKILL-BUILDER 4–16*

A hot water heater has a resistance of 10.47 Ω and draws a current of 22.9 A. What is the power rating?

GIVEN:
  $I = 22.9\text{ A}$
  $R = 10.47\ \Omega$

FIND:
  P

SOLUTION:

$P = I^2R$	select formula
$= 22.9^2 \, A \times 10.47 \, \Omega$	substitute values
$= 524.41 \times 10.47$	square
$= 5,490 \, W$	multiply
$= 5.49 \, kW$	apply metric prefix

Repeat the Skill-Builder using the following values. You should obtain the results that follow.

Given:

I	17.6 A	38.5 A	21.3 A	6.8 A	9.4 A
R	7.91 Ω	2.14 Ω	2.51 Ω	74.61 Ω	46.17 Ω

Find:

P	2.45 kW	3.17 kW	1.14 kW	3.45 kW	4.08 kW

● *SKILL-BUILDER 4–17*

A 1,500-W toaster oven operates on 117 V. What is the resistance?

GIVEN:
$V = 117 \, V$
$P = 1,500 \, W$

FIND:
R

SOLUTION:

$R = V^2 \div P$	select formula
$= 117^2 \, V \div 1,500 \, W$	substitute values
$= 13,689 \div 1,500$	square
$= 9.13 \, \Omega$	divide

Repeat the Skill-Builder using the following values. You should obtain the results that follow.

Given:

V	135 V	24.7 V	476 V	233 V	11.8 V
P	1.45 kW	65.7 W	12.53 kW	8.75 kW	31.4 W

Find:

R	12.57 Ω	9.29 Ω	18.08 Ω	6.2 Ω	4.43 Ω

● *SKILL-BUILDER 4–18*

The heater of a small dishwasher has a resistance of 14.46 Ω and a power of 996 W. What is the current?

GIVEN:
$P = 996 \, W$
$R = 14.46 \, \Omega$

FIND:
I

SOLUTION:

$I = \sqrt{(P \div R)}$ select formula

    $= \sqrt{(996\text{ W} \div 14.46\ \Omega)}$ substitute values

    $= \sqrt{68.88}$ divide

    $= 8.3$ A   square root

     Repeat the Skill-Builder using the following values. You should obtain the results that follow.

Given:

P	1,435 W	2,283 W	1,576 W	2,199 W	1,347 W
R	21.62 Ω	13.49 Ω	17.56 Ω	61.08 Ω	23.98 Ω

Find:

I	8.15 A	13.01 A	9.47 A	6 A	7.49 A

● *SKILL-BUILDER 4–19*

An 1,100-W hair dryer has a resistance of 13.09 Ω. What is the voltage?

GIVEN:

$P = 1,100$ W
$R = 13.09\ \Omega$

FIND:

V

SOLUTION:

$V = \sqrt{(PR)}$ select formula

    $= \sqrt{(1,100\text{ W} \times 13.09\ \Omega)}$ substitute values

    $= \sqrt{14,399}$ multiply

    $= 120$ V   square root

     Repeat the Skill-Builder using the following values. You should obtain the results that follow.

Given:

P	12.56 kW	3,428 W	5,348 W	2,214 W	3,743 W
R	18.42 Ω	22.71 Ω	2.6 Ω	24.73 Ω	11.78 Ω

Find:

V	481 V	279 V	118 V	234 V	210 V

→ *Self-Test*

8. What is the Formula Wheel?
9. How many electrical quantities must be known to find the remaining quantities using the Formula Wheel?

SECTION 4–4: Multimeters

In this section you will learn the following:

→ A multimeter is a portable instrument that measures voltage, current, and resistance.

→ A VOM (volt-ohm-milliammeter) is an analog multimeter with a needle-and-scale display.

→ A DMM is a digital multimeter with a numerical display.

A *multimeter* is a portable meter that measures several electrical quantities. There are two basic styles of multimeter: analog and digital.

The Analog Multimeter

An analog multimeter has a moveable pointer and several numeric scales printed on its dial. Figure 4–10 shows a VOM (volt-ohm-milliammeter), a typical analog multimeter used to measure voltage, current, and resistance.

An analog multimeter contains a mechanism called a *meter movement,* which uses a current of less than 50 μA in a coil of thin wire to produce magnetism, moving the pointer against the force of a spring. We measure current, voltage, or resistance by looking at the position of the pointer on the appropriate scale.

The Digital Multimeter

Figure 4–11 shows a digital multimeter, or DMM. The digital multimeter has a window called a display. Numbers appear in the window, measuring voltage, current, and resistance. Many DMMs can also measure other quantities, such as frequency, temperature, capacitance, and so on.

An important feature of a DMM is the number of digits it can display. Often the first digit in the display can be only a zero or one and is counted as half a digit. Thus, a 4½-digit DMM can display values up to 19999, but a 3½-digit DMM can display only to 1999.

FIGURE 4–10
Volt-ohm-milliammeter (VOM).

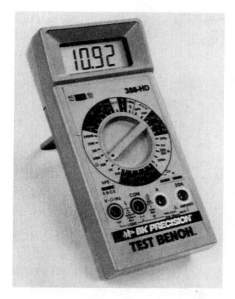

FIGURE 4–11
Digital Multimeter (DMM).

Setting Up a Multimeter

Figure 4–12 illustrates the parts of a multimeter. The *test leads* are the wires that connect the meter to the circuit or component to be measured. The *probes* are metal points or clips mounted in plastic at one end of each test lead. The probes allow you to touch the leads temporarily to test points. Alligator clips or grippers allow you to attach the leads to test points so you may work with your hands free. The banana plugs allow the test leads to be inserted into the *jacks* (holes) in the meter. The *function switches* select the type of measurement to be made, and the range switches select the sensitivity of the meter.

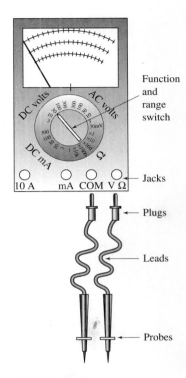

FIGURE 4–12
Parts of a multimeter.

Never connect a meter to a circuit until the test leads are inserted into the proper jacks and the function and range switches are properly set. You could damage either the meter or the circuit, and possibly injure yourself, if the settings are not correct.

The first step in setting up a multimeter is to insert the leads into the correct jacks. The black test lead plug goes in the COM (common) jack. The red test lead plug goes into the appropriate jack for the quantity you want to measure. Look for unit abbreviations printed near the jack: V (volts), Ω (ohms), A (amps), or mA (milliamps). Often one jack is used for several quantities. Special jacks are used for measuring high values, and may be labeled 10 A, 1,000 VDC, and so on.

The second step is to select the proper function switch setting. All meters have settings for measuring dc voltage, dc current, ac voltage, ac current, and resistance. DMMs may also have settings for measuring continuity, capacitance, frequency, temperature, or for testing diodes and transistors.

Some multimeters have a single rotary switch that selects both function and range. The function settings are grouped together and are indicated by brackets or colored areas.

Other multimeters have separate function and range switches. These meters often use push-button switches. Functions are selected by pressing the push-buttons in various combinations according to diagrams printed on the meter.

The third step is to select the proper range. Near each range switch position is a number that indicates the limit (maximum value) of the range. In other words, all ranges begin at zero and end at the indicated maximum value.

Lower ranges are more sensitive and, thus, the measurements are more precise. Therefore, select the lowest range whose limit is greater than the expected measurement. For example, if you intend to measure the voltage of an automobile battery, you would expect the measurement to be approximately 12 V. If you have available meter ranges of 3 V, 10 V, 30 V, and 100 V, you would select the 30-V range as the lowest (most sensitive) range with a limit greater than the expected measurement (12 V).

If you do not know what value to expect, select the highest range for your first measurement; then select a lower range for your final measurement.

To set up a multimeter:

1. Insert the black test lead plug in the COM jack and the red test lead plug in the appropriate quantity jack.
2. Select the proper function switch setting.
3. Select the lowest range whose limit is greater than the expected measurement.

Reading an Analog Multimeter

Analog multimeters have several *scales,* the numbers and marks printed on the face of the meter, as shown in Figure 4–13. The marks are called *divisions.* Longer marks are called major divisions, and shorter marks are called minor divisions. Notice that some scales have several sets of numbers for the major divisions. Thus, to read an analog multimeter, you must make three decisions:

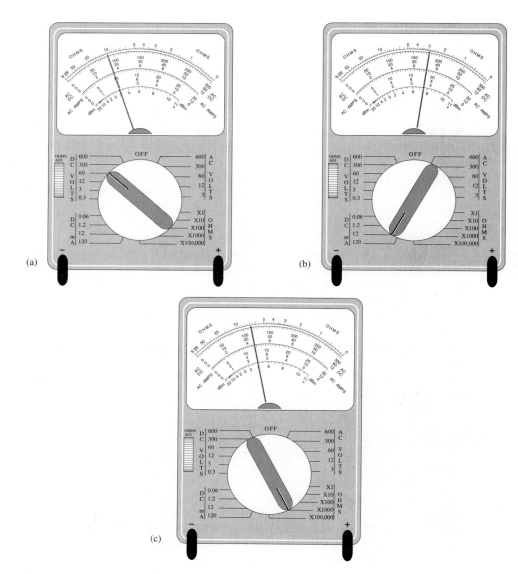

FIGURE 4–13
Reading analog multimeters.

    1. which scale to read;
    2. which set of scale numbers to use;
    3. how much each major and minor division is worth.

The meters in Figure 4–13 have four scales: ohms, ac/dc, ac amps, and dBm. The dBm scale is used for measuring audio power in decibel-milliwatts. We will not use this scale for now.

    Choose the scale according to the function switch setting. In Figure 4–13(a), the function switch is set for dc volts. We are not measuring ohms or amps, so we will not use these scales. The only scale left to use is the ac/dc scale, and this is our choice.

    Choose the set of scale numbers according to the range switch setting. In Figure 4–13(a), the top set of scale numbers ends in 300. It is used for ranges that are multiples of three, such as 0.3, 3, and 300. The middle set of scale numbers ends in 60 and is used for ranges that are multiples of six, such as 0.06, 60, and 600. The bottom set of scale numbers ends in 12 and is used for ranges that are multiples of 12, such as 1.2, 12, and 120. Because our range switch setting is 60 V, we will use the set of scale numbers ending in 60.

Choose the value of the major and minor divisions according to the range switch setting. In Figure 4–13, the range switch setting is 60 V. This is the value of the endmark on our scale. Therefore, each major division is worth 10 V, and each minor division is worth 1 V. Thus, the meter in Figure 4–13 is measuring 18 VDC.

In Figure 4–13(b), the function switch is set for dc mA. Therefore, we use the ac/dc scale again (we are not measuring ohms or ac amps). The range switch is set on 12, so we use the bottom set of scale numbers ending in 12. We give the endmark a value of 12 mA, which makes each major division worth 2 mA and each minor division worth 0.2 mA. Thus, the meter in Figure 4–13(b) is measuring 7.2 mA.

In Figure 4–13(c), the function switch is set for ohms. Therefore, we use the ohms scale. Notice that this scale is divided unevenly, and zero is on the right side of the scale. The pointer is between major divisions 5 and 10. There are ten minor divisions between these major divisions, and each minor division is worth 0.5 ($10 - 5 = 5$, and $5 \div 10 = 0.5$). Thus, the pointer is indicating a value of 7.0.

The range setting for ohms is different than for volts or amps. The ohms range setting is called a multiplier. The scale reading should be multiplied by the range setting to get the true reading. Our multiplier is × 1,000. Therefore, our true reading is $7.0 \times 1,000 = 7,000\ \Omega$.

Often the range settings will include unit prefixes, such as mA, mV, kΩ, or MΩ, which determine the worth of the divisions.

Analog Off-scale Readings

When measuring dc voltage or current with an analog meter, you must be careful about the polarity of the probes. Reverse polarity will produce a backwards force on the pointer, causing it to go *off scale* (past the end mark of the scale). A strong off-scale movement can damage the pointer by bending it against its zero stop.

If you select a range that is too low, the pointer will go off scale at the high end and could be damaged by bending against its end stop.

Reading a Digital Multimeter

Digital multimeters are easier and quicker to read than analog meters because digital meters have no scales. However, you must still examine the range setting to determine the units or multiplier. Some DMMs are autoranging, selecting the appropriate range automatically and displaying the units along with the numerical value.

Resolution

The *resolution* of a DMM is the smallest change in value that the DMM can display. Resolution depends on the number of digits in the display, the position of the decimal according to the range setting, and the internal circuits of the meter.

For example, if a DMM displays a reading of 3.5 V, the next lowest reading would be 3.4 V, and the next highest reading would be 3.6 V. Thus, the resolution is 0.1 V (the smallest change the DMM can display). If a more sensitive range is selected, the 3.5-V reading may change to 3.53 V. Now the next lowest reading is 3.52 V, and the next highest reading is 3.54 V. The resolution has increased to 0.01 V. Notice that we say the reso-

FIGURE 4–14
Typical DMM
off-scale
reading.

lution has increased, although the value of the resolution is a smaller number. Resolution is the ability to look or measure closely. Thus, the resolution is higher when the minimum change is smaller.

Sampling

The electronic circuits of digital instruments do not measure continuously. Instead they take a measurement, process it, and then take another measurement. Since there are moments of time when no measurement is being taken, the process is called *sampling*. The number of samples per second is called the sampling rate. If a change occurs between samples, the instrument may not notice the change. Therefore, instruments with higher resolution require higher sampling rates.

The sampling process of DMMs is one reason why a DMM display may be unsteady, with the last digit flickering or changing. Technicians learn to ignore this, or choose a range with lower resolution so they don't have to watch it.

Digital Off-scale Readings

A DMM typically indicates an off-scale reading by displaying a single-digit one, with no decimal point or zeros, as shown in Figure 4–14. A DMM has no pointer to be damaged by an off-scale reading, but extreme off-scale conditions can damage the DMM's electronic circuits.

In measuring resistance with a DMM, do not confuse a reading of *infinity* (the resistance of an open circuit) with an off-scale reading. For example, if you have selected a low-resistance range and get an off-scale reading, do not quickly say that you have infinity and an open circuit. You may get a normal resistance reading when you go to a higher range. However, if you are at the highest resistance range and still get an off-scale reading, then you may reasonably conclude that you have an open circuit and a reading of infinity.

→ *Self-Test*

10. What is a multimeter?
11. What is a VOM?
12. What is a DMM?

SECTION 4–5: Measuring Voltage

In this section you will learn the following:

- → An electrical point is a conductor or group of conductors with no difference in electrical potential between any of the physical points. All voltage measurements are taken between electrical points.
- → To measure voltage with a multimeter, place the two probes across two electrical points in a circuit.
- → When no current is flowing, the potential is the same on both sides of a resistive load. Therefore, the difference in potential is zero.
- → If a current path is open between two points, and complete between all other points, the potential across the open points equals the potential of the voltage source.
- → In simple circuits with resistive loads, the normal voltmeter reading is full source voltage across an open switch and zero volts across a closed switch.

Physical and Electrical Points

When a conductor is short (less than 100 feet), its resistance is generally less than 1 Ω. Therefore, for practical purposes, we ignore the small resistance and consider the resistance of a conductor to be almost zero ohms. This means that we also ignore the slight loss in voltage that occurs in a conductor. In other words, we consider the potential energy (voltage) to be the same at all points along the conductor. This way of thinking leads to an important concept: the *electrical point*.

In Figure 4–15, suppose the current path, represented by the lines in the schematic, consists of bare wire. Thus, physical points *A, B,* and *C* represent the two physical ends and the physical middle of one wire. Similarly, points *D, E,* and *F* represent physical points on another wire.

Because we ignore resistance in short conductors, we consider physical points *A, B,* and *C* to have the same electrical potential. Therefore, we say that *A, B,* and *C* are all the same electrical point. Likewise, physical points *D, E,* and *F* are another electrical point. In other words, the circuit in Figure 4–15 has only two electrical points.

To measure voltage, choose two electrical points in a circuit, and touch or connect the black probe to one point and the red probe to the other. The meter will read the voltage across the two points, which means the voltage difference between the two points.

In Figure 4–15, one probe would be placed at either of points *A, B,* or *C.* The other probe would be placed at either of points *D, E,* or *F.* In simple language, the voltage is the same at either end of a wire. Therefore, when taking voltage measurements, it doesn't matter whether you place a probe at one end of a wire or the other.

All voltmeters are designed to measure the voltage difference between the two probes, comparing the red probe to the black one. Figure 4–16 will help us understand voltage measurements. In schematics, the symbol for a voltmeter is a circle with a V inside it.

In Figure 4–16(a), two voltmeters are measuring a 12-V battery. V_1 is across physical points *A* and *B,* and V_2 is across physical points *C* and *D.* However, points *A* and *C* are the same electrical point, and so are points *B* and *D.* In other words, both meters are across the same two electrical points, even though they are touching different physical points. Thus, both voltmeters are measuring the voltage of the battery, although their probes are not physically touching the battery terminals.

Meter V_1 has its black probe on the battery's negative terminal and its red probe on the battery's positive terminal. It measures a voltage difference of 12 V, with the red probe more positive than the black. Therefore, the reading of V_1 is a positive voltage reading: +12 V.

On the other hand, meter V_2 has its black probe on the battery's positive terminal and its red probe on the battery's negative terminal. It measures a voltage difference of 12 V, with the red probe more negative than the black. Therefore, the reading of V_2 is a negative voltage reading: −12 V.

Many people are confused about positive and negative voltage readings and sometimes think that a negative voltage reading means a value less than zero. This is incorrect. There is no such thing as a voltage less than zero.

Figure 4–16(a) shows the true meaning of negative voltage readings. The voltage of the battery is 12 V in both cases. The only meaning of the negative reading of V_2 is that the red probe is at a point more negative than the black probe. Often it doesn't matter whether a point is truly positive or negative, only whether one point is more or less positive or negative than another point.

Voltmeter V_2 has its probes placed improperly. When the polarity of the power source is known, the black probe should be placed on the negative side of the source, and the red probe on the positive side. (There are exceptions to this rule, which we will learn

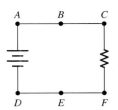

FIGURE 4–15
Physical and electrical points.

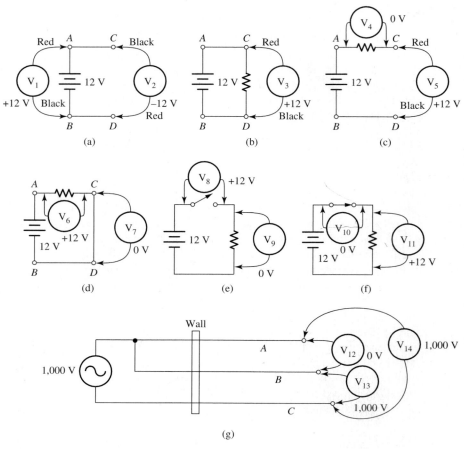

FIGURE 4–16
Measuring voltage.

in later chapters.) Analog meters can be damaged if the probes are placed improperly, causing the pointer to bend as it tries to move in the wrong direction. Digital meters will not be damaged if the probes are placed improperly but may indicate a reverse polarity value, which may confuse you.

In Figure 4–16(b), a resistive load has been connected between terminals *C* and *D*, and current is now flowing in the circuit. However, the battery's electromotive force is able to maintain a 12-V potential across the terminals, as measured by voltmeter V_3. The probes are positioned correctly, and the voltmeter shows a positive polarity.

V_3 is measuring the voltage across the battery and also measuring the voltage across the load. There is really only one voltage across terminals *C* and *D*, and it is merely a choice of words to call it the source voltage or the load voltage. In this simple circuit, source voltage and load voltage are one and the same.

In Figure 4–16(c), the load has been placed between terminals *A* and *C*. No current is flowing because the path is open between terminals *C* and *D*. Because no current is flowing, the potential energy at point *A* does not become kinetic energy and is not dissipated by the load. However, the resistive load completes the path between terminals *A* and *C*, and allows the potential at *A* to be felt at *C*. In other words, since no current is flowing, the potential is the same at *A* and *C*. Thus, there is no difference in potential across *A* and *C*, and voltmeter V_4 measures 0 V.

However, since C has the same potential as A, there is a difference in potential across C and D equal to the potential of the battery. Voltmeter V_5 is actually measuring the source voltage: +12 V.

Figure 4–16(c) applies only to resistive loads. Remember that there are other kinds of loads: reactive, electromechanical, and electronic. These loads might not complete the current path between A and C and would cause both voltmeters V_4 and V_5 to measure zero volts.

In Figure 4–16(d), the path is complete between C and D, and current is flowing in the load. C and D are now the same electrical point because of the connection between them. Since they are the same electrical point, they must have the same potential. Therefore, the difference in potential is zero, as shown by voltmeter V_7.

Voltmeter V_6 is actually taking the same measurement as voltmeter V_3 in part (b). Do not be confused by the way the schematic is drawn. You must learn to recognize electrical points, not physical points.

In Figure 4–16(e), voltmeter V_8 is across an open switch and reads the same as voltmeter V_5 in part (c). Voltmeter V_9, across the load with no current flowing, reads the same as voltmeter V_4.

In Figure 4–16(f), voltmeter V_{10} is across a closed switch and reads the same as voltmeter V_7 in part (d). Current is flowing in the load, and voltmeter V_{11} reads the same as voltmeter V_6.

In Figure 4–16(g), a 1,000-V ac generator in one room is connected to wires passing through a wall into another room. Notice that the top wire has been split into two wires so terminals A and B are actually the same electrical point. Voltmeters V_{13} and V_{14} both show the 1,000-V potential between the top wire and the bottom wire. However, voltmeter V_{12} correctly measures zero volts across A and B because there is no difference in potential between them.

Suppose an inexperienced person measures zero volts across A and B in Figure 4–16(g). Now suppose the person thinks that the reading of zero volts means that there is no voltage on the wires and touches A or B, and then touches C. The person would be badly hurt or killed by the current produced by a thousand volts.

This example is important to your safety in working with electricity. Always remember that a voltmeter reading of zero only means there is no difference in voltage between those two points. You need to take more readings across several other points to get the full picture before doing something to the circuit.

To measure voltage with a multimeter, follow these steps:

1. Set the jacks and plugs, function switches, and range switches properly.
2. In dc circuits, observe polarity by placing the black probe on the more negative of the two points to be measured and the red probe on the more positive point. In ac circuits, the meter will not indicate polarity, and the placement of the probes will not make a difference in the reading.
3. Take the reading.

→ *Self-Test*

13. What is an electrical point?
14. What is the procedure for measuring voltage with a multimeter?
15. Describe the potential of a resistive load when no current is flowing.
16. Describe the potential across an open.
17. Describe typical voltmeter readings across a switch.

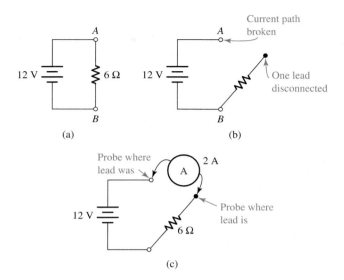

FIGURE 4–17
Measuring current.

SECTION 4–6: Measuring Current

In this section you will learn the following:

→ To measure current with a multimeter, open the circuit and insert the meter into the current path.

To measure current with a multimeter, you must first create an open in the circuit. This means you must disconnect the current path at some point, as shown in Figure 4–17(b), where one lead of the resistive load has been disconnected at terminal *A*. We describe this by saying the current path has been broken at point *A*. Breaking a current path simply means making an open in the path by disconnecting something. In schematics, the symbol for an ammeter is a circle with an A inside it.

When a multimeter is set to measure current, the leads are connected inside the meter so the two leads act like a single piece of wire. Therefore, when the probes are connected across the break, the meter completes the current path, and current flows in the circuit again. Since the meter becomes part of the current path, it accurately measures the current in the circuit.

In Figure 4–17(c), one ammeter probe is placed where the disconnected lead now is, and the other probe placed where the lead was before disconnection. Thus, the probes act as the two ends of a single wire to reconnect the lead to point *A* and allow current to flow again.

To measure current with a multimeter, follow these steps:

1. Set the jacks and plugs, function switches, and range switches properly.
2. Turn off the power to the circuit.
3. Break the current path: disconnect a component or conductor, and move the lead out of position.
4. Connect the meter properly: one probe where the loose lead now is, the other probe where the loose lead was.
5. In dc circuits, observe polarity by placing the black probe on the more negative point, and the red probe on the more positive point. In ac circuits, the meter will not indicate polarity, and the placement of the probes will not make a difference in the reading.
6. Turn the power back on.
7. Take the reading.

PRACTICAL TIP

It takes more time to measure current than to measure voltage because power must be turned off, a lead in the circuit must be disconnected and moved out of position, and the meter probes attached. After the reading, the lead must be reconnected to the circuit. Moreover, there is a risk of breaking or damaging the component in handling it.

Also remember that the probes act as a single piece of wire when measuring current. If you put the probes in the wrong place, you could bypass the load and draw a high amount of current that could damage the meter or the circuit, and possibly harm you if you are holding the probes. Multimeters contain a small internal fuse to protect against high current, but this may not be enough protection when the voltage is high, or when the current in the probes is enough to damage the circuit but not enough to blow the fuse.

For all these reasons, experienced technicians prefer to learn as much as possible from voltage readings, and take current readings only when necessary.

→ Self-Test

18. What is the procedure for measuring current with a multimeter?

SECTION 4–7: Measuring Resistance

In this section you will learn the following:

→ To measure resistance with a multimeter, turn off the power to the circuit, disconnect one or more leads of the component to be measured, and place the probes across the component.

→ Continuity is a resistance reading of nearly zero ohms taken between two physical points of a current path with no current flowing. Continuity indicates there is no break or open between the two points.

→ Infinity is a resistance reading that is off scale on the highest range, taken between two physical points of a current path with no current flowing. Infinity indicates there is a break or open between the two points.

Figure 4–18 shows how ohmmeters measure resistance. When a component is connected between the probes of an ohmmeter, a battery in the meter sends current through the component. The meter measures the current and interprets the amount of current as the resistance of the component, according to Ohm's Law. In schematics, the symbol for an ohmmeter is a circle with an Ω inside it.

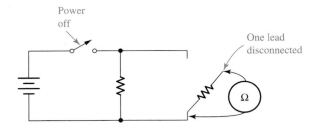

FIGURE 4–18
Measuring resistance.

For example, suppose the battery in the ohmmeter produces 3 V across the probes, and the component between the probes has a resistance of 60 Ω. According to Ohm's Law, the current would be 3 V ÷ 60 Ω = 50 mA. The meter actually measures 50 mA but indicates 60 Ω, based on the fact that its battery produces 3 V.

This means that if the battery voltage is not exactly what it should be, the ohmmeter reading will be incorrect. Analog ohmmeters compensate for incorrect battery voltage with an ohms-adjust control, usually a small knob or thumbwheel. After setting the jacks, function, and range correctly, touch the probe tips to each other and adjust the control until the scale reads exactly zero ohms; then measure the component.

Digital meters do not have an ohms-adjust control. The electronic circuits of a digital meter automatically correct variations in battery voltage to produce an accurate reading. When the battery runs down and the circuits cannot correct for the low voltage, a message appears in the display indicating the battery should be replaced.

The ohmmeter reading will be accurate only if the battery alone is producing the current, and only if the current is flowing just in the component being measured.

To measure resistance with a multimeter, follow these steps:

1. Set the jacks and plugs, function switches, and range switches properly.
2. Touch the probe tips together to check the leads and battery.
3. If necessary, adjust the meter to read zero ohms.
4. If necessary, replace the battery.
5. Turn off the power to the circuit.
6. Disconnect one or more leads of the component to isolate it from other components.
7. Place the probes on either side of the component.
8. Take the reading.

 PRACTICAL TIP

Resistance measurements are slow and require disturbing the circuit. Experienced technicians take resistance measurements only when voltage readings indicate a problem in the component, or when voltage readings alone do not give enough information.

Remember that many components such as heater elements have low resistance when cold and high resistance when hot. Since you cannot take resistance readings in a live circuit, you may be confused by cold resistance readings.

For example, suppose you have a 240-V, 5,500-W hot water heater that is not working. You remove the heater element and measure a resistance of 3 Ω. Your calculations tell you the resistance should be 10.47 Ω (240^2 V ÷ 5,500 W). You expect the cold resistance to be less than the hot resistance, but you don't know whether 3 Ω is a normal cold reading. You decide to take the element with you and go to a parts store, where you can measure the cold resistance of a new element.

In components such as heater elements, it is normal for the cold resistance to vary by as much as plus or minus 20%. Suppose the new element has a cold resistance of 2.1 Ω or 3.9 Ω. This is a 30% difference in cold resistance and is too close to the 20% normal difference for you to be sure whether your element is bad. However, if the new element measures 0.8 Ω or 5.3 Ω, far outside the 20% to 30% range, you have stronger evidence that your element is bad.

Testing Conductors with Ohmmeters

We test conductors by measuring the resistance from one end to the other with an ohmmeter. A good conductor with good connectors and contacts will normally have a resistance of nearly zero ohms. Thus, we say we have *continuity* between the two ends. However, if there is an open between the probes, the ohmmeter cannot measure resistance because its battery cannot produce any current. We say the meter reads *infinity,* symbolized ∞.

An ohms-adjust check also tests the continuity of the leads and probes, which often break inside from normal handling. It is a good idea to perform an ohms-adjust check before using a multimeter for other measurements.

Testing Switches and Relays with Ohmmeters

To test a switch or relay, you must disconnect it from the circuit with all wires removed from all terminals. Set a switch to each of its positions and use the ohmmeter to determine which pairs of terminals have continuity and which pairs are open. If the switch is unfamiliar to you, careful testing will reveal how the contacts operate inside the switch. If you know how the switch should operate, the test will verify whether the switch is good.

Since relay contacts are held in normal positions by springs, the first set of ohmmeter readings will not give a complete picture of how the contacts work. You must take another set of ohmmeter readings while holding the contacts in their momentary positions by pressing the armatures with a finger or screwdriver.

A strong word of warning: you must *never* move relay contacts manually when power is on. Any machine controlled by relays and contactors operates in a complex fashion. Do not experiment with such equipment without proper training and supervision.

Testing Fuses with Ohmmeters

Since a fuse is a piece of wire or metal, it can be tested out of the circuit with an ohmmeter. A good fuse will show continuity; a blown fuse will show infinity. It is often difficult to be sure whether a fuse is good or blown just by looking at it. Therefore, if you suspect a fuse is blown, remove it from the holder and test it with an ohmmeter.

→ *Self-Test*

19. What is the procedure for measuring resistance with a multimeter?
20. What is continuity?
21. What is infinity?

SECTION 4–8: Other Electrical Meters and Measurements

Panel Meters

Panel meters measure a single electrical quantity and are permanently mounted. The quantity measured by the meter is often indicated on the meter face, as in Figure 4–19. Panel meters may be analog or digital. A modern version of the analog panel meter is the

FIGURE 4–19
Panel meters.

row of green and red indicator lights commonly found in audio equipment. When voltage or current increases by a certain amount, the next indicator in the row lights up.

Clamping Ammeter

Figure 4–20(a) illustrates a *clamping ammeter,* which measures ac current without breaking the current path. To use the meter, open the spring-loaded clamp and place it around a

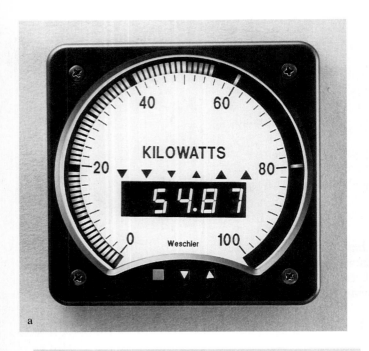

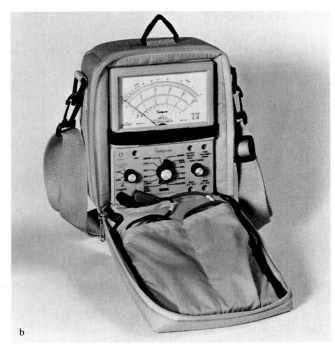

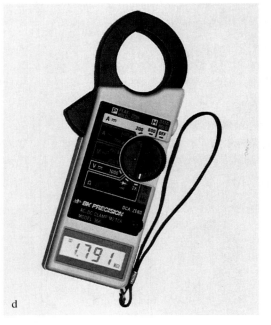

FIGURE 4–20
(a) Amprobe, (b) megger, (c) wattmeter, (d) kilowatt-hour meter.

conductor carrying an ac current. Wires inside the clamp act like an antenna to pick up the pulsating magnetic field around the conductor. The clamping ammeter measures the strength of the magnetic field and interprets this as the amount of current flowing in the conductor.

Because clamping ammeters are fast, easy, and safe to use, they are widely used by industrial maintenance electricians. However, they only work on ac currents of 5 amps or more and are generally not as accurate as ammeters connected directly to the current path.

Insulation Testers (Meggers)

Figure 4–20(b) illustrates an insulation tester, often nicknamed a *megger.* The insulation tester is a high-voltage ohmmeter. No insulator is perfect, and enough voltage can force some current through any insulator. In industrial electricity, a common problem is a crack or weak spot in the insulation of a conductor inside a long metal conduit. When high voltage is applied to the conductor, a leakage current flows from the conductor through the weak insulation into the metal conduit, blowing fuses and shutting down equipment connected to the conductor.

When power is removed and the insulation between the conductor and the conduit is tested with an ordinary ohmmeter, the insulation seems to be good. The ohmmeter reads off scale on its highest range, indicating that the insulation has infinite resistance. The problem is that ordinary ohmmeters with low-voltage internal batteries cannot measure resistance above 2 MΩ, and the flaws in the insulation may have a resistance of 10 MΩ to 50 MΩ.

When the same conductor is tested with a megger, the problem is revealed. The high voltage of the megger produces enough current to give an accurate reading of the insulation's resistance.

Older insulation testers contain a hand-cranked generator able to produce pulses of several thousand volts, enough to produce a measurable current through the flaws in the insulation. Modern testers use electronic circuits to produce the high-voltage pulses. An insulation tester must be handled carefully to avoid personal injury.

Wattmeters

Figure 4–20(c) illustrates a *wattmeter,* which measures power by combining the readings of a voltmeter and an ammeter, according to Watt's Law. Wattmeters are often permanently installed near a main power panel.

Kilowatt-Hour Meter

Figure 4–20(d) illustrates a *kilowatt-hour meter* used by electric utility companies to measure the energy consumed by individual customers. The meter is actually a small motor driving gears that turn a group of dials, indicating the total kilowatt-hours consumed since the meter was first installed. Each month an electric company employee reads the meter and subtracts the last reading from the current reading to determine the month's consumption of energy.

Formula List

Power:

$$P = VI$$
$$P = I^2R$$

$$P = \frac{V^2}{R}$$

Resistance:

$$R = \frac{V}{I}$$

$$R = \frac{P}{I^2}$$

$$R = \frac{V^2}{P}$$

Current:

$$I = \frac{V}{R}$$

$$I = \frac{P}{V}$$

$$I = \sqrt{\frac{P}{R}}$$

Voltage:

$$V = IR$$

$$V = \frac{P}{I}$$

$$V = \sqrt{PR}$$

● TROUBLESHOOTING: Applying Your Knowledge

If your mother's microwave oven normally draws more current than the circuit breaker can handle, the circuit breaker will trip open. In this case, there is probably nothing wrong with the new oven. The solution to the problem would be to install a larger circuit breaker.

The circuit breaker stays closed when the toaster and coffee maker are on, which suggests there is nothing wrong with the circuit breaker or the wiring. Nonetheless, it would be a good idea to calculate the current these appliances draw and compare this value to the current rating of the breaker.

MICROWAVE OVEN:

GIVEN:
  $P = 2.6\ \text{kW}$
  $V = 120\ \text{V}$

FIND:
  $I = ?$

SOLUTION:

$I = P \div V$	select formula
$\quad = 2.6\ \text{kW} \div 120\ \text{V}$	substitute values
$\quad = 2.6\ \text{EXP}\ 3 \div 120\ \text{V}$	enter metric prefix
$\quad = 21.67\ \text{A}$	divide

TOASTER and COFFEE MAKER:

$P = 1{,}000\ \text{W} + 1{,}200\ \text{W}$	find total power
$\quad = 2{,}200\ \text{W}$	add
$I = P \div V$	select formula
$\quad = 2{,}200\ \text{W} \div 120\ \text{V}$	substitute values
$\quad = 18.33\ \text{A}$	divide

CONCLUSION: The normal combined current of the toaster and coffee maker is 18.33 A, less than the 20-A limit of the circuit breaker. The normal current of the microwave oven is 21.67 A, more than the rating of the breaker. The breaker is operating normally, and there is no reason to suspect any problem with the oven. The best recommendation would be to replace the 20-A breaker with a 30-A breaker, if the wires and outlets can safely handle a 30-A current. A licensed electrician should be consulted to be certain no codes or other regulations would be violated by using the larger breaker.

CHAPTER SUMMARY: ANSWERS TO SELF-TESTS

1. Ohm's Law expresses the relationship between voltage, current, and resistance.
2. Ohm's Law states that resistance is the ratio of voltage and current: $R = V \div I$.
3. If resistance is constant, current is directly proportional to voltage: An increase in voltage produces an increase in current.
4. If voltage is constant, current is inversely proportional to resistance: An increase in resistance produces a decrease in current.
5. Watt's Law expresses the relationship between voltage, current, and power.
6. Watt's Law states that power is the product (combination) of voltage and current: $P = V \times I$.
7. A low-voltage circuit needs high current to produce large amounts of power. A high-voltage circuit can produce the same amount of power with a low current.
8. Ohm's Law and Watt's Law may be rearranged and combined to produce 12 formulas known as the Formula Wheel.
9. If any two basic electrical quantities are known, the other two can be found using the Formula Wheel.
10. A multimeter is a portable instrument that measures voltage, current, and resistance.
11. A VOM (volt-ohm-milliammeter) is an analog multimeter with a needle-and-scale display.
12. A DMM is a digital multimeter with a numerical display.
13. An electrical point is a conductor or group of conductors with no difference in electrical potential between any of the physical points. All voltage measurements are taken between electrical points.
14. To measure voltage with a multimeter, place the two probes across two electrical points in a circuit.
15. When no current is flowing, the potential is the same on both sides of a resistive load. Therefore, the difference in potential is zero.
16. If a current path is open between two points and complete between all other points, the potential across the open points equals the potential of the voltage source.
17. In simple circuits with resistive loads, the normal voltmeter reading is full source voltage across an open switch, and zero volts across a closed switch.
18. To measure current with a multimeter, open the circuit and insert the meter into the current path.
19. To measure resistance with a multimeter, turn off the power to the circuit, disconnect one or more leads of the component to be measured, and place the probes across the component.
20. Continuity is a resistance reading of nearly zero ohms, taken between two physical points of a current path with no current flowing. Continuity indicates there is no break or open between the two points.
21. Infinity is a resistance reading that is off scale on the highest range, taken between two physical points of a current path with no current flowing. Infinity indicates there is a break or open between the two points.

CHAPTER REVIEW QUESTIONS

Determine whether the following questions are true or false.

4–1. Ohms equals volts divided by amps.
4–2. Watts equals volts divided by amps.
4–3. A low-voltage circuit needs low current to produce large amounts of power.
4–4. Only high-voltage circuits can produce large amounts of power.
4–5. Any power level can be produced by any source voltage, if the source can also produce a sufficient amount of current.
4–6. If voltage changes, resistance also changes.
4–7. Resistance is a ratio between voltage and current.
4–8. Two circuits can draw different amounts of current and yet produce the same amount of power.
4–9. If resistance changes, voltage also changes.
4–10. Power is the product of voltage and current.
4–11. If current changes, resistance also changes.
4–12. Voltage is the product of power and current.
4–13. If voltage changes, current also changes.
4–14. A change in current can be caused only by a change in voltage.
4–15. Two loads can have different values of voltage and current and still have the same value of power.
4–16. An AMM is an analog multimeter.
4–17. A VOM contains a mechanism called a movement.
4–18. A DMM has a window called a display.
4–19. An alligator clip and a banana plug cannot be used on the same test lead.
4–20. The function and range switches may be combined into a single switch.
4–21. An off-scale reading is the same as a reading of infinity.
4–22. We consider the potential energy (voltage) to be the same at all points along a conductor.
4–23. To measure voltage, choose an electrical point in a circuit and touch both probes to the point.
4–24. To measure current with a multimeter, break the circuit and insert the meter into the current path.

4–25. To measure resistance with a multimeter, turn off the power to the circuit and place the probes across the component.

4–26. A blown fuse produces an ohmmeter reading of infinity.

4–27. A megger measures current in an ac conductor by detecting the strength of the magnetic field around the conductor.

Select the response that best answers the question or completes the statement.

4–28. If resistance is constant,
   a. voltage is directly proportional to current.
   b. current is directly proportional to voltage.
   c. voltage is inversely proportional to current.
   d. current is inversely proportional to voltage.

4–29. If voltage is constant,
   a. resistance is directly proportional to current.
   b. current is directly proportional to resistance.
   c. resistance is inversely proportional to current.
   d. current is inversely proportional to resistance.

4–30. The power of a circuit is
   a. the sum of voltage and current.
   b. the sum of voltage and resistance.
   c. the product of voltage and current.
   d. the product of voltage and resistance.

4–31. All multimeters have a common
   a. range   **c.** jack
   b. resolution   **d.** probe tip

4–32. Special jacks are used for
   a. autoranging
   b. measuring high values
   c. adjusting ohms
   d. connecting a VOM to a DMM

4–33. What is the meter reading in Figure 4–21 if the switch is set
   a. as shown   **d.** 1.2 mA dc
   b. × 1,000 ohms   **e.** 0.3 V dc
   c. 0.06 mA dc   **f.** 300 V ac

4–34. In Figure 4–22, which DMM display indicates an off-scale reading?

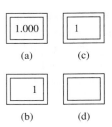

(a)   (c)

(b)   (d)

FIGURE 4–22

4–35. The resistance of an open circuit is called
   a. continuity   **c.** off scale
   b. infinity   **d.** conductivity

4–36. The resistance between two physical points of the same electrical point is called
   a. continuity   **c.** zero
   b. infinity   **d.** conductivity

4–37. How many electrical points are in Figure 4–23(a)?
   a. 1   **d.** 4
   b. 2   **e.** 7
   c. 3   **f.** 8

4–38. How many electrical points are in Figure 4–23(b)?
   a. 1   **d.** 4
   b. 2   **e.** 7
   c. 3   **f.** 8

Solve the following problems.

4–39. What is the current when the voltage is 9 V and the resistance is 3 Ω?

4–40. What is the voltage when the current is 2.5 A and the resistance is 6 Ω?

4–41. What is the current when the voltage is 12 V and the resistance is 8 Ω?

4–42. What is the power when the voltage is 7 V and the current is 300 mA?

4–43. What is the voltage when the power is 600 W and the current is 5 A?

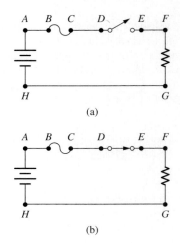

FIGURE 4–21

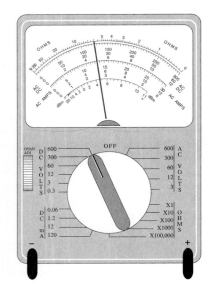

FIGURE 4–23

4–44. What is the current when the voltage is 12 V and the power is 240 W?

4–45. What is the power when the voltage is 75 V and the resistance is 3.7 Ω?

4–46. What is the resistance when the power is 1,500 W and the voltage is 117 V?

4–47. What is the power when the current is 12.3 A and the resistance is 3.7 Ω?

4–48. What is the current when the power is 60 W and the resistance is 2.4 Ω?

4–49. What is the voltage when the power is 200 W and the resistance is 72 Ω?

Refer to the DMM illustrated in Figure 4–24. What should the switch position be for each of the following situations?

4–50. Current in a circuit with a 36-V dc source and a 2-kΩ load.

4–51. Voltage drop across a 470-Ω resistor with a current of 38 mA dc.

4–52. Resistance of a resistor with bands colored orange-blue-yellow-gold.

The following problems are more challenging.

4–53. A 345-kV transmission line carries 325 A. What is the power?

4–54. An antenna picks up 30 μV and produces 6 nW of power. What is the current?

4–55. An average home consumes power at a rate of 1.35 kW. How much power must an electric utility provide for a city of 235,000 homes?

4–56. What is the current produced by a 12-V power supply when connected to each of the following resistors?

    a. 360 Ω   **c.** 2 kΩ

    b. 470 kΩ   **d.** 5.1 MΩ

4–57. What is the maximum resolution of a 4½-digit DMM?

4–58. Suppose you intend to use a VOM to test the voltage at a wall outlet, but you mistakenly set up the meter as an ammeter. What will happen when you insert the probes into the outlet?

Refer to the DMM illustrated in Figure 4–24. What should the switch position be for each of the following situations?

4–59. Current in an automobile circuit that dissipates 1.8 W.

4–60. Resistance of a night light that draws 62.5 mA from a 120-V wall outlet, and with a hot resistance 150 times greater than its cold resistance.

GLOSSARY

clamping ammeter A meter that measures ac current by detecting the strength of the magnetic field around the conductor.

continuity A resistance reading of nearly zero ohms, taken between two physical points of a current path.

division A mark on a scale of an analog meter.

electrical point A conductor or group of conductors with no difference in electrical potential between any of the physical points.

Formula Wheel Twelve formulas derived from Ohm's Law and Watt's Law.

function switch The switch that selects the electrical quantity a multimeter measures.

infinity A resistance reading that is off scale on the highest range, taken between two physical points of a current path.

jack A hole in a panel or meter intended to receive a plug.

kilowatt-hour meter A meter used by electric utility companies to measure the energy consumed by individual customers.

megger A high-voltage ohmmeter used to test insulation.

meter movement A coil of thin wire that produces magnetism from current to move the pointer of an analog meter against the force of a spring.

multimeter A portable meter that measures several electrical quantities.

off scale A reading beyond the scale or display setting of a multimeter.

Ohm's Law The relationship between current, voltage, and resistance.

probes Metal points or clips mounted in plastic at one end of each test lead.

resolution The smallest change in value that a DMM can display.

sampling The process by which the electronic circuits of a DMM take repeated measurements at brief intervals of time.

scale The numbers and marks printed on the face of an analog meter.

test leads The wires that connect the meter to the circuit or component to be measured.

Watt's Law The relationship between power, voltage, and current.

wattmeter A meter that measures power by combining the functions of a voltmeter and an ammeter.

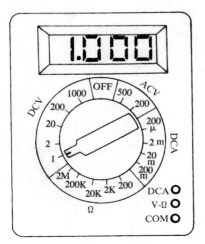

FIGURE 4–24

CHAPTER

5 SERIES CIRCUITS

SECTION OUTLINE

5–1   Definition of Series Circuits

5–2   Resistance in Series Circuits

5–3   Current in Series Circuits

5–4   Voltage in Series Circuits

5–5   Power in Series Circuits

5–6   Polarity in Series Circuits

5–7   Multiple Voltage Sources in Series Circuits

5–8   The *RIPE* Table in Series Circuits

5–9   Ground-Referenced Voltage Readings

5–10   Shorts and Opens in Series Circuits

5–11   Practical Examples of Series Circuits

TROUBLESHOOTING: Applying Your Knowledge

Two weeks before Christmas, you receive a phone call from your aunt. The string of Christmas tree lights she bought in 1953 for 79 cents are not lighting up this year (she only burns them two hours a year, on Christmas Eve, which is why they have lasted this long).

Your aunt tells you there is a "short" in the wire, and since you are going to school to learn electricity, you wouldn't mind driving 75 miles to her house this weekend and fixing them for her, would you? Because she is not about to go out and spend good money for a new set of lights.

As you drive through the snowstorm to her house, you wonder what is wrong with her Christmas lights. What could be the problem?

OVERVIEW AND LEARNING OBJECTIVES

The series circuit is the most basic type. Understanding the rules and relationships of voltage, current, resistance, and power in the series circuit is the foundation of understanding all other circuits.

The concept of ground is universal in electrical circuits and widely misunderstood. Most test voltage readings in actual circuits are ground-referenced. You must understand what these readings mean and how they relate to the voltage drops of the components in the circuit.

Opens and shorts are the most common electrical problems. You must be able to distinguish them clearly and locate them through voltage measurements.

Solving electrical problems requires a systematic method. Organizing your information effectively on paper is your best resource for a successful solution. In this chapter, you will learn a unique method, the *RIPE* table, which will reinforce your fundamental knowledge of electrical laws and rules and greatly simplify your work in solving theoretical problems.

After completing this chapter, you will be able to:

1.   Recognize components connected in series.
2.   Calculate total resistance, current, voltage drops, and power in a series circuit.
3.   Determine polarity of components in a series circuit.
4.   Calculate ground-referenced voltage readings in series circuits.
5.   Identify and locate shorts and opens using voltmeters.

SECTION 5–1: Definition of Series Circuits

In this section you will learn the following:

→ A series circuit has two or more loads, connected so there is only one path for current.

→ Series circuit visual recognition rule: If two components are connected to each other on only one side and no other component is connected between them, then the two components are in series.

The word *series* means one thing following after another. For example, in baseball we speak of the World Series, which is one game played after another.

An electrical circuit with several loads connected one after another is called a *series circuit*. The current in the circuit has to flow through all of the loads. Thus, a series circuit has only one path for current, which means the current has to go through all the loads.

The circuit in Figure 5–1 has three resistors. To follow the current path, put your finger on the battery's negative (−) terminal and move your finger along the lines around to the battery's positive (+) terminal. Your finger moved through all three resistors, one after another. The circuit is clearly a series circuit: two or more loads connected so there is only one path for current.

If two components are connected to each other on only one side and no other component is connected between them, then the two components are in series. We can use this rule to recognize a series connection in a schematic diagram. We will call this the Series Circuit Visual Recognition Rule.

For example, in Figure 5–1, R_1 and R_2 are in series because they are connected on only one side and no other component is connected between them. R_2 and R_3 are also connected in series.

Although R_1 and R_3 are not connected directly to each other, they are still considered to be in series because the same current flows through both of them. In other words, since the current that flows through R_1 must also flow through R_3, we say they are in series.

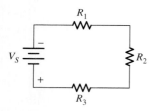

FIGURE 5–1
A series circuit.

→ *Self-Test*
- - - - - - -

1. Describe a series circuit.
2. State the Series Circuit Visual Recognition Rule.

SECTION 5–2: Resistance in Series Circuits

In this section you will learn the following:

→ The total load resistance of a series circuit is equal to the sum of the individual load resistances in the circuit.

In Figure 5–2(a) we see a single resistor with a value of 100 ohms. In Figure 5–2(b) we see two resistors in series, each one with a value of 50 ohms. It is easy to see that the two circuits have the same total resistance. Resistance in a series circuit is like a total resistance broken up into pieces, then put back together.

Series Resistance Rule

The total load resistance of a series circuit is equal to the sum of the individual load resistances in the circuit. We will call this the Series Resistance Rule:

$$R_T = R_1 + R_2 + \ldots + R_n$$

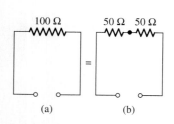

FIGURE 5–2
Total resistance in series circuits.

● SKILL-BUILDER 5–1

What is the total resistance in the circuit in Figure 5–3?

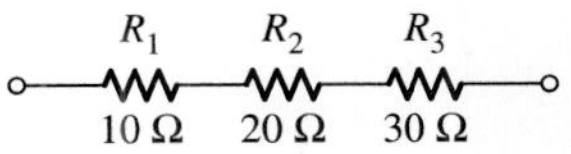

GIVEN:
$R_1 = 10\ \Omega$
$R_2 = 20\ \Omega$
$R_3 = 30\ \Omega$

FIND: R_T

SOLUTION:

$R_T = R_1 + R_2 + R_3$ Series Resistance Rule
$= 10\ \Omega + 20\ \Omega + 30\ \Omega$
$= 60\ \Omega$

FIGURE 5–3

   Repeat the Skill-Builder using the following values. You should obtain the results that follow.

Given:

R_1	470 Ω	150 Ω	68 Ω	10 kΩ	2.2 kΩ
R_2	360 Ω	75 Ω	30 Ω	4.7 kΩ	5.1 kΩ
R_3	200 Ω	200 Ω	22 Ω	560 Ω	1.8 kΩ

Find:

R_T	1,030 Ω	425 Ω	120 Ω	15.26 kΩ	9.1 kΩ

→ Self-Test

3. Describe the total load resistance of a series circuit.

SECTION 5–3: Current in Series Circuits

In this section you will learn the following:

→ Total current in a series circuit equals source voltage divided by total resistance.
→ There is only one current in a series circuit. Therefore, current in a series circuit has the same value in all components.

   Since a series circuit acts like a big circuit broken into pieces, the source voltage feels the total resistance all at once. This determines the current, according to Ohm's Law.

Ohm's Law for Total Values

Total current equals source voltage divided by total resistance.

$$\frac{V_S}{I_T \mid R_T}$$

Series Current Rule

There is only one current in a series circuit. Therefore, current in a series circuit has the same value in all components. We will call this the Series Current Rule.

$$I_T = I_1 = I_2 = .\ .\ . = I_n$$

If the current is known at any location in a series circuit, then the current is known for all locations in the circuit.

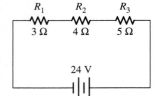

FIGURE 5–4

● *SKILL-BUILDER 5–2*

Find the current in the circuit in Figure 5–4.

GIVEN:
$V_S = 24$ V
$R_1 = 3$ Ω
$R_2 = 4$ Ω
$R_3 = 5$ Ω

FIND:
I_T

STRATEGY:
1. Use the Series Resistance Rule to find R_T. R_1, R_2, and R_3 are required. All these values are given.
2. Use Ohm's Law to find I_T. V_S and R_T are required. V_S is given.

SOLUTION:

1. $R_T = R_1 + R_2 + R_3$ Series Resistance Rule
$= 3\ \Omega + 4\ \Omega + 5\ \Omega$
$= 12\ \Omega$

2. $I_T = \dfrac{V_S}{R_T}$ Ohm's Law

$= \dfrac{24\text{ V}}{12\ \Omega}$

$= 2$ A

Repeat the Skill-Builder using the following values. You should obtain the results that follow.

Given:

V_S	9 V	12 V	15 V	27 V	36 V
R_1	300 Ω	1 kΩ	4.7 kΩ	360 Ω	5.1 kΩ
R_2	470 Ω	1.8 kΩ	3.3 kΩ	200 Ω	7.5 kΩ
R_3	150 Ω	2.7 kΩ	3.9 kΩ	300 Ω	6.8 kΩ

Find:

R_T	920 Ω	5.5 kΩ	11.9 kΩ	860 Ω	19.4 kΩ
I	9.78 mA	2.18 mA	1.26 mA	31.4 mA	1.86 mA

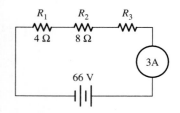

FIGURE 5–5

● *SKILL-BUILDER 5–3*

The current is 3 A in the circuit in Figure 5–5. What is the resistance of R_3?

GIVEN:
$V_S = 66$ V
$I_T = 3$ A
$R_1 = 4$ Ω
$R_2 = 8$ Ω

FIND:
  R_3

STRATEGY:
1. Use Ohm's Law to find R_T. V_S and I_T are required, and both are given.
2. Use the Series Resistance Rule to find R_3 by subtracting. R_1, R_2 and R_T are required. R_1 and R_2 are given.

SOLUTION:

1. $R_T = \dfrac{V_S}{I_T}$ Ohm's Law

    $= \dfrac{66\text{ V}}{3\text{ A}}$

    $= 22\ \Omega$

2. $R_3 = R_T - R_1 - R_2$ Series Resistance Rule
    $= 22\ \Omega - 4\ \Omega - 8\ \Omega$
    $= 10\ \Omega$

    Repeat the Skill-Builder using the following values. You should obtain the results that follow.

Given:

V_S	21 V	5 V	38 V	29 V	17 V
I_T	446 µA	4.81 mA	3.62 mA	2.07 mA	243 µA
R_1	10 kΩ	510 Ω	4.7 kΩ	3.3 kΩ	10 kΩ
R_2	22 kΩ	300 Ω	2.2 kΩ	3.9 kΩ	33 kΩ

Find:

R_T	47,085 Ω	1,040 Ω	10,497 Ω	14,010 Ω	69,959 Ω
R_3	15,085 Ω	230 Ω	3,597 Ω	6,810 Ω	26,959 Ω

→ *Self-Test*

4. Describe the total current in a series circuit.
5. How many currents are there in a series circuit, and what effect does this have on the value of current in the various components?

SECTION 5–4: Voltage in Series Circuits

In this section you will learn the following:

→ Kirchhoff's Voltage Law states that in any closed loop of any circuit, the algebraic sum of all voltage sources and voltage drops in that loop equals zero.

→ A voltage drop is a difference in electrical potential across a load component.

→ In a series circuit, the sum of the voltage drops of all load components equals the source voltage.

→ Ohm's Law may be used to calculate values for any part of a circuit, as long as the values used in the formula apply to that part of the circuit.

→ The voltage drop of each load component in a series circuit is proportional to the resistance of that component.

Kirchhoff's Voltage Law

Kirchhoff's Voltage Law states that in any closed loop of any circuit, the algebraic sum of all voltage sources and voltage drops in that loop equals zero.

The series circuits in this chapter are the simplest forms of a closed loop. Thus, a series circuit offers the simplest example of Kirchhoff's Voltage Law. The term *algebraic sum* refers to assigning positive and negative values to the various voltages in the loop. When this is done correctly, the voltages in the loop will add up to zero. As we study more complex circuits in later chapters, we will understand this more clearly.

As it applies to series circuits with only one voltage source, Kirchhoff's Voltage Law tells us that all the voltage created by the source will be used up by the load. We have seen this in circuits with only one load, where the load voltage equals the source voltage.

However, a series circuit behaves as if the load has been broken up into pieces. Therefore, it makes sense that each part of the load uses up part of the source voltage, and the sum of these voltage losses equals the source voltage.

Voltage Drops

A *voltage drop* is a difference in electrical potential across a load component. In other words, a voltage drop is a difference in energy between two locations because electrons in the current have more energy in one location than in the other. As current moves through each load component in the circuit, electron energy is used up by the resistance in the component. The amount of energy used up depends on the amount of resistance. The loss in energy is measured by a voltmeter placed across the component and is called a voltage drop.

Series Voltage Rule

In a series circuit, the sum of the voltage drops of all load components equals the source voltage. We will call this the Series Voltage Rule:

$$V_S = V_1 + V_2 + \ldots + V_n$$

Figure 5–6 shows us all we have learned so far about series circuits. The total resistance is $8\ \Omega + 4\ \Omega = 12\ \Omega$. The 24-V source feels the total $12\ \Omega$. Therefore, the current is $24\ V \div 12\ \Omega = 2\ A$. 16 V is dropped across R_1, and 8 V is dropped across R_2. The sum of the voltage drops equals the source voltage: $16\ V + 8\ V = 24\ V$, following the Series Voltage Rule and Kirchhoff's Voltage Law.

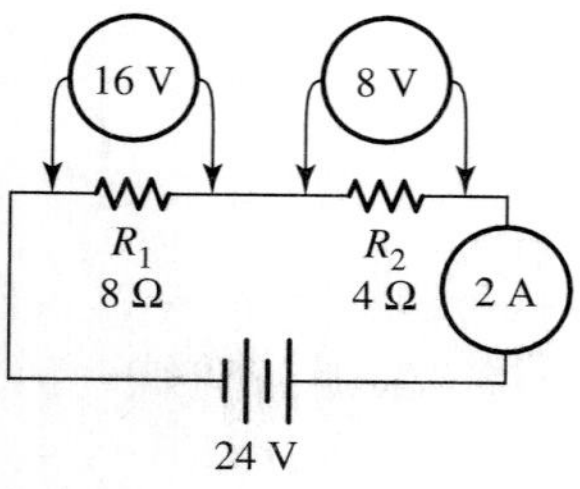

FIGURE 5–6
Kirchhoff's Voltage Law in series circuits.

Voltage Drops and Ohm's Law

Ohm's Law may be used to calculate values for any part of a circuit, as long as the values used in the formula apply to that part of the circuit.

The exact amount of each voltage drop can be calculated by Ohm's Law, as long as we use the correct values of current and resistance for each component. The current in both R_1 and R_2 is 2 A, according to the Series Current Rule. R_1 has a resistance of $8\ \Omega$, and R_2 has $4\ \Omega$. Voltage equals current multiplied by resistance, according to Ohm's Law. Putting all this together, we calculate that the voltage drop across R_1 is $2\ A \times 8\ \Omega = 16\ V$. For R_2, $2\ A \times 4\ \Omega = 8\ V$.

● *SKILL-BUILDER 5–4*

What is the source voltage in the circuit in Figure 5–7?

GIVEN:
$V_1 = 3$ V
$V_2 = 2$ V
$V_3 = 7$ V

FIND:
V_S

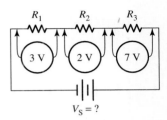

FIGURE 5–7

SOLUTION:

$V_S = V_1 + V_2 + V_3$ Series Voltage Rule
   $= 3$ V $+ 2$ V $+ 7$ V
   $= 12$ V

Repeat the Skill-Builder using the following values. You should obtain the results that follow.

Given:

V_1	7.5 V	13.2 V	6.8 V	3.9 V
V_2	3.7 V	14.8 V	4.2 V	8.6 V
V_3	2.2 V	16.4 V	3.4 V	6.3 V

Find:

V_S	13.4 V	44.4 V	14.4 V	18.8 V

● *SKILL-BUILDER 5–5*

What is the voltage drop of R_2 in the circuit in Figure 5–8?

GIVEN:
$V_S = 16$ V
$V_1 = 5$ V
$V_3 = 7$ V

FIND:
V_2

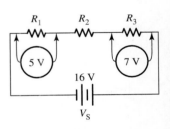

FIGURE 5–8

SOLUTION:

$V_2 = V_S - V_1 - V_3$ Series Voltage Rule
   $= 16$ V $- 5$ V $- 7$ V
   $= 4$ V

Repeat the Skill-Builder using the following values. You should obtain the results that follow.

Given:

V_S	17 V	11 V	16 V	29.8 V	1.26 V
V_1	3 V	3.3 V	4.75 V	9.8 V	0.35 V
V_3	6 V	4.6 V	3.28 V	7.6 V	0.68 V

Find:

V_2	8 V	3.1 V	7.97 V	12.4 V	0.23 V

Voltage Drops and Proportionality

The voltage drop of each load component in a series circuit is proportional to the resistance of that component.

In a series circuit, each load component drops its fair share of the source voltage, according to that component's share of total resistance. Thus, large resistors drop more of the total voltage than small resistors because large resistors have a larger share of the total resistance. This fair-share relationship is called a proportional relationship. Thus, the voltage drops can also be calculated with formulas based on proportionality.

Proportionality is similar to percentages. For any load component, its percentage of total voltage dropped equals its percentage of total resistance. This is expressed in Formula 5–1, where V_X is the voltage drop of a load component, and R_X is the resistance of that component.

Formula 5–1: Proportionality of Voltage Drops

$$\frac{V_X}{V_S} = \frac{R_X}{R_T}$$

This formula can be rearranged to give us Formula 5–2 for the voltage drop, V_X.

Formula 5–2: Voltage Divider Formula

$$V_X = \frac{R_X}{R_T} \times V_S$$

● SKILL-BUILDER 5–6

Find V_1 in Figure 5–6 using the voltage divider formula.

GIVEN:
  $V_S = 24$ V
  $R_1 = 8\ \Omega$
  $R_2 = 4\ \Omega$

FIND:
  V_1

STRATEGY:
  1. Use the voltage divider formula to find V_1. V_S and R_1 are given. R_T is not given.
  2. Use the Series Resistance Rule to find R_T. R_1 and R_2 are given.

SOLUTION:
  1. $R_T = R_1 + R_2$ Series Resistance Rule
     $= 8\ \Omega + 4\ \Omega$
     $= 12\ \Omega$

  2. $V_1 = \dfrac{R_1}{R_T} \times V_S$ voltage divider formula

     $= \dfrac{8\ \Omega}{12\ \Omega} \times 24$ V

     $= 16$ V

Repeat the Skill-Builder using the following values. You should obtain the results that follow.

Given:

V_S	15 V	28 V	3 V	12.6 V	4.35 V
R_1	27 Ω	5.1 kΩ	150 Ω	10 kΩ	27.2 Ω
R_2	38 Ω	3.6 kΩ	270 Ω	22 kΩ	35.8 Ω

Find:

R_T	65 Ω	8.7 kΩ	420 Ω	32 kΩ	63 kΩ
V_1	6.23 V	16.41 V	1.07 V	3.94 V	1.88 V

→ *Self-Test*

6. State Kirchhoff's Voltage Law.
7. What is a voltage drop?
8. What is the relationship between voltage drops and source voltage in a series circuit?
9. How may Ohm's Law be used to calculate values within a circuit?
10. What is the relationship between the voltage drop of a load component in a series circuit and the resistance of that component?

SECTION 5–5: Power in Series Circuits

In this section you will learn the following:

→ The total power of a series circuit is the sum of the power developed in each component.

→ In series circuits, the component with the highest resistance dissipates the most power (produces the most heat).

Series Power Rule

The total power of a series circuit is the sum of the power developed in each component. We will call this the Series Power Rule:

$$P_T = P_1 + P_2 + \ldots + P_n$$

Power equals voltage times current. In a series circuit, all components have the same current. The voltage divider formula informs us that the series component with the greatest resistance will have the largest voltage drop. Therefore, in series circuits, the component with the highest resistance dissipates the most power (produces the most heat).

● *SKILL-BUILDER 5–7*

In the circuit in Figure 5–9, find the power dissipation of each component and the total power level of the circuit.

GIVEN:
   $V_S = 120$ V
   $R_1 = 10\ \Omega$
   $R_2 = 20\ \Omega$
   $R_3 = 30\ \Omega$

FIND:
   P_T, P_1, P_2, P_3

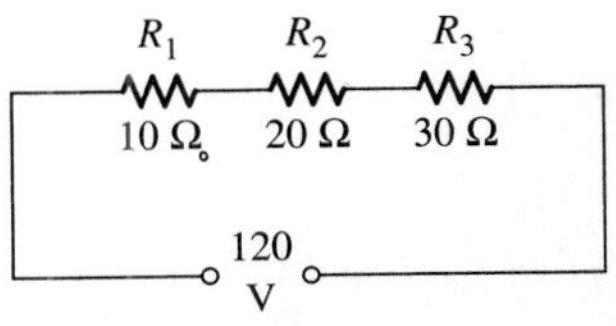

FIGURE 5–9

STRATEGY:

1. Use the power formula $P = I^2 R$ to find each power level. All values of R are given. I is not given.
2. Use Ohm's Law to find I. V_S and R_T are required. V_S is given. R_T is not given.
3. Use the Series Resistance Rule to find R_T. All required values of R are given.

SOLUTION:

1. $R_T = R_1 + R_2 + R_3$ Series Resistance Rule
   $= 10\ \Omega + 20\ \Omega + 30\ \Omega$
   $= 60\ \Omega$

2. $I = V_S \div R_T$ Ohm's Law
   $= 120\ \text{V} \div 60\ \Omega$
   $= 2\ \text{A}$

3. $P_T = I^2 R_T$ Watt's Law
   $= 2^2\ \text{A} \times 60\ \Omega$
   $= 240\ \text{W}$

   $P_1 = I^2 R_1$ Watt's Law
   $= 2^2\ \text{A} \times 10\ \Omega$
   $= 40\ \text{W}$

   $P_2 = I^2 R_2$ Watt's Law
   $= 2^2\ \text{A} \times 20\ \Omega$
   $= 80\ \text{W}$

   $P_3 = I^2 R_3$ Watt's Law
   $= 2^2\ \text{A} \times 30\ \Omega$
   $= 120\ \text{W}$

R_3, the largest resistor, dissipates the greatest amount of power.

Repeat the Skill-Builder using the following values. You should obtain the results that follow.

Given:

V_S	37 V	51 V	26 V	1.63 V	208 V
R_1	180 Ω	6.5 Ω	2,700 Ω	1.3 Ω	40 Ω
R_2	220 Ω	3.9 Ω	3,300 Ω	0.8 Ω	32 Ω
R_3	360 Ω	2.8 Ω	1,500 Ω	0.4 Ω	19 Ω

Find:

R_T	760 Ω	13.2 Ω	7,500 Ω	2.5 Ω	91 Ω
I_T	48.68 mA	3.864 mA	3.47 mA	0.652 A	2.286 A
P_T	1,801 mW	197 W	90 mW	1.06 W	475 W
P_1	427 mW	97 W	32 mW	0.55 W	209 W
P_2	521 mW	58 W	40 mW	0.34 W	167 W
P_3	853 mW	42 W	18 mW	0.17 W	99 W

→ *Self-Test*

11. Describe the total power in a series circuit.
12. What is the relationship between the resistance of a component in a series circuit and the power dissipated by that component?

SECTION 5–6: Polarity in Series Circuits

In this section you will learn the following:

→ Electrical polarity is caused by a difference in charge or potential between any two points in a circuit, with one point more negative and the other point more positive.

→ A resistive load component has polarity only when current flows through it.

Batteries always have polarity because the chemicals inside them remove electrons from one terminal and force those electrons onto the other terminal. However, resistive load components have polarity only when current flows through them.

The negative end of a resistor is the one closer to the negative terminal of the battery. Likewise, the positive end of a resistor is the one closer to the positive end of the battery.

The words positive and negative can be misleading. It is best to include the words more and less to make the meaning clear.

In the circuit of Figure 5–10, it is easy to see that point *A* is connected directly to the negative terminal of the battery, and point *D* is connected directly to the positive terminal. Therefore, it seems correct to say that point *A* is negative and point *D* is positive.

However, this is not the best way to describe the two points. It is better to say that point *A* is more negative than point *D*, or that point *D* is more positive than point *A*.

In describing polarity, we compare one point to another. To be clear, we include the words more or less. In fact, we can describe the polarity of points *A* and *D* by saying that point *A* is less positive than point *D*, or that point *D* is less negative than point *A*.

In Figure 5–10, how should we describe the polarity of point *B*? The polarity markings (+ and −) shown on the schematic can confuse us. Point *B* has a plus on the left and a minus on the right. What does this mean?

We cannot say anything about point *B* by itself. We can only describe point *B* by comparing it to another point. The following statements about point *B* are all true:

Point *B* is more positive than point *A*.
Point *B* is less negative than point *A*.
Point *B* is more negative than point *C*.
Point *B* is less positive than point *C*.

For each of the three resistors in Figure 5–10, the left side is closer to the battery's negative terminal, and the right side is closer to the positive terminal. Thus, electron current enters on the left and exits on the right of each resistor. This is why the left sides are marked negative and the right sides marked positive. The polarity markings refer to the resistors, not to the points.

In summary, when describing polarity, we speak of two points, and speak of the first point compared to the second point. We can describe any polarity as either positive or negative, if we also include the words more or less. Some components such as batteries have *intrinsic polarity*. This means that one pole (end) is always more negative and the other always more positive. Other components such as resistors have polarity only

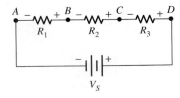

FIGURE 5–10
Polarity in series circuits.

when current is flowing. Since electrons actually flow from the negative pole of a voltage source toward the positive pole, the negative side of the resistor is the side closer to the negative pole of the source, and vice versa for the positive side.

→ *Self-Test*

13. What causes electrical polarity?
14. When does a resistive load component have polarity?

SECTION 5–7: Multiple Voltage Sources in Series Circuits

In this section you will learn the following:

→ If two voltage sources in series are connected positive to negative, they aid each other by attempting to produce current in the same direction. The total voltage is the sum of the individual voltages.

→ If two voltage sources in series are connected positive to positive or negative to negative, they oppose each other by attempting to produce current in opposite directions. The total voltage is the difference between the individual voltages.

Figure 5–11 shows a flashlight with a long handle containing several dry cells (batteries). This is a simple example of a series circuit with multiple voltage sources. Each cell has a voltage of 1.5 V. We start at the positive terminal of the first cell (the one touching the bulb). The electrons leave that cell's negative terminal with a potential of 1.5 V. The electrons enter the second cell and gain another 1.5 V, giving a total of 3 V. Going through the third cell increases their potential to 4.5 V.

Referring to the schematic for the circuit, we see the three batteries connected so that their polarities are in the same direction. This allows the batteries to help each other, since they are all pushing the electrons in the same direction. This connection of voltage sources is called *series aiding.* The total voltage is the sum of the individual voltages, and the load must drop the total voltage of 4.5 V.

If batteries in series are turned so that their polarities are opposite, they fight each other by trying to push the electrons in opposite directions. This is called *series opposing,* as shown in Figure 5–12. The total voltage is the difference between the sources, found

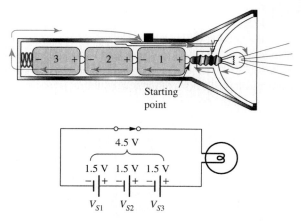

FIGURE 5–11
Multiple voltage sources in series aiding.

by subtracting the voltage in one direction from the voltage in the other direction. The electrons flow in the direction of the stronger voltage and actually flow backwards through the weaker battery.

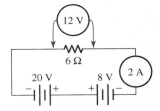

FIGURE 5–12
Multiple voltage sources in series opposing.

 PRACTICAL TIP

There are many small electrical and electronic devices which use batteries, such as flashlights, portable radios, and tape players, and so on. Many of these use more than one battery. Now you can understand why all the batteries must be put in facing the same direction. If some of the batteries are turned the wrong way, their voltage opposes the other batteries, instead of aiding. As a result, there is not enough total voltage for the device to operate properly.

● SKILL-BUILDER 5–8

What is the total voltage dropped by the load resistor in the circuit in Figure 5–13?

GIVEN:
$V_A = 3$ V
$V_B = 9$ V

FIND:
V_L

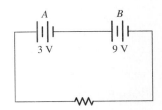

FIGURE 5–13

STRATEGY:
  By inspecting the schematic, we see the batteries are connected in series aiding. Therefore, the total source voltage is the sum of the two sources.

SOLUTION:
$V_L = V_A + V_B$ sources in series aiding
$\quad = 3$ V $+ 9$ V
$\quad = 12$ V

● SKILL-BUILDER 5–9

What is the total voltage dropped by the load resistor in the circuit in Figure 5–14?

GIVEN:
$V_A = 3$ V
$V_B = 9$ V

FIND:
V_L

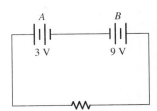

FIGURE 5–14

STRATEGY:
  By inspecting the schematic, we see the batteries are connected in series opposing. Therefore, the total source voltage is the difference of the two sources.

SOLUTION:
$V_L = V_B - V_A$ multiple sources series opposing
$\quad = 9$ V $- 3$ V
$\quad = 6$ V

15. What is the relationship between two voltage sources in series connected positive to negative?
16. What is the relationship between two voltage sources in series connected positive to positive or negative to negative?

SECTION 5–8: The *RIPE* Table in Series Circuits

The *RIPE* table is a simple organization method for solving circuit problems. The table has five columns: component, resistance, current, power, and voltage. In the table, the voltage column is labeled with the letter *E,* which stands for electromotive force, another name for voltage. There is a row for each load component and a row for total values. We will use the circuit problem in Figure 5–15 to illustrate the use of the table.

PROBLEM: In the circuit in Figure 5–15, find the value of R_3 and all voltage drops.

Begin by drawing the *RIPE* table. The table has five columns. The first column identifies the component. The remaining columns represent the four basic electrical quantities. Record in the table all values given in the problem. *Note:* The battery voltage is the total voltage.

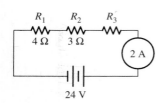

FIGURE 5–15

COMPONENT	R	I	P	E
R_1	4 Ω			
R_2	3 Ω			
R_3				
Total		2 A		24 V

To use the table, we follow a few simple rules:

1. Place all given values in the *RIPE* table. The total voltage (battery voltage) always goes in the bottom row of the table.
2. Look for a row that contains two values.
3. If you find such a row, use Ohm's Law to find the third value in the row.
4. When you find a third value in a row, think of the rule for that value. Then think of a way to use that value in another row.
5. If you can't find a row with two values, look at each column, one at a time. Think of the rule for that column. Look at the information in the column. Think of a way to get another value to use in a row.
6. Repeat steps 2–5 until the table is complete.

After placing all given values in the table (step 1), we see (step 2) that we have two values in the Total row: voltage and current. Therefore, we can find total resistance with Ohm's Law (step 3):

$$R_T = V_S \div I \qquad \text{Ohm's Law}$$
$$= 24 \text{ V} \div 2 \text{ A}$$
$$= 12 \text{ Ω}$$

We record this new value in the table:

COMPONENT	R	I	P	E
R_1	4 Ω			
R_2	3 Ω			
R_3				
Total	12 Ω	2 A		24 V

We now repeat our steps. This time, we cannot follow step 2 because we do not see a row that has two values. Therefore, we must skip steps 3 and 4. We go to step 5, look at each column, think of the series rules, and figure out another value from the information we have.

The first column is resistance. We remember the Series Resistance Rule, which says that the total resistance is equal to the sum of the component values. With this in mind, we look at our table and see that in the resistance column, we have all the cells filled in except for R_3. Now we can see what to do next and how to do it. The Series Resistance Rule says to add the component values to find the total. Since we already know the total, we can use subtraction to work backwards and find R_3.

$$R_3 = R_T - R_1 - R_2 \qquad \text{Series Resistance Rule}$$
$$= 12\ \Omega - 4\ \Omega - 3\ \Omega$$
$$= 5\ \Omega$$

This is half the answer to our problem. Let's put it in the table:

COMPONENT	R	I	P	E
R_1	4 Ω			
R_2	3 Ω			
R_3	5 Ω			
Total	12 Ω	2 A		24 V

Now let's complete the other cells to find the voltage drops. This will show how helpful the table really is. As long as you follow the steps, the table fills itself in almost automatically, and you can answer any question about the values in the circuit.

At this point, we cannot follow step 2 (no row with two values), so we skip steps 3 and 4. Following step 5, we notice the Current column and remember the series rule for current, which tells us that in a series circuit, the current is the same value everywhere. This means that we can copy the 2-A value into the other rows. The table now looks like this:

COMPONENT	R	I	P	E
R_1	4 Ω	2 A		
R_2	3 Ω	2 A		
R_3	5 Ω	2 A		
Total	12 Ω	2 A		24 V

Now we have two values in three rows. We can follow step 3 and use Ohm's Law to complete these rows.

$$V_1 = I_1 \times R_1$$
$$= 2\,\text{A} \times 4\,\Omega$$
$$= 8\,\text{V}$$

Ohm's Law

$$V_2 = I_2 \times R_2$$
$$= 2\,\text{A} \times 3\,\Omega$$
$$= 6\,\text{V}$$

Ohm's Law

$$V_3 = I_3 \times R_3$$
$$= 2\,\text{A} \times 5\,\Omega$$
$$= 10\,\text{V}$$

Ohm's Law

When we put these values into the table, we have the answers we were asked to find:

COMPONENT	R	I	P	E
R_1	4 Ω	2 A		8 V
R_2	3 Ω	2 A		6 V
R_3	5 Ω	2 A		10 V
Total	12 Ω	2 A		24 V

We can double-check our voltage drops by remembering the Series Voltage Rule: The voltage drops add up to equal the source voltage. Looking at the voltage column, it is easy to see this is true.

Although the problem did not ask for power, we can easily calculate these values from the information in the table to be sure that all resistors are operating within their power rating.

$$P_1 = I_1 \times V_1$$
$$= 2\,\text{A} \times 8\,\text{V}$$
$$= 16\,\text{W}$$

Watt's Law

$$P_2 = I_2 \times V_2$$
$$= 2\,\text{A} \times 6\,\text{V}$$
$$= 12\,\text{W}$$

Watt's Law

$$P_3 = I_3 \times V_3$$
$$= 2\,\text{A} \times 10\,\text{V}$$
$$= 20\,\text{W}$$

Watt's Law

$$P_T = I_T \times V_T$$
$$= 2\,\text{A} \times 24\,\text{V}$$
$$= 48\,\text{W}$$

Watt's Law

These are large power levels, so the resistors need to be large wire-wound or ceramic types, capable of dissipating at least 20 W. We now complete the table by adding the power values.

COMPONENT	R	I	P	E
R_1	4 Ω	2 A	16 W	8 V
R_2	3 Ω	2 A	12 W	6 V
R_3	5 Ω	2 A	20 W	10 V
Total	12 Ω	2 A	48 W	24 V

When you solve circuit problems, you may have difficulty knowing where to start and what to do next. Using the *RIPE* table and following the steps eliminates these difficulties.

There is another advantage to using the table. You should never do any calculations without a clear reason. Using the table and following the steps reminds you of the reasons for each calculation: the laws and the rules. This teaches you how to think logically and work systematically with electrical circuits. When you work on real equipment, you may not have to do circuit calculations. However, the way of thinking will be exactly the same. Let's summarize this most important mental skill.

1. You must have a reason for taking the first step. Why did you start there?
2. You must have a reason for taking the next step. Generally, it will be because of something you just discovered in the last step.
3. You should write down everything you do in a organized manner that is easy to read. This keeps you from getting confused and makes it easy for other people to help you if you have difficulty.

This last point is very important. People cannot read your mind or make sense of a few scribbled numbers and words. You also will not remember everything you did after you have been working on a problem for a long time. Writing everything down, step by step as you do it, seems time-consuming at first. The truth is, it usually saves a lot more time when you run into trouble and need help. Get into the habit now, and stay with it!

Often there is more than one way to fill in the table. The sequence you choose is not as important as having a good reason for doing so.

→ *Self-Test*

17. What are the five columns in the *RIPE* table?
18. What is the first step in using the *RIPE* table?
19. Where does the total voltage (battery voltage) go in the *RIPE* table?
20. What should you do when you find a row in the *RIPE* table that contains two values?
21. What should you do after you find a third value in a row in the *RIPE* table?
22. What should you do when you cannot find a row in the *RIPE* table that contains two values?

SECTION 5–9: Ground-Referenced Voltage Readings

In this section you will learn the following:

→ Ground is the electrical point in a circuit where the voltage is considered to be zero. Therefore, ground is the starting point or reference point for all voltage measurements in the circuit.
→ In many machines that include electrical circuits, ground is the metal frame or chassis of the machine, used as a common return path for current.
→ In electric utility company systems, ground is the earth itself.

The first definition is always true. The second and third definitions apply only in certain situations.

Reference Ground

We have learned that voltage is a difference in energy or charge between two points. Whenever we take a voltage reading between two points, we must choose one of them as

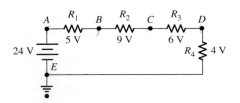

FIGURE 5–16
Ground at point *E*.

the starting or reference point. If we choose one point to be the common reference point for the entire circuit, we call this reference point *ground,* and we read the voltage at the other points compared to reference ground.

We can choose any point in a circuit as ground. There is no natural law which decides where ground should be. However, it is a common practice among electrical engineers to choose one terminal of the voltage source as ground. We will follow this custom in our first example.

In Figure 5–16, we have chosen point *E* to be ground, and the symbol for ground is shown at point *E*. This point is the negative terminal of the battery. The circuit has four resistors in series. The voltage drops are shown for each resistor. Four other points are identified, *A* through *D*. Now we will find the ground-referenced voltage at each point.

Technicians usually begin checking a circuit at the power supply and work their way to ground, watching the voltage readings decrease to zero as they move along.

The 24-V battery is located between ground and point *A*. Therefore, the ground-referenced voltage at *A* is 24 V.

What about the polarity of point *A?* Since ground has been placed on the negative terminal of the battery, point *A* is clearly more positive than ground. Therefore, the voltage at *A* is +24 V (positive with respect to ground).

R_1 is between point *A* and point *B*. R_1 shows a voltage drop (loss) of 5 V. Therefore, the ground-referenced voltage at point *B* is +24 V − 5 V = +19 V.

R_2 is between point *B* and point *C*. R_2 drops 9 V. Therefore, the ground-referenced voltage at *C* is +19 V − 9 V = +10 V.

R_3 is between point *C* and point *D*. R_3 drops 6 V. Therefore, the ground-referenced voltage at *D* is +10 V − 6 V = +4 V.

TABLE 5–1
Ground-referenced voltages for Figure 5–16.

POINT	GROUND-REFERENCED VOLTAGE	VOLTAGE DROP	COMPONENT
A	+24 V		Battery
		5 V	R_1
B	+19 V		
		9 V	R_2
C	+10 V		
		6 V	R_3
D	+4 V		
		4 V	R_4
Ground *(E)*	0 V		

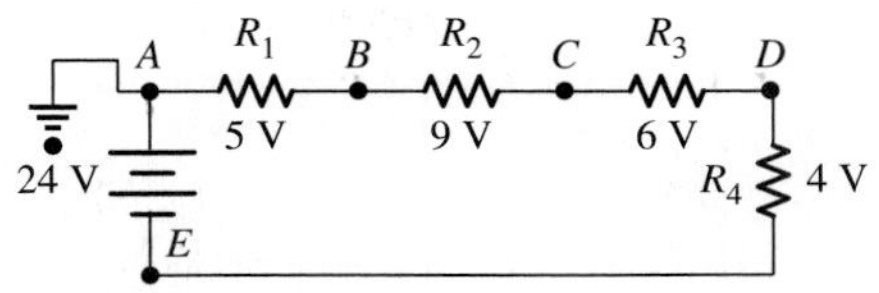

FIGURE 5–17
Ground at point A.

The last 4 V is dropped by R_4. When we return to ground at the bottom of R_4, we have +4 V – 4 V = 0 V.

Table 5–1 summarizes our readings. It shows how the ground-referenced voltage readings decrease according to the voltage drops of each component.

Remember, there is nothing natural about the location of ground in this circuit. It is simply the place we decided to call zero volts. To illustrate this idea, let's put ground at point A, as shown in Figure 5–17.

The battery is still between point E and point A. The difference in voltage between these points is still 24 V. However, since ground is on the positive terminal of the battery, we now must say that the ground-referenced voltage at point E is –24 V.

Working our way toward ground, we cross over R_4 to point D. R_4 drops 4 V. Therefore, the ground-referenced voltage at D is –24 V – 4 V = –20 V.

Plus and minus signs can be confusing in electricity because they have two different meanings: polarity and arithmetic. In the previous step, we had 24 V and we lost (subtracted) 4 V, leaving us with 20 V. Since our 24 V had a negative polarity, the 20 V also has a negative polarity.

We cross R_3 to point C. R_3 drops 6 V. Therefore, at point C we have –20 V – 6 V = –14 V.

We cross R_2 to point B. R_2 drops 9 V. Therefore, the voltage reading at point B is –14 V – 9 V = –5 V.

We cross R_1 to point A (ground). R_1 drops the last 5 V, leaving us at zero volts.

Notice that nothing has changed electrically. All the components are the same. We have simply put ground in a different place. Now our table looks like Table 5–2.

As our last example, let's put ground in the middle of the circuit at point C, as shown in Figure 5–18. This will cause our ground-referenced voltage readings at the other points to be a combination of positive and negative polarities.

TABLE 5–2
Ground-referenced voltages for Figure 5–17.

POINT	GROUND-REFERENCED VOLTAGE	VOLTAGE DROP	COMPONENT
E	–24 V		Battery
		4 V	R_4
D	–20 V		
		6 V	R_3
C	–14 V		
		9 V	R_2
B	–5 V		
		5 V	R_1
Ground (A)	0 V		

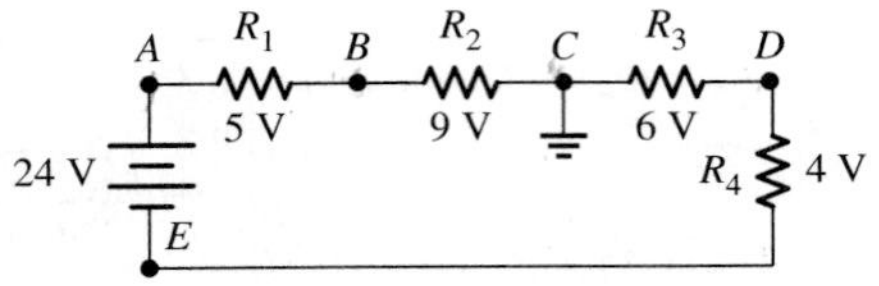

FIGURE 5–18
Ground at point C.

If we start at ground (point C) and move toward either terminal of the battery, we will appear to gain voltage. The polarity of the voltage depends on which terminal we move toward.

R_2 causes a 9-V difference between ground and point B. Point B is closer to the positive terminal of the battery. Therefore, the ground-referenced voltage at point B is +9 V.

R_1 causes a 5-V difference between point B and point A. Point A is closer to the positive terminal of the battery. Therefore, the ground-referenced voltage at point A is +14 V.

R_3 causes a 6-V difference between ground and point D. Point D is closer to the negative terminal of the battery. Therefore, the ground-referenced voltage at point D is −6 V.

R_4 causes a 4-V difference between point D and point E. Point E is closer to the negative terminal of the battery. Therefore, the ground-referenced voltage at point E is −10 V.

Now our table looks like Table 5–3:

TABLE 5–3
Ground-referenced voltages for Figure 5–18.

POINT	GROUND-REFERENCED VOLTAGE	VOLTAGE DROP	COMPONENT
A	+14 V		
		5 V	R_1
B	+9 V		
		9 V	R_2
Ground (C)	0 V		
		6 V	R_3
D	−6 V		
		4 V	R_4
E	−10 V		

Notice that in this table we do not see the battery voltage anywhere.

Chassis Ground

In many machines which include electrical circuits, the metal frame or chassis of the machine is used as a common return path for current. This is done to save wire. Therefore, it makes sense to choose the chassis as the ground point for all circuits in the machine. This is called *chassis ground*.

A good example of this is an automobile. In most automobiles, the negative terminal of the battery is connected directly to the steel frame. The positive terminal is con-

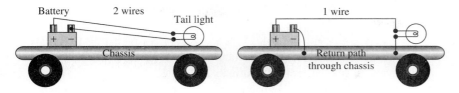

FIGURE 5–19
Chassis ground provides a current return path.

nected to all the electrical devices: starter motor, horn, lights, heater, radio, wipers, and so on.

How does this save wire? The tail lights in an automobile could be 15 feet away from the battery. By using the steel frame as a conductor (one side of the current path), we need only one long wire instead of two. This saves time and money in building the automobile. We see this illustrated in Figure 5–19.

Any metal part of the automobile's frame or body can be used as a reference point when taking voltage readings with a voltmeter. Simply touch the black probe to the frame or body and the red probe to the electrical connector you are examining. You must be certain that the black probe is touching bare metal. You may have to scratch through dirt, grease, or paint to get a good contact.

Chassis ground is also used in televisions, sound systems, and computers. The aluminum frame which holds the various circuit boards together is used as a common return path for current, and also chosen as the ground reference point.

Earth Ground

A utility company generates electricity for millions of people spread out over large areas. In a circuit this big and complicated, the earth itself is the only place that makes sense to be the reference point for measuring voltage. This is called *earth ground*.

At thousands of locations within the system, the utility company will put copper rods into the soil at a depth of several feet and make electrical connections to these rods. The electricians who install wiring in buildings and homes will do the same thing to match the utility company's ground system.

Summary of Ground

We see now why the word *ground* has several different meanings. To the student or experimenter building small circuits, ground means a certain location in the circuit. To the technician, ground means the metal frame of the machine. To the electrician, ground means the copper rod buried in the earth. Yet for all of these, ground means the starting point for measuring voltages. In other words, ground is the chosen place where we think of the voltage as being zero.

→ *Self-Test*

23. What is ground?
24. What is chassis ground?
25. What is earth ground?

SECTION 5–10: Shorts and Opens in Series Circuits

In this section you will learn the following:

→ A short causes more current than normal.

→ An open causes less current than normal.

→ A short is a low-resistance connection that causes current to flow around a normal load instead of through it.

→ In a series circuit, a short will produce a voltage drop of zero. All the other voltage drops will be higher than normal.

→ A short will cause the ground-referenced voltage readings to be the same on both sides of the short, when normally they should be different.

→ An open is a break in the current path.

→ In a series circuit, an open causes all voltage drops to be zero. Across the open, a voltmeter reads full source voltage.

→ When there is an open in a series circuit, all ground-referenced voltage readings will be either zero or full source voltage. The open is located between the two points where the readings change.

The most common problems in all electrical circuits are either shorts or opens. However, the two problems are totally different. People often speak of shorts incorrectly to mean any kind of electrical problem. As we shall see, shorts and opens have opposite meanings. We will learn how to recognize the two problems with current and voltage readings.

Shorts

A *short* is a low-resistance connection that causes current to flow around a normal load instead of through it. Since the resistance is much less, the current is much higher. Often the high current produces enough heat to melt the circuit and start a fire. Therefore, shorts are damaging and dangerous. Shorts are caused by wires connected around a component, accidental contact between wires or metal parts, faulty insulation, poor workmanship, pinched conductors, and so on.

A short is a bypass or detour for current, so that the current goes around a load instead of through it. A simple way to create a short is to connect a piece of wire around a component.

Figure 5–20(a) shows a normal series circuit of three resistors. The total resistance is $3 \, \Omega + 4 \, \Omega + 5 \, \Omega$, which equals $12 \, \Omega$. The total voltage is 12 V. The normal current is found by Ohm's Law:

$$I = V \div R$$
$$= 12 \, \text{V} \div 12 \, \Omega$$
$$= 1 \, \text{A}$$

Figure 5–20(b) shows a piece of wire connected around the second resistor. This creates a short. The current will go through the wire now instead of through the resistor. The wire's resistance is almost zero ohms. Therefore, the total resistance has changed. The current feels only the resistance of the loads it goes through. It doesn't feel the middle resistor any more. The total resistance is now $3 \, \Omega + 5 \, \Omega = 8 \, \Omega$.

A short causes the total resistance to be less than normal. This causes the total current to be more than normal. With only $8 \, \Omega$ in the circuit, the current is now $12 \, \text{V} \div 8 \, \Omega = 1.5 \, \text{A}$.

Figure 5–20(c) illustrates the effect the short has on the circuit. Since the current no longer feels the second resistor, we think of it as being gone from the circuit.

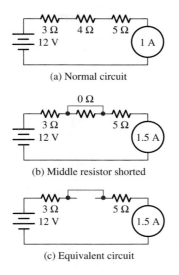

FIGURE 5–20
Effect of a short in a series circuit.

Shorts may be caused by connecting a piece of wire around a load, as we just saw. However, shorts are more often caused by accidental contact. For example, if you are working on a wall outlet or switch with a screwdriver, you will cause a short if the screwdriver slips and touches both a live terminal and the metal box at the same time. Because the metal box is connected to ground outside the building, the screwdriver has created a short between the live terminal and ground. Current will flow through the screwdriver instead of through the normal load. Since the screwdriver has almost no resistance, the current will be very high, causing a tremendous spark and producing enough heat to melt the screwdriver blade. Thus, we can see that shorts are dangerous because of the high current they produce.

Shorts often occur when wire is so old that its insulation cracks and falls off, exposing the bare wire inside. This can cause a danger of both fire and electrocution. For example, a wire inside an old electric drill might lose its insulation and touch the case of the drill. Many older drills have metal cases. A person holding the drill could be badly shocked. To avoid this danger, most modern electric drills have cases made of plastic instead of metal.

Another example is the wiring in the walls of old buildings. If these wires lose their insulation and short by touching each other, the high current might create enough heat to

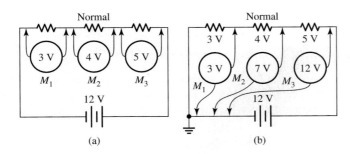

(a) (b)

FIGURE 5–21
Normal voltage readings.

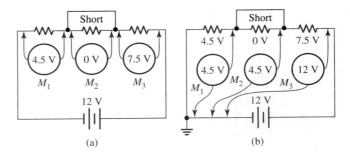

FIGURE 5–22
Voltage readings with middle component shorted.

start a fire and burn down the building. When working with wires in old buildings, electricians must be very careful not to crack the insulation when handling the wires.

Shorts can also occur by poor workmanship. Using too much solder, burning insulation with a soldering iron, stripping away too much insulation, using insulating tape improperly, and using connectors improperly can create shorts.

Another common cause of shorts occurs with electrical equipment that is contained inside a cabinet. When technicians close doors or put covers back in place, wires can be pinched or terminals forced into contact with the frame, causing a short.

Finding Shorts with Voltmeters

Voltmeters are quick, safe, and easy to use. We prefer to find electrical problems with voltmeters whenever we can. How can we use a voltmeter to find a short?

Figure 5–21 shows the normal voltmeter readings in our circuit. The drawing on the left shows the individual voltage drops. The drawing on the right shows ground-referenced voltage readings.

Figure 5–22 shows the voltage readings with the middle resistor shorted. On the left, we notice that the other two resistors have voltage drops higher than normal. This is because of the higher current flowing through them. The shorted resistor shows a zero voltage drop. This is the characteristic pattern that reveals a short.

In summary, a short will produce a voltage drop of zero. All the other voltage drops will be higher than normal.

A short will cause the ground-referenced voltage readings to be the same on both sides of the short, when normally they should be different. We see this in the readings on the right.

Troubleshooting Exercises in Series Shorts

The following examples will help us understand series shorts and will also teach us some important principles of electrical troubleshooting.

In all troubleshooting problems, there are three steps. We must learn:

1. What the circuit should be.
2. What the circuit is.
3. Why the circuit is not what it should be.

● *SKILL-BUILDER 5–10*

- -

Which resistor is shorted in the circuit in Figure 5–23?

GIVEN:

$$V_S = 12 \text{ V}$$
$$I = 1.5 \text{ A}$$
$$R_1 = 3 \text{ } \Omega$$
$$R_2 = 4 \text{ } \Omega$$
$$R_3 = 5 \text{ } \Omega$$

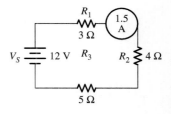

FIGURE 5–23

FIND:

Which resistor is shorted?

STRATEGY:

1. Consider what the circuit should be, if all components were normal. Find normal R_T with the Series Resistance Rule, and normal I with Ohm's Law.
2. Consider what the circuit actually is, accepting the given values as correct. The actual current and actual source voltage are given. Find actual R_T using Ohm's Law.
3. Compare actual values with normal values. The values are different because a component is defective. Use the comparison of values to determine which component is at fault.

SOLUTION:

1. Normal $R_T = R_1 + R_2 + R_3$ Series Resistance Rule
   $$= 3 \text{ } \Omega + 4 \text{ } \Omega + 5 \text{ } \Omega$$
   $$= 12 \text{ } \Omega$$

2. Actual $R_T = V_S \div I$ Ohm's Law
   $$= 12 \text{ V} \div 1.5 \text{ A}$$
   $$= 8 \text{ } \Omega$$

3. The difference between normal
   R_T and actual R_T is
   $12 \text{ } \Omega - 8 \text{ } \Omega = 4 \text{ } \Omega$ compare normal to actual

DEDUCTION: We were told that one resistor was shorted. The difference between normal and actual R_T is 4 Ω. Shorting a resistor removes that resistor's value from the total resistance. Therefore, R_2 is the shorted resistor.

Repeat the Skill-Builder using the following values. You should obtain the results that follow.

Given:					
V_S	25 V	48 V	16 V	37 V	53 V
Actual I_T	71.43 mA	11.43 mA	14.04 mA	26.81 mA	1.04 mA
R_1	100 Ω	1.5 kΩ	680 Ω	470 Ω	18 kΩ
R_2	200 Ω	2.7 kΩ	750 Ω	1.2 kΩ	33 kΩ
R_3	150 Ω	3.3 kΩ	390 Ω	910 Ω	22 kΩ
Find:					
Normal R_T	450 Ω	7.5 kΩ	1,820 Ω	2,580 Ω	73 kΩ
Actual R_T	350 Ω	4.2 kΩ	1,140 Ω	1,380 Ω	51 kΩ
Difference	100 Ω	3.3 kΩ	680 Ω	1,200 Ω	22 kΩ
Short	R_1	R_3	R_1	R_2	R_3

- -

● SKILL-BUILDER 5–11

Which resistor is shorted in the circuit in Figure 5–24?

GIVEN:
$V_S = 36$ V
$V_B = 20.57$ V
$R_1 = 12$ Ω
$R_2 = 16$ Ω
$R_3 = 22$ Ω

FIND:
Shorted resistor.

STRATEGY:
    Use a process of elimination. Assume each resistor is the one that is shorted, and calculate the ground-referenced voltage at point *B*. Match the calculated values with the actual value to determine which assumption is correct.

SOLUTION:
  1. Assume R_1 is shorted. Then point *B* is electrically the same as point *A:*
      $V_B = V_A = V_S = 36$ V   does not match data

  2. Assume R_2 is shorted.
      $R_T = R_1 + R_3$
          $= 12$ Ω $+ 22$ Ω
          $= 34$ Ω

    If R_2 is shorted, point *B* is electrically the same as point *C,* and $V_B = V_C = V_3$. Calculate V_3 using the voltage divider formula.

$$V_3 = \frac{R_3}{R_T} \times V_S$$

$$= \frac{22\ \Omega}{34\ \Omega} \times 36\ \text{V}$$

$$= 23.29\ \text{V} \qquad\qquad\qquad \text{does not match data}$$

  3. Assume R_3 is shorted.
      $R_T = R_1 + R_2$
          $= 12$ Ω $+ 16$ Ω
          $= 28$ Ω

    If R_3 is shorted, point *C* is electrically the same as point *D* (ground), and $V_B = V_2$. Calculate V_2 using the voltage divider formula.

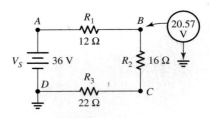

FIGURE 5–24

$$V_2 = \frac{R_2}{R_T} \times V_S$$

$$= \frac{16\ \Omega}{28\ \Omega} \times 36\ \text{V}$$

$$= 20.57\ \text{V} \qquad\qquad \text{this is a match!}$$

DEDUCTION: Assumption 3 produces a calculated value for V_B that matches the actual value. Therefore, it appears that assumption 3 is correct and R_3 is the shorted resistor.

- -

Opens

An *open* is a break in the current path. In a series circuit, an open causes all current to stop. An open is caused by conductors not touching or by contamination such as dirt or corrosion, which prevents current from flowing.

An open is the opposite of a short. When there is an open, no current can flow at all. The resistance is so high that we say the ohms are infinite, which means the ohms are larger than we can measure.

In Figure 5–25(a), the middle resistor has been disconnected from the circuit, producing an open. No current can flow in the circuit. Therefore, in the drawing on the left, the voltage drops of M_1 and M_3 are zero volts. However, you may be surprised to see that voltmeter M_2 reads 12 volts. In fact, this meter is feeling the battery terminals through resistors 1 and 3. Remember that resistors are conductors: They allow electrons to move through them. When no current is flowing, the negative battery terminal can push electrons all the way through the resistor up to the end of the wire. Similarly, the positive terminal can pull electrons away from the other end of the wire through the resistor.

In effect, the ends of the wire become extensions of the battery terminals. This makes it easy to find an open with a voltmeter. When you try to measure voltage drops in the circuit, your readings will be zero everywhere except across the open. Here you will read the full value of the source voltage.

In summary, an open causes all voltage drops to be zero. Across the open, a voltmeter reads full source voltage.

The ground-referenced readings for an open are shown in Figure 5–25(b). When there is an open in a series circuit, all ground-referenced voltage readings will be either zero or full source voltage. The open is located between the two points where the readings change.

We cannot do Ohm's Law calculations with open series circuits because there is no current. Therefore, we will not do any troubleshooting exercises here.

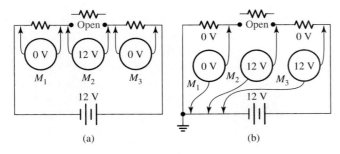

FIGURE 5–25
Voltage readings with middle component open.

Checking Switches with Voltmeters

The purpose of a switch is to produce an open when the switch is off and a low-resistance path (similar to a short) when the switch is on. Therefore, we can use a voltmeter to test a switch, as shown in Figure 5–26.

A good switch in its off position will produce the same voltmeter readings as an open. This is shown in Figure 5–26(a). The voltmeter reads full source voltage across the terminals of the open switch.

A good switch in its on position produces the same voltmeter readings as a short. This is shown in Figure 5–26(b). The voltmeter reads a drop of zero volts across the terminals of the closed switch.

If the switch produces voltmeter readings other than these, there is cause to suspect that the switch is bad.

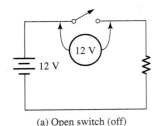

(a) Open switch (off)

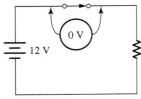

(b) Closed switch (on)

FIGURE 5–26
Checking a switch with a voltmeter.

→ *Self-Test*

26. What effect does a short have on current?
27. What effect does an open have on current?
28. Describe a short.
29. What is the effect of a short on the voltage drops in a series circuit?
30. What is the effect of a short on ground-referenced voltage readings?
31. Describe an open.
32. What is the effect of an open on the voltage drops in a series circuit?
33. What is the effect of an open on ground-referenced voltage readings?

SECTION 5–11: Practical Examples of Series Circuits

Switches, Fuses, and Circuit Breakers

Any device that controls current will be in series with the load. Switches control current by opening to stop the current, or closing to allow the current to flow. Thus, switches are in series with the load that they control.

Fuses and circuit breakers are overcurrent protection devices. They open to stop the current when the current becomes dangerously high. All overcurrent protection devices must be in series with the load that they protect.

Voltage Dividers

Figure 5–27 is a typical transistor circuit. The transistor is the circle with lines and an arrow inside. There are several types of transistors. This type has a lead (wire) called the base. The base is the control wire. The base must have a certain voltage before the transistor can work properly. This voltage is not the same for all transistor circuits. The engineer who designs the circuit calculates exactly what this voltage should be.

In our example, the engineer has decided the base should have a voltage of 3 volts. However, the circuit needs a 9-V battery for other reasons. The engineer has created a series circuit of R_1 and R_2 to divide the 9 volts. R_1 drops 6 volts, and R_2 drops the remaining 3 volts. Therefore, at the top of R_2, the ground-referenced voltage is 3 volts. The base of the transistor is connected to this point to give the base the exact voltage the engineer wants it to have. In this circuit, R_1 and R_2 would be described as a *voltage divider*. Formula 5–2 is often used for these calculations and, thus, is called the voltage divider formula.

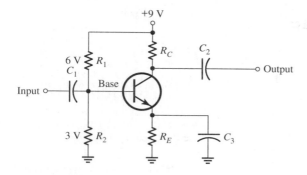

FIGURE 5–27
R_1 and R_2 act as a voltage divider.

Christmas Tree Lights

We all know the frustration of Christmas tree lights, where none of the lights will burn if any bulb is bad. This means the bulbs are wired in series. We know that in a series circuit, there is only one current. If any bulb is loose, burned out, broken, or dirty, this creates an open (a break in the current path), stopping the current for all bulbs.

Formula List

$$R_T = R_1 + R_2 + \ldots + R_n$$
$$I_T = I_1 = I_2 = \ldots = I_n$$
$$V_S = V_1 + V_2 + \ldots + V_n$$

$$\frac{V_X}{V_S} = \frac{R_X}{R_T}$$

$$V_X = \frac{R_X}{R_T} \times V_S$$

$$P_T = P_1 + P_2 + \ldots + P_n$$

● TROUBLESHOOTING: Applying Your Knowledge

If your aunt's string of Christmas tree lights had a short in it, most likely it would be a dead short between the hot and neutral wires. That would have been an exciting moment for your aunt when she plugged in the lights. The short would have produced sparks, bits of molten copper, and burning plastic. Since your aunt didn't speak of it, most likely it didn't happen, and there is probably no short. Still, as old as the wires are, it is possible the insulation finally crumbled off when she brought them down from the attic.

You happen to know that in 1953, Christmas tree lights were wired in series. Therefore, if one bulb burned out, it would open the current path for all other bulbs. Your common sense tells you this is most likely what is wrong. Rather than spend hours trying to locate and replace the defective bulb, you wonder whether you should just stop at a store and buy a new string of lights. Hmmm . . . looks like a shopping center just ahead. . . .

CHAPTER SUMMARY:
ANSWERS TO SELF-TESTS

1. A series circuit has two or more loads connected so there is only one path for current.
2. Series Circuit Visual Recognition Rule: If two components are connected to each other on only one side and no other component is connected between them, then the two components are in series.
3. The total load resistance of a series circuit is equal to the sum of the individual load resistances in the circuit.
4. Total current in a series circuit equals source voltage divided by total resistance.
5. There is only one current in a series circuit. Therefore, current in a series circuit has the same value in all components.
6. Kirchhoff's Voltage Law states that in any closed loop of any circuit, the algebraic sum of all voltage sources and voltage drops in that loop equals zero.
7. A voltage drop is a difference in electrical potential across a load component.
8. In a series circuit, the sum of the voltage drops of all load components equals the source voltage.
9. Ohm's Law may be used to calculate values for any part of a circuit, as long as the values used in the formula apply to that part of the circuit.
10. The voltage drop of each load component in a series circuit is proportional to the resistance of that component.
11. The total power of a series circuit is the sum of the power developed in each component.
12. In series circuits, the component with the highest resistance dissipates the most power (produces the most heat).
13. Electrical polarity is caused by a difference in charge or potential between any two points in a circuit, with one point more negative and the other point more positive.
14. A resistive load component has polarity only when current flows through it.
15. If two voltage sources in series are connected positive to negative, they aid each other by attempting to produce current in the same direction. The total voltage is the sum of the individual voltages.
16. If two voltage sources in series are connected positive to positive or negative to negative, they oppose each other by attempting to produce current in opposite directions. The total voltage is the difference between the individual voltages.
17. The five columns in the *RIPE* table are component, resistance, current, power, and voltage.
18. The first step in using the *RIPE* table is to place all given values in the table.
19. The total voltage (battery voltage) always goes in the bottom row of the *RIPE* table.
20. If you find a row in the *RIPE* table that contains two values, use Ohm's Law to find the third value in the row.

21. When you find a third value in a row, think of the rule for that value. Then think of a way to use that value in another row.
22. If you can't find a row with two values, look at each column, one at a time. Think of the rule for that column. Look at the information in the column. Think of a way to get another value to use in a row.
23. Ground is the electrical point in a circuit where the voltage is considered to be zero. Therefore, ground is the starting point or reference point for all voltage measurements in the circuit.
24. In many machines which include electrical circuits, ground is the metal frame or chassis of the machine, used as a common return path for current.
25. In electric utility company systems, ground is the earth itself.
26. A short causes more current than normal.
27. An open causes less current than normal.
28. A short is a low-resistance connection that causes current to flow around a normal load instead of through it.
29. In a series circuit, a short will produce a voltage drop of zero. All the other voltage drops will be higher than normal.
30. A short will cause the ground-referenced voltage readings to be the same on both sides of the short, when normally they should be different.
31. An open is a break in the current path.
32. In a series circuit, an open causes all voltage drops to be zero. Across the open, a voltmeter reads full source voltage.
33. When there is an open in a series circuit, all ground-referenced voltage readings will be either zero or full source voltage. The open is located between the two points where the readings change.

CHAPTER REVIEW QUESTIONS

Determine whether the following questions are true or false.

5–1. A series circuit has only one path for current.
5–2. A series circuit can be thought of as a total resistance broken up into pieces, then put back together.
5–3. The total load resistance of a series circuit is equal to the largest individual load resistance in the circuit.
5–4. To find current in a series circuit, divide source voltage by the value of each resistor.
5–5. There is only one voltage drop in a series circuit.
5–6. If the current is known at any location in a series circuit, then the current is known for all locations in the circuit.
5–7. In a series circuit, all the voltage created by the source will be used up by the load.
5–8. A voltage drop is a difference in electrical potential across a load component.
5–9. In a series circuit, the sum of the currents of all load components equals the total current.
5–10. Ohm's Law may be used only to calculate total values.

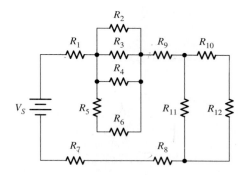

FIGURE 5–28

5–11. The total power of a series circuit equals the power developed by the largest component.
5–12. We can choose only certain points in a circuit as ground.
5–13. A short causes more current than normal.
5–14. An open causes more current than a short.
5–15. A short is a high-resistance connection that causes current to flow around a normal load instead of through it.
5–16. An open has an ohmic value of zero.
5–17. An open in a series circuit causes all voltage drops to be zero.
5–18. Across an open, a voltmeter reads full source voltage.

Select the response that best answers the question or completes the statement.

5–19. Decide whether or not the following resistors are in series in Figure 5–28.
   a. R_7 and R_8 **e.** R_1 and R_3
   b. R_9 and R_{11} **f.** R_5 and R_6
   c. R_9 and R_{10} **g.** R_1 and R_9
   d. R_4 and R_5 **h.** R_9 and R_{12}
5–20 Identify the following statement: In any closed loop of any circuit, the algebraic sum of all voltage sources and voltage drops in that loop equals zero.
   a. Series Resistance Rule
   b. Series Voltage Rule
   c. Kirchhoff's Voltage Law
   d. Ohm's Law for Total Values
5–21. The voltage drop of each load component in a series circuit is proportional to the _____ of that component.
   a. current   **c.** power
   b. resistance   **d.** voltage
5–22. Electrical polarity is caused by a difference in _____ _____ between any two points in a circuit.
   a. voltage   **c.** current
   b. charge or potential   **d.** power

5–23. Resistive load components have polarity only when
   a. current flows through them.
   b. voltage is applied.
   c. they touch each other on only one side.
   d. connected in series aiding.
5–24. Ground is the electrical point in a circuit where the voltage is considered to be
   a. positive   **c.** zero
   b. negative   **d.** neutral
5–25. Chassis ground is
   a. the earth.
   b. the point of least resistance.
   c. the metal frame of the machine.
   d. the point of least polarity.
5–26. Shorts and opens are
   a. exactly the same.
   b. the most common problems in electrical circuits.
   c. difficult and dangerous to find.
   d. the result of polarity.
5–27. An electrical fire can be caused by
   a. a short.   **c.** high resistance.
   b. an open.   **d.** excessive insulation.
5–28. A short will produce a voltage drop equal to
   a. source voltage.
   b. zero.
   c. the largest resistor in the circuit.
   d. the smallest resistor in the circuit.
5–29. A short will cause the ground-referenced voltage readings to be _____ on both sides of the short.
   a. zero   **c.** equal to source voltage
   b. the same   **d.** different
5–30. In a series circuit, an open causes
   a. dangerously high current.
   b. a loss of ground.
   c. all current to stop.
   d. a reversal of polarity.
5–31. An open in a series circuit causes all ground-referenced voltage readings to be
   a. zero.
   b. equal to source voltage.
   c. either zero or full source voltage.
   d. higher than normal.

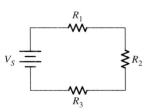

FIGURE 5–29

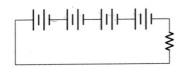

FIGURE 5–30

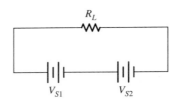

FIGURE 5–31

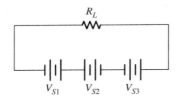

FIGURE 5–32

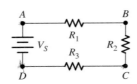

FIGURE 5–33

Solve the following problems.

Problems 5–32 through 5–40 refer to Figure 5–29.

5–32. $R_1 = 27\ \Omega$, $R_2 = 15\ \Omega$, $R_3 = 36\ \Omega$. Find total resistance.

5–33. $R_T = 85\ \Omega$, $R_1 = 16\ \Omega$, $R_3 = 47\ \Omega$. Find R_2.

5–34. $V_S = 90$ V, $R_1 = 7\ \Omega$, $R_2 = 15\ \Omega$, $R_3 = 12\ \Omega$. Find current.

5–35. $V_S = 105$ V, $I = 2.5$ A, $R_1 = 12.5\ \Omega$, $R_2 = 19.8\ \Omega$. Find R_3.

5–36. $I = 0.8$ A, $R_1 = 16.3\ \Omega$, $R_2 = 27.9\ \Omega$, $R_3 = 38.3\ \Omega$. Find V_S.

5–37. $V_1 = 17$ V, $V_2 = 23$ V, $V_3 = 6$ V. Find V_S.

5–38. $V_S = 90$ V, $V_1 = 33$ V, $V_3 = 19$ V. Find V_2.

5–39. $V_S = 72$ V, $R_1 = 12\ \Omega$, $R_2 = 18\ \Omega$, $R_3 = 6\ \Omega$. Find V_1 using the voltage divider formula.

5–40. $V_S = 36$ V, $R_1 = 8\ \Omega$, $R_2 = 6\ \Omega$, $R_3 = 10\ \Omega$. Find total power and power of each component.

5–41. What is the total voltage dropped across the load in Figure 5–30? All cells are 1.5 V.

5–42. In Figure 5–31, $V_{S1} = 12$ V and $V_{S2} = 8$ V. What is the voltage drop across R_L?

5–43 If the polarity of V_{S1} is reversed in Figure 5–31, and the voltages remain at $V_{S1} = 12$ V and $V_{S2} = 8$ V, what is the voltage drop across R_L?

5–44. In Figure 5–32, $V_{S1} = 17$ V, $V_{S2} = 20$ V, and $V_{S3} = 12$ V. What is the voltage drop across R_L?

5–45. In Figure 5–33, $V_S = 70$ V, $R_1 = 15\ \Omega$, $R_2 = 38\ \Omega$, and $R_3 = 24\ \Omega$.
   a. If point D is ground, what is the ground-referenced voltage reading at point B?
   b. If point A is ground, what is the ground-referenced voltage reading at point C?

5–46. In Figure 5–33, $V_S = 85$ V, $R_1 = 12\ \Omega$, $R_2 = 38\ \Omega$, $R_3 = 21\ \Omega$.
   a. What is the voltage drop across R_1 if R_3 is shorted?
   b. What is the voltage reading between points B and C if R_2 is open?
   c. What is the voltage reading between points A and B if R_3 is open?
   d. If point D is ground and R_3 is open, what is the ground-referenced voltage reading at point B?
   e. If point A is ground and R_1 is shorted, what is the ground-referenced voltage reading at point C?

The following problems are more challenging.

5–47. In Figure 5–33, $V_S = 48$ V and $R_2 = 23\ \Omega$. The voltage reading from points A to C is 26.68 V. The voltage reading from points B to D is 36.685 V. What is the resistance of R_3?

5–48. In Figure 5–33, $V_S = 117$ V, $R_2 = 34.6\ \Omega$, and $R_3 = 28.4\ \Omega$. The voltage drop across R_1 is 26.778 V. What is the resistance of R_1?

5–49. In Figure 5–33, $V_S = 90$ V, $R_1 = 22\ \Omega$, $R_2 = 36\ \Omega$, $R_3 = 47\ \Omega$, and $I = 1.304$ A. Which resistor is shorted?

5–50. In Figure 5–33, $V_S = 24$ V, $R_1 = 17\ \Omega$, $R_2 = 28\ \Omega$, and $R_3 = 39\ \Omega$. The voltage reading between points A and C is 10.03 V. Which resistor is shorted?

GLOSSARY

chassis ground   The frame of an electrical machine used both as a return path for current and as the voltage reference point (zero-volt point).

earth ground   A connection from utility company transformers or generators into the earth, establishing it as the point of zero voltage for the power circuit.

ground   The electrical point in a circuit where the voltage is considered to be zero.

intrinsic polarity   A polarity determined by the internal structure of a component rather than by the direction of current flow through the device.

Kirchhoff's Voltage Law   The fundamental relationship between voltage sources and voltage drops: In any closed loop of any circuit, the algebraic sum of all voltage sources and voltage drops in that loop equals zero.

open   A break in the current path.

series aiding Voltage sources connected in series so that the total voltage is the sum of the individual source voltages.

series circuit A circuit with two or more loads connected so there is only one path for current.

series opposing Voltage sources connected in series so that the total voltage is the difference of the individual source voltages.

short A low-resistance connection that causes current to flow around a normal load instead of through it.

voltage divider A series circuit used to provide specific voltages to another device, usually to a transistor or integrated circuit.

voltage drop A difference in electrical potential across a load component.

6

PARALLEL CIRCUITS

● SECTION OUTLINE

6–1   Definition of Parallel Circuits

6–2   Current in Parallel Circuits

6–3   Resistance in Parallel Circuits

6–4   Voltage in Parallel Circuits

6–5   Power in Parallel Circuits

6–6   Multiple Voltage Sources in Parallel Circuits

6–7   The *RIPE* Table in Parallel Circuits

6–8   Shorts and Opens in Parallel Circuits

6–9   Practical Examples of Parallel Circuits

● TROUBLESHOOTING: Applying Your Knowledge

Your friend bought a CB radio for his truck. He mounted it in the dashboard and looked for a way to connect the power wires. He saw a heavy red wire connect to the AM-FM radio. He cut this red wire in the middle, connected the red wire from the CB radio to one of the cut ends, and the black wire from the CB radio to the other cut end. To his dismay, neither radio works now, although the lights in the radios glow dimly. What do you think is wrong? What should he do to correct the problem?

● OVERVIEW AND LEARNING OBJECTIVES

Parallel circuits are quite common. For example, all the wall outlets in your home are wired in parallel, and so are all the light bulbs in your car. You must understand the theory of parallel circuits to troubleshoot common electrical circuits.

After completing this chapter, you will be able to:

1. Recognize components connected in parallel.
2. Calculate resistance, current, voltage, and power in a parallel circuit.
3. Apply Kirchhoff's Current Law to understand values of current in parallel branches and feeders.
4. Solve troubleshooting problems involving open branches in parallel circuits.

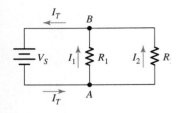

FIGURE 6–1
A parallel circuit.

SECTION 6–1: Definition of Parallel Circuits

In this section you will learn the following:

→ A parallel circuit has two or more current paths (branches).
→ Parallel Circuit Visual Recognition Rule: If two components are connected to each other on both sides, the two components are in parallel.

A parallel circuit has points where total current can split up into smaller currents and flow in several paths before returning to the source. Each parallel path is called a *branch* and, thus, the smaller currents are called branch currents.

The parallel circuit in Figure 6–1 has two branches. The total current I_T leaves the negative terminal of the voltage source and flows along a single current path until it reaches point A, where it has a choice of two current paths. The total current splits into two smaller branch currents, I_1 and I_2, which flow through the branches. At point B, I_1 and I_2 come together again as total current, which returns to the positive terminal of the voltage source on a single current path.

If two components are connected to each other on both sides, the two components are in parallel. We will call this the Parallel Circuit Visual Recognition Rule.

In Figure 6–1, R_1 and R_2 are in parallel because they are connected to each other on both sides.

→ *Self-Test*

1. Describe a parallel circuit.
2. State the Parallel Circuit Visual Recognition Rule.

SECTION 6–2: Current in Parallel Circuits

In this section you will learn the following:

→ Kirchhoff's Current Law states that the sum of all currents entering an electrical point equals the sum of all currents leaving that point.
→ The total current of a parallel circuit equals the sum of the branch currents.

Each branch of a parallel circuit acts like a series circuit. Thus, a parallel circuit acts like a group of series circuits all connected to the same voltage source, as shown in Figure 6–2.

R_1 is a 6–Ω resistor connected directly to a 12-V source. Therefore, by Ohm's Law, I_1 (the current in R_1) equals 2 A (12 V ÷ 6 Ω). Similarly, R_2 is also directly connected to the same 12-V source, and I_2 equals 3 A (12 V ÷ 4 Ω). Since the source must produce both I_1 and I_2 at the same time, the total current I_T leaving the source must equal $I_1 + I_2$, which is 5 A.

R_1 and R_2 are connected in parallel according to the Parallel Visual Recognition Rule: They touch each other on both sides. It doesn't matter that the voltage source is located at the points of connection because the Parallel Visual Recognition Rule does not speak of other components.

In Figure 6–3 we have redrawn Figure 6–2 to look like Figure 6–1. R_1 and R_2 are still connected to each other on both sides, and both are connected directly to the source. The only difference is the current path between points A and B and the source, which is now a single current path shared by both branches. For this reason, we see the total current of 5 A in this section. It is common practice to save wire by having parallel branches share common current paths wherever possible.

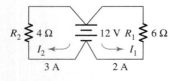

FIGURE 6–2
Parallel branches share a common voltage source.

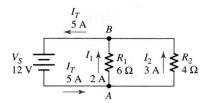

FIGURE 6–3
Standard schematic for a
parallel circuit.

Kirchhoff's Current Law

Kirchhoff's Current Law states that the sum of all currents entering an electrical point
equals the sum of all currents leaving that point.

Figure 6–3 illustrates Kirchhoff's Current Law. At point *A*, the sum of all currents
entering the point is 5 A *(I_T)*, and the sum of all currents leaving the point is also 5 A
($I_1 + I_2$). The same is true for point *B:* 5 A enters *($I_1 + I_2$)* and 5 A leaves *(I_T)*. Kirchhoff's
Current Law is very helpful in solving parallel circuit problems.

Parallel Current Rule

The total current of a parallel circuit equals the sum of the branch currents. We will call
this the Parallel Current Rule.

Formula 6–1

$$I_T = I_1 + I_2 + \dots + I_n$$

Since all branches of a parallel circuit are connected to the same voltage source,
each branch current can be found by Ohm's Law. The branch currents are then added to
find the total current.

● *SKILL-BUILDER 6–1*
- -

Find branch currents and total current in the circuit in Figure 6–4.

GIVEN:
   $V_S = 24$ V
   $R_1 = 2\ \Omega$
   $R_2 = 6\ \Omega$
   $R_3 = 8\ \Omega$

FIND:
   I_1, I_2, I_3, I_T

STRATEGY:
   1. Use Ohm's Law to find I_1, I_2, and I_3. V_S, R_1, R_2, and R_3 are required. All these
      values are given.
   2. Use the Parallel Current Rule to find I_T. All branch currents are required, and
      none are given.

SOLUTION:
   1. $I_1 = V_S \div R_1$ Ohm's Law
          $= 24$ V $\div 2\ \Omega$
          $= 12$ A

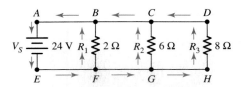

FIGURE 6–4

$$I_2 = V_S \div R_2 \qquad\qquad\qquad \text{Ohm's Law}$$
$$= 24 \text{ V} \div 6 \text{ } \Omega$$
$$= 4 \text{ A}$$

$$I_3 = V_S \div R_3 \qquad\qquad\qquad \text{Ohm's Law}$$
$$= 24 \text{ V} \div 8 \text{ } \Omega$$
$$= 3 \text{ A}$$

2. $I_T = I_1 + I_2 + I_3$ Parallel Current Rule
$$= 12 \text{ A} + 4 \text{ A} + 3 \text{ A}$$
$$= 19 \text{ A}$$

Repeat the Skill-Builder using the following values. You should obtain the results that follow.

Given:

V_S	15 V	48 V	61 V	28 V	37 V
R_1	270 Ω	3 kΩ	15 kΩ	330 Ω	2.7 kΩ
R_2	360 Ω	4.7 kΩ	22 kΩ	510 Ω	3.9 kΩ
R_3	470 Ω	5.1 kΩ	18 kΩ	270 Ω	6.8 kΩ

Find:

I_1	55.56 mA	16 mA	4.07 mA	84.85 mA	13.70 mA
I_2	41.67 mA	10.21 mA	2.77 mA	54.90 mA	9.49 mA
I_3	31.91 mA	9.41 mA	3.39 mA	103.70 mA	5.44 mA
I_T	129.14 mA	35.62 mA	10.23 mA	243.45 mA	28.63 mA

● **SKILL-BUILDER 6–2**

In Figure 6–4, find the current between all adjacent points.

STRATEGY: This example is an exercise in Kirchhoff's Current Law. No other rule is needed, since we have just calculated all branch currents.

1. The section between E and F connects to a terminal of the voltage source. Therefore, I_{EF} (the current between E and F) must equal the total current: $I_{EF} = 19$ A.
2. The section between F and B is the branch containing R_1. Therefore, $I_{FB} = I_1 = 12$ A.
3. To find I_{FG}, we must apply Kirchhoff's Current Law. Since 19 A entered point F, 19 A must leave point F. We know 12 A left point F flowing toward point B. Therefore, the current that left point F flowing toward point G must be the rest of the 19 A. Thus, $I_{FG} = 19$ A $- 12$ A $= 7$ A.
4. I_{GC} equals I_2: 4 A.
5. Sections GH, HD, and DC are all part of the same current path through R_3. Therefore, the current in these three sections equals I_3: 3 A.
6. Point C is where I_2 and I_3 flow together again. Thus, the sum of the currents entering point C equals 4 A $+ 3$ A $= 7$ A. By Kirchhoff's Current Law, the current leaving point C toward point B must also equal 7 A. Thus, $I_{CB} = 7$ A.
7. Point B is where I_1 and I_{CB} flow together. Thus, the current entering point B equals

12 A + 7 A = 19 A. By Kirchhoff's Current Law, the current leaving point B must also equal 19 A. Thus, $I_{BA} = 19$ A. This is actually I_T, flowing back to the positive terminal of the source.

● SKILL-BUILDER 6–3

What is the value of R_1 in the circuit in Figure 6–5?

GIVEN:
   $V_S = 36$ V
   $I_{CD} = 18$ A
   $R_2 = 6$ Ω

FIND:
   R_1

STRATEGY:
   1. Use Ohm's Law to find I_2. V_S and R_2 are required. Both are given.
   2. Use Kirchhoff's Current Law to find I_1. I_{CD} and I_2 are required. I_{CD} is given. I_2 is not given.
   3. Use Ohm's law to find R_1. V_S and I_1 are required. V_S is given. I_1 is not given.

SOLUTION:

1. $I_2 = V_S \div R_2$ Ohm's Law
    $= 36$ V $\div 6$ Ω
    $= 6$ A

2. $I_1 = I_{CD} - I_2$ Kirchhoff's Current Law
    $= 18$ A $- 6$ A
    $= 12$ A

3. $R_1 = V_S \div I_1$ Ohm's Law
    $= 36$ V $\div 12$ A
    $= 3$ Ω

Repeat the Skill-Builder using the following values. You should obtain the results that follow.

Given:

V_S	56 V	37 V	45 V	96 V	23 V
I_{CD}	51.85 mA	18.15 mA	50.45 mA	12.36 mA	64.49 mA
R_2	1.8 kΩ	4.7 kΩ	2.2 kΩ	12 kΩ	680 Ω

Find:

I_2	31.11 mA	7.87 mA	20.45 mA	8.00 mA	33.82 mA
I_1	20.74 mA	10.28 mA	30.00 mA	4.36 mA	30.67 mA
R_1	2.7 kΩ	3.6 kΩ	1.5 kΩ	22 kΩ	750 Ω

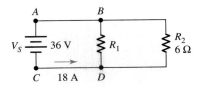

FIGURE 6–5

→ *Self-Test*
- - - - - - -
3. State Kirchhoff's Current Law.
4. Describe current in a parallel circuit.
- -

SECTION 6–3: Resistance in Parallel Circuits

In this section you will learn the following:

→ The equivalent (total) resistance of a parallel circuit equals the reciprocal of the sum of the reciprocals of the branch resistances.

→ The equivalent resistance of parallel branches will always be smaller than the resistance of any single branch.

Ohm's Law for total values states that total resistance equals source voltage divided by total current. If we apply this rule to the circuit in Figure 6–6, we have:

$$R_T = V_S \div I_T$$
$$= 60 \text{ V} \div 5 \text{ A}$$
$$= 12 \text{ }\Omega$$

This means the circuit in Figure 6–6 acts like it has a total resistance of 12 Ω. We have already learned that we must think of a circuit in two ways: what it really is, and what it acts like it is. The circuit in Figure 6–6 is really two resistors of 20 Ω and 30 Ω; but because they are connected in parallel, they act like 12 Ω, drawing 5 A from a 60-V source.

Equivalent Circuits

Figure 6–7 is the *equivalent circuit* of Figure 6–6. An equivalent circuit is a mathematical substitute for a real circuit. It has fewer components than the real circuit yet produces identical currents and voltage drops.

The equivalent circuit is an important concept in electrical theory. Although equivalent circuits are not real, we treat them as if they were because calculations based on the equivalent circuit give the same results as the real circuit.

Parallel Resistance Rule

The equivalent (total) resistance of a parallel circuit equals the reciprocal of the sum of the reciprocals of the branch resistances. We will call this the Parallel Resistance Rule.

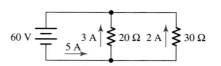

FIGURE 6–6
Actual parallel circuit.

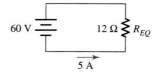

FIGURE 6–7
Equivalent circuit.

We have learned that the reciprocal of a number is 1 divided by the number:

$$\text{Reciprocal of } X = \frac{1}{X}$$

The Parallel Resistance Rule has three steps:

1. Calculate the reciprocal of the resistance of each branch.
2. Add the branch reciprocals.
3. Reciprocate again.

The reciprocal of resistance is defined as *conductance*. Conductance measures how easy it is for current to pass through a load. The unit of conductance is the *siemen*, abbreviated S. Because the Series Resistance Rule involves the reciprocals of resistance, it is often called the conductance formula.

Formula 6–2

$$R_{EQ} = \frac{1}{\dfrac{1}{R_1} + \dfrac{1}{R_2} + \ldots + \dfrac{1}{R_n}}$$

● **SKILL-BUILDER 6–4**

Use the Parallel Resistance Rule to find the equivalent resistance of the circuit in Figure 6–6.

GIVEN:
  $R_1 = 20\ \Omega$
  $R_2 = 30\ \Omega$

FIND:
  R_{EQ}

STRATEGY:
    We already know from Ohm's Law that this circuit's equivalent resistance is 12 Ω. Use the Parallel Resistance Rule and see if the value is the same.

    1. Calculate the reciprocals of the resistance of each branch.

$1 / 20\ \Omega = 0.0500$
$1 / 30\ \Omega = 0.0333$

    2. Add the branch reciprocals.

$0.0500 + 0.0333 = 0.0833$

    3. Reciprocate again.

$1 / 0.0833 = 12\ \Omega$ \qquad the value is the same

    Repeat the Skill-Builder using the following values. You should obtain the results that follow.

Given:
R_1	5.6 kΩ	8.2 kΩ	150 Ω	1.5 MΩ
R_2	2.2 kΩ	15 kΩ	910 Ω	3 MΩ

Find:
R_T	1,579 Ω	5,302 Ω	129 Ω	1 MΩ

MATH TIP

Using the Parallel Resistance Rule on a scientific calculator is quite easy:
    1. Enter each branch resistance and press the reciprocal key (1/*x*).
    2. Press the addition key (+) before repeating step 1 for the next branch.
    3. After repeating step 1 for the last branch, press the equal key (=) to complete the additions of step 2.
    4. Press the reciprocal key (1/*x*) one more time.
    The keystroke sequence for the circuit in Figure 6–6 is shown in Figure 6–8.

FIGURE 6–8
Calculator keystrokes for Skill-Builder 6–4.

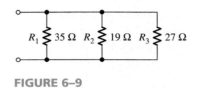

FIGURE 6–9

● *SKILL-BUILDER 6–5*

Find the equivalent resistance in the circuit in Figure 6–9.

GIVEN:
   $R_1 = 35\ \Omega$
   $R_2 = 19\ \Omega$
   $R_3 = 27\ \Omega$

FIND:
   R_{EQ}

SOLUTION:
   $1 / R_1 = 1 / 35\ \Omega = 0.02857$ use the Parallel Resistance Rule
   $1 / R_2 = 1 / 19\ \Omega = 0.05263$
   $1 / R_3 = 1 / 27\ \Omega = \underline{0.03704}$
   $ 0.11824$

   $1 / 0.11824 = 8.46\ \Omega$

   Repeat the Skill-Builder using the following values. You should obtain the results that follow.

Given:

R_1	68 Ω	1.5 kΩ	3.7 Ω	47 kΩ	27 Ω
R_2	75 Ω	2.2 kΩ	2.5 Ω	68 kΩ	13 Ω
R_3	96 Ω	3.6 kΩ	1.2 Ω	1.5 kΩ	5 Ω

Find:

R_{EQ}	26 Ω	715 Ω	0.67 Ω	1423 Ω	3.19 Ω

Special Formula for Two Branches

If the parallel circuit contains only two branches, the following formula may be used:

Formula 6–3

$$R_{EQ} = \frac{R_1 \times R_2}{R_1 + R_2}$$

   This formula is called the product-over-sum formula.

● *SKILL-BUILDER 6–6*

Use the product-over-sum formula to calculate the equivalent resistance of the circuit in Figure 6–6.

GIVEN:

   $R_1 = 20 \ \Omega$

   $R_2 = 30 \ \Omega$

FIND:

   R_{EQ}

SOLUTION:

   $R_{EQ} = \dfrac{R_1 \times R_2}{R_1 + R_2}$ voltage divider formula

   $= \dfrac{20 \ \Omega \times 30 \ \Omega}{20 \ \Omega + 30 \ \Omega}$

   $= \dfrac{600}{50}$

   $= 12 \ \Omega$

Repeat the Skill-Builder using the following values. You should obtain the results that follow.

Given:

R_1	27 Ω	150 Ω	68 Ω	200 Ω	33 Ω
R_2	51 Ω	270 Ω	75 Ω	360 Ω	39 Ω

Find:

R_T	17.6 Ω	96.4 Ω	35.7 Ω	128.6 Ω	17.9 Ω

- -

Special Formula for Equal Branches

If the resistance is the same for each branch in a parallel circuit, the following formula diagram may be used:

Formula Diagram 6–4

Branch Resistance	
Number of Branches	Total Resistance

● SKILL-BUILDER 6–7
- -

What is the total resistance of a parallel circuit with eight branches? Each branch has a resistance of 48 Ω.

SOLUTION:

Using the special formula for equal branches.

   $R_T = \dfrac{R_B}{N}$

   $= \dfrac{48 \ \Omega}{8}$

   $= 6 \ \Omega$

Repeat the Skill-Builder using the following values. You should obtain the results that follow.

Given:

R_B	48 Ω	84 Ω	150 Ω	44 Ω
N	3	4	6	11

Find:

R_T	16 Ω	21 Ω	25 Ω	4 Ω

A Handy Rule for Parallel Equivalent Resistance

The equivalent resistance of parallel branches will always be smaller than the resistance of any single branch.

→ *Self-Test*

5. Describe total resistance in a parallel circuit.
6. What is the relationship between the equivalent resistance of parallel branches and the resistance of any branch?

SECTION 6–4: Voltage in Parallel Circuits

In this section you will learn the following:

→ All parallel branches have the same voltage drop.

Parallel Voltage Rule

All parallel branches have the same voltage drop. We will call this the Parallel Voltage Rule.

Formula 6–5

$$V_1 = V_2 = \ldots = V_n$$

Because parallel branches are connected to each other on both sides, they share the same two electrical points. Therefore, they share the same voltage drop.

● *SKILL-BUILDER 6–8*

What is the source voltage in the circuit in Figure 6–10?

GIVEN:
 $R_1 = 4\ \Omega$
 $R_2 = 6\ \Omega$
 $I_T = 5\ \text{A}$
 $I_2 = 2\ \text{A}$

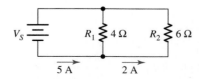

FIGURE 6–10

FIND:
  V_S

SOLUTION:
  1. $V_2 = I_2 \times R_2$ Ohm's Law
        $= 2\,A \times 6\,\Omega$
        $= 12\,V$

  2. $V_S = V_2$ Parallel Voltage Rule
        $= 12\,V$

CHECK:
  We may check our answer by finding the voltage drop across R_1.
  1. Use Ohm's Law to find V_1. R_1 is given. I_1 is not given.
  2. Use the Parallel Current Rule to find I_1. I_T and I_2 are both given.

SOLUTION:
  1. $I_1 = I_T - I_2$
        $= 5\,A - 2\,A$
        $= 3\,A$

  2. $V_1 = I_1 \times R_1$
        $= 3\,A \times 4\,\Omega$
        $= 12\,V$

  Repeat the Skill-Builder using the following values. You should obtain the results that follow.

Given:

R_1	15 Ω	150 Ω	39 Ω	680 Ω	1.5 kΩ
R_2	47 Ω	330 Ω	68 Ω	470 Ω	2 kΩ
I_T	2.55 A	359 mA	1.78 A	198 mA	15.2 mA
I_2	0.62 A	112 mA	0.65 A	117 mA	6.5 mA

Find:

I_1	1.93 A	247 mA	1.13 A	81 mA	8.7 mA
V_S	29 V	37 V	44 V	55 V	13 V

● SKILL-BUILDER 6–9

What is the source voltage in the circuit in Figure 6–11?

GIVEN:
  $R_2 = 8\,\Omega$
  $I_T = 4\,A$
  $I_1 = 2.5\,A$

FIND:
  V_S

STRATEGY:
  According to the Parallel Voltage Rule, the voltage drop across either R_1 or R_2 equals the source voltage. Ohm's Law will give us the voltage drop for either branch if we also know the resistance and current for that branch. We are given R_2, and we can use Kirchhoff's Current Law to find I_2.
  1. Use Kirchhoff's Current Law to find I_2. Both I_T and I_1 are given.
  2. Use Ohm's Law to find V_2. R_2 is given.
  3. Use the Parallel Voltage Rule to prove that $V_2 = V_S$.

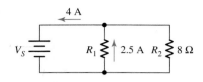

FIGURE 6–11

SOLUTION:

1. $I_2 = I_T - I_1$ Kirchhoff's Current Law
   $= 4\,A - 2.5\,A$
   $= 1.5\,A$

2. $V_2 = I_2 \times R_2$ Ohm's Law
   $= 1.5\,A \times 8\,\Omega$
   $= 12\,V$

3. $V_S = V_2$ Parallel Voltage Rule
   $= 12\,V$

Repeat the Skill-Builder using the following values. You should obtain the results that follow.

Given:

R_2	360 Ω	27 Ω	470 Ω	51 Ω	150 Ω
I_T	271.0 mA	538.7 mA	78.5 mA	203.6 mA	404.4 mA
I_1	168.2 mA	242.4 mA	44.5 mA	115.4 mA	144.4 mA

Find:

I_2	102.8 mA	296.3 mA	34.0 mA	88.2 mA	260.0 mA
V_S	37 V	8 V	16 V	4.5 V	39 V
R_1	220 Ω	33 Ω	360 Ω	39 Ω	270 Ω

→ *Self-Test*

7. Describe voltage in a parallel circuit.

SECTION 6–5: Power in Parallel Circuits

In this section you will learn the following:

→ The total power of a parallel circuit is the sum of the power developed in each component.

→ In parallel branches, the branch with the lowest resistance dissipates the most power (produces the most heat).

Parallel Power Rule

The total power of a parallel circuit is the sum of the power developed in each component. We will call this the Parallel Power Rule.

Formula 6–6

$$P_T = P_1 + P_2 + \ldots + P_n$$

Notice that the Parallel Power Rule is exactly the same as the Series Power Rule. In fact, the power rule is the same for all circuits.

Power equals voltage times current. In a group of parallel branches, all branches have the same voltage drop. The branch with the lowest resistance will draw the highest current. Therefore, in parallel branches, the branch with the lowest resistance dissipates the most power (produces the most heat).

● **SKILL-BUILDER 6–10**

In the circuit in Figure 6–12, find the power dissipation of each component and the total power level of the circuit.

GIVEN:
$V_S = 24$ V
$R_1 = 30\ \Omega$
$R_2 = 40\ \Omega$
$R_3 = 60\ \Omega$

FIND:
P_T, P_1, P_2, P_3

STRATEGY:
1. Use the Parallel Voltage Rule to find V_1, V_2, and V_3. V_S is given.
2. Use the power formula $P = E^2 \div R$ to find P_1, P_2, and P_3. R_1, R_2, and R_3 are given. V_1, V_2, and V_3 are not given.
3. Use the Parallel Power Rule to find P_T. P_1, P_2, and P_3 are not given.

SOLUTION:
1. $V_S = V_1 = V_2 = V_3$ Parallel Voltage Rule
    $= 24$ V

2. $P_1 = V_1^2 \div R_1$ Watt's Law
    $= 24^2$ V $\div 30\ \Omega$
    $= 576 \div 30$
    $= 19.2$ W

   $P_2 = V_2^2 \div R_2$ Watt's Law
    $= 24^2$ V $\div 40\ \Omega$
    $= 576 \div 40$
    $= 14.4$ W

   $P_3 = V_3^2 \div R_3$ Watt's Law
    $= 24^2$ V $\div 60\ \Omega$
    $= 576 \div 60$
    $= 9.6$ W

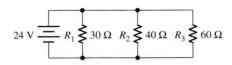

FIGURE 6–12

3. $P_T = P_1 + P_2 + P_3$ Parallel Power Rule
     $= 19.2\ \text{W} + 14.4\ \text{W} + 9.6\ \text{W}$
     $= 43.2\ \text{W}$

R_1, the branch with the lowest resistance, dissipates the greatest amount of power.

Repeat the Skill-Builder using the following values. You should obtain the results that follow.

Given:

V_S	36 V	14 V	53 V	22 V	41 V
R_1	270 Ω	150 Ω	12 kΩ	6.8 kΩ	7.5 kΩ
R_2	510 Ω	200 Ω	22 kΩ	5.1 kΩ	8.6 kΩ
R_3	680 Ω	300 Ω	18 Ω	4.7 Ω	91 Ω

Find:

P_1	4.8 W	1.3 W	234 mW	71 mW	224 mW
P_2	2.5 W	1.0 W	128 mW	95 mW	195 mW
P_3	1.9 W	0.7 W	156 mW	103 mW	185 mW
P_T	9.2 W	2.9 W	518 mW	269 mW	604 mW

→ *Self-Test*

8. Describe total power in a parallel circuit.
9. What is the relationship between the resistance of a parallel branch and the power dissipated in that branch?

SECTION 6–6: Multiple Voltage Sources in Parallel Circuits

In this section you will learn the following:

→ To increase current, multiple sources may be connected in parallel. All sources should have the same voltage rating.

→ Voltage sources connected in parallel should have the same voltage and should be connected positive to positive, negative to negative.

Figure 6–13 shows an automobile with a dead battery being jump-started by another automobile. This is an example of multiple voltage sources in parallel. The elec-

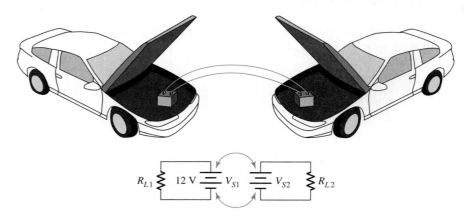

FIGURE 6–13
Multiple voltage sources in parallel.

trical loads of both cars (R_{L1} and R_{L2}) require 12 V. Both V_{S1} and V_{S2} have a potential of 12 V. However, V_{S2} has lost its ability to produce current.

To provide current at the proper voltage to the electrical system of the second car, we have jumper cables connected to V_{S1} and V_{S2} in parallel. This illustration emphasizes two important points: Voltage sources connected in parallel should have the same voltage, and should be connected positive to positive, negative to negative.

All voltage sources are rated for the maximum amount of current they produce. If a load needs more current than a single source can produce, then multiple sources must be connected in parallel to increase the current-producing capacity of the system.

dc voltage sources such as batteries are rated for current either according to cell size or ampere-hour capacity. Cell sizes for small batteries range from AAA (the smallest current rating) to D (the largest current rating). Automobile batteries are rated in ampere-hours, which represent the amount of charge in the battery.

In summary, to increase voltage, multiple sources may be connected in series aiding. All sources should have the same current rating. To increase current, multiple sources may be connected in parallel. All sources should have the same voltage rating.

→ Self-Test

10. How can multiple voltage sources be connected to increase current?
11. How should voltage sources in parallel be connected?

SECTION 6–7: The *RIPE* Table in Parallel Circuits

We will use the circuit problem in Figure 6–14 to illustrate the use of the *RIPE* table in parallel circuits.

We draw the *RIPE* table and insert the given values. Notice that there is no place in the table for the 1.25-A feeder current. However, we will use this information later in the problem.

COMPONENT	*R*	*I*	*P*	*E*
R_1	6 Ω			
R_2				
R_3	12 Ω		6.75 W	
Total				

We see two values for R_3 and, thus, we begin there. We use Ohm's Law and Watt's Law formulas to complete the row.

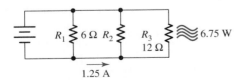

FIGURE 6–14
Using the *RIPE* table in parallel circuits.

$$I_3 = \sqrt{(P_3 \div R_3)} \qquad \text{current formula}$$
$$= \sqrt{(6.75\ \text{W} \div 12\ \Omega)}$$
$$= \sqrt{0.5625}$$
$$= 0.75\ \text{A}$$

$$V_3 = I_3 \times R_3 \qquad \text{voltage formula}$$
$$= 0.75\ \text{A} \times 12\ \Omega$$
$$= 9\ \text{V}$$

We now insert these values into the table:

COMPONENT	R	I	P	E
R_1	6 Ω			
R_2				
R_3	12 Ω	0.75 A	6.75 W	9 V
Total				

We have inserted values in the current and voltage columns. Therefore, we think of the rules for these columns, looking for ways to use these new values elsewhere.

The Parallel Voltage Rule states that the voltage drop is the same across all parallel branches. Therefore, V_3 is also V_1, V_2, and V_S.

We now insert these values into the table:

COMPONENT	R	I	P	E
R_1	6 Ω			9 V
R_2				9 V
R_3	12 Ω	0.75 A	6.75 W	9 V
Total				9 V

Looking back at Figure 6–14, we see the feeder current of 1.25 A is actually the combined current of I_2 and I_3. Therefore, according to Kirchhoff's Current Law,

$$I_2 = 1.25\ \text{A} - I_3 \qquad \text{Kirchhoff's Current Law}$$
$$= 1.25\ \text{A} - 0.75\ \text{A}$$
$$= 0.50\ \text{A}$$

We now insert this value into the table:

COMPONENT	R	I	P	E
R_1	6 Ω			9 V
R_2		0.50 A		9 V
R_3	12 Ω	0.75 A	6.75 W	9 V
Total				9 V

Now we have two rows (R_1 and R_2) each with two values in the row. We use Ohm's Law and Watt's Law formulas to complete these rows.

$$I_1 = V_1 \div R_1 \qquad \text{current formula}$$
$$= 9\,\text{V} \div 6\,\Omega$$
$$= 1.5\,\text{A}$$

$$P_1 = I_1 \times V_1 \qquad \text{power formula}$$
$$= 1.5\,\text{A} \times 9\,\text{V}$$
$$= 13.5\,\text{W}$$

$$R_2 = V_2 \div I_2 \qquad \text{resistance formula}$$
$$= 9\,\text{V} \div 0.5\,\text{A}$$
$$= 18\,\Omega$$

$$P_2 = I_2 \times V_2 \qquad \text{power formula}$$
$$= 0.5\,\text{A} \times 9\,\text{V}$$
$$= 4.5\,\text{W}$$

We now insert these values into the table:

COMPONENT	R	I	P	E
R_1	6 Ω	1.50 A	13.5 W	9 V
R_2	18 Ω	0.50 A	4.5 W	9 V
R_3	12 Ω	0.75 A	6.75 W	9 V
Total				9 V

We can find total current with the Parallel Current Rule and total power with the Parallel Power Rule.

$$I_T = I_1 + I_2 + I_3 \qquad \text{Parallel Current Rule}$$
$$= 1.50\,\text{A} + 0.50\,\text{A} + 0.75\,\text{A}$$
$$= 2.75\,\text{A}$$

$$P_T = P_1 + P_2 + P_3 \qquad \text{Parallel Power Rule}$$
$$= 13.5\,\text{W} + 4.5\,\text{W} + 6.75\,\text{W}$$
$$= 24.75\,\text{W}$$

We now insert these values into the table:

COMPONENT	R	I	P	E
R_1	6 Ω	1.50 A	13.5 W	9 V
R_2	18 Ω	0.50 A	4.5 W	9 V
R_3	12 Ω	0.75 A	6.75 W	9 V
Total		2.75 A	24.75 W	9 V

We can find R_{EQ} either by Ohm's Law using the values on the totals row or by the Parallel Resistance Rule using the values in the resistance column.

Method 1:

$$R_T = V_S \div I_T \qquad \text{resistance formula}$$
$$= 9 \text{ V} \div 2.75 \text{ A}$$
$$= 3.27 \text{ }\Omega$$

Method 2:

$$R_T = \cfrac{1}{\cfrac{1}{R_1} + \cfrac{1}{R_2} + \cfrac{1}{R_3}}$$

$$= \cfrac{1}{\cfrac{1}{6 \text{ }\Omega} + \cfrac{1}{18 \text{ }\Omega} + \cfrac{1}{12 \text{ }\Omega}}$$

$$= \cfrac{1}{0.1667 + 0.0556 + 0.0833}$$

$$= \cfrac{1}{0.3056}$$

$$= 3.27 \text{ }\Omega$$

When we insert this value, the table is complete.

COMPONENT	R	I	P	E
R_1	6 Ω	1.50 A	13.5 W	9 V
R_2	18 Ω	0.50 A	4.5 W	9 V
R_3	12 Ω	0.75 A	6.75 W	9 V
Total	3.27 Ω	2.75 A	24.75 W	9 V

SECTION 6–8: Shorts and Opens in Parallel Circuits

Shorts

In this section you will learn the following:

→ A short across any parallel branch is also a short across all other branches in that group.

→ When parallel branches are connected to a voltage source, a short across any branch is also a short across the voltage source. This is a dangerous condition, producing high current and possible fire.

→ An open in a parallel branch does not affect the operation of any other branch. However, the open does reduce total current and thus increases equivalent resistance.

In Figure 6–15, R_3 is shorted. The short is actually between electrical points *A* and *B*, which are shared by the other branches. Therefore, R_1 and R_2 are also shorted, and the equivalent resistance between points *A* and *B* is nearly zero ohms.

Figure 6–16 is electrically identical to Figure 6–15. The short is in a different physical location but still connects electrical points *A* and *B*.

If a voltage source is added, as in Figure 6–17, we see the voltage source is also shorted out. This is a very dangerous condition. There is a current path of nearly zero

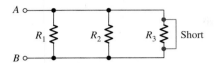

FIGURE 6–15
If one branch is shorted, all
branches are shorted.

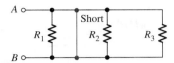

FIGURE 6–16
Physical location of the
short does not matter in
parallel branches.

ohms between the terminals of the voltage source. The result will be an abnormally high
current, often enough to melt and destroy the circuit, or start a fire.

Practical Application

When electrical power lines enter a building, they are connected to a circuit breaker box.
Each breaker protects and controls several light fixtures and wall outlets. All the fixtures
and outlets on each breaker are wired in parallel, as shown in Figure 6–18.

If an electrical appliance plugged into one of the outlets develops a short, the high
current should trip the circuit breaker and stop all current in all branch outlets until the
problem has been corrected.

Opens

An open in a parallel branch does not affect the operation of any other branch. However,
the open reduces total current and thus increases equivalent resistance.

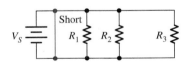

FIGURE 6–17
The short across the branches
also shorts the voltage
source.

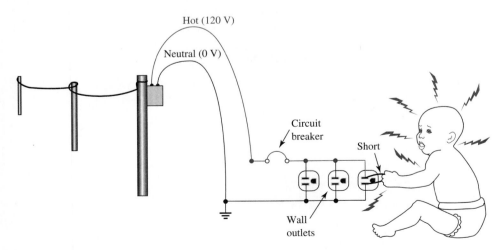

FIGURE 6–18
Wall outlets are in parallel with the voltage source.

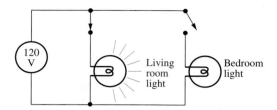

FIGURE 6–19
An open in one branch does not affect
another branch.

We wire our buildings, automobiles, and so on in parallel so that the branches can operate independently. You know that the light switch in one room does not affect the lights in another room. This is because the two rooms are wired in parallel, as shown in Figure 6–19. The open switch in the bedroom branch keeps the bedroom light off, while the closed switch in the living room branch keeps the living room light on.

● *SKILL-BUILDER 6–11*

Which branch is open in the circuit in Figure 6–20?

GIVEN:
$V_S = 36$ V
$I_T = 10$ A
$R_1 = 6\ \Omega$
$R_2 = 4\ \Omega$
$R_3 = 9\ \Omega$

FIND:
Which branch is open?

STRATEGY:
Since we know that one branch is open, we know that the actual total current will be less than the normal total current. Thus, we will calculate the normal branch currents and the normal total current. Compare these calculations to the actual total current to determine which branch is open.

SOLUTION:

$$I_{1norm} = V_S \div R_1 \qquad\qquad \text{Ohm's Law}$$
$$= 36\text{ V} \div 6\ \Omega$$
$$= 6\text{ A}$$

$$I_{2norm} = V_S \div R_2 \qquad\qquad \text{Ohm's Law}$$
$$= 36\text{ V} \div 4\ \Omega$$
$$= 9\text{ A}$$

$$I_{3norm} = V_S \div R_3 \qquad\qquad \text{Ohm's Law}$$
$$= 36\text{ V} \div 9\ \Omega$$
$$= 4\text{ A}$$

$$I_{Tnorm} = I_{1norm} + I_{2norm} + I_{3norm} \qquad\qquad \text{Parallel Current Rule}$$
$$= 6\text{ A} + 9\text{ A} + 4\text{ A}$$
$$= 19\text{ A}$$

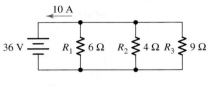

FIGURE 6–20

Now compare normal total current with actual total current by subtracting.

$$I_{Tnorm} - I_{Tactual} = 19\,A - 10\,A$$
$$= 9\,A$$

The difference between normal and actual current equals the normal current of R_2. Therefore, R_2 is the open branch.

Repeat the Skill-Builder using the following values. You should obtain the results that follow.

Given:

V_S	45 V	15 V	87 V	16 V	37 V
R_1	150 Ω	68 Ω	2.7 kΩ	47 Ω	560 Ω
R_2	220 Ω	91 Ω	3.9 kΩ	20 Ω	750 Ω
R_3	330 Ω	36 Ω	4.7 kΩ	30 Ω	680 Ω
Actual I_T	505 mA	582 mA	51 mA	1333 mA	115 mA

Find:

Normal branch currents

I_1	300 mA	221 mA	32 mA	340 mA	66 mA
I_2	205 mA	165 mA	22 mA	800 mA	49 mA
I_3	136 mA	417 mA	19 mA	533 mA	54 mA

Normal total current

I_T	641 mA	802 mA	73 mA	1674 mA	170 mA

Difference between normal and actual total current

136 mA	220 mA	22 mA	341 mA	55 mA

Open branch:

R_3	R_1	R_2	R_1	R_3

→ **Self-Test**

12. What effect does a short across a parallel branch have on the other branches?
13. When parallel branches are connected to a voltage source, what effect does a short across a branch have on the voltage source?
14. What effect does an open in a parallel branch have on the other branches?

SECTION 6–9: Practical Examples of Parallel Circuits

Interior Wiring

Figure 6–21 shows four sets of lights controlled by two switches and one circuit breaker, illustrating practical series-parallel relationships.

Each set of lights has two lamps, wired in parallel so one lamp will continue to operate if the other lamp is open. Next each pair of sets is wired in parallel, for the same reason: One set will continue to operate if the other set opens. SW_1 is in series with set A

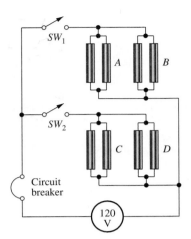

FIGURE 6–21
Parallel relationships in
interior wiring.

and *B*, so it can control both sets. Similarly, *SW₂* is in series with sets *C* and *D*. However, both switches are in parallel on the feeder controlled by the circuit breaker in series. When both switches are closed, the entire circuit consists of only two electrical points and is a parallel circuit with eight branches: Each lamp is a separate branch.

In general, if an open in one device causes current to stop in another device, then the two devices are in series. For example, in Figure 6–21, opening *SW₁* stops current in light sets *A* and *B*. Therefore, these devices are in series.

On the other hand, if two devices share the same voltage source, yet an open in one device has no effect on current in the other device, then these devices are in parallel. For example, in Figure 6–21, if any lamp is removed from any light set, this has no effect on all the other lamps. Therefore, the lamps are in parallel.

Automotive Circuits

All the electrical devices of an automobile are wired in parallel with switches and fuses inserted at appropriate points. The negative terminal of the battery is connected to the metal frame (chassis) of the automobile, which serves as ground.

To install an electrical device in an automobile, you must first determine where to connect the positive power wire of the device. Check the fuse box for empty or unused fuse connections. Having the device on its own branch protected by its own fuse is often the ideal solution. However, if this is impossible or undesirable, you must locate a wire or terminal where +12 V to ground is available, and connect your positive power wire there.

For example, in Figure 6–22, a citizen's band radio has been installed in a car, connected to the positive power wire of the regular AM-FM radio. The negative power wire can be connected to the metal body of the car at any location. Be sure you have good contact with bare metal. Dirt or paint may interfere with the ground connection.

Remember that the new device will now be sharing a fuse with other devices. It is a good idea to make your connections temporary until you have determined which fuse is in the circuit. The new device will draw additional current through this fuse, which may exceed the fuse's rating. Therefore, test the system by turning on all devices connected to the fuse to see if the fuse can take the extra current.

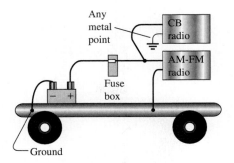

FIGURE 6–22
Parallel devices in automotive circuits.

Formula List

Formula 6–1

$$I_T = I_1 + I_2 + \ldots + I_n$$

Formula 6–2

$$R_{EQ} = \cfrac{1}{\cfrac{1}{R_1} + \cfrac{1}{R_2} + \ldots + \cfrac{1}{R_n}}$$

Formula 6–3

$$R_{EQ} = \frac{R_1 \times R_2}{R_1 + R_2}$$

Formula Diagram 6–4

Branch Resistance	
Number of Branches	Total Resistance

Formula 6–5

$$V_1 = V_2 = \ldots = V_n$$

Formula 6–6

$$P_T = P_1 + P_2 + \ldots + P_n$$

● TROUBLESHOOTING: Applying Your Knowledge

Your friend has wired the two radios in series. Thus, the radios act as a voltage divider, each getting only part of the 12 V supplied by the battery. The radios need to be connected in parallel, so both can get the full 12 V.

The CB radio needs its own fuse. Your friend should remove the CB radio wires and reconnect the two cut ends of the AM-FM radio red wire. The CB radio red wire should be connected to an empty fuse holder and the proper size fuse installed. The black CB radio wire should be grounded by connecting it to the metal frame or chassis of the truck.

CHAPTER SUMMARY: ANSWERS TO SELF-TESTS

1. A parallel circuit has two or more current paths (branches).
2. Parallel Circuit Visual Recognition Rule: If two components are connected to each other on both sides, the two components are in parallel.
3. Kirchhoff's Current Law states that the sum of all currents entering an electrical point equals the sum of all currents leaving that point.
4. The total current of a parallel circuit equals the sum of the branch currents.
5. The equivalent (total) resistance of a parallel circuit equals the reciprocal of the sum of the reciprocals of the branch resistances.
6. The equivalent resistance of parallel branches will always be smaller than the resistance of any single branch.
7. All parallel branches have the same voltage drop.
8. The total power of a parallel circuit is the sum of the power developed in each component.
9. In parallel branches, the branch with the lowest resistance dissipates the most power (produces the most heat).
10. To increase current, multiple sources may be connected in parallel. All sources should have the same voltage rating.
11. Voltage sources connected in parallel should have the same voltage and should be connected positive to positive, negative to negative.
12. A short across any parallel branch is also a short across all other branches in that group.
13. When parallel branches are connected to a voltage source, a short across any branch is also a short across the voltage source. This is a dangerous condition, producing high current and possible fire.
14. An open in a parallel branch does not affect the operation of any other branch. However, the open does reduce total current and thus increases equivalent resistance.

CHAPTER REVIEW QUESTIONS

Determine whether the following questions are true or false.

6–1. A parallel circuit has only one path for current.
6–2. Two components can be in parallel without being physically connected to each other.
6–3. The equivalent resistance of a parallel circuit is smaller than the smallest branch resistance in the circuit.
6–4. There is only one voltage drop across parallel branches.
6–5. If the current is known at any location in a parallel circuit, then the current is known for all locations in the circuit.
6–6. In a parallel circuit, the sum of the branch currents equals the total current.
6–7. An open in a parallel branch increases total current.
6–8. An open in a parallel branch does not affect the other branches.

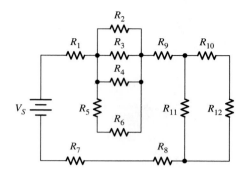

FIGURE 6–23

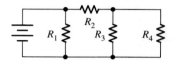

FIGURE 6–24

Select the response that best answers the question or completes the statement.

6–9. In Figure 6–23, which resistors are in parallel?
   a. R_2 and R_3 c. R_2 and R_4
   b. R_{11} and R_{12} d. R_4 and R_6
6–10. In Figure 6–24, which resistors are in parallel?
   a. R_1 and R_2 c. R_1 and R_3
   b. R_2 and R_3 d. R_3 and R_4
6–11. Identify the following statement: The sum of all currents entering an electrical point equals the sum of all currents leaving that point.
   a. Ohm's Law   c. Kirchhoff's Voltage Law
   b. Watt's Law   d. Kirchhoff's Current Law
6–12. In a parallel circuit, an open branch causes
   a. dangerously high current.
   b. a loss of ground.
   c. all current to stop.
   d. a decrease in total current.

Solve the following problems, referring to Figure 6–25.

6–13. $R_1 = 27\ \Omega$, $R_2 = 15\ \Omega$, $R_3 = 36\ \Omega$. Find total resistance.
6–14. $V_S = 35\ \text{V}$, $R_1 = 7\ \Omega$, $R_2 = 10\ \Omega$, $R_3 = 20\ \Omega$. Find all branch currents and total current.
6–15. $V_S = 25\ \text{V}$, $I_T = 7.25\ \text{A}$, $I_{BC} = 5.25\ \text{A}$, $R_3 = 20\ \Omega$. Find R_1 and R_2.

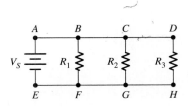

FIGURE 6–25

The following problems are more challenging.

6–19. $R_T = 7.734\ \Omega$, $R_1 = 16\ \Omega$, $R_3 = 47\ \Omega$. Find R_2.

6–20. R_1, R_2, and R_3 are equal in value. I_T is 12 A. What is the value of I_T if R_2 is opened?

6–21. $V_S = 30$ V. R_1 has twice the resistance of R_2, and the same resistance as R_3. The current is 6 A. Find R_1, R_2, and R_3.

GLOSSARY

branch   A parallel current path.

conductance   The reciprocal of resistance; the ease of current passing through a material.

equivalent circuit   An imaginary mathematical substitute for a real circuit.

Kirchhoff's Current Law   The sum of all currents entering an electrical point equals the sum of all currents leaving that point.

siemen   The unit of conductance.

6–16. $I_{FG} = 12.5$ A, $I_{GH} = 4.5$ A, $R_2 = 3\ \Omega$. Find V_S.

6–17. $V_S = 24$ V, $R_1 = 8\ \Omega$, $R_2 = 6\ \Omega$, $R_3 = 10\ \Omega$. Find total power and power of each component.

6–18. $V_S = 36$ V, $R_1 = 12\ \Omega$, $R_2 = 18\ \Omega$, $R_3 = 9\ \Omega$, $I_T = 5$ A. Which branch is open?

SERIES-PARALLEL CIRCUITS

● SECTION OUTLINE

7–1   Definition of Series-Parallel Circuits

7–2   The *RIPE* Table in Series-Parallel Circuits

7–3   Opens and Shorts in Series-Parallel Circuits

7–4   Loading Effects of Meters

● TROUBLESHOOTING: Applying Your Knowledge

Your classmate has constructed the circuit shown in Figure 7–1 and now is having trouble measuring the voltage drops. He reads 4.84 V across R_1 and 4.46 V across R_2. He knows these reading are incorrect because they add up to 9.30 V, and by Kirchhoff's Voltage Law the readings should add to 15 V. The meter reads accurately on other circuits.

Your classmate has spent hours checking and rechecking the circuit and cannot find the trouble. What would you suggest, and why?

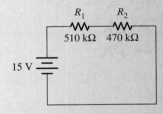

FIGURE 7–1

● OVERVIEW AND LEARNING OBJECTIVES

Most electronic circuits have some components connected in series and others connected in parallel. Thus, we call these circuits series-parallel circuits. In this chapter, you will master your ability to analyze these circuits, using the process of circuit simplification and the tools you have developed thus far: the basic formulas derived from Ohm's Law and Watt's Law, the rules for series and parallel circuits, and the *RIPE* table.

Although meters are supposed to measure electrical quantities in a circuit without affecting the circuit, the truth is that meters become part of the circuit while taking measurements and cause a change in the values they measure. This is called loading effect, and you must understand it clearly to interpret meter readings correctly.

After completing this chapter, you will be able to:

1. Simplify a series-parallel circuit.
2. Solve for all component voltage drops and currents using the *RIPE* table.
3. Analyze the effects of opens and shorts in series-parallel circuits.
4. Use ground-referenced voltage readings to estimate the possible existence and location of opens and shorts in series-parallel circuits.
5. Analyze and reduce the effects of meter loading in measuring voltage drops across circuit components.

SECTION 7–1: Definition of Series-Parallel Circuits

In this section you will learn the following:

→ A series-parallel circuit is one where all current paths can be determined by examining the connection of the components without reference to the value of the components.

A series-parallel circuit is one where all current paths can be determined by examining the connection of the components without reference to the value of the components. In Figure 7–2, the amount of current in each part of the circuit depends on voltage and resistance values, but the direction of current depends only on the way the components are connected to each other. When the direction of current in each component can be determined from connections alone without knowing the voltage or resistance values, the circuit is a series-parallel circuit.

In a typical series-parallel circuit, some components have a series relationship and other components have a parallel relationship. For example, in Figure 7–2, R_2, R_4, and R_6 all have the same current and, thus, are in series. This group of three resistors is in parallel with R_3. Also R_1 and R_5 are in series because the same current flows through them.

Circuit Simplification

We simplify series-parallel circuits to make the series and parallel relationships easier to see. We use the Series and Parallel Visual Recognition Rules to help us in this process, by combining groups of components into a single component with a value equivalent to the value of the group.

In Figure 7–2, R_2, R_4, and R_6 are clearly in series because they connect to each other on only one side with no other component connected between them, according to the Series Visual Recognition Rule. Therefore, we can create an imaginary equivalent component, R_7 (the next available subscript) and let R_7 equal R_2, R_4, and R_6 in series. The pluses in the expression indicate a series relationship.

$R_7 = R_2 + R_4 + R_6$ Series Resistance Rule

We now remove R_2, R_4, and R_6 from the circuit and replace the group with the equivalent component, R_7, as shown in Figure 7–3.

Now when we look at Figure 7–3, we can see that R_3 and R_7 are clearly in parallel because they touch each other on both sides, according to the Parallel Visual Recognition Rule. Notice that it does not matter that R_3 is an actual component and R_7 is an equivalent component. When simplifying a circuit, we treat actual and equivalent components the same way.

Since R_3 and R_7 are in parallel, they can be replaced with an equivalent component, R_8. The double slash in the expression indicates a parallel relationship.

$R_8 = R_3 \,|\, | \, R_7$ Parallel Resistance Rule

We now remove R_3 and R_7 from the circuit and replace them with the equivalent component, R_8, as shown in Figure 7–4.

In Figure 7–4, we see that R_1, R_8, and R_5 are clearly in series according to the Series Visual Recognition Rule. Therefore, we can create an equivalent component, R_9.

$R_9 = R_1 + R_8 + R_5$ Series Resistance Rule

We remove R_1, R_8, and R_5, replace them with R_9, and now have the simple circuit shown in Figure 7–5. R_9 is the equivalent of the entire circuit, and we cannot simplify any further.

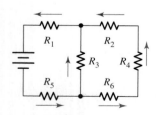

FIGURE 7–2
In series-parallel circuits, the direction of current depends only on the connections of the components.

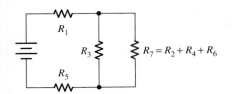

FIGURE 7–3
Step 1 in simplifying Figure 7–2.

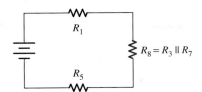

FIGURE 7–4
Step 2 in simplifying Figure 7–2.

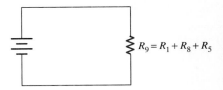

FIGURE 7–5
Final step in simplifying Figure 7–2.

Now we can see how circuit simplification helps us understand series and parallel relationships. For example, in the original circuit of Figure 7–2, we could not use our Recognition Rules to describe the relationship between R_3 and R_4. However, in Figure 7–3, we can see that R_4 is part of a group of components in parallel with R_3.

Similarly, in Figure 7–2, we had no rule to describe the relationship between R_1 and R_5. However, when we simplified the circuit down to Figure 7–4, we could see that R_1 and R_5 are in series.

→ *Self-Test*

1. Describe a series-parallel circuit.

SECTION 7–2: The *RIPE* Table in Series-Parallel Circuits

Since we have already simplified Figure 7–2, let's assign values to the components and use the *RIPE* table to determine the current and voltage drop of each component, as shown in Figure 7–6.

We begin by setting up the table for the actual and equivalent components. Notice that the 9-V source is assigned to the bottom row, which is total equivalent resistance. In using the *RIPE* table, the source voltage will always be assigned to the bottom row.

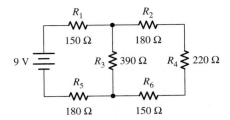

FIGURE 7–6
Circuit of Figure 7–2 with
component values.

COMPONENT	$R\,\Omega$	$I\,$mA	$P\,$mW	$E\,$V
R_1	150 Ω			
R_2	180 Ω			
R_3	390 Ω			
R_4	220 Ω			
R_5	180 Ω			
R_6	150 Ω			
$R_7 = R_2 + R_4 + R_6$				
$R_8 = R_3 \,\|\, R_7$				
$R_9 = R_1 + R_8 + R_5$				9.00 V

We calculate resistance values for R_7, R_8, and R_9, using the Series Resistance Rule and the Parallel Resistance Rule:

$$R_7 = R_2 + R_4 + R_6 \qquad \text{Series Resistance Rule}$$
$$= 180\ \Omega + 220\ \Omega + 150\ \Omega$$
$$= 550\ \Omega$$

$$R_8 = R_3 \,\|\, R_7 \qquad \text{Parallel Resistance Rule}$$

$$= \cfrac{1}{\cfrac{1}{R_3} + \cfrac{1}{R_7}}$$

$$= \cfrac{1}{\cfrac{1}{390\ \Omega} + \cfrac{1}{550\ \Omega}}$$

$$= 228\ \Omega$$

$$R_9 = R_1 + R_8 + R_5 \qquad \text{Series Resistance Rule}$$
$$= 150\ \Omega + 228\ \Omega + 180\ \Omega$$
$$= 558\ \Omega$$

We now record the values for R_7, R_8, and R_9 in the table:

COMPONENT	$R\,\Omega$	$I\,$mA	$P\,$mW	$E\,$V
R_1	150 Ω			
R_2	180 Ω			
R_3	390 Ω			
R_4	220 Ω			
R_5	180 Ω			
R_6	150 Ω			
$R_7 = R_2 + R_4 + R_6$	550 Ω			
$R_8 = R_3 \,\|\, R_7$	228 Ω			
$R_9 = R_1 + R_8 + R_5$	558 Ω			9.00 V

Remember that in using the *RIPE* table, we look for a row that contains two values. We have resistance and voltage values for R_9 and, thus, we find current with Ohm's law:

$$I_9 = 9\ \text{V} \div 558\ \Omega \qquad \text{Ohm's Law}$$
$$= 0.0161\ \text{A}$$
$$= 16.1\ \text{mA}$$

We now record the value of I_9 in the table:

COMPONENT	$R\,\Omega$	I mA	P mW	E V
R_1	150 Ω			
R_2	180 Ω			
R_3	390 Ω			
R_4	220 Ω			
R_5	180 Ω			
R_6	150 Ω			
$R_7 = R_2 + R_4 + R_6$	550 Ω			
$R_8 = R_3 \,\|\, R_7$	228 Ω			
$R_9 = R_1 + R_8 + R_5$	558 Ω	16.1 mA		9.00 V

Since R_9 is the equivalent of a group of components in series, the current of R_9 is also the current of R_1, R_8, and R_5 (the Series Current Rule). Therefore, we can assign the value of I_9 to each of these rows.

COMPONENT	$R\,\Omega$	I mA	P mW	E V
R_1	150 Ω	16.1 mA		
R_2	180 Ω			
R_3	390 Ω			
R_4	220 Ω			
R_5	180 Ω	16.1 mA		
R_6	150 Ω			
$R_7 = R_2 + R_4 + R_6$	550 Ω			
$R_8 = R_3 \,\|\, R_7$	228 Ω	16.1 mA		
$R_9 = R_1 + R_8 + R_5$	558 Ω	16.1 mA		9.00 V

We now have values for resistance and current in three rows (R_1, R_5, and R_9). Thus, we can find voltage in these rows using Ohm's Law and record these values in the table:

$$V_1 = I_1 \times R_1 \qquad \text{Ohm's Law}$$
$$= 16.1 \text{ mA} \times 150 \ \Omega$$
$$= 2.42 \text{ V}$$

$$V_5 = I_5 \times R_5 \qquad \text{Ohm's Law}$$
$$= 16.1 \text{ mA} \times 180 \ \Omega$$
$$= 2.90 \text{ V}$$

$$V_8 = I_8 \times R_8 \qquad \text{Ohm's Law}$$
$$= 16.1 \text{ mA} \times 228 \ \Omega$$
$$= 3.67 \text{ V}$$

COMPONENT	$R\,\Omega$	I mA	P mW	E V
R_1	150 Ω	16.1 mA		2.42 V
R_2	180 Ω			
R_3	390 Ω			
R_4	220 Ω			
R_5	180 Ω	16.1 mA		2.90 V
R_6	150 Ω			
$R_7 = R_2 + R_4 + R_6$	550 Ω			
$R_8 = R_3 \,\|\, R_7$	228 Ω	16.1 mA		3.67 V
$R_9 = R_1 + R_8 + R_5$	558 Ω	16.1 mA		9.00 V

Since R_8 is the equivalent of a group of components in parallel, the voltage drop of R_8 is also the voltage drop of R_3 and R_7 (the Parallel Voltage Rule). Therefore, we can assign the value of V_8 to each of these rows.

COMPONENT	$R\ \Omega$	$I\ \text{mA}$	$P\ \text{mW}$	$E\ \text{V}$
R_1	150 Ω	16.1 mA		2.42 V
R_2	180 Ω			
R_3	390 Ω			3.67 V
R_4	220 Ω			
R_5	180 Ω	16.1 mA		2.90 V
R_6	150 Ω			
$R_7 = R_2 + R_4 + R_6$	550 Ω			3.67 V
$R_8 = R_3\,\|\,R_7$	228 Ω	16.1 mA		3.67 V
$R_9 = R_1 + R_8 + R_5$	558 Ω	16.1 mA		9.00 V

We now can use Ohm's Law to find the current in these rows and record these values in the table:

$$I_3 = 3.67\ \text{V} \div 390\ \Omega \qquad \text{Ohm's Law}$$
$$= 9.4\ \text{mA}$$

$$I_7 = 3.67\ \text{V} \div 550\ \Omega$$
$$= 6.7\ \text{mA}$$

COMPONENT	$R\ \Omega$	$I\ \text{mA}$	$P\ \text{mW}$	$E\ \text{V}$
R_1	150 Ω	16.1 mA		2.42 V
R_2	180 Ω			
R_3	390 Ω	9.4 mA		3.67 V
R_4	220 Ω			
R_5	180 Ω	16.1 mA		2.90 V
R_6	150 Ω			
$R_7 = R_2 + R_4 + R_6$	550 Ω	6.7 mA		3.67 V
$R_8 = R_3\,\|\,R_7$	228 Ω	16.1 mA		3.67 V
$R_9 = R_1 + R_8 + R_5$	558 Ω	16.1 mA		9.00 V

Since R_7 is the equivalent of a group of components in series, the current of R_7 is also the current of R_2, R_4, and R_6 (the Series Current Rule). Therefore, we can assign the value of I_7 to each of these rows.

COMPONENT	$R\ \Omega$	$I\ \text{mA}$	$P\ \text{mW}$	$E\ \text{V}$
R_1	150 Ω	16.1 mA		2.42 V
R_2	180 Ω	6.7 mA		
R_3	390 Ω	9.4 mA		3.67 V
R_4	220 Ω	6.7 mA		
R_5	180 Ω	16.1 mA		2.90 V
R_6	150 Ω	6.7 mA		
$R_7 = R_2 + R_4 + R_6$	550 Ω	6.7 mA		3.67 V
$R_8 = R_3\,\|\,R_7$	228 Ω	16.1 mA		3.67 V
$R_9 = R_1 + R_8 + R_5$	558 Ω	16.1 mA		9.00 V

We can now find voltage in these rows with Ohm's Law and record these values in the table:

$$V_2 = I_2 \times R_2$$
$$= 6.7 \text{ mA} \times 180 \text{ }\Omega$$
$$= 1.20 \text{ V}$$
Ohm's Law

$$V_4 = I_4 \times R_4$$
$$= 6.7 \text{ mA} \times 220 \text{ }\Omega$$
$$= 1.47 \text{ V}$$
Ohm's Law

$$V_6 = I_6 \times R_6$$
$$= 6.7 \text{ mA} \times 150 \text{ }\Omega$$
$$= 1.00 \text{ V}$$
Ohm's Law

COMPONENT	R Ω	I mA	P mW	E V
R_1	150 Ω	16.1 mA		2.42 V
R_2	180 Ω	6.7 mA		1.20 V
R_3	390 Ω	9.4 mA		3.67 V
R_4	220 Ω	6.7 mA		1.47 V
R_5	180 Ω	16.1 mA		2.90 V
R_6	150 Ω	6.7 mA		1.00 V
$R_7 = R_2 + R_4 + R_6$	550 Ω	6.7 mA		3.67 V
$R_8 = R_3 \| R_7$	228 Ω	16.1 mA		3.67 V
$R_9 = R_1 + R_8 + R_5$	558 Ω	16.1 mA		9.00 V

As a last exercise, we can calculate power for each actual component and for the entire circuit using Watt's Law. The results are included in the following table.

COMPONENT	R Ω	I mA	P mW	E V
R_1	150 Ω	16.1 mA	39 mW	2.42 V
R_2	180 Ω	6.7 mA	8 mW	1.20 V
R_3	390 Ω	9.4 mA	35 mW	3.68 V
R_4	220 Ω	6.7 mA	10 mW	1.47 V
R_5	180 Ω	16.1 mA	47 mW	2.90 V
R_6	150 Ω	6.7 mA	7 mW	1.00 V
$R_7 = R_2 + R_4 + R_6$	550 Ω	6.7 mA		3.68 V
$R_8 = R_3 \| R_7$	228 Ω	16.1 mA		3.68 V
$R_9 = R_1 + R_8 + R_5$	558 Ω	16.1 mA	145 mW	9.00 V

There is no need to calculate power of R_7 and R_8, since these are not actual components.

Thus, the *RIPE* table gives us the current and voltage drop of each component in the circuit. By applying the various rules and laws, we can double-check our work.

$$I_5 = I_3 + I_6$$
$$16.1 \text{ mA} = 9.4 \text{ mA} + 6.7 \text{ mA}$$
$$16.1 \text{ mA} = 16.1 \text{ mA}$$
Kirchhoff's Current Law

$$V_3 = V_2 + V_4 + V_6$$
$$3.67 \text{ V} = 1.20 \text{ V} + 1.47 \text{ V} + 1.00 \text{ V}$$
$$3.67 \text{ V} = 3.67 \text{ V}$$
Kirchhoff's Voltage Law

$$V_S = V_1 + V_3 + V_5$$
$$9.00 \text{ V} = 2.42 \text{ V} + 3.68 \text{ V} + 2.90 \text{ V}$$
$$9.00 \text{ V} = 9.00 \text{ V}$$
Kirchhoff's Voltage Law

The key step in using the *RIPE* table in series-parallel circuits is in assigning the current of a series-equivalent group, or the voltage of a parallel-equivalent group, to the components in that group.

SECTION 7–3: Opens and Shorts in Series-Parallel Circuits

In this section you will learn the following:

→ In a series-parallel circuit, if a single component's resistance value changes, the voltage, current, and power values of all other components may also change.
→ A short anywhere in a series-parallel circuit will decrease total equivalent resistance, and thereby increase total current and total power.
→ An open anywhere in a series-parallel circuit will increase total equivalent resistance, and thereby decrease total current and total power.
→ A slight current still flows in a shorted component. However, the current is so small that the component cannot function normally.
→ A ground-referenced voltage reading higher than normal may be caused by a short between the point and the source or by an open between the point and ground.
→ A ground-referenced voltage reading lower than normal may be caused by an open between the point and the source or by a short between the point and ground.

In a series-parallel circuit, if a single component's resistance value changes, the voltage, current, and power values of all other components may also change. In other words, if we change the resistance value of any actual component in a series-parallel circuit, the entire *RIPE* table may need to be recalculated.

If the change is an open or shorted component, there is even more work to do. The circuit simplification must be redone, and then the *RIPE* table recalculated.

Shorts

Let's begin by shorting R_4 in Figure 7–6. We now have the circuit shown in Figure 7–7. Part (b) of this figure emphasizes the fact that shorting R_4 has the same effect as replacing it with a conductor. Thus, we have a different circuit with five components instead of six. This will affect the simplification: R_7 will now equal $R_2 + R_6$, leaving out R_4.

To complete the *RIPE* table, follow the same sequence of steps given in the previous section for the table of the normal circuit. The complete *RIPE* table for Figure 7–7 follows. Carefully compare the values with those in the table for the normal circuit. You will discover that for the actual components, every value of current, power, and voltage has changed.

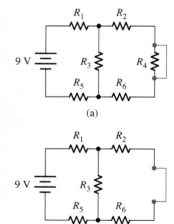

FIGURE 7–7
A short in a series-parallel circuit.

COMPONENT	R Ω	I mA	P mW	E V
R_1	150 Ω	17.7 mA	47 mW	2.65 V
R_2	180 Ω	9.6 mA	17 mW	1.72 V
R_3	390 Ω	8.1 mA	26 mW	3.16 V
R_4 shorted	—	—	—	—
R_5	180 Ω	17.7 mA	56 mW	3.18 V
R_6	150 Ω	9.6 mA	14 mW	1.44 V
$R_7 = R_2 + R_6$	330 Ω	9.6 mA		3.16 V
$R_8 = R_3 \| R_7$	179 Ω	17.7 mA		3.16 V
$R_9 = R_1 + R_8 + R_5$	509 Ω	17.7 mA	159 mW	9.00 V

A short anywhere in a series-parallel circuit will decrease total equivalent resistance, and thereby increase total current and total power.

From the *RIPE* tables, we can compare the values of R_9 (total equivalent resistance) for the normal and shorted circuits, and see that the short has indeed caused total equivalent resistance to decrease and, thus, total current and power to increase.

Normal: $R_T = 558\ \Omega$ $I_T = 16.1$ mA $P_T = 145$ mW
R_4 shorted: $R_T = 509\ \Omega$ $I_T = 17.7$ mA $P_T = 159$ mW

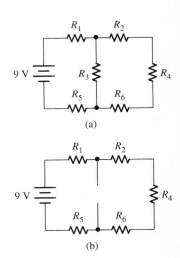

FIGURE 7–8
An open in a series-parallel circuit.

Opens

If we open R_3 in Figure 7–6, we will have the circuit shown in Figure 7–8(b). In this case, the circuit has become a simple series circuit. The complete *RIPE* table follows.

COMPONENT	R Ω	I mA	P mW	E V
R_1	150 Ω	10.2 mA	16 mW	1.53 V
R_2	180 Ω	10.2 mA	19 mW	1.84 V
R_3 open	—	—	—	—
R_4	220 Ω	10.2 mA	23 mW	2.25 V
R_5	180 Ω	10.2 mA	19 mW	1.84 V
R_6	150 Ω	10.2 mA	16 mW	1.53 V
R_T	880 Ω	10.2 mA	92 mW	9.00 V

An open anywhere in a series-parallel circuit will increase total equivalent resistance, and thereby decrease total current and total power.

From the *RIPE* tables, we can compare the values of R_9 (total equivalent resistance) for the normal and opened circuits and see that the open has indeed caused total equivalent resistance to increase and, thus, total current and power to decrease.

Normal: $R_T = 558\ \Omega$ $I_T = 16.1$ mA $P_T = 145$ mW
R_3 opened: $R_T = 880\ \Omega$ $I_T = 10.2$ mA $P_T = 92$ mW

Shorts as a Parallel Branch

If a component is shorted, the component and the short actually form two parallel branches. In the previous examples, we have considered the resistance value of a short to be zero ohms. However, a piece of wire has some resistance, and we can use the methods for series-parallel circuits to understand the details of what happens when a component is shorted. Let's consider the circuit shown in Figure 7–9.

We would simplify the circuit by defining R_3 as R_2 in parallel with the short, and R_T as R_1 plus R_3.

We can calculate R_3 using the usual Parallel Resistance Rule formulas:

$R_3 = R_2 \,||\, \text{short}$ Parallel Resistance Rule

$$= \cfrac{1}{\cfrac{1}{R_2} + \cfrac{1}{\text{short}}}$$

$$= \cfrac{1}{\cfrac{1}{220\ \Omega} + \cfrac{1}{0.02\ \Omega}}$$

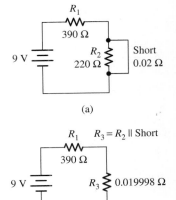

FIGURE 7–9
A short acts as a parallel branch.

$$= \frac{1}{0.004545 + 50}$$

$$= \frac{1}{50.004545}$$

$$= 0.019998 \ \Omega$$

In the simplified circuit, notice that the value of R_3 is slightly smaller than the value of the short, since the equivalent resistance of parallel branches must be smaller than the resistance of the smallest branch.

The total resistance of the circuit is $R_1 + R_3$, which equals 390.019998 Ω. Thus, the current equals 9 V ÷ 390.019998 Ω, which is 23.075739 mA. This give us a slight voltage drop across R_3: 23.075739 mA × 0.019998 Ω equals 461.468629 μV. Such a small voltage drop could not be measured with an ordinary voltmeter, and thus we say that for all practical purposes, the voltage drop is zero across a short.

If we apply the voltage drop across both branches (R_2 and the short), we can calculate the current in each branch:

$I_2 = 461.46829 \ μV ÷ 220 \ \Omega$ Ohm's Law
$\quad = 0.00209758 \ mA$

$I_{short} = 461.46829 \ μV ÷ 0.02 \ \Omega$
$\quad\quad = 23.073431 \ mA$

When we add the two branch currents together, we have

$I_T = 23.073431 \ mA + 0.00209758 \ mA$
$\quad = 23.075529 \ mA$

This value is slightly different from the total current of 23.075739 mA, calculated previously. The difference is due to round-off error and can be ignored.

Normally, we would not bother with such detailed calculations to six decimal places. However, the example proves an important point: Although most of the current (23.073 mA) flows through the short, a slight current (about 2 μA) still flows through the shorted resistor R_2.

In summary, a slight current still flows in a shorted component. However, the current is so small that the component cannot function normally.

You may have heard people say that "electricity takes the path of least resistance." This common belief is not correct. Current takes every available path. However, when one path has a normal resistance and the other has almost zero resistance (a short), almost all of the current flows through the short. Since the component receives so little current that it cannot function, it appears that no current is flowing through it.

Effects of Opens and Shorts on Ground-Referenced Voltage Readings

We will use Figure 7–10 to illustrate the effects of opens and shorts on ground-referenced voltage readings. The principles we will learn here are useful in practical troubleshooting.

Following is the *RIPE* table for the normal circuit.

COMPONENT	$R \ \Omega$	$I \ mA$	$P \ mW$	$E \ V$
R_1	100 Ω	369.23 mA		36.92 V
R_2	120 Ω	369.23 mA		44.31 V
R_3	100 Ω	176.23 mA		17.62 V

(continued)

COMPONENT	$R\,\Omega$	$I\,\mathrm{mA}$	$P\,\mathrm{mW}$	$E\,\mathrm{V}$
R_4	200 Ω	193.85 mA		38.77 V
R_5	120 Ω	176.23 mA		21.15 V
$R_6 = R_3 + R_5$	220 Ω	176.23 mA		38.75 V
$R_7 = R_4 \| R_6$	105 Ω	369.23 mA		38.77 V
$R_8 = R_1 + R_2 + R_7$	325 Ω	369.23 mA		120.00 V

The normal ground-referenced voltage at point A is equal to the normal voltage drop across R_4: 38.77 V. We will now examine the ground-referenced voltage at point A under four different cases involving opens and shorts.

Case 1: Short between Point A and Ground

If we place a short across R_3, we have the circuit shown in Figure 7–11.
Here is the *RIPE* table for this circuit.

COMPONENT	$R\,\Omega$	$I\,\mathrm{mA}$	$P\,\mathrm{mW}$	$E\,\mathrm{V}$
R_1	100 Ω	406.78 mA		40.68 V
R_2	120 Ω	406.78 mA		48.81 V
R_4	200 Ω	152.54 mA		30.51 V
R_5	120 Ω	254.24 mA		30.51 V
$R_6 = R_4 \| R_5$	75 Ω	406.78 mA		30.51 V
$R_7 = R_1 + R_2 + R_6$	295 Ω	406.78 mA		120.00 V

The ground-referenced voltage at point A equals the voltage drop across R_4: 30.51 V. The short across R_3 has caused the voltage at point A to be lower than the normal value of 38.77 V.

Case 2: Open Between Point A and Ground

If R_3 is opened, we have the circuit shown in Figure 7–12. The open at R_3 has also disconnected R_5 from the circuit. Thus, the circuit now consists of R_1, R_2, and R_4 in series.
Here is the *RIPE* table for this circuit.

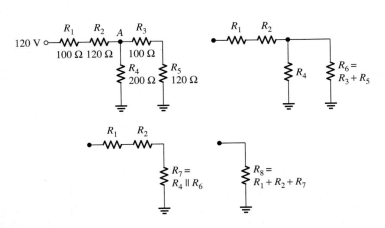

FIGURE 7–10
Normal series-parallel circuit with simplification.

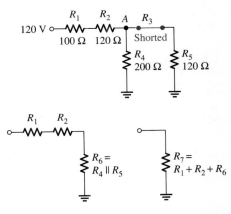

FIGURE 7–11
Circuit of Figure 7–10 with shorted component between point A and ground.

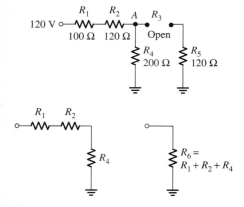

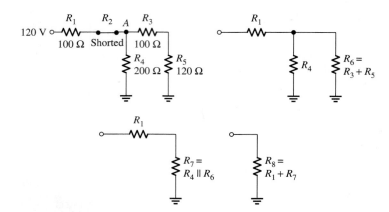

FIGURE 7–12
Circuit of Figure 7–10 with open component between point A and ground.

FIGURE 7–13
Circuit of Figure 7–10 with shorted component between point A and voltage source.

COMPONENT	R Ω	I mA	P mW	E V
R_1	100 Ω	285.71 mA		28.57 V
R_2	120 Ω	285.71 mA		34.29 V
R_4	200 Ω	285.71 mA		57.14 V
$R_6 = R_1 + R_2 + R_4$	420 Ω	285.71 mA		120.00 V

The ground-referenced voltage at point A equals the voltage drop across R_4: 57.14 V. The open at R_3 has caused the voltage at point A to be higher than the normal value of 38.77 V.

Case 3: Short Between Point A and the Voltage Source

If R_2 is shorted, we have the circuit shown in Figure 7–13.
Here is the *RIPE* table for this circuit.

COMPONENT	R Ω	I mA	P mW	E V
R_1	100 Ω	585.37 mA		58.54 V
R_3	100 Ω	279.36 mA		27.94 V
R_4	200 Ω	307.30 mA		61.46 V
R_5	120 Ω	279.36 mA		33.52 V
$R_6 = R_3 + R_5$	220 Ω	279.36 mA		61.46 V
$R_7 = R_4 \| R_6$	105 Ω	585.37 mA		61.46 V
$R_8 = R_1 + R_7$	205 Ω	585.37 mA		120.00 V

The ground-referenced voltage at point A equals the voltage drop across R_4: 61.46 V. The short across R_2 has caused the voltage at point A to be higher than the normal value of 38.77 V.

Case 4: Open Between Point A and the Voltage Source

If R_2 is opened, we have the circuit shown in Figure 7–14. No current can flow between the voltage source and ground. Since the current through R_4 is zero, the voltage drop is also zero. Thus, the ground-referenced voltage at point A is zero. The open at R_2 has caused the voltage at point A to be lower than its normal value.

FIGURE 7–14
Circuit of Figure 7–10 with open component between point A and voltage source.

Summary of the Four Cases

Ground-Referenced
Voltage at Point A

Location	*Short*	*Open*
between point *A* and source	higher than normal	lower than normal
between point *A* and ground	lower than normal	higher than normal

When troubleshooting real circuits, technicians often take ground-referenced voltage readings at designated points. By noticing whether the reading is higher or lower than normal and by referring to the preceding table, you may get an idea of the possible problem.

In summary, a ground-referenced voltage reading higher than normal may be caused by a short between the point and the source or by an open between the point and ground. A ground-referenced voltage reading lower than normal may be caused by an open between the point and the source or by a short between the point and ground.

→ *Self-Test*

2. If the resistance value of a component changes in a series-parallel circuit, what effect does this have on the voltage and current values of the other components?
3. What is the effect of a short in a series-parallel circuit?
4. What is the effect of an open in a series-parallel circuit?
5. Describe current in a shorted component.
6. What could cause a ground-referenced voltage reading to be higher than normal?
7. What could cause a ground-referenced voltage reading to be lower than normal?

SECTION 7–4: Loading Effects of Meters

In this section you will learn the following:

→ The internal resistance of a voltmeter is caused by the circuit inside the meter.

→ When a voltmeter measures a voltage drop across a component, the internal resistance of the meter is placed in parallel with the component. Therefore, the equivalent resistance of the meter and component is lower than the resistance of the component alone.

→ The meter's internal resistance causes the measured voltage drop across a component to be less than the true voltage drop. This effect is called meter loading.

→ Meter loading produces large, unacceptable errors in voltage measurement when the internal resistance of the meter is similar or smaller in value compared to the resistance of the circuit component. The measurement error is greatly reduced by switching the meter to the next higher range setting.

When a voltmeter measures the voltage drop of a component, the meter is actually in parallel with the component. The circuits inside the voltmeter produce an internal resistance that allows a small sample current to flow into the meter. By measuring this sample current, the meter determines the voltage drop across the component.

We know that placing one component in parallel with another creates an equivalent resistance that is smaller than the smallest branch. This means that placing a voltmeter across a component has the effect of changing the equivalent resistance of the component, which changes the voltage drop across the component. In other words, whenever we place a voltmeter across a component, the voltage drop decreases slightly because of the parallel resistance of the meter.

In summary, the internal resistance of a voltmeter is caused by the circuit inside the meter. When the internal resistance of the meter is placed in parallel with a circuit component, the equivalent resistance is lower than the resistance of the component alone. The meter's internal resistance causes the measured voltage drop across a component to be less than the true voltage drop. This effect is called meter loading.

The amount of error in the voltage reading due to meter loading depends on the internal resistance of the meter and the resistance of the circuit component. The internal resistance of the meter depends on the type of meter. There are three types of multimeters: VOMs, VTVMs, and DMMs.

The VOM (volt-ohm-milliammeter) is an analog (needle-type) meter that samples enough current from the circuit to move the needle. The internal resistance of a VOM can vary widely from 6 MΩ to as little as 40 kΩ. The voltage readings of the VOM are accurate enough for ordinary troubleshooting. However, because of the VOM's relatively low internal resistance, its voltage readings are the least accurate of the multimeter types.

The VTVM (vacuum-tube-volt-milliammeter) is also an analog-type meter. The VTVM contains vacuum-tube electronic circuits which sample a much smaller current than the VOM, and amplify the sample current into a stronger current which can move the needle. Since the VTVM draws very little current from the circuit, it has a high internal resistance, typically 10 MΩ. This high resistance when placed in parallel with circuit components does not change the equivalent resistance very much. Thus, the VTVM gives a more accurate voltage reading than the VOM.

The DMM also contains electronic circuits to amplify the sample current from the circuit. However, the DMM does not have a large needle to move against a spring. Instead it has a digital display, which requires very little current to operate. The electronic circuits inside the DMM are small transistorized chips. The DMM has a high internal resistance, typically 10 MΩ and, thus, gives more accurate voltage readings than the VOM.

All VOMs have a rating called a *sensitivity factor,* which allows us to calculate the internal resistance at any range setting. The sensitivity factor is expressed in units of ohms per volt. A typical sensitivity factor is 20,000 Ω/V. To find the internal resistance, simply multiply the sensitivity factor and the range setting. For example, if the range setting is 12 V, then the internal resistance is 12 V $\times$ 20,000 Ω/V = 240,000 Ω, or 240 kΩ. We will use this value in the following examples.

Figure 7–15 compares the true voltage drop of a component with the voltage readings of a DMM and a VOM. In part (a), the voltage divider formula gives a true voltage drop across R_2 of 7.714 V.

In part (b), a DMM with an internal resistance of 10 MΩ is placed across R_2. The slight loading of the meter's resistance lowers the equivalent resistance of R_2 to 3.5987 kΩ. According to the voltage divider formula, the drop across R_2 is now 7.713 V, and this is the voltage reading that would appear on the DMM's display. The error is a mere 0.001 V (1 mV), which is not enough to matter to us.

In part (c), a VOM is placed across R_2. Assuming the internal resistance is 240 kΩ, the equivalent resistance of R_2 becomes 3.5468 kΩ and the measured voltage drop is 7.673 V, or 0.4 V lower than the true reading. The error of the VOM is much greater than that of the DMM, yet the reading is still fairly accurate.

Figure 7–15 is typical of a practical situation. The slightly greater error of the VOM is probably not enough to matter. However, the error becomes significant when the size of the circuit components are approximately equal to the meter's internal resistance, as shown in Figure 7–16.

The components are now 100 times larger in resistance. Notice, however, that the true voltage drop in part (a) remains the same.

In part (b), the DMM's measured voltage is 7.616 V, which is an error of 0.098 V. Although the DMM still gives a fairly accurate reading, the error is nearly 100 times larger than it was in Figure 7–15.

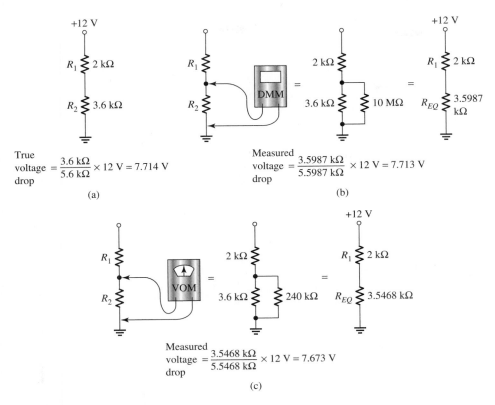

True voltage drop $= \dfrac{3.6 \text{ k}\Omega}{5.6 \text{ k}\Omega} \times 12 \text{ V} = 7.714 \text{ V}$

(a)

Measured voltage drop $= \dfrac{3.5987 \text{ k}\Omega}{5.5987 \text{ k}\Omega} \times 12 \text{ V} = 7.713 \text{ V}$

(b)

Measured voltage drop $= \dfrac{3.5468 \text{ k}\Omega}{5.5468 \text{ k}\Omega} \times 12 \text{ V} = 7.673 \text{ V}$

(c)

FIGURE 7–15
(a) True voltage drop. (b) Measured voltage drop with DMM. (c) Measured voltage drop with VOM.

Part (c) illustrates the gross error produced by meter loading. The VOM's internal resistance is actually smaller than the value of R_2. According to the rules of parallel branches, the equivalent resistance of R_2 has been reduced from 360 kΩ to 144 kΩ, and the voltage reading has dropped from the true value of 7.714 V to a measured value of 5.023 V. The difference of 2.691 V is an error of nearly 35%, which is not acceptable.

However, there is a simple remedy to the problem: Simply set the VOM to the next higher range setting. Remember that the internal resistance of a VOM equals the range setting times the sensitivity factor. If we suppose that the next higher setting is a 60-V range, then the internal resistance becomes

$$60 \text{ V} \times 20{,}000 \text{ }\Omega/\text{V} = 1.2 \text{ M}\Omega$$

Part (d) shows that at the 60-V range setting, the equivalent resistance becomes 276.92 kΩ and the measured voltage is 6.968 V. The difference between true and measured voltage is now 0.746 V, which is an error of 9.6%. Although the error is still considerable, it is much less at the 60-V range setting than it was at the 12-V setting.

We have learned in taking meter measurements to use the lowest possible range setting in order to get a more accurate reading. This procedure is always correct when using a DMM or VTVM because these meters have a high internal resistance that causes little error due to meter-loading effects. It is also correct when using VOMs with components whose resistance is fairly low.

However, when using VOMs to measure components whose resistance is fairly high, the meter-loading effects produce large errors in the voltage readings. By switching the meter to a higher range setting, we actually get a more accurate reading.

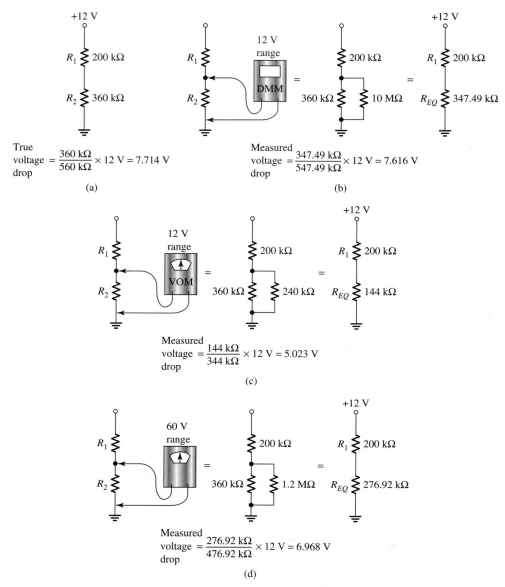

FIGURE 7–16
(a) True voltage drop. R_2 has high resistance. (b) Measured voltage drop with DMM. Greater error in reading due to high resistance of R_2. (c) Large error in voltage measured with VOM across high resistance. (d) Error is reduced by switching VOM to higher range setting.

Meter-loading effects are most noticeable in voltage readings, but loading errors also occur in current readings. In most cases the errors are small and can be disregarded. The exception to the rule is when measuring the small currents flowing in the control wires of certain transistors. Even DMMs cause large errors in this case. However, the remedy is the same: Use a higher range setting. The reading may be less sensitive, but it will be more accurate.

In summary, meter loading produces large, unacceptable errors in voltage measurement when the internal resistance of the meter is similar or smaller in value compared to the resistance of the circuit component. The measurement error is greatly reduced by switching the meter to the next higher range setting.

→ *Self-Test*

8. What causes the internal resistance of a voltmeter?
9. What effect does the internal resistance of a voltmeter have when the voltmeter measures a voltage drop?
10. How does a voltmeter's internal resistance affect the accuracy of voltage measurements? What is this effect called?
11. Under what conditions does meter loading produce large, unacceptable errors in voltage measurement? How may these errors be reduced?

● TROUBLESHOOTING: Applying Your Knowledge

You notice that the resistors in the circuit shown in Figure 7–1 are relatively large in value, and you suspect the meter's internal resistance is loading the circuit and producing inaccurate readings. Upon inquiry, you learn that your classmate is using an inexpensive VOM with a sensitivity of 20 kΩ/V, set on the 20-V range. This gives the meter an internal resistance of 400 kΩ, smaller than either resistor. This internal resistance would produce an equivalent resistance of 224 kΩ when placed in parallel with R_1, and 216 kΩ in parallel with R_2. A little calculation shows the voltmeter readings to be accurate, given the severe loading effect.

You suggest that your classmate set the meter on a higher voltage range, or even better, use a DMM or VTVM with high internal resistance.

CHAPTER SUMMARY: ANSWERS TO SELF-TESTS

1. A series-parallel circuit is one where all current paths can be determined by examining the connection of the components without reference to the value of the components.
2. In a series-parallel circuit, if a single component's resistance value changes, the voltage, current, and power values of all other components may also change.
3. A short anywhere in a series-parallel circuit will decrease total equivalent resistance, and thereby increase total current and total power.
4. An open anywhere in a series-parallel circuit will increase total equivalent resistance, and thereby decrease total current and total power.
5. A slight current still flows in a shorted component. However, the current is so small that the component cannot function normally.
6. A ground-referenced voltage reading higher than normal may be caused by a short between the point and the source or by an open between the point and ground.
7. A ground-referenced voltage reading lower than normal may be caused by an open between the point and the source or by a short between the point and ground.
8. The internal resistance of a voltmeter is caused by the circuit inside the meter.
9. When a voltmeter measures a voltage drop across a component, the internal resistance of the meter is placed in parallel with the component. Therefore, the equivalent resistance of the meter and component is lower than the resistance of the component alone.

10. The meter's internal resistance causes the measured voltage drop across a component to be less than the true voltage drop. This effect is called meter loading.
11. Meter loading produces large, unacceptable errors in voltage measurement when the internal resistance of the meter is similar or smaller in value compared to the resistance of the circuit component. The measurement error is greatly reduced by switching the meter to the next higher range setting.

CHAPTER REVIEW QUESTIONS

Determine whether the following questions are true or false.

7–1. We simplify series-parallel circuits to make the series and parallel relationships easier to see.

Questions 7–2 through 7–10 refer to Figure 7–17.

7–2. R_1 and R_2 are in series.
7–3. R_3 and R_4 are in parallel.
7–4. R_4 and R_5 are in parallel.

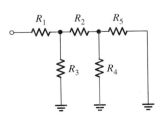

FIGURE 7–17

7–5. If R_4 is open, R_2 and R_5 are in series.

7–6. The group consisting of R_2, R_4, and R_5 is in parallel with R_3.

7–7. If R_2 is shorted, then R_3, R_4, and R_5 are in parallel.

7–8. If R_4 is shorted, R_2 and R_5 are in series.

7–9. If R_5 is shorted, R_2 and R_3 are in parallel.

7–10. If R_2 is open, R_1 and R_3 are in series.

7–11. If a group of components are in series, they all have the same current value.

7–12. If a group of components are in parallel, they all have the same voltage value.

7–13. In a series-parallel circuit, a single component may change value without affecting the voltage, current, and power values of the other components.

7–14. No current flows in a shorted component.

7–15. The internal resistance of a voltmeter is caused by the circuits inside the meter.

7–16. Sensitivity factor is a meter rating that allows us to calculate the internal resistance at any range setting.

7–17. Sensitivity factor is expressed in units of volts per ohm.

7–18. Meter loading produces large, unacceptable errors in voltage measurement when the internal resistance of the meter is much larger than the resistance of the circuit component.

Select the response that best answers the question or completes the statement.

7–19. A series-parallel circuit is one where
   a. some but not all current paths can be determined by examining the connection of the components.
   b. all current paths can be determined by examining the connections of the components.
   c. the values of the components must be known to determine current paths.
   d. it is impossible to determine current paths without the *RIPE* table.

7–20. In a typical series-parallel circuit,
   a. either all components have a series relationship or else all components have a parallel relationship.
   b. some components have a series relationship and other components have a parallel relationship.
   c. components near the source are in series and components near ground are in parallel.
   d. the series-parallel relationships depend on the direction of current flow.

7–21. A short anywhere in a series-parallel circuit will
   a. increase total equivalent resistance, and thereby decrease total current and total power.
   b. have no effect on total equivalent resistance, current, or power.
   c. decrease total equivalent resistance, and thereby increase total current and total power.
   d. cause total equivalent resistance, current, and power to be zero.

7–22. An open anywhere in a series-parallel circuit will
   a. increase total equivalent resistance, and thereby decrease total current and total power.
   b. have no effect on total equivalent resistance, current, or power.

c. decrease total equivalent resistance, and thereby increase total current and total power.
d. cause total equivalent resistance, current, and power to be zero.

7–23. A ground-referenced voltage reading higher than normal may be caused by
   a. an open between the point and the source or a short between the point and ground.
   b. an open or a short between the point and the source.
   c. a short between the point and the source or an open between the point and ground.
   d. an open or a short between the point and ground.

7–24. A ground-referenced voltage reading lower than normal may be caused by
   a. an open between the point and the source or a short between the point and ground.
   b. an open or a short between the point and the source.
   c. a short between the point and the source or an open between the point and ground.
   d. an open or a short between the point and ground.

7–25. When the internal resistance of the meter is placed in parallel with a circuit component, the equivalent resistance
   a. is lower than the resistance of the component alone.
   b. is higher than the resistance of the component alone.
   c. does not change.
   d. changes only if the meter is a VOM.

7–26. Meter loading causes the measured voltage drop across a component to be
   a. equal to the true voltage drop.
   b. more than the true voltage drop.
   c. less than the true voltage drop.
   d. unpredictable.

7–27. The internal resistance of a VOM voltmeter equals sensitivity factor _____ range setting.
   a. plus   **c.** multiplied by
   b. minus   **d.** divided by

7–28. Measurement error due to meter loading can be greatly reduced by
   a. using a VOM instead of a DMM or VTVM.
   b. measuring components with very high resistance values.
   c. switching the meter to the next lower range setting.
   d. switching the meter to the next higher range setting.

7–29. What is the best simplification step in Figure 7–18?
   a. $R_6 = R_4 \,\|\, R_5$ **c.** $R_6 = R_1 + R_3$
   b. $R_6 = R_2 \,\|\, R_3$ **d.** $R_6 = R_2 \,\|\, R_3 + R_5$

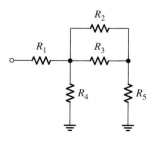

FIGURE 7–18

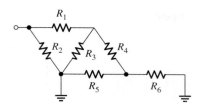

FIGURE 7–19

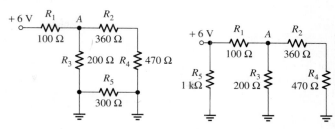

FIGURE 7–24   **FIGURE 7–25**

7–30. What is the best simplification step in Figure 7–19?
a. $R_7 = R_2 \| R_3$ **c.** $R_7 = R_1 + R_4$
b. $R_7 = R_3 \| R_4$ **d.** $R_7 = R_5 \| R_6$

Solve the following problems.

7–31. Find the voltage drop and current for all components in Figure 7–20.

7–32. In Figure 7–21, the current in R_1 is 57.6923 mA. What is the resistance of R_4?

7–33. What is the total resistance of the circuit in Figure 7–22?

7–34. In Figure 7–23, the ground-referenced voltage at point A is 3.7026 V. What is the source voltage?

7–35. In Figure 7–24, V_1 (the voltage drop across R_1) is 2.2974 V. What will V_1 be if R_5 is removed?

7–36. In Figure 7–25, I_3 (the current through R_3) is 18.513 mA. What is the ground-referenced voltage at point A if R_5 is removed?

7–37. A voltmeter has a sensitivity of 25,000 Ω/V. Find its internal resistance on each of the following ranges:
a. 300 mV   **c.** 10 V
b. 3 V   **d.** 30 V

7–38. In Figure 7–26, the VOM has a sensitivity of 18,000 Ω/V and is set on the 300-V range. What is the true voltage drop of R_2 and the measured voltage drop?

The following problems are more challenging.

Refer to Figure 7–27 for the following problems.

7–39. V_A and V_B both equal 10.125 V. Which resistor is shorted?

7–40. $V_A = 18.000$ V. $V_B = 10.286$ V. Which resistor is shorted?

7–41. V_A and V_B both equal 14.087 V. Which resistor is open?

7–42. $V_A = 14$ V. $V_B = 8$ V. Which resistor is open?

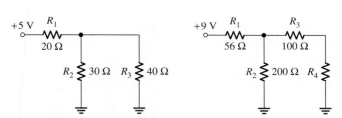

FIGURE 7–20   **FIGURE 7–21**

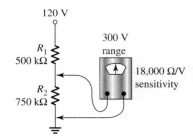

FIGURE 7–26

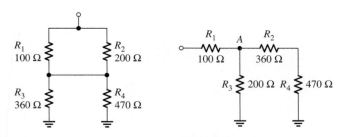

FIGURE 7–22   **FIGURE 7–23**

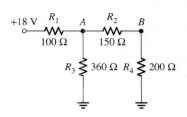

FIGURE 7–27

CIRCUIT THEOREMS

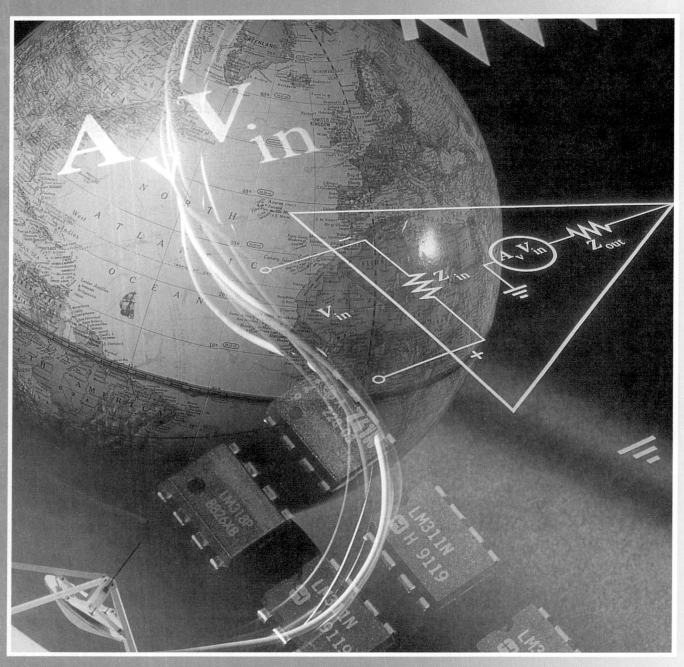

● SECTION OUTLINE

8–1   Viewing a Circuit from a Terminal Pair

8–2   Internal Resistance

8–3   Thevenin's Theorem

8–4   Networks

8–5   Bridges

8–6   Superposition Theorem

● TROUBLESHOOTING: Applying Your Knowledge

Your younger brother dreams of becoming a rock superstar and has put together a little band of musicians. They have bought a lot of second-hand amplifiers, keyboards, mixers, speakers, and so on, and are trying to get the whole thing hooked up and working. They are having lots of problems with microphones. Sometimes they get a roaring hum when the volume control is just barely turned up, and sometimes they can barely hear anything when the volume is all the way up. What do you think might be wrong with their sound system?

● OVERVIEW AND LEARNING OBJECTIVES

All advanced circuit analysis rests on Ohm's Law and the two Kirchhoff Laws. Several important theorems have been developed from such analysis. Thevenin's Theorem is widely used to understand the effect of a load upon a voltage source. Bridge theorems are applied to many test instruments and sensors. The superposition theorem applies to the effects of multiple voltage sources in the same circuit.

After completing this chapter, you will be able to:

1. Calculate the resistance of a simple circuit from any terminal pair.
2. Calculate the value of internal resistance of a practical voltage source based on measured unloaded output voltage, loaded output voltage, and load resistance.
3. Determine the Thevenin equivalent circuit for a practical voltage source and use the Thevenin circuit to predict the results of various loads on the output voltage.
4. Match load impedance to source impedance to produce maximum power transfer.
5. Apply Thevenin's Theorem to simple bridge circuits to determine voltage drop, current, and polarity of the bridge component.

SECTION 8–1: Viewing a Circuit from a Terminal Pair

In this section you will learn the following:

→ Viewing a circuit from a terminal pair means determining the total resistance the circuit would have if a voltage source were connected across the terminal pair.

A terminal is a point in a circuit where a physical connection can be made. Terminals are usually grouped as *terminal pairs.* In most practical circuits, the terminal pair of the load is called the *input,* and the terminal pair of the voltage source is called the *output.* In other words, the voltage source is connected to the input terminals of the load, and the load is connected to the output terminals of the voltage source.

For example, you may have a stereo system consisting of a separate CD player, amplifier, and loudspeaker, as shown in Figure 8–1. If we think of the amplifier as a circuit, then the CD player acts as a voltage source and the loudspeaker acts as a load. Thus, the CD player is connected to the amplifier's input terminals, and the loudspeakers are connected to the amplifier's output terminals.

Now consider the circuit shown in Figure 8–2(a). Suppose we choose terminal pair A and B as the input and connect a voltage source, as shown in Figure 8–2(b). Current flows through resistors R_1, R_3, and R_4, and the total resistance of the circuit is 80 Ω.

If we now choose terminal pair C and D as the input and connect the voltage source as shown in Figure 8–2(c), current flows through resistors R_2, R_3, and R_5, and the total resistance is 100 Ω.

We would express this by saying that the total resistance of the circuit is 80 Ω viewed from terminal pair AB, and 100 Ω viewed from terminal pair CD. In other words, the total resistance of the circuit changes if you look at it from different terminal pairs. We have already learned that a circuit's resistance depends on the values of the components and the way in which the components are connected. Now we see that the total resistance also depends on the choice of terminal pairs.

When we describe a circuit using expressions such as "viewed," "seen," or "looking" from a terminal pair, we are referring to what the circuit's resistance would be if a voltage source were connected between that terminal pair.

In summary, viewing a circuit from a terminal pair means determining the total resistance the circuit would have if a voltage source were connected across the terminal pair.

→ *Self-Test*

1. What does it mean to view a circuit from a terminal pair?

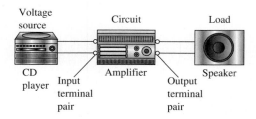

FIGURE 8–1
A circuit often has multiple terminal pairs.

SECTION 8–2: Internal Resistance

In this section you will learn the following:

→ An ideal device has perfect characteristics. A practical, real-world device has imperfect characteristics.

→ The output voltage of an ideal voltage source remains constant regardless of load current.

→ The output voltage of a practical voltage source decreases as load current increases.

→ Internal resistance is a concept used to explain a voltage source's decrease in output voltage with an increase in load current by supposing that the source has a certain amount of resistance inside itself in series with the load.

→ When current flows, a practical voltage source drops some of its voltage across its own internal resistance.

→ A practical voltage source is the equivalent of an ideal voltage source in series with a specific amount of internal resistance.

→ Internal resistance may be calculated from loaded and unloaded output voltage and load resistance.

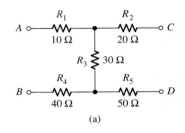

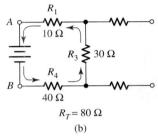

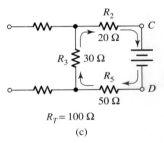

FIGURE 8–2
(a) A circuit with several terminals.
(b) The circuit as seen from terminal pair *AB*.
(c) The circuit as seen from terminal pair *CD*.

Ideal and Practical Devices and Circuits

The word *ideal* means perfect. When we speak of an ideal device or circuit, we mean one that has perfect behavior and characteristics. Since no real device or circuit is perfect, an ideal device or circuit is imaginary. However, we often speak of ideal devices and circuits and draw schematic diagrams of them to give us an idea of what a perfect device or circuit would do.

The word *practical* means real. When we speak of a practical device or circuit, we mean a real-world device or circuit whose behavior and characteristics are not perfect. By comparing the practical device or circuit to the ideal, we see how close we have come to perfection.

In summary, an ideal device has perfect characteristics. A practical, real-world device has imperfect characteristics.

Ideal and Practical Voltage Sources

An *ideal voltage source* maintains constant output voltage under all load conditions. A *practical voltage source* has a decrease in output voltage as load current increases.

Part (a) of Figure 8–3 shows the effect of a load on the terminal (output) voltage of an ideal voltage source. The important thing to notice is that the load has no effect at all. In this example, the terminal voltage remains equal to the source voltage of 12 volts regardless of the load. Thus, the output voltage of an ideal voltage source remains constant regardless of load current.

However, part (b) shows a different result, which we have not seen before. In practical voltage sources, the only time the terminal voltage will equal the source voltage is when the terminals are open and there is no load (no current). When a load is connected and current begins to flow, the terminal voltage decreases below the value of the source voltage.

As more current flows, the terminal voltage drops even more. In this example, the 100-Ω load draws a small current, and so the terminal voltage decreases only a small amount, going down to 11.765 V. When the load is 10 Ω, the current is larger, and the terminal voltage decrease is also larger, going down to 10 V. Thus, the output voltage of a practical voltage source decreases as load current increases.

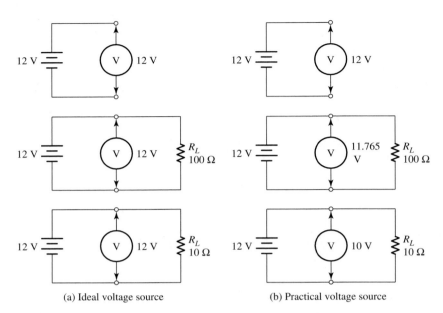

(a) Ideal voltage source   (b) Practical voltage source

FIGURE 8–3
Effect of various loads on ideal and practical voltage sources.

We know from our previous studies that the voltage drops in a series circuit must add up to equal the source voltage. If the terminal voltage has decreased to 10 V, the question is what has happened to the other 2 volts? The answer is explained next.

Internal Resistance

Internal resistance is a concept used to explain a voltage source's decrease in output voltage with an increase in load current by supposing that the source has a certain amount of resistance inside itself, in series with the load.

When current flows, a practical voltage source drops some of its voltage across its own internal resistance. This explains the terminal voltage losses we just discussed. For example, batteries contain chemicals, and these chemicals cause a certain amount of resistance to the current flowing through the battery. Generators contain windings with lots of wire, which also cause resistance to current. Therefore, we explain the decrease in terminal voltage by saying that the voltage source has a certain amount of internal resistance.

Figure 8–4 helps us imagine the idea of internal resistance. The box represents the practical voltage source, a battery, for example. We imagine that inside the battery is an ideal electromotive force (emf) of 12 V in series with an internal resistance of 2 Ω. The circuits in Figure 8–4 show what the battery acts like, and are called equivalent circuits.

Equivalent circuits are widely used in electricity and electronics. We have already seen examples of them when we learned to simplify series-parallel circuits. An equivalent circuit shows what a practical device acts like without explaining why. In other words, if you cut open a battery, you will not find another battery and a little resistor inside. Yet somehow the battery is acting as if this were true. Therefore, we use equivalent circuits to get a good mental picture of what a device acts like, and we don't worry about what's really inside, making it act like that.

A practical voltage source is the equivalent of an ideal voltage source in series with a specific amount of internal resistance. If we imagine that our 12-V battery has 2 Ω of internal resistance, it is easy to explain the decreases in output voltage shown in Figure 8–3(b). When the 100-Ω load resistor is placed across the terminals, current flows through a total resistance of 102 Ω. Following the rules of series circuits, 0.235 volts are dropped across the internal resistance, and the remaining 11.765 volts are dropped across

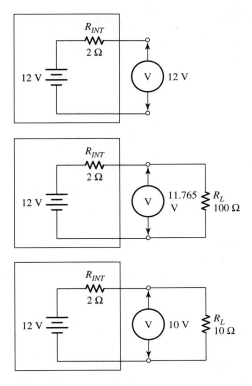

FIGURE 8–4
Internal resistance.

the load connected to the terminals. If the load becomes 10 Ω, the current is much higher, and the internal voltage drop is greater. Now 2 volts are dropped internally, leaving 10 volts across the load.

Another way to look at the problem is to think of the fair-share voltage relationship between the internal resistance and the external load. When the external load is many times greater than the internal resistance, most of the voltage is dropped across the load. However, when the load resistance is only a little larger than the internal resistance, a larger share of the total voltage is dropped internally.

Determining the Value of Internal Resistance

Internal resistance may be calculated from loaded and unloaded output voltage and load resistance. In other words, the value of internal resistance may be determined by Ohm's Law. The steps are:

1. Determine the load current with Ohm's Law: load voltage divided by load resistance.

$I = V_L \div R_L$ Ohm's Law

2. Determine the internal voltage drop by subtracting the load voltage from the unloaded output voltage.

$V_{INT} = V_{OUT} - V_L$ Kirchhoff's Voltage Law

3. Determine the internal resistance with Ohm's Law: the internal voltage drop divided by the current.

$R_{INT} = V_{INT} \div I$ Ohm's Law

● SKILL-BUILDER 8–1

- -

Find the value of the internal resistance in Figure 8–3(b).

GIVEN:
$$V_{OUT} = 12 \text{ V}$$
$$V_L = 11.765 \text{ V}$$
$$R_L = 100 \text{ }\Omega$$

FIND:
$$R_{INT}$$

STRATEGY:
　Use the three-step procedure given previously.

SOLUTION:
1. $I = V_L \div R_L$ 　　　　　　　　　Ohm's Law
   $= 11.765 \text{ V} \div 100 \text{ }\Omega$
   $= 117.65 \text{ mA}$

2. $V_{INT} = V_{OUT} - V_L$ 　　　　　Kirchhoff's Voltage Law
   $= 12 \text{ V} - 11.765 \text{ V}$
   $= 0.235 \text{ V}$

3. $R_{INT} = V_{INT} \div I$ 　　　　　　Ohm's Law
   $= 0.235 \text{ V} \div 117.65 \text{ mA}$
   $= 1.997 \text{ }\Omega$, or approximately 2 Ω

　　　Repeat the Skill-Builder using the following values. You should obtain the results that follow.

Given:

V_{OUT}	24 V	12.53 V	16.7 V	115.6 V	6.35 V
V_L	23.19 V	11.94 V	14.22 V	110.3 V	6.23 V
R_L	120 Ω	68 Ω	37 Ω	11 Ω	2361 Ω

Find:

I	193.3 mA	175.6 mA	384.3 mA	10.03 A	2.6 mA
V_{INT}	0.81 V	0.59 V	2.48 V	5.30 V	0.12 V
R_{INT}	4.19 Ω	3.36 Ω	6.45 Ω	0.53 Ω	45.48 Ω

- -

　　　The preceding three steps may be combined into a single formula:

Formula 8–1

$$R_{INT} = R_L \times \frac{(V_{OUT} - V_L)}{V_L}$$

Using our preceding example,

$$R_{INT} = 100 \text{ }\Omega \times \frac{(12 \text{ V} - 11.765 \text{ V})}{11.765 \text{ V}}$$

$$= 100 \text{ }\Omega \times \frac{0.235 \text{ V}}{11.765 \text{ V}}$$

$$= 100 \text{ }\Omega \times 0.0199745$$

$$= 1.997 \text{ }\Omega, \text{ or approximately 2 }\Omega$$

→ *Self-Test*

2. What is the difference between an ideal device and a practical device?
3. Describe the output voltage of an ideal voltage source.
4. Describe the output voltage of a practical voltage source.
5. What is internal resistance?
6. What happens when current flows in a practical voltage source?
7. What is the equivalent of a practical voltage source?
8. How may internal resistance be calculated?

SECTION 8–3: Thevenin's Theorem

In this section you will learn the following:

→ Thevenin's Theorem states that when viewed from any terminal pair, any circuit is equivalent to an ideal voltage source in series with an internal resistance.

→ A practical voltage source is normally unregulated, which means the output voltage varies with load current.

→ A voltage regulator is a device that may be added to a practical voltage source to keep output voltage constant within a given accuracy for a given range of load currents.

→ A practical voltage source with an added voltage regulator is called a regulated voltage source or regulated power supply.

→ Maximum power is transferred from a voltage source to a load when the ohmic value of the load equals the internal resistance of the voltage source.

→ The overall ohmic value of a device is called impedance.

→ Matching the output impedance of a voltage source with the input impedance of a load produces maximum power transfer.

A theorem is a mathematical idea that is not obvious at first but can be proven to be true. Therefore, a theorem is accepted as if it were a law. A theorem often has the name of the person who first proved it. In this section we will learn *Thevenin's Theorem,* one of the most important in electrical theory.

Thevenin was an electrical engineer who worked with the French telegraph service between 1870 and 1900. He developed a theorem to help predict exactly how much the voltage and current would change when a new telegraph station was added to the system. Since then, his theorem has become one of the most useful ways for electrical engineers to predict how voltage and current will change when two circuits are connected together.

In simplified form, Thevenin's Theorem states that when viewed from any terminal pair, any circuit is equivalent to an ideal voltage source in series with an internal resistance.

Thevenin's Theorem is a concept or idea. Figure 8–5 will help us understand it. Suppose a complex circuit is acting as a voltage source with two output terminals as shown in Figure 8–5(a). The output is unloaded (open), and a voltmeter measures a voltage across the terminals.

The circuit shown in Figure 8–5(b) is the Thevenin equivalent of the complex circuit: an ideal voltage source in series with an internal resistance. V_{TH}, the value of the ideal voltage source, equals V_{OUT}, the unloaded output voltage of the actual source. R_{INT}, the value of the internal resistance, may be determined as described in the previous section.

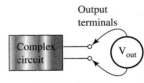

(a) Actual circuit.

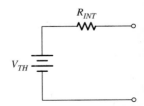

(b) Thevenin equivalent circuit.

FIGURE 8–5

● SKILL-BUILDER 8–2

What is the Thevenin equivalent circuit of the circuit shown in Figure 8–6?

Thevenin Voltage

The Thevenin voltage is the unloaded output voltage of an actual circuit. In this example, the output terminals are across the 8-Ω resistor. Therefore, the unloaded output voltage is the voltage drop across the 8-Ω resistor with the output terminals unloaded (open). Using the methods learned in previous chapters, we determine this voltage drop to be 8 V.

$4\,\Omega + 8\,\Omega = 12\,\Omega$	Series Resistance Rule
$12\,V \div 12\,\Omega = 1\,A$	Ohm's Law
$1\,A \times 8\,\Omega = 8\,V$	Ohm's Law

Determining the Value of Internal Resistance

To determine the internal resistance, we connect a load resistor to the output terminals, measure the voltage drop across the load, and apply the formula for internal resistance.

The load resistor could be any value, and 100 Ω is a handy size for this circuit. Using the methods previously learned, we determine that the drop across the load resistor is 7.792 V.

$8\,\Omega \;\|\; 100\,\Omega = 7.407\,\Omega$	Parallel Resistance Rule
$4\,\Omega + 7.407\,\Omega = 11.407\,\Omega$	Series Resistance Rule
$12\,V \div 11.407\,\Omega = 1.052\,A$	Ohm's Law
$1.052\,A \times 7.407\,\Omega = 7.792\,V$	Ohm's Law

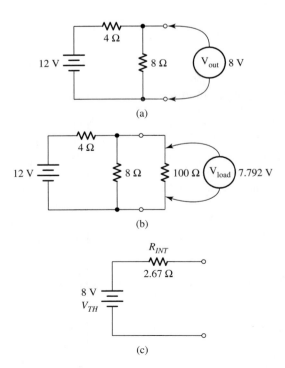

FIGURE 8–6
Determining the Thevenin equivalent circuit.

We now apply the internal resistance formula.

$8 \text{ V} - 7.792 \text{ V} = 0.208 \text{ V}$
$100 \,\Omega \times (0.208 \text{ V} \div 7.792 \text{ V}) = 2.67 \,\Omega$

Therefore, the Thevenin equivalent circuit has an ideal voltage of 8 V in series with an internal resistance of 2.67 Ω.

Repeat the Skill-Builder using the following values. You should obtain the results that follow.

Given:

V_S	16 V	8 V	22 V	14 V	7.3 V
R_1	5.2 Ω	2.5 Ω	1.15 Ω	3.24 Ω	0.48 V
R_2	7.3 Ω	3.1 Ω	1.85 Ω	4.06 Ω	0.73 Ω
R_L	150 Ω	180 Ω	68 Ω	75 Ω	28 Ω

Find:

$V_{TH} = V_S \times R_2/(R_1 + R_2)$

9.344 V	4.429 V	13.567 V	7.786 V	4.404 V

$R_{EQ} = R_2 \parallel R_L$

6.961 Ω	3.048 Ω	1.801 Ω	3.852 Ω	0.711 Ω

$V_L = V_S \times R_{EQ}/(R_1 + R_{EQ})$

9.159 V	4.395 V	13.427 V	7.604 V	4.359 V

$I_L = V_L/R_L$

61.06 mA	24.42 mA	197.45 mA	101.38 mA	155.68 mA

$V_{INT} = V_{TH} - V_L$

0.185 V	0.034 V	0.140 V	0.183 V	0.045 V

$R_{TH} = V_{INT}/I_L$

3.03 Ω	1.39 Ω	0.71 Ω	1.80 Ω	0.29 Ω

The Use of the Thevenin Equivalent Circuit

The Thevenin circuit may be used to calculate the effects of applying a different load to the real circuit. It is one of the most important and widely used methods for designing and understanding electronic circuits.

● SKILL-BUILDER 8–3

In Figure 8–7, we see the same actual circuit as in Figure 8–6 and the Thevenin equivalent circuit of the actual circuit. This time we have connected a different load (50 Ω) to the output terminals.

Let's begin by using the actual circuit to calculate the load voltage drop and current.

$8 \,\Omega \parallel 50 \,\Omega = 6.9 \,\Omega$	Parallel Resistance Rule
$4 \,\Omega + 6.9 \,\Omega = 10.9 \,\Omega$	Series Resistance Rule
$12 \text{ V} \div 10.9 \,\Omega = 1.101 \text{ A}$	Ohm's Law
$1.101 \text{ A} \times 6.9 \,\Omega = 7.597 \text{ V } (V_L)$	Ohm's Law
$7.597 \text{ V} \div 50 \,\Omega = 0.152 \text{ A } (I_L)$	Ohm's Law

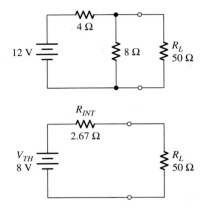

FIGURE 8–7
The Thevenin equivalent circuit is used to calculate the effect of a different load on the actual circuit.

Now let's use the Thevenin equivalent circuit and see what we get.

$2.67\ \Omega + 50\ \Omega = 52.67\ \Omega$	Series Resistance Rule
$8\ V \div 52.67\ \Omega = 0.152\ A\ (I_L)$	Ohm's Law
$0.152\ A \times 50\ \Omega = 7.6\ V\ (V_L)$	Ohm's Law

Repeat the Skill-Builder using the following values. You should obtain the results that follow.

Given:

V_{TH}	9.344 V	4.429 V	13.567 V	7.786 V	4.404 V
R_{TH}	3.037 Ω	1.384 Ω	0.709 Ω	1.802 Ω	0.290 Ω
New R_L	125 Ω	163 Ω	81 Ω	66 Ω	37 Ω

Find:

$$V_L = V_{TH} \times R_L/(R_{TH} + R_L)$$

9.122 V	4.391 V	13.449 V	7.579 V	4.370 V

$$I_L = V_L/R_L$$

72.98 mA	26.94 mA	166.04 mA	114.84 mA	118.11 mA

Notice that the Thevenin circuit gave the same results as the actual circuit, allowing for round-off error. However, working with the Thevenin equivalent circuit was quicker and easier, since it is a simple series circuit.

When the actual circuit is complex or unknown, using the Thevenin equivalent circuit becomes the only practical method of predicting the effects of a load, as shown in the following example.

● *SKILL-BUILDER 8–4*

An automobile battery has an unloaded output voltage of 11.75 V. The output voltage drops to 11.20 V when a load draws a current of 10 A. What will the output voltage be with a load current of 17 A? The circuit is shown in Figure 8–8.

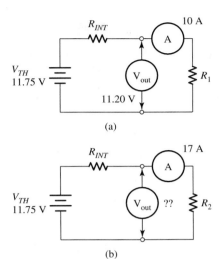

(a)

(b)

FIGURE 8–8
What will the output voltage be
when the current increases from
10 A to 17 A?

GIVEN:
$V_{TH} = 11.75$ V
$V_L = 11.20$ V
$I_L = 10$ A

FIND:
V_L when $I_L = 17$ A

STRATEGY:
First, we must recognize that the unloaded output voltage, 11.75 V, is by definition the Thevenin voltage. Then rather than use the internal resistance formula, we will use the series circuit voltage rule to find the internal voltage drop (V_{INT}), Ohm's Law to find R_{INT} at 10 A, and Ohm's Law again to find V_L at 17 A.

SOLUTION:
1. $V_{INT} = V_{TH} - V_L$
      $= 11.75$ V $- 11.20$ V
      $= 0.55$ V

   $R_{INT} = V_{INT} \div I$
      $= 0.55$ V $\div 10$ A
      $= 0.055$ Ω

2. $V_{INT} = I \times R_{INT}$
      $= 17$ A $\times 0.55$ Ω
      $= 0.935$ V

   $V_L = V_{TH} - V_{INT}$
      $= 11.75$ V $- 0.935$ V
      $= 10.815$ V

Repeat the Skill-Builder using the following values. You should obtain the results that follow.

Given:

V_{TH}	16.23 V	21.6 V	53.8 V	116.5 V	2.36 V
V_L	15.47 V	19.2 V	48.4 V	110.3 V	2.19 V
I_{L1}	9.3 A	2.35 A	9.3 A	15.4 A	6.42 A
I_{L2}	13.6 A	4.08 A	8.7 A	21.3 A	3.17 A

Find:

$V_{INT} = V_{TH} - V_L$

0.76 V	2.4 V	5.4 V	6.2 V	0.17 V

$R_{INT} = V_{INT} / I_{L1}$

0.082 Ω	1.021 Ω	0.581 Ω	0.403 Ω	0.026 Ω

$V_{INT} = I_{L2} \times R_{INT}$

1.111 V	4.167 V	5.052 V	8.575 V	0.084 V

$V_{L2} = V_{TH} - V_{INT}$

15.119 V	17.433 V	48.748 V	107.925 V	2.276 V

- -

Voltage Regulation

If output voltage remains steady at all normal load currents, we say the voltage source is regulated. If output voltage changes with load current, we say the voltage source is unregulated.

Thevenin's Theorem and Skill-Builder 8–4 show us that practical voltage sources have an unsteady output voltage that changes with load current. Thus, practical voltage sources are normally unregulated. This is generally not what we want. We prefer to have an ideal voltage source with a steady output voltage at all normal current levels. In other words, we want our voltage source to be regulated.

To do this, we must design some sort of device to automatically notice and correct any changes in output voltage due to changes in load current. Such a device is called a *voltage regulator*.

There are several different ways to regulate voltage and so there are many different types of voltage regulators. Most of them use electronic circuits to keep the output voltage at a steady level. The general idea of all voltage regulators is to hold some extra voltage in reserve, and then automatically make more or less voltage available at the output terminals to make up for changes in voltage due to changes in current. This happens so quickly that it seems like the output voltage never changes.

In summary, practical voltage sources are normally unregulated, which means the output voltage varies with load current. A voltage regulator is a device that may be added to a practical voltage source to keep output voltage constant within a given accuracy for a given range of load currents. A practical voltage source with an added voltage regulator is called a regulated voltage source or regulated power supply.

Maximum Power Transfer Theorem

The *maximum power transfer theorem* states that maximum power is transferred from a voltage source to a load when the ohmic value of the load equals the internal resistance of the voltage source.

Figure 8–9 illustrates this important and useful theorem. Notice that the maximum load power of 4.5 W is developed when R_L equals R_{INT}.

The overall ohmic value of a device is called *impedance*. We will learn more about impedance in a later chapter. For now, impedance simply means the ohmic value of the load or the source.

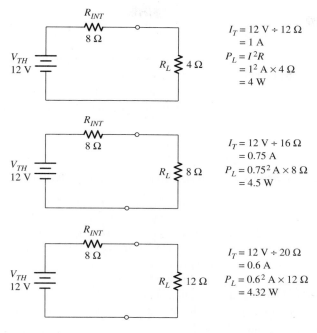

$I_T = 12 \text{ V} \div 12 \text{ } \Omega$
$= 1 \text{ A}$
$P_L = I^2 R$
$= 1^2 \text{ A} \times 4 \text{ } \Omega$
$= 4 \text{ W}$

$I_T = 12 \text{ V} \div 16 \text{ } \Omega$
$= 0.75 \text{ A}$
$P_L = 0.75^2 \text{ A} \times 8 \text{ } \Omega$
$= 4.5 \text{ W}$

$I_T = 12 \text{ V} \div 20 \text{ } \Omega$
$= 0.6 \text{ A}$
$P_L = 0.6^2 \text{ A} \times 12 \text{ } \Omega$
$= 4.32 \text{ W}$

FIGURE 8–9
Maximum power is transferred when $R_L = R_{INT}$.

Matching the output impedance of a voltage source with the input impedance of a load produces maximum power transfer. For example, if an audio amplifier output requires 8-Ω loudspeakers and we connect 4-Ω speakers, maximum power cannot be transferred from the amplifier to the speakers because of impedance mismatching. The internal impedance of the amplifier is 8 Ω, and the impedance of the loudspeakers should match this value.

Another example of impedance matching involves connecting microphones to recorders and sound systems. Many recorders and sound systems will have separate inputs for high-Z (high-impedance) and low-Z (low-impedance) microphones. A low-Z mike has an impedance of about 150 Ω to 600 Ω. A high-Z mike's impedance is typically 10 kΩ or above. If a mike is plugged into the wrong input, the impedance mismatch will produce undesirable results: Either the audio output is weak at high-volume settings, or else the output is so strong that the amplifier howls and squeals even at low-volume settings.

→ Self-Test

9. What is Thevenin's Theorem?
10. What do we mean when we say that a practical voltage source is normally unregulated?
11. What is a voltage regulator?
12. What is a regulated power supply?
13. What conditions cause maximum power to be transferred from a source to a load?
14. What is impedance?
15. What is the relationship between maximum power transfer and impedance matching?

SECTION 8–4: Networks

In this section you will learn the following:

→ In both series and parallel circuits, the direction of current flow does not depend on the value of the components.

→ A network is a circuit that contains at least one component where the direction of current flow cannot be predicted without knowing the values of other components and voltage sources.

Direction of Current in Series and Parallel Circuits

So far, we have studied two types of circuits: series and parallel. Series and parallel circuits have an important common characteristic: in both series and parallel circuits, the direction of current flow does not depend on the value of the components.

Notice that we are speaking only of the direction of current flow, not the amount. Obviously, we must know the value of the components to calculate the amount of current. But if all we are interested in is the direction of current, it is not necessary to know the component values. The direction of current flow in all components can be determined from examining the polarity of the input terminals and the connections of the components. Figure 8–10 illustrates this idea.

In each of these circuits, we do not know the ohmic value of any component, nor the value of the voltage source. Yet we are certain the currents will flow as shown by the arrows because we know the polarity of the voltage source and the connections of the components.

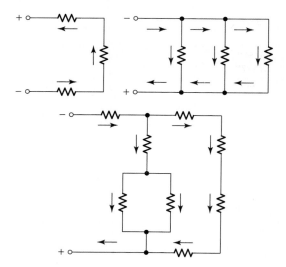

FIGURE 8–10
The direction of current flow in series and parallel circuits does not depend on the values of the components.

Networks

A *network* is a circuit that contains at least one component where the direction of current flow cannot be predicted without knowing the values of other components and voltage sources.

Figure 8–11(a) is an example of a simple network. The direction of current flow in R_5 cannot be predicted simply by looking at the polarity and connections. We must also know something about the ohmic values of the components.

If R_5 is removed and the component values are as shown in Figure 8–11(b), the voltage drop across R_2 is 8 V and the drop across R_4 is 6.86 V. Since the negative terminal of the source is grounded, the voltage readings are +8 V at point A and +6.86 V at point B.

We usually think of current flowing from a negative point toward a positive point. However, if both points are positive, electrons will move from the less positive point toward the more positive point. Similarly, if both points are negative, electrons will move from the more negative point toward the less negative point.

In Figure 8–11(b), both points A and B are positive compared to ground. However, point B is less positive than point A, and current will flow from point B (less positive) to point A (more positive).

If the component values are changed as in Figure 8–11(c), the voltage readings become +4 V at point A and +6.86 V at point B. Now point A is less positive than point B, and current would flow from point A to point B. Thus, the direction of current between the two points depends on the value of the components around R_5.

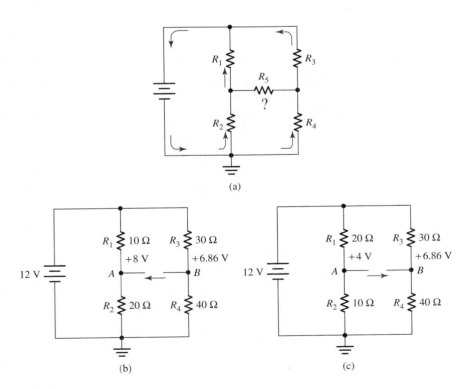

(a)

(b)

(c)

FIGURE 8–11

A simple network. The direction of current flow in R_5 depends on the values of the other components.

When R_5 is connected to either circuit, the voltages at both points will change from the values shown in Figure 8–11 because the total resistance of the circuit will change. Notice that the connection of R_5 does not match either of the circuit recognition rules. In other words, with R_5 connected, the circuit is no longer either series or parallel. R_5 makes the circuit a network, and the rules of series or parallel circuits do not apply.

Figure 8–12 illustrates a complicated network. There are ways to calculate the voltages and currents in networks like this, but the mathematics is beyond the scope of this book. The series and parallel circuit simplification and calculation methods we learned in earlier chapters cannot be used in evaluating networks.

Many large circuits are networks. For example, the power lines of electric utility companies are interconnected to make networks that often include several cities and rural areas. Utility companies are also interconnected with other utility companies to form giant networks covering several states. Telephones and computers are also usually connected in networks.

→ *Self-Test*

16. What is the relationship between the direction of current flow and the values of components in series and parallel circuits?
17. What is a network?

SECTION 8–5: Bridges

In this section you will learn the following:

→ A bridge consists of five components. Four components form two branches. The fifth component, called the bridge component, interconnects the two branches.

The circuit shown in Figure 8–11 is a simple example of a *bridge* because component R_5 bridges (interconnects) the two branches. Bridges are widely used in instruments to make highly accurate electrical measurements or to control equipment such as motors very precisely. Usually, one of the other four resistors is adjustable or is replaced by a transistor that is electronically adjustable until the current is zero in the bridge component. This type of bridge is often called a Wheatstone or balanced bridge.

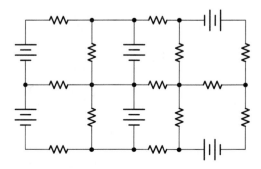

FIGURE 8–12
A complicated network.

Special Procedure for Calculating Bridge Voltage and Current

Thevenin's Theorem can be used to predict the voltage drop and current of a bridge component, following these steps:

1. Thevenin voltage equals the difference in ground-referenced voltages at points A and B.
2. To find internal resistance, calculate the resistance of each branch as if the branch components were in parallel. Then add the two values together.

● SKILL-BUILDER 8-5
- -
Find the Thevenin equivalent circuit for the bridge shown in Figure 8–13.

Step 1.
$$V_{TH} = V_A - V_B$$
$$= 8\text{ V} - 6.86\text{ V}$$
$$= 1.14\text{ V}$$

Step 2.
$$10\ \Omega\ ||\ 20\ \Omega = 6.67\ \Omega$$
$$30\ \Omega\ ||\ 40\ \Omega = 17.14\ \Omega$$
$$R_{INT} = 6.67\ \Omega + 17.14\ \Omega$$
$$= 23.81\ \Omega$$

Since point A is more positive than point B in the original circuit, point A is shown toward the positive side of the Thevenin voltage in the Thevenin circuit.

Repeat the Skill-Builder using the following values. You should obtain the results that follow.

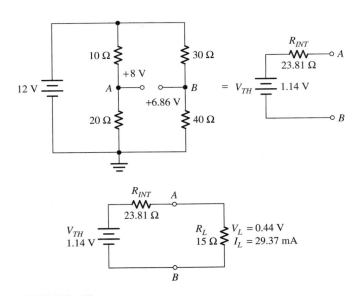

FIGURE 8–13
A bridge may be converted into a Thevenin circuit to calculate load voltage and current.

Given:

V_S	14 V	16 V	22 V	7 V	35 V
R_1	17 Ω	47 Ω	150 Ω	6.3 Ω	2.2 kΩ
R_2	14 Ω	51 Ω	200 Ω	5.1 Ω	2.7 kΩ
R_3	7 Ω	36 Ω	390 Ω	8.3 Ω	3.6 kΩ
R_4	12 Ω	22 Ω	270 Ω	15.7 Ω	1.8 kΩ

Find:

$V_A = V_S \times R_2/(R_1 + R_2)$

6.323 Ω	8.327 Ω	12.571 Ω	3.132 Ω	19.286 Ω

$V_B = V_S \times R_4/(R_3 + R_4)$

8.842 Ω	6.069 Ω	9.000 Ω	4.579 Ω	11.667 Ω

$V_{TH} = |V_A - V_B|$

2.520 V	2.258 V	3.571 V	1.448 V	7.619 V

$R_A = R_1 \,||\, R_2$

7.677 Ω	24.459 Ω	85.714 Ω	2.818 Ω	1,212 Ω

$R_B = R_3 \,||\, R_4$

4.421 Ω	13.655 Ω	159.545 Ω	5.430 Ω	1,200 Ω

$R_{TH} = R_A + R_B$

12.098 Ω	38.114 Ω	245.260 Ω	8.248 Ω	2,412 Ω

- -

● *SKILL-BUILDER 8–6*
- -

In Figure 8–13, find the voltage and current in a 15-Ω bridge resistor.

STRATEGY: Now that a Thevenin circuit has been created, the bridge resistor R_5 is treated like a load resistor in the Thevenin circuit, and the load voltage and current are easy to calculate.

SOLUTION:

$R_T = R_{TH} + R_L$ Series Resistance Rule
   $= 23.81\ Ω + 15\ Ω$
   $= 38.81\ Ω$

$I = 1.44\ V \div 38.81\ Ω$ Ohm's Law
   $= 29.37\ mA$

$V_L = 29.37\ mA \times 15\ Ω$ Ohm's Law
   $= 0.44\ V$

    Repeat the Skill-Builder using the following values. You should obtain the results that follow.

Given:

V_{TH}	2.52 V	2.258 V	3.571 V	1.448 V	7.619 V
R_{TH}	12.098 Ω	38.114 Ω	245.26 Ω	8.248 Ω	2412 Ω
R_L	150 Ω	200 Ω	1500 Ω	75 Ω	10 kΩ

Find:

R_T	162 Ω	238 Ω	1745 Ω	83 Ω	12,412 Ω
I	15.55 mA	9.48 mA	2.05 mA	17.39 mA	614 μA
V_L	2.33 V	1.9 V	3.07 V	1.3 V	6.14 V

- -

→ *Self-Test*

18. Describe a bridge.

SECTION 8–6: Superposition Theorem

In this section you will learn the following:

→ The Superposition Theorem states that the current in any component is the algebraic sum of the currents produced by each voltage source in the circuit acting alone, with all other voltage sources replaced with their internal resistances.

There are many circuits where multiple voltage sources produce separate currents that combine together into a single current. For example, in a pop music concert, many microphones are used on stage. Each microphone produces a complex ac voltage and current. These many currents are mixed into a single current which flows to the loudspeakers. Even though only one current flows to the loudspeaker, the audience clearly hears all the separate voices and instruments. In other words, the one current behaves as if it were many separate currents.

Cable television is another similar situation. The currents of many different television stations are mixed together into a single current in the cable. Although the cable has only one current, the television is able to tune in the different channels as if the currents were still separate.

This behavior of current is unique and remarkable. The fact that a single complex current behaves like a mixture of many simple currents allows us to design many advanced electronic devices. For example, a space probe can send radar signals to the surface of a planet and receive back billions of scattered echoes combined into a single signal. A computer can analyze the complex signal and separate out the many echoes to

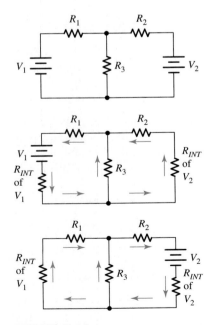

FIGURE 8–14
Superposition theorem.

create a map of the mountains and valleys on the planet's surface. A similar process is used in magnetic resonance imaging (MRI) equipment, which gives doctors an extremely detailed picture of a patient's internal organs and tissue.

The *Superposition Theorem* states that the current in any component is the algebraic sum of the currents produced by each voltage source in the circuit acting alone, with all other voltage sources replaced with their internal resistances. This theorem provides a mathematical basis for current analysis in the preceding examples.

The words *algebraic sum* simply mean some currents add to each other and others subtract from each other, as shown in the following example.

Figure 8–14(a) is a simple network. In part (b), we see the currents produced by V_1 acting alone, with V_2 replaced with its internal resistance. Notice that V_1's internal resistance is also included for greater accuracy. Similarly in part (c), we see the currents produced by V_2 acting alone, with V_1 replaced with its internal resistance. The actual currents in R_1, R_2, and R_3 are the results (algebraic sums) of the separate currents produced by V_1 and V_2. In R_1 and R_3, the two currents flow in opposite directions. Thus, the actual currents in these resistors would be the difference between the separate currents. However, in R_3 the two currents flow in the same direction, and the actual current would be the sum of the separate currents.

● *SKILL-BUILDER 8–7*

In Figure 8–14, find the voltage drops and currents if the components have the following values:

$$V_1 = 16 \text{ V}$$
$$R_{INT} \text{ of } V_1 = 0.4 \text{ } \Omega$$
$$V_2 = 18 \text{ V}$$
$$R_{INT} \text{ of } V_2 = 0.2 \text{ } \Omega$$
$$R_1 = 4 \text{ } \Omega$$
$$R_2 = 6 \text{ } \Omega$$
$$R_3 = 8 \text{ } \Omega$$

STRATEGY: Use the series-parallel rules and the *RIPE* table method to determine the currents produced by each voltage source acting alone. Then combine the currents algebraically to determine the actual currents, and use Ohm's Law to determine the actual voltage drops.

SOLUTION: The complete tables for the two voltage sources acting alone follow.

Note: In superposition problems, we recommend you do not round off your figures as you work through the table. In the following tables, values are shown to four decimal places, but were actually computed to nine decimal places.

For V_1:

COMPONENT	R	I	E
R_1	4.0 Ω	2.0271 A	8.1085 V
R_2	6.0 Ω	1.1420 A	6.8522 V
R_3	8.0 Ω	0.8851 A	7.0807 V
R_{INT} of V_1	0.4 Ω	2.0271 A	0.8108 V
R_{INT} of V_2	0.2 Ω	1.1420 A	0.2284 V
$R_7 = R_2 + R_{INT} V_2$	6.2000 Ω	1.1420 A	7.0807 V
$R_8 = R_3 \mid\mid R_7$	3.4930 Ω	2.0271 A	7.0807 V
$R_9 = R_1 + R_{INT} V_1 + R_8$	7.8930 Ω	2.0271 A	16.0000 V

For V_2:

COMPONENT	R	I	E
R_1	4.0 Ω	1.2848 A	5.1392 V
R_2	6.0 Ω	1.9914 A	11.9486 V
R_3	8.0 Ω	0.7066 A	5.6531 V
R_{INT} of V_1	0.4 Ω	1.2848 A	0.5139 V
R_{INT} of V_2	0.2 Ω	1.9914 A	0.3983 V
$R_7 = R_1 + R_{INT} V_1$	4.4000 Ω	1.2848 A	5.6531 V
$R_8 = R_3 \mid\mid R_7$	2.8387 Ω	1.9914 A	5.6531 V
$R_9 = R_2 + R_{INT}V_2 + R_8$	9.0387 Ω	1.9914 A	18.0000 V

At this point in the solution, it is better to look at the two sets of currents rather than the various voltage drops. Figure 8–15 shows the two sets of currents flowing in the circuit. We reduce these to a combined current for each component by addition if the two currents flow in the same direction. We use subtraction if the currents flow in opposite directions, with the combined current flowing in the direction of the stronger of the two currents. The results are calculated next.

Current in R_1: 2.0271 A − 1.2848 A = 0.7423 A
Current in R_2: 1.9914 A − 1.1420 A = 0.8494 A
Current in R_3: 0.8851 A + 0.7066 A = 1.5917 A
Current in R_{INT} of V_1 = Current of R_1
Current in R_{INT} of V_2 = Current of R_2

These current values are used to calculate the actual voltage drops.

COMPONENT	R	I	E
R_1	4.0 Ω	0.7423 A	2.9623 V
R_2	6.0 Ω	0.8494 A	5.0964 V
R_3	8.0 Ω	1.5917 A	12.7338 V
R_{INT} of V_1	0.4 Ω	0.7423 A	0.2969 V
R_{INT} of V_2	0.2 Ω	0.8494 A	0.1699 V

Let's think about the meaning of the voltage drops across the internal resistances in Figure 8–16. Remember that internal resistance is located inside the physical voltage source. Thus, in an actual circuit, the terminal voltage of each source would equal the unloaded voltage minus the voltage drop due to internal resistance. In our example, the

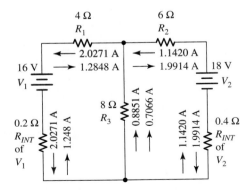

FIGURE 8–15
The currents produced by V_1 and V_2.

terminal voltage of V_1 would be 16.0000 V − 0.2969 V, which equals 15.7031 V. Similarly, the terminal voltage of V_2 would be 18.0000 V − 0.1699 V, which equals 17.8301 V.

Applying Kirchhoff's Current Law

Kirchhoff's Current Law states that the sum of all currents entering a junction equals the sum of all currents leaving the junction. We can double-check our work by applying Kirchhoff's Current Law at points *A* and *B* in Figure 8–18:

Currents entering point *A*	Currents leaving point *A*
1.5917 A	0.7423 A
	0.8494 A
1.5917 A	1.5917 A

Currents entering point *B*	Currents leaving point *B*
0.7423 A	1.5917 A
0.8494 A	
1.5917 A	1.5917 A

Applying Kirchhoff's Voltage Law

Kirchhoff's Voltage Law states that the algebraic sum of all voltage drops and rises around a closed circuit loop is zero.

The circuit in Figure 8–16 gives us a chance to apply Kirchhoff's Voltage Law, stated earlier.

Before we begin, we need to remember a few details about the polarity of components. The polarity of a dc voltage source does not change. Thus, we have marked the positive and negative terminals of the two batteries in Figure 8–16.

On the other hand, the polarity of a resistor depends on the direction of current flow. Where the electron current enters the resistor is the more negative end, and where the electron current comes out of the resistor is the more positive end. Now that we have determined the direction of the combined current in the circuit, we have marked the polarity of the resistors in Figure 8–16.

Kirchhoff's Voltage Law speaks of closed circuit loops. A circuit loop is a simple idea. You can identify a circuit loop by placing your finger on any point in the cir-

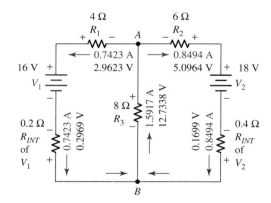

FIGURE 8–16
The actual current is the algebraic sum of the separate currents.

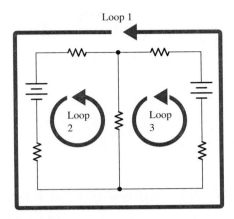

FIGURE 8–17
Circuit loops.

cuit and moving your finger along any current path until you come back to your starting point. The path your finger has taken is called a closed circuit loop. In tracing out the loop, you may move your finger in any direction, but once you start, you cannot change the direction of movement. The circuit in Figure 8–16 has three loops, as shown in Figure 8–17.

When applying Kirchhoff's Voltage Law to a circuit loop, it does not matter where you start in the loop or which way you go around the loop. The procedure is simple: Whenever you go through a component, notice the polarity of the end where you come out. Let the sign of the polarity become the sign of the voltage value. This will give you a mixture of positive and negative numbers. When you add them up (the algebraic sum), the result will be a value of zero, as predicted by the voltage law.

● *SKILL-BUILDER 8–8*

Using the values of the circuit in Figure 8–16, begin at point *A* and follow the three loops shown in Figure 8–17 to determine the algebraic sums of the voltages in each loop.

Loop 1	Loop 2	Loop 3
+ 2.9623 V	+ 2.9623 V	−12.7338 V
−16.0000 V	−16.0000 V	− 0.1699 V
+ 0.2969 V	+ 0.2969 V	+18.0000 V
− 0.1699 V	+12.7338 V	− 5.0964 V
+18.0000 V		
− 5.0964 V		
− 0.0071 V	− 0.0070 V	− 0.0001 V

The sums are not zero because of round-off error. In actual circuits, the sum of the voltages in a loop is precisely zero.

If we had gone around the loops in the other direction, all the signs of the numbers would have been reversed and the algebraic sums would have been the same value.

→ *Self-Test*

- - - - - - -

19. State the Superposition Theorem.

- -

Formula List

Formula 8–1

$$R_{INT} = R_L \times \frac{(V_{OUT} - V_L)}{V_L}$$

● TROUBLESHOOTING: Applying Your Knowledge

Your brother's problems with microphones and amplifiers suggest an impedance mismatching problem. In other words, he may be plugging high-impedance mikes into low-impedance inputs, or vice versa. Since the source and load impedances do not match, the voltage produced by the microphone is either too high or too low for the input circuits. If the equipment does not have the correct jacks to accommodate the mikes, your brother can solve the problem by using special impedance matching transformers available from audio and electronics stores.

CHAPTER SUMMARY: ANSWERS TO SELF-TESTS

- - - - - - - - - - - - - - -

1. Viewing a circuit from a terminal pair means determining the total resistance the circuit would have if a voltage source were connected across the terminal pair.

2. An ideal device has perfect characteristics. A practical, real-world device has imperfect characteristics.

3. The output voltage of an ideal voltage source remains constant regardless of load current.

4. The output voltage of a practical voltage source decreases as load current increases.

5. Internal resistance is a concept used to explain a voltage source's decrease in output voltage with an increase in load current by supposing that the source has a certain amount of resistance inside itself in series with the load.

6. When current flows, a practical voltage source drops some of its voltage across its own internal resistance.

7. A practical voltage source is the equivalent of an ideal voltage source in series with a specific amount of internal resistance.

8. Internal resistance may be calculated from loaded and unloaded output voltage and load resistance.

9. Thevenin's Theorem states that when viewed from any terminal pair, any circuit is equivalent to an ideal voltage source in series with an internal resistance.

10. A practical voltage source is normally unregulated, which means the output voltage varies with load current.

11. A voltage regulator is a device that may be added to a practical voltage source to keep output voltage constant within a given accuracy for a given range of load currents.

12. A practical voltage source with an added voltage regulator is called a regulated voltage source or regulated power supply.

13. Maximum power is transferred from a voltage source to a load when the ohmic value of the load equals the internal resistance of the voltage source.

14. The overall ohmic value of a device is called impedance.

15. Matching the output impedance of a voltage source with the input impedance of a load produces maximum power transfer.

16. In both series and parallel circuits, the direction of current flow does not depend on the value of the components.

17. A network is a circuit that contains at least one component where the direction of current flow cannot be predicted without knowing the values of other components and voltage sources.

18. A bridge consists of five components. Four components form two branches. The fifth component, called the bridge component, interconnects the two branches.

19. The Superposition Theorem states that the current in any component is the algebraic sum of the currents produced by each voltage source in the circuit acting alone, with all other voltage sources replaced with their internal resistances.

CHAPTER REVIEW QUESTIONS

- - - - - - - - - - - - - - - -

Determine whether the following questions are true or false.

8–1. A terminal is a point where a circuit ends.

8–2. A circuit has different values of resistance when viewed from different terminal pairs.

8–3. An ideal circuit is an imaginary circuit with perfect characteristics.

8–4. A Thevenin circuit is an ideal circuit.

8–5. The output voltage of an ideal voltage source changes with load current.

8–6. The output voltage of a practical voltage source decreases as load current increases.

8–7. When the output terminals are open, a practical voltage source drops some of its voltage across its own internal resistance.

8–8. A voltage regulator is a device that may be added to a practical voltage source to keep output voltage constant within a given accuracy for a given range of load currents.

8–9. Matching the output impedance of a voltage source with the input impedance of a load produces maximum power transfer.

8–10. In both series and parallel circuits, the direction of current flow does not depend on the value of the components.

8–11. A network is a circuit that contains at least one component where the direction of current flow cannot be predicted without knowing the values of other components and voltage sources.

Select the response that best answers the question or completes the statement.

8–12. The voltage source is connected to the _____ terminals.
   a. input
   b. Thevenin
   c. output
   d. maximum power

8–13. The load is connected to the _____ terminals.
   a. input
   b. Thevenin
   c. output
   d. maximum power

8–14. Internal resistance is
   a. an actual resistor inside a voltage source.
   b. a concept used to explain a voltage source's decrease in output voltage with an increase in load current.
   c. the opposite of external resistance.
   d. found only in dc voltage sources.

8–15. Which of the following is the best statement of Thevenin's theorem?
   a. Viewed from only one particular terminal pair, any circuit is equivalent to an ideal voltage source in series with an internal resistance.
   b. Viewed from any terminal pair, any circuit is equivalent to an ideal voltage source in parallel with an internal resistance.
   c. Viewed from the output terminal, any practical voltage source is equivalent to an ideal voltage source in series with an internal resistance.
   d. Viewed from any terminal pair, any circuit is equivalent to an ideal voltage source in series with an internal resistance.

8–16. A voltage source is said to be regulated if
   a. output voltage changes with load current.
   b. it has a Thevenin equivalent.
   c. output voltage remains steady at all normal load currents.
   d. it has been approved by the National Electrical Association.

8–17. A voltage source is said to be unregulated if
   a. output voltage changes with load current.
   b. it does not have a Thevenin equivalent.
   c. output voltage remains steady at all normal load currents.
   d. it has not been approved by the National Electrical Association.

8–18. Maximum power will be transferred from a voltage source to a load when the ohmic value of the load
   a. is greater than the internal resistance of the voltage source.
   b. equals the internal resistance of the voltage source.
   c. is less than the internal resistance of the voltage source.
   d. the ohmic value of the load approaches zero.

8–19. The minimum number of components in a bridge is
   a. three
   b. four
   c. one
   d. five

Solve the following problems.

8–20. Find the total resistance for the circuit in Figure 8–18 when viewed from terminal pairs *AB,* and when viewed from terminal pairs *CD.*

8–21. A voltage source has an unloaded output voltage of 18 V. When connected to a 200-Ω load, the voltage drops to 17.6471 V. Find the value of the internal resistance.

8–22. What is the Thevenin equivalent circuit of the circuit shown in Figure 8–19?

8–23. A Thevenin circuit has an unloaded output voltage of 6.9 V and an internal resistance of 3.6 Ω. What load resistance would cause the output voltage to drop to 6.3 V?

8–24. Use the Thevenin circuit for Figure 8–19 to calculate the voltage drop and current for a 75-Ω load.

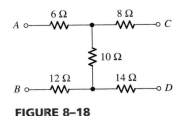

FIGURE 8–18

The following problems are more challenging.

8–25. For the bridge shown in Figure 8–20, use the special procedure based on Thevenin's theorem to determine the voltage drop and current of a 9-Ω bridge resistor.

8–26. Use the Superposition Theorem to find the currents in R_1, R_2, and R_3 in Figure 8–21.

8–27. Use the voltage drops for the circuit in Figure 8–21 to demonstrate the truth of Kirchhoff's Voltage Law for all loops in the circuit.

8–28. In the circuit shown in Figure 8–22, what ohmic value of R_6 will cause R_6 to produce the most heat?

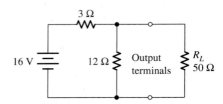

FIGURE 8–19

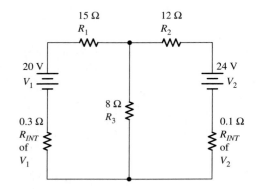

FIGURE 8–21

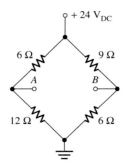

FIGURE 8–20

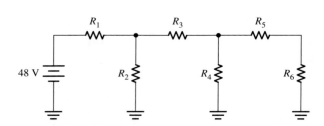

FIGURE 8–22

GLOSSARY

bridge   A circuit consisting of five components, one of which interconnects parallel branches formed by pairs of the remaining resistors.

ideal voltage source   One that maintains constant voltage under all load conditions.

impedance   The overall ohmic value of a device.

impedance matching   Any step taken to ensure that the output impedance of a source matches as nearly as possible the input impedance of a load.

input   The terminal pair of a load.

internal resistance   A concept used to explain a voltage source's decrease in output voltage with an increase in load current by supposing that the source has a certain amount of resistance inside itself in series with the load.

maximum power transfer theorem   The fact that maximum power will be transferred from a voltage source to a load when the ohmic value of the load equals the internal resistance of the voltage source.

network   A circuit that contains at least one component where the direction of current flow cannot be predicted without knowing the values of other components and voltage sources.

output   The terminal pair of a voltage source.

practical voltage source   One whose output voltage decreases as load current increases.

Superposition Theorem   The mathematical proof that the current in any component is the algebraic sum of the currents produced by each voltage source in the circuit acting alone, with all other voltage sources replaced with their internal resistances.

terminal pair   Two terminals in a circuit selected either as input or output.

Thevenin's Theorem   The mathematical proof that when viewed from any terminal pair, any circuit is equivalent to an ideal voltage source in series with an internal resistance.

voltage regulator   A device which may be added to a practical voltage source to keep output voltage constant within a given accuracy for a given range of load currents.

9

CIRCUIT COMPONENTS II

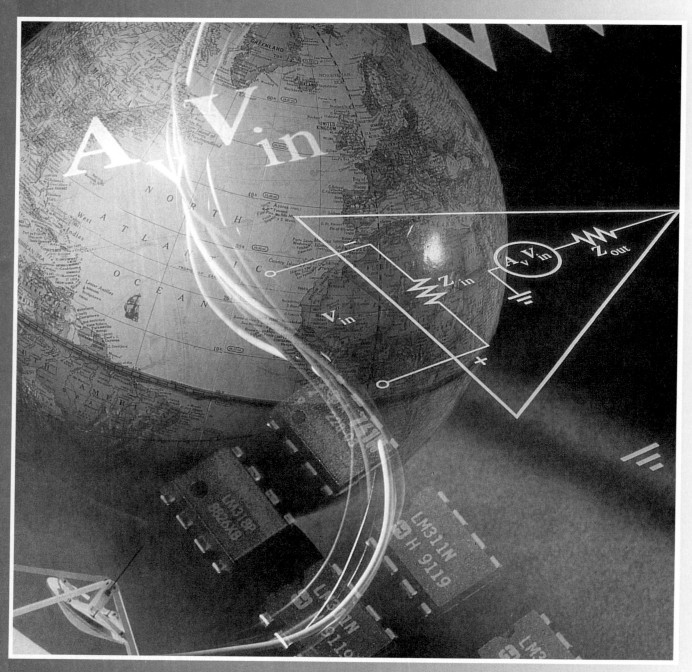

● SECTION OUTLINE

9–1 Cells and Batteries
9–2 Wire and Cable
9–3 Insulators
9–4 Circuit Breakers and Residential Wiring

● TROUBLESHOOTING:
Applying Your Knowledge

Your neighbor is upset about an electrical problem in his house. Last week his 240-V hot water heater quit working. He checked the circuit breaker panel and found that the breaker had tripped. He reset the breaker, but soon it tripped again. He noticed the breaker looked old and worn, so he replaced the breaker. The hot water heater still didn't work, although the breaker remained on. He ran back and forth to the hardware store, replacing first the heating element, then the thermostat, and finally replacing the entire water heater. Nothing worked, and he still has no hot water. He wonders if he now must rip the old wiring out of the wall and replace it too. What would you suggest he do?

● OVERVIEW AND LEARNING OBJECTIVES

As a technician, you should be able to deal with routine electrical problems as well as high-tech electronic problems. In this chapter you will learn about common electrical devices such as batteries, wire and cable, and the basics of residential wiring. You will find this information useful both in your technical career and in your personal life.

After completing this chapter, you should be able to:

1. Describe the basic construction and principle of operation of any cell or battery.
2. Estimate the AWG wire size appropriate to common electrical applications, and explain the concept of ampacity.
3. Describe how to connect 120-V and 240-V equipment to residential circuit breakers.
4. Describe the purpose and operation of a grounding wire and a ground-fault circuit interrupter.

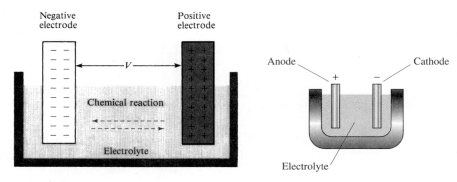

FIGURE 9–1
A basic cell.

SECTION 9–1: Cells and Batteries

In this section you will learn the following:

→ A cell is a dc voltage source consisting of two electrodes immersed in an electrolyte.
→ A battery is a group of cells connected in series and/or parallel.
→ An ampere-hour rating is the continuous discharge current of a battery multiplied by a six-hour discharge cycle.

A *cell* consists of two electrodes of different metals placed in a liquid or paste electrolyte, as shown in Figure 9–1. Chemical reactions between the electrolyte and the electrodes transfer electrons from one electrode to the other, producing a voltage. Sometimes gases are released, and the cell may need to be vented to allow the gases to escape.

All chemicals are classified as being either acid or alkaline. Thus, you will see these words used to describe cells and batteries, according to the chemical used as the electrolyte.

You can make a simple cell by using two wires or nails of different metals as electrodes and inserting them into a lemon. The sour juice acts as an electrolyte, and a small voltage will develop between the electrodes.

The voltage of a cell is a natural value determined by the metals used for the electrodes and the chemical used for the electrolyte. Cell voltages are low, generally less than 4 V. Each cell can produce a limited amount of current. To create higher voltages or currents, several cells are connected together to make a *battery,* as shown in Figure 9–2.

The amount of current a cell can produce partly depends on the amount of electrolyte inside the cell. Thus, larger cells produce more current for longer periods of time than smaller cells. Eventually, all of the original chemicals in the electrolyte will be changed into other chemicals, and the cell cannot produce any more current. We say the cell is "run down."

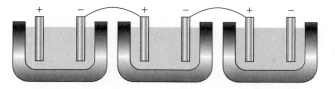

FIGURE 9–2
Several cells connected to form a battery.

Cells are classified as wet cells (liquid electrolyte) or dry cells (paste electrolyte). Automobile batteries are wet cells, and flashlight cells are dry.

In some cells, once the chemical reactions are completed, the cell is useless and cannot be recharged. A cell that can be used only once is called a primary cell.

In other cells, we can recharge the cell by applying an external voltage, forcing current to flow backward through the cell. This process reverses the chemical reaction and restores the cell to its original chemical form. A cell that can be recharged is called a secondary cell.

Each time a secondary cell is charged and discharged, the cell deteriorates slightly. Eventually, the secondary cell wears out and cannot be recharged. Secondary cells are rated according to the number of charge-discharge cycles they can endure.

Ampere-Hour Rating

Cells and batteries are often rated in ampere-hours, abbreviated Ah. The *ampere-hour rating* is based on a six-hour discharge cycle. Thus, the ampere-hour rating divided by six gives the estimated maximum continuous current of the battery, as shown in Formula Diagram 9–1. For example, suppose a battery is rated at 200 ampere-hours. We divide this by 6 hours and get 33 A. Therefore, this battery can provide a continuous current of up to 33 A.

Automobile and marine batteries are often rated in cranking amps. This is the amount of current the battery can provide for brief periods of time to start (crank) an engine. Cranking currents of 200 A to 400 A are common.

Formula Diagram 9–1

$$\frac{\text{Ampere-hour rating}}{6 \;\bigg|\; \text{Maximum continuous current}}$$

In the following sections, we will summarize information about many common types of cells used in batteries.

Lead-Acid Cell

The lead-acid cell, shown in Figure 9–3, is a wet cell widely used for automotive and marine batteries. Each cell produces 2.1 V. The electrodes are made of lead and lead peroxide (spongy lead), and the electrolyte is sulfuric acid. The cell can be recharged.

Lead and sulfuric acid are inexpensive materials capable of providing large amounts of currents. The cell has poor performance at high and low temperatures, which is why you may have trouble starting your car on a cold morning.

In automobile batteries, six cells are connected internally to make a 12.6-V battery. Each cell has its own chamber of electrolyte with a filler cap to provide access to the chamber.

Automobile and marine batteries are often mounted in an engine compartment where heat produced by the engine evaporates the water. The loss of water raises the specific gravity, exposes the plates, and reduces the performance and life of the battery. Filler caps allow the user to add water to each cell to maintain normal electrolyte volume. An instrument called a hydrometer is used to measure the mixture of acid and water in the electrolyte.

UNIQUE MANIFOLD VENT •

EXCLUSIVE TRP • CONTAINER

PREMIUM ENVELOPED • SEPARATOR DESIGN

COMPUTER DESIGNED • CALCIUM GRIDS

PROTECTIVE BOTTOM • DESIGN

• TORQUE RESISTANT TERMINALS

• INBOARD LUG DESIGN

• THROUGH-THE-PARTITION INTERNAL WELDS

• ARMORED PLATE CELL BONDING

FIGURE 9–3
Lead-acid cell.

The fluffy green-white material you may have seen around the terminals of your auto battery is lead sulfate, produced by the chemical reaction in the electrolyte. The buildup of lead sulfate prevents proper electrical contact between the battery terminals and cable clamps. An alkaline solution of water and baking soda is a good way to wash off the lead sulfate and neutralize any corrosive activity on or around the battery.

Lead-acid cells can be dangerous. Lead is poisonous, and sulfuric acid is highly corrosive. If the acid gets on your skin, it can cause severe burns. If it gets in your eyes, it can cause blindness. Therefore, you should wear protective clothing, rubber gloves, and safety goggles when working with lead-acid batteries.

When lead-acid cells produce current, the chemical reaction releases explosive hydrogen gas near the negative terminal. For this reason, you should be careful not to produce electrical sparks or place a flame or lighted cigarette near an automobile battery. When jump-starting automobile batteries, follow the safety precautions provided by the manufacturer of the jumper cables.

Carbon-Zinc Cell

The carbon-zinc cell, shown in Figure 9–4, is a dry cell widely used in flashlights, portable radios, and so on. Each cell produces about 1.5 V. The electrodes are zinc and manganese dioxide, and the electrolyte is a solution of ammonium chloride and zinc chloride. The cell cannot be recharged.

The carbon-zinc cell was the first successful cell using a paste electrolyte. It is inexpensive and available in an assortment of sizes.

The body of a carbon-zinc cell is called the can. The bottom of the can is the negative terminal. The metal button at the top of the can is the positive terminal.

A 6-V lantern battery consists of four cells connected internally, and a 9-V transistor battery consists of six cells.

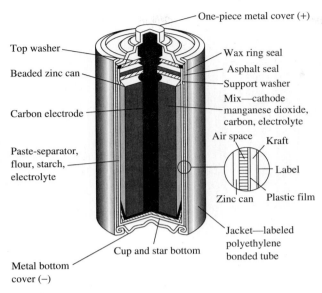

One-piece metal cover (+)

Top washer

Beaded zinc can

Carbon electrode

Paste-separator,
flour, starch,
electrolyte

Wax ring seal

Asphalt seal

Support washer

Mix—cathode
manganese dioxide,
carbon, electrolyte

Air space   Kraft

Label

Zinc can   Plastic film

Jacket—labeled
polyethylene
bonded tube

Cup and star bottom

Metal bottom
cover (−)

FIGURE 9–4
Carbon-zinc cell.

Zinc Chloride Cell

The zinc chloride cell is a modified version of the carbon-zinc cell and is often called a heavy-duty cell. By using only zinc chloride as the electrolyte, voltage regulation is improved. Also, the zinc chloride cell becomes completely dry when it is fully discharged, so there is no concern about leaking.

Alkaline Cell

The alkaline cell, shown in Figure 9–5, is a dry cell that is more expensive than a carbon-zinc cell but has much better performance and longer life. The electrolyte is potassium hydroxide, an alkaline chemical. Most alkaline cells are not rechargeable.

Nickel-Cadmium Cell

The nickel-cadmium (ni-cad) cell, shown in Figure 9–6, is a rechargeable dry cell. The cell has good voltage regulation and can be recharged many times. The electrodes are made of nickel and cadmium, and the electrolyte is an alkaline mixture of potassium hydroxide and lithium hydroxide. The cell does not release gas and thus does not require venting.

Nickel-Iron Cell

The nickel-iron cell was developed by Thomas Edison and is used in industrial equipment. It is similar in chemistry and performance to the nickel-cadmium cell. However, the nickel-iron cell is heavier than the ni-cad and releases gas.

CHAPTER 9

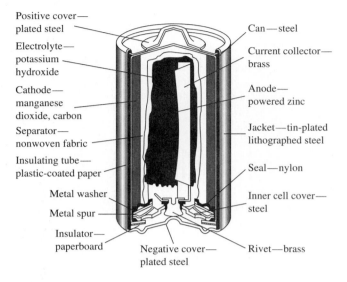

FIGURE 9–5
Alkaline cell.

FIGURE 9–6
Nickel-cadmium cell.

Mercury Cell

The mercury cell, shown in Figure 9–7, is a type of miniature alkaline cell widely used in computers and VCRs to power clock/calendar circuits when the main ac power is off. The cell produces 1.4 V and is not rechargeable.

Lithium Cell

The lithium cell, shown in Figure 9–8, provides a much higher voltage and more energy than all other nonrechargeable types with a shelf life of over ten years. Its disadvantages are that it is expensive and highly toxic.

The lithium cell is a wet cell. The electrolyte is kept in a liquid state with solvents under high pressure, which the capsule must withstand.

Miscellaneous Cells

A wide variety of other cell types exists, some in production with limited commercial applications, others as experimental prototypes. Some of these are described next.

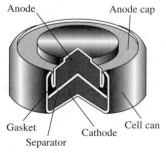

• Anodes are a gelled mixture of amalgamated zinc powder and electrolyte.

• Cathodes
  Silver cells: AgO_2, MnO_2, and conductor
  Mercury cells: HgO and conductor (may contain MnO_2)
  Manganese dioxide cells: MNO_2 and conductor

FIGURE 9–7
Generic button cell: silver, mercury, or manganese dioxide cathode.

FIGURE 9–8
Lithium cells.

Caustic soda cells use sodium hydroxide as the electrolyte. It is inexpensive and gives long service. Caustic soda batteries have current ratings as high as 1,000 ampere-hours but are not suitable for use in vehicles because the container is glass. The batteries are used to activate railroad signals in rural areas.

A reserve cell is a type that is activated immediately before use. Before activation, a reserve cell has an indefinite shelf life; but after activation, it must be used at once until it is completely discharged. A reserve cell generally uses chemicals too strong to be used in ordinary cells. The cell may be activated by heat, gas, water, or an electrolyte. One type of reserve cell uses seawater as the electrolyte. Because seawater is an electrolyte, ocean-going ships have a large block of zinc mounted below the waterline, so the electrolytic action of the seawater will eat away at the zinc instead of the steel hull of the ship. The block is thus called the sacrificial zinc.

The fuel cell uses gases as its electrodes. The gases are continuously added during operation. The advantage of the fuel cell is that its electrodes never deteriorate, since the gases are constantly being replenished. One gas is always oxygen. The other gas is hydrogen or some hydrocarbon compound and is called the fuel. Potassium hydroxide is a common electrolyte. A fuel cell operates at high temperatures (240°C or 464°F), produces 0.6 to 0.9 V, and has an extremely high energy/volume ratio of up to 10 kilowatts per cubic foot. Fuel cells are typically used on manned space vehicles and similar high-technology applications.

A solar cell, properly called a photovoltaic cell, does not involve any chemical reactions. Two pieces of semiconductor material are formed into layers and bonded together, as shown in Figure 9–9. Light shining on the surface of one layer produces free electrons, which migrate to the other layer, producing a charge difference (voltage) between the two layers. The free electrons migrate because of differences in quantum energy levels among the atoms of the two layers. A solar cell produces less than 1 V, and current capacity is usually only a few microamperes. Thus, the solar cell is suited only for low-power applications, such as small calculators. However, the solar cell is a good source of power in satellites and spacecraft because the intensity of the sun's radiation in space greatly increases the output of the cell.

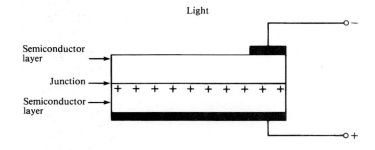

FIGURE 9–9
Solar cell.

Nuclear batteries use radioactive materials to release high-energy charged particles which strike electrodes, producing voltage. The batteries produce an unusual combination of high voltage (up to 10,000 V) and low current (50 μA). They are used to power electronic circuits on deep-space probes and similar vehicles.

→ *Self-Test*

1. What is a cell?
2. What is a battery?
3. What is an ampere-hour rating?

SECTION 9–2: Wire and Cable

In this section you will learn the following:

- → The ampacity of a conductor is the maximum current the conductor may carry under specific conditions according to the National Electrical Code.
- → A circular mil is the cross-sectional area of a round conductor with a diameter of 1 mil (0.001 inch).
- → The American Wire Gauge (AWG) and kilo-circular-mil (kcmil) systems measure the size of conductors. Both systems are based on the circular mil.
- → Soldering creates permanent electrical connections by melting a tin/lead alloy into the surfaces of conductors.
- → The characteristic impedance of a wire or cable is its overall opposition to current, expressed in ohms.
- → Power transmission lines carry power at high voltage to minimize power losses in the lines.

Ampacity

The word *ampacity* means ampere-carrying capacity. As more current flows in a conductor, more heat is produced. An excessive current would produce enough heat to melt the insulation, exposing the bare conductor and creating a serious danger of shorts and fire. Therefore, for safety reasons, we set a limit on the current a conductor may carry under specific conditions.

Ampacity is an engineering standard or limit, not a natural law. Table 9–1 is an example of an ampacity table, reprinted from the National Electrical Code. The National Electrical Code (NEC) is a book of electrical safety standards published by the National Fire Protection Association, a nonprofit organization supported by insurance companies and other groups interested in fire safety. The NEC is not law but is widely used as the basis of city and county laws for electrical wiring in local buildings.

As Table 9–1 shows, the ampacity of a conductor depends on five factors:

1. The conductor material: copper or aluminum.
2. The way the conductor is mounted or installed: open air or enclosed.
3. The air temperature around the conductor.
4. The electrical insulation around the conductor.
5. The cross-sectional area (size) of the conductor.

TABLE 9–1
Ampacity table.

COPPER

AMPACITIES OF NOT MORE THAN THREE SINGLE INSULATED CONDUCTORS RATED 0-2,000 VOLTS						
IN RACEWAY, CABLE, OR EARTH, BASED ON AMBIENT TEMPERATURE OF 30°C				IN RACEWAY OR CABLE, BASED ON AMBIENT TEMPERATURE OF 40°C		
SIZE	60°C	75°C	85°C	90°C	150°C	200°C	250°C
AWG OR Kcmil	TW* UF*	FEPW* RH* RHW* ZW* THW* THWN* XHHW* THHW* USE*	V	TA, TBS SA, RHH* SIS, FEP* FEPB* THHW* THHN* XHHW*	Z	FEP FEPB PFA	PFAH TFE NICKEL OR NICKEL-COATED COPPER
14	20*	20*	25	25*	34	36	39
12	25*	25*	30	30*	43	45	54
10	30	35*	40	40*	55	60	73
8	40	50	55	55	76	83	93
6	55	65	70	75	96	110	117
4	70	85	95	95	120	125	148
3	85	100	110	110	143	152	166
2	95	115	125	130	160	171	191
1	110	130	145	150	186	197	215
0	125	150	165	170	215	229	244
00	145	175	190	195	251	260	273
000	165	200	215	225	288	297	308
0000	195	230	250	260	332	346	361
250	215	255	275	290	–	–	–
300	240	285	310	320	–	–	–
350	260	310	340	350	–	–	–
400	280	335	365	380	–	–	–
500	320	380	415	430	–	–	–
600	355	420	460	475	–	–	–
700	385	460	500	520	–	–	–
750	400	475	515	535	–	–	–
800	410	490	535	555	–	–	–
900	435	520	565	585	–	–	–
1000	455	545	590	615	–	–	–
1250	495	590	640	665	–	–	–
1500	520	625	680	705	–	–	–
1750	545	650	705	735	–	–	–
2000	560	665	725	750	–	–	–

Overcurrent protection for conductor types marked (*) will not exceed 15 amperes for size 14 AWG. 20 amperes for size 12 AWG, and 30 amperes for size 10 AWG. (–) for dry locations only. See 75°C column for wet locations.

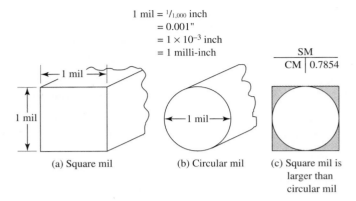

FIGURE 9–10
Comparison of a square mil and a circular mil.

The Circular Mil

Wire cross-sectional area is measured in a special unit called the *circular mil*. As shown in Figure 9–10, a mil is one thousandth of an inch. A circle with a diameter of 1 mil has an area of 1 circular mil.

The circular mil area of a round conductor is calculated by squaring the diameter in mils, as shown in Figure 9–11.

Formula 9–2

$$\text{Circular mil area} = (\text{Diameter in mils})^2$$

A square that measures 1 mil on all sides has an area of 1 square mil. It is important to understand that a square mil is a larger value than a circular mil, as shown in Figure 9–10(c). Formula Diagram 9–3 gives the relationship between the two units.

Formula Diagram 9–3

$$\frac{\text{Square mils}}{\text{Circular mils} \mid 0.7854}$$

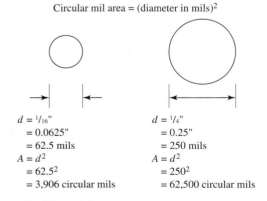

FIGURE 9–11
Calculating circular mil area.

Wire Gauge Systems

The American Wire Gauge *(AWG)* and kilo-circular-mil *(kcmil)* systems measure the size of conductors. Both systems are based on the circular mil.

The word *gauge* means size. Thus, a wire gauge system is a method of specifying wire size. There are two wire-gauge systems in common use, and both are based on the circular mil area of wire.

The American Wire Gauge System

The American Wire Gauge (AWG) is used for smaller wires, from the fine wires used in delicate electronic components to the medium-sized wires used to bring electrical power into homes and offices.

Like many gauge systems, the AWG uses dimensionless numbers to indicate size. A dimensionless number is one that has no units. For example, you may have a kitchen sink made of 18-gauge steel, or you may own a 12-gauge shotgun. These gauge numbers have no units such as inches, and so the numbers do not measure the actual thickness of the steel or the barrel of the shotgun. However, in any gauge system, some table of gauge standards exists, which defines the gauge sizes in exact dimensions.

Table 9–2 is a table from the National Electrical Code listing AWG gauge numbers and defining them in terms of circular mil area. For example, a #18 wire has a cross-sectional area of 1,620 circular mils, and a #16 wire has an area of 2,580 cm. It is, therefore, obvious that a #16 wire is larger than a #18. This feature is characteristic of most gauge systems: Larger sizes have smaller gauge numbers.

If smaller gauge numbers indicate larger wire sizes, what happens when we reach gauge #1? The AWG has an interesting solution to this problem. Wire sizes larger than #1 are designated by groups of zeros. In other words, #0 is a size larger than #1. The series of gauge numbers would be: #1, #0, #00, #000, and #0000. At this point, the AWG system stops.

As a shortcut, the zero gauges are written as follows: #0 = #1/0, #00 = #2/0, #000 = #3/0, #0000 = #4/0. The word *aught,* an Old English word for zero, is used in speaking these gauge numbers. Thus, #1/0 wire is spoken of as "one-aught" or "single-aught" wire; #2/0 is spoken of as "two aught" or "double aught"; and so on.

Although the AWG system is somewhat arbitrary, there is a pattern to it. A decrease of three gauge numbers approximately doubles the cross-sectional area. For example, #4 wire has an area of 41,740 cm, and #1 wire (three numbers smaller) has an area of 83,690 cm, approximately twice as large.

Table 9–2 includes data on the dc resistance of the various gauges. Since resistance depends on temperature and metal, three columns are given. The term *ohm/MFT* means ohms per thousand feet. Thus, #12 single-strand uncoated copper wire at 75°C has a resistance of 1.93 ohms per thousand feet of length.

Table 9–3 lists various AWG wire sizes and their typical uses in homes.

The Kilo-Circular-Mil (kcmil) System

The kilo-circular-mil (kcmil) system is used for larger conductors. The kcmil system begins where the AWG system ends. In Table 9–2, notice that the largest AWG wire, #4/0, has an area of 211,600 cm. The next wire size is 250 kcmil, which means an area of 250,000 cm. Thus, the table does not give cm areas for the kcmil wires, since the wire size is based on the area. For example, a 900-kcmil conductor has an area of 900,000 cm.

The kcmil system was formerly called the *MCM* system, and many people still speak of it this way. For many years, the term *mil* was used loosely in America to mean

TABLE 9–2
AWG and kcmil wire table.

CONDUCTOR PROPERTIES

SIZE AWG/ Kcmil	AREA CIR. MILS	CONDUCTORS				DC RESISTANCE AT 75°C (167°F)		
		STRANDING		OVERALL		COPPER		ALUMI-NUM
		QUAN-TITY	DIAM. IN.	DIAM. IN.	AREA IN.^{2}	UNCOATED ohm/MFT	COATED ohm/MFT	ohm/ MFT
18	1620	1	—	0.040	0.001	7.77	8.08	12.8
18	1620	7	0.015	0.046	0.002	7.95	8.45	13.1
16	2580	1	—	0.051	0.002	4.89	5.08	8.05
16	2580	7	0.019	0.058	0.003	4.99	5.29	8.21
14	4110	1	—	0.064	0.003	3.07	3.19	5.06
14	4110	7	0.024	0.073	0.004	3.14	3.26	5.17
12	6530	1	—	0.081	0.005	1.93	2.01	3.18
12	6530	7	0.030	0.092	0.006	1.98	2.05	3.25
10	10380	1	—	0.102	0.008	1.21	1.26	2.00
10	10380	7	0.038	0.116	0.011	1.24	1.29	2.04
8	16510	1	—	0.128	0.013	0.764	0.786	1.26
8	16510	7	0.049	0.146	0.017	0.778	0.809	1.28
6	26240	7	0.061	0.184	0.027	0.491	0.510	0.808
4	41740	7	0.077	0.232	0.042	0.308	0.321	0.508
3	52620	7	0.087	0.260	0.053	0.245	0.254	0.403
2	66360	7	0.097	0.292	0.067	0.194	0.201	0.319
1	83690	19	0.066	0.332	0.087	0.154	0.160	0.253
1/0	105600	19	0.074	0.373	0.109	0.122	0.127	0.201
2/0	133100	19	0.084	0.419	0.138	0.0967	0.101	0.159
3/0	167800	19	0.094	0.470	0.173	0.0766	0.0797	0.126
4/0	211600	19	0.106	0.528	0.219	0.0608	0.0626	0.100
250	—	37	0.082	0.575	0.260	0.0515	0.0535	0.0847
300	—	37	0.090	0.630	0.312	0.0429	0.0446	0.0707
350	—	37	0.097	0.681	0.364	0.0367	0.0382	0.0605
400	—	37	0.104	0.728	0.416	0.0321	0.0331	0.0529
500	—	37	0.116	0.813	0.519	0.0258	0.0265	0.0424
600	—	61	0.099	0.893	0.626	0.0214	0.0223	0.0353
700	—	61	0.107	0.964	0.730	0.0184	0.0189	0.0303
750	—	61	0.111	0.998	0.782	0.0171	0.0176	0.0282
800	—	61	0.114	1.03	0.834	0.0161	0.0166	0.0265
900	—	61	0.122	1.09	0.940	0.0143	0.0147	0.0235
1000	—	61	0.128	1.15	1.04	0.0129	0.0132	0.0212
1250	—	91	0.117	1.29	1.30	0.0103	0.0106	0.0169
1500	—	91	0.128	1.41	1.57	0.00858	0.00883	0.0141
1750	—	127	0.117	1.52	1.83	0.00735	0.00756	0.0121
2000	—	127	0.126	1.63	2.09	0.00643	0.00662	0.0106

TABLE 9–3
AWG sizes and applications.

AWG	TYPICAL USES
24–18	Speaker Wire Audio/Stereo Cables Doorbell & Thermostat
18–16	Indoor Extension Cords
16–12	Outdoor Extension Cords
14–12	Interior 120V Outlets
10–6	Interior 240V Outlets
4	Grounding Wire
1–3/0	200 AMP Service Conductors

either thousand (1,000) or thousandth (0.001). This naturally caused confusion, especially in the term *MCM,* which meant mil circular mil (thousand circular thousandths). To eliminate this confusion, the term *mil* now means only thousandths, similar to the metric prefix *milli,* and the MCM system was renamed as kcmil.

The following examples make use of what we have just learned about ampacity and gauge systems.

● *SKILL-BUILDER 9–1*
- -

What is the circular mil area of a solid wire with a diameter of 1/16"?

GIVEN:
  $d = 1/16''$

FIND:
  A

STRATEGY:
Convert 1/16 to a decimal fraction. Convert the decimal fraction to mils. Square the diameter in mils.

SOLUTION:
  $d = 1/16''$
    $= 0.0625''$ 1 divided by 16
    $= 62.5$ mils   move decimal three places
  $A = d^2$ Formula Diagram 9–2
    $= 62.5^2$
    $= 3,906$ cm

Repeat the Skill-Builder using the following values. You should obtain the results that follow.

Given:

Diameter	3/32"	9/64"	1/32"	5/64"	7/32"

Find:

Diameter	0.09375"	0.140625"	0.03124"	0.078125"	0.21875"
Diameter	93.75 mils	140.63 mils	31.25 mils	78.13 mils	218.75 mils
Area	8,789 cm	19,775 cm	977 cm	6,104 cm	47,852 cm

- -

● *SKILL-BUILDER 9–2*

What is the circular mil area of a #10 conductor?

GIVEN:
   Conductor size = #10 AWG

FIND:
   Circular mil area

STRATEGY:
   Use Table 9–2 to determine the cm area.

SOLUTION:
   A #10 conductor has a cm area of 10,380 cm.

   Repeat the Skill-Builder using the following values. You should obtain the results that follow.

Given:

AWG	16	8	14	1/0	6

Find:

Area	2,580 cm	16,510 cm	4,110 cm	105,600 cm	26,240 cm

Solid and Stranded Conductors

Wire may be solid (a single large conductor) or stranded (many small conductors twisted together). Solid wire is generally preferred for electronic hookup, breadboarding, and ac power wiring in the walls of buildings.

Stranded wire is preferred when flexibility is important. Cords are combinations of two or three conductors made of stranded wire, insulated separately and either bonded together or enclosed in a plastic or rubber sheath. Cords are used to conduct ac power to equipment and thus are commonly called either power cords or extension cords.

Magnet Wire

As we will learn in a later chapter, a current flowing in a conductor produces a magnetic field. This principle is the basis for many electrical machines, including generators, motors, transformers, relays, and solenoids. These machines contain large amounts of wire wrapped into coils or windings.

The wire used for coils or windings is called magnet wire. It is not magnetic itself. The name refers to the way it will be used. Magnet wire is insulated with a thin coating of varnish, which gives it a dark red color. When making connections with magnet wire, you must scrape the varnish off the ends to expose the bare copper.

Cables

A combination of two or more conductors is a *cable,* although the term is often used for large stranded single conductors as well.

Figure 9–12 illustrates several types of cable. Twisted-pair cable consists of even numbers of ordinary conductors encased in a common plastic sheath, as shown in Figure 9–12(a).

Shielded cable is shown in Figure 9–12(b). The particular type of shielded cable shown here is called coaxial cable, or simply coax (pronounced co-ax). You are probably

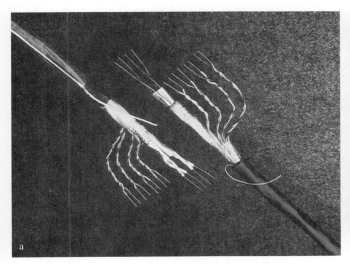

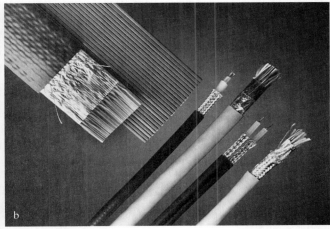

Belden

TV and Amateur Radio Cables

Maxi-Color® TV Lead-In Wire

Part No. 9085

Product Description: Bare copper-covered steel, 2 conductors parallel. Foam Brown polyethylene oval jacket.
Recommended for use in FM radio and VHF, UHF, Color, and Black/White television installations in uncongested or fringe areas where interference is not a problem. Designed to deliver more signal to the receiver under adverse environmental conditions. The Brown foam polyethylene insulation provides the major portion of the signal energy fields around the conductors and resists ultraviolet degradation.

300 ohms
80°C
c

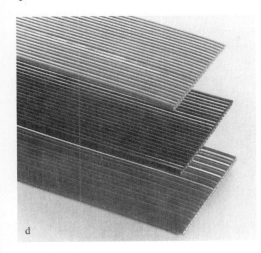

FIGURE 9–12
Cables: (a) twisted pair, (b) shielded coaxial, (c) twin-lead, and (d) ribbon.

familiar with coax, since it is used in cable television and VCR connections. The solid inner conductor carries the signal and the woven metallic braid or aluminum foil jacket acts both as the return path for signal current and also as the shield.

As we will learn in a later chapter, magnetic fields can create voltages in wires. Our environment is full of magnetic fields created by radio and television transmitters, and by the current flowing in power lines. These magnetic fields create unwanted voltages and currents in signal wires. The unwanted currents are called noise because they produce distortions in the audio or video signal.

In coaxial cable, the noise currents are created in the shield, not in the signal wire. The shield is grounded to carry away the noise currents. Thus, the signal wire is free of noise and distortion.

Figure 9–12(c) shows twin-lead cable, widely used as antenna wire for television and FM radio receivers. Twin lead is unshielded so the magnetic fields of the transmitters can create voltage in it. Of course, the twin lead also picks up noise from unwanted magnetic fields as well.

Figure 9–12(d) shows ribbon cable, a flat type of unshielded cable widely used in computers and related equipment. Special connectors allow this cable to be easily connected to metal strips on the edges of printed circuits, as shown. The removable printed circuits used in computers are called cards or boards. The connectors used with ribbon cable are called edge-card connectors.

Soldering

Permanent electrical connections are often made by soldering. *Solder* (pronounced "sodder") is an alloy of lead and tin that melts at temperatures below 800°F and adheres to the surface of other metals, such as copper, gold, and silver. Solder joints have low mechanical strength and serve only to provide an electrical connection.

The surfaces to be soldered must be completely clean. All surface dirt and grease should be removed with a solvent. Surface oxidation may be removed by lightly scraping the surfaces with a blade or rubbing them with emery cloth. The parts are joined mechanically by inserting, bending, or twisting them, as appropriate. Heat is then applied with a soldering gun or iron. A liquid or paste cleaning agent called *flux* is applied to the heated surfaces to remove any remaining layers of oxide, and to allow the solder to flow freely into the joint. The hot flux produces a slight smoke. The solder is then applied to the heated surfaces, where it melts into the surfaces and cools to form a solid connection.

Soldering is used in plumbing, jewelry making, and sheet-metal work as well as in electrical work. Each of these applications uses its own kind of solder and flux. You must use the correct kind of solder and flux for electrical work. Other types can damage the components and circuits. The solder for electrical work is a mixture of lead and tin. The flux for electrical work is rosin, an organic material derived from pine trees.

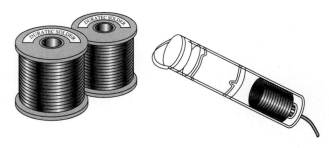

FIGURE 9–13
Solder.

Electrical solder is usually made in a form resembling wire and is sold in spools or dispensers, as shown in Figure 9–13. Electronic solder is rosin core, which means the solder contains the rosin flux inside itself in hollow tunnels, as shown in Figure 9–14. Rosin-core solder is convenient, since the flux applies itself during the soldering operation. However, solder is also available in solid-core form, with the flux in a separate dispenser.

The composition of the solder is indicated by a pair of numbers. For example, 60/40 solder is 60% lead and 40% tin. A *eutectic alloy* is an alloy that has the lowest possible melting temperature. The eutectic alloy of solder is 63/37. Electronic solder is available in alloys of 40/60, 50/50, 63/37, and 60/40.

An important characteristic of eutectic solder is that it changes physical state directly between solid and liquid. The other alloys have an intermediate semisoft state. If the component leads or wires move while the solder is cooling and in the semisoft state, the resulting joint will be weak and unacceptable. Therefore, eutectic solder is used when it is critical that no movement occur when cooling down.

Figure 9–15 shows a variety of tools and accessories used for soldering. Soldering guns and irons are rated in watts, indicating the amount of heat they can produce. Soldering guns, shown in Figure 9–15(a), are rated from 75 W to over 200 W and are used to solder large wires and components. They develop tip temperatures of over 1,000°F and should not be used on delicate electronic components. Soldering irons, shown in Figure 9–15(b), are used for working with small components on printed circuit boards. These irons have heating elements which range from 15 W to 55 W and produce tip temperatures between 600°F and 850°F. Soldering irons may have interchangeable handles, heating elements, and tips for various jobs. A new tip must always be tinned (coated with solder) before it is used.

A holder for a soldering iron is shown in Figure 9–15(c). The holder includes a small sponge, which is kept wet with water. Each time the soldering iron is taken from the holder, its tip is wiped against the sponge to remove the crust of oxide that forms on the hot tip from exposure to air.

The tools shown in Figure 9–15(d) are used to scrape, file, and clean leads and terminals before soldering. An important aid to soldering is the heat sink, shown enlarged in Figure 9–15(e). This is a simple aluminum clip placed on the lead of a delicate compo-

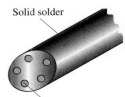

Solid solder

Hollow tunnels filled with rosin flux

FIGURE 9–14
Strand of rosin-core solder.

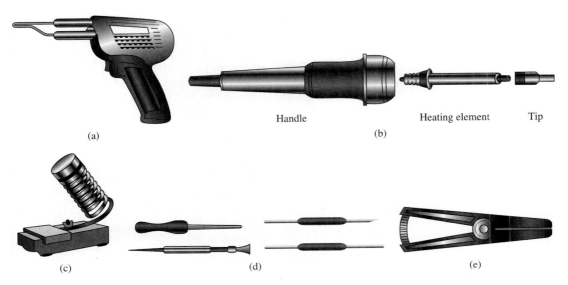

Handle   Heating element   Tip

(a)   (b)

(c)   (d)   (e)

FIGURE 9–15
Soldering equipment: (a) gun; (b) iron: holder, heating element, tips; (c) stand with sponge; (d) scrapers, picks, brushes; (e) heat sink.

nent between the body of the component and the tip of the soldering iron. Its purpose is to absorb excess heat from the iron and thus protect the component during soldering.

In repair work, it is often necessary to desolder (remove solder). Figure 9–16 shows a variety of desoldering equipment, including a solder bulb (a) and solder pump (b). These are used to suck up melted solder. Solder wick (c) is a mesh or braid of fine copper wire coated with a material which absorbs melted solder from a heated joint. A desoldering iron (d) has the bulb mounted to the iron, and a desoldering station (e) uses a vacuum pump to remove melted solder.

After both soldering and desoldering, solvents such as denatured alcohol are applied to the joint to dissolve any flux residue, which is then brushed away.

Characteristic Impedance

The *characteristic impedance* of a conductor or cable is its overall opposition to current, expressed in ohms. Impedance includes resistance plus other electrical characteristics, which we will study in later chapters.

Twin-lead cable has a characteristic impedance of 300 Ω. Coaxial cable used for television and VCRs has a characteristic impedance of 75 Ω. There are other varieties of coaxial cable with other impedances, such as 50-Ω coax.

Characteristic impedance cannot be measured with an ohmmeter. It is determined by laboratory tests using special instruments. However, it is important to match the impedance of the cable to the impedance of the terminals. For example, a television

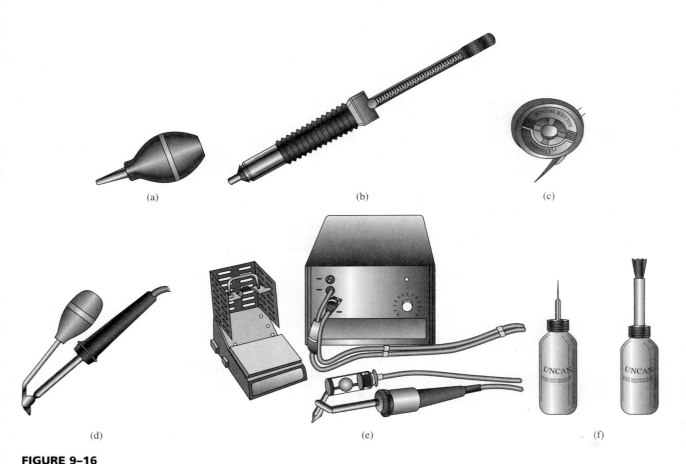

(a)   (b)   (c)

(d)   (e)   (f)

FIGURE 9–16
Desoldering equipment: (a) solder bulb, (b) solder sucker, (c) solder wick, (d) desoldering iron, (e) desoldering station with vacuum pump, (f) flux solvent dispenser and brush.

receiver has at least two sets of input terminals: one set for 300-Ω twin-lead cable and another for 75-Ω coax. If you need to change from one impedance to the other, you can use an impedance-matching transformer, available in stores that sell video accessories.

Power Loss in Transmission Lines

Because wire and cable have impedance, current flowing in transmission lines produces a power loss, which is undesirable. In the case of electric utility company power lines, which furnish large amounts of power to consumers, the power losses can be enormous, as illustrated in the following example.

Suppose that an electric utility company has 1,500 kcmil aluminum transmission lines between its generating station and a subdivision of homes 30 miles away. The subdivision consists of 200 homes, with an average power demand of 12 kilowatt per home. If the utility company decides to transmit the power at 240 V, what will be the power loss in the transmission lines?

The total power requirement is 12 kW times 200 homes, which equals 2.4 MW. If the voltage in the lines is 240 V, then the current is 2.4 MW divided by 240 V, which equals 10,000 A.

The resistance of the 1,500-kcmil aluminum lines is 0.0141 Ω per thousand feet, from Table 9–2. A mile is 5,280 feet, and so the resistance of 1 mile of line is 5.28 times 0.0141 Ω, which equals 0.074448 Ω per mile. The lines are 30 miles long, and so the total resistance is 30 times 0.074448 Ω, which equals 2.233 Ω.

Now we have a problem. If the lines have a resistance of 2.233 Ω and the current in the lines is 10,000 A, then according to Ohm's Law the voltage drop in the lines is 2.233 Ω times 10,000 A, which equals 22,330 V. This means that the generator must produce 22,570 V to have 240 V reach the customer. This doesn't make sense.

There is a bigger problem. According to Watt's Law, power equals resistance times the square of current: $P = I^2R$. Applying this formula to the lines, we get $(10,000 \text{ A})^2$ times 2.233 Ω, which equals 223.3 MW. This means the light company must burn enough fuel to produce over 223 million watts of power just to sell 2.4 million watts to its customers. This is not practical. No one could pay that much money for electricity!

Remember that it is current that causes voltage drops and power losses in lines. The only way we can send a large amount of electrical power across a long distance is to keep the current so low that the line loss is not a problem. This is why the voltage in a power line is so high: to keep the current low and still provide a large amount of power.

Suppose the utility company raises the voltage on the transmission lines to 345,000 V (345 kV). By repeating the previous calculations, we find that the current is now 6.96 A, the voltage drop in the lines is 15.5 V, and the power loss is 108 W. Obviously, these drops and losses are acceptably small.

From this example you can see that in power transmission lines, when the distance is long and the current high, even a small amount of resistance causes tremendous voltage and power losses. This is why the main goal of all power distribution systems is to keep the current as low as possible by keeping the voltage high, right up to the final transformer at the customer's location.

→ *Self-Test*

- - - - - - -

4. What is ampacity?
5. What is a circular mil?
6. What are the AWG and kcmil systems?
7. How does soldering create an electrical connection?
8. What is characteristic impedance?
9. Why do power transmission lines carry power at high voltage?

SECTION 9–3: Insulators

In this section you will learn the following:

→ When voltage is applied across any insulator, a small leakage current flows through the insulator.

→ The dielectric strength of an insulator is the voltage required to produce dielectric breakdown in a layer of the material of a specific thickness.

An insulator is a material whose molecules do not readily give up or receive electrons. The most common electrical insulators are various forms of rubber, plastic, glass, or ceramic.

Oil is a good liquid insulator and is often used in high-voltage equipment, where it also serves as a coolant. In equipment such as a large transformer, the oil is pumped through the equipment and then through radiators with fans to release the heat, similar to the cooling system of an automobile engine.

Pure water is a fairly good insulator. However, many chemicals such as salt dissolve into ions in water. The ions act as charge carriers, and the water solution becomes a good conductor. Tap water in the home contains chlorine and other ions, and sweat on the skin contains various organic salts dissolved in ionic forms. Rainwater and seawater contain various ions as well. Thus, water in general is considered a conductor and a safety hazard to electrical equipment.

Wood and paper are good insulators when dry. However, because wood is an organic material, it absorbs water from the air. Thus, wood is an unreliable insulator.

Many rocks are good insulators, particularly a crystalline rock called mica, which is used in electronic components.

Leakage Current

Because insulator materials do contain a few charge carriers, when a voltage is applied across an insulator, a small amount of current will pass through the insulator. This is called *leakage current.* The leakage current is extremely small, perhaps a few microamps or nanoamps. Therefore, leakage current is usually ignored for practical purposes. However, in the case of biomedical equipment or sensitive, high-performance circuits, leakage currents may become important.

The word *dielectric* is the technical term for an insulator. When a voltage is applied across an insulator, the molecules are stressed and distorted as their valence electrons respond to the electrical field. The distortion of the molecules, called dielectric stress, absorbs and stores energy from the electric field. Thus, the insulator acquires potential energy, which can be released into the circuit at a later time.

Dielectric Strength

If the electric field is strong enough, electrons are ripped off the valence shells. This breaks down the chemical bonds within the material and causes it to crack. An electric arc (a spark) often passes through the material, producing enough heat to melt the material in the region of the crack. The *dielectric strength* is the voltage needed to break down the insulator. Dielectric strength is measured in volts per mil. Table 9–4 lists the dielectric strength of several materials.

TABLE 9–4
Dielectric strength.

MATERIAL	DIELECTRIC STRENGTH (VOLTS/MIL)
Air	80
Oil	375
Ceramic	1000
Paper (paraffined)	1200
Teflon®	1500
Mica	1500
Glass	2000

● *SKILL-BUILDER 9–3*
-- --

How much voltage is necessary to produce an arc across a 1/4″ air gap?

GIVEN:
$$\text{Gap} = 1/4''$$
$$\text{Dielectric} = \text{Air}$$

FIND:
$V_{\text{breakdown}}$

STRATEGY:
Determine the gap in mils. Multiply by the dielectric strength of air given in Table 9–4.

SOLUTION:
Gap = 1/4″
= 0.25″
= 250 mils
Dielectric strength for air
= 80 V/mil   from Table 9–4
$V_{\text{breakdown}} = 80 \times 250$
= 20 kV

Repeat the Skill-Builder using the following values. You should obtain the results that follow.

Given:

Gap	5/8″	3/16″	1/2″	9/32″
Dielectric	Mica	Glass	Oil	Ceramic

Find:

Gap	625 mils	188 mils	500 mils	281 mils
Strength	1.5 kV/mil	2 kV/mil	0.375 kV/mil	1 kV/mil
Breakdown	937.5 kV	375 kV	187.5 kV	281.25 kV

You may be surprised at how much voltage it takes to arc across a 1/4-inch air gap. The sparks you occasionally see when you plug in or turn on 120-V equipment occur when the contacts are about 1.5 mils apart, practically touching.

→ *Self-Test*
- - - - - - -
10. What is leakage current?
11. What is dielectric strength?
- -

SECTION 9–4: Circuit Breakers and Residential Wiring

In this section you will learn the following:

> → A residential wiring system consists of two hot wires, a neutral wire, a grounding wire, and a variety of circuit breakers, ground fault circuit interrupters, and surge suppressors.

Circuit Breakers

Circuit breakers are widely used as overcurrent protection devices in buildings. The panel and enclosure containing the circuit breakers is called a load center or service panel, and nicknamed a breaker box. A typical residential load center is shown in Figure 9–17.

A main breaker controls the entire load center. It disconnects the utility company's supply lines from the *buss bars* inside the load center. The buss bars are bare metal strips which act as conductors. The word *service* is used to describe the current rating of the main breaker and the buss bars. For example, a 200 amp service means a 200 amp main breaker installed in a load center whose buss bars are also rated at 200 amps.

The branch breakers are the numerous smaller breakers arranged in a double row below the main breaker. Figure 9–17 shows one branch breaker mounted on the right buss. The main breaker feeds power into the buss bars, and the branch breakers snap onto the buss bars to distribute power to branch loads throughout the building. A large machine or appliance such as a range or hot water heater may be the only load on its branch and thus may have its own branch breaker. Other branches may provide power to several lights and wall outlets. In this case, loads in several rooms may be on the same branch breaker.

The rating of the circuit breaker is printed on its case. Typical ratings for branch breakers are 15 A, and 20 A to 70 A in 10-A increments. The breaker will trip (turn off the

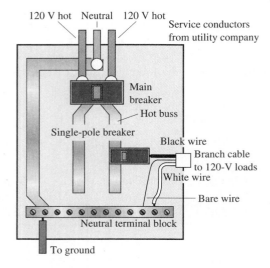

FIGURE 9–17
Residential load center, 120-V loads.

current) when the current value exceeds 125% of the breaker's rating for more than two or three seconds.

Circuit breakers contain both thermal and magnetic tripping devices. The thermal cutout device trips the breaker if the temperature of the breaker exceeds a certain value, presumably because of a current overload. The magnetic cutout device responds to extremely high currents caused by shorts and trips instantly.

The handle of a circuit breaker has three positions. The on position is toward the buss bar. The manual off position is away from the buss bar. Thus, the circuit breaker can be used as a branch switch to disconnect power before working on the branch. If the circuit breaker trips due to an overload, the handle goes to a middle off position.

The three-position design allows you to determine whether a breaker has tripped or has been manually turned off. If the breaker has tripped and the overload has been corrected, the handle should be moved to the manual off position, then back to the on position.

Circuit breakers deteriorate with age and after several years will begin to trip irregularly at currents below the rated value. This is more likely to happen if the breaker panel is outside and exposed to sunlight for most of the day. Often the plastic case of the breaker will show signs of physical deterioration. When this happens, the breaker should be replaced.

When replacing circuit breakers, you should know the name of the manufacturer. Breakers by different manufacturers are generally not interchangeable. Some manufacturers make both full-size and half-size breakers in certain current ratings. The purpose is to allow more breakers in the load center.

Many appliances and power strips contain miniature circuit breakers called a reset button.

Residential Wiring

The wiring system used for homes and small commercial buildings is called residential wiring. Three supply wires enter the load center and three buss bars are used. One of the buss bars is grounded into the earth through metal water pipes or a special copper rod. Therefore, the buss potential is zero volts, and the buss is called neutral. The wire connected to this buss is called the *neutral wire*. The other two buss bars are called hot because they carry a voltage relative to ground. The *hot wires* are connected to these busses.

In residential wiring, the voltage between neutral and either hot buss is between 110 V and 125 V. Ordinary light fixtures and wall outlets are connected between neutral and one of the hot busses. The branch circuit breaker must be installed on the hot wire. Since a 120-V branch has only one hot wire, each branch uses a single-pole breaker that connects to only one buss bar, as shown in Figure 9–17.

The voltage between both hot busses is between 220 V and 250 V. Large appliances such as ranges, ovens, water heaters, and large air conditioners operate at this higher voltage. The neutral wire is not used. Instead, the load is connected between the two hot wires.

Therefore, 240-V circuits require either two single-pole breakers, as shown in Figure 9–18, or else a double-pole breaker, shown in Figure 9–19. A double-pole breaker contains two breakers in the same case with two handles tied together. Double-pole breakers are designed so one breaker connects to one hot buss, and the other breaker to the other hot buss. The buss-bar extension shown in Figure 9–19 is highly simplified and exaggerated. Actual buss bars are more complex in shape, allowing both single-pole and double-pole breakers to be intermixed freely on either side of the panel.

Some appliances require both 120 V and 240 V, such as a clothes dryer that uses 120 V for the motor and 240 V for the heater element. In this case, a double-pole breaker will be used and a neutral wire also supplied to the appliance. Inside the appliance, the 240-V heater element will be connected between the two hot wires, and the 120-V motor between either hot wire and the neutral.

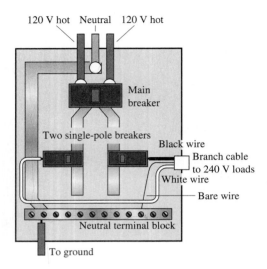

FIGURE 9–18
240-V load on two single-pole breakers.

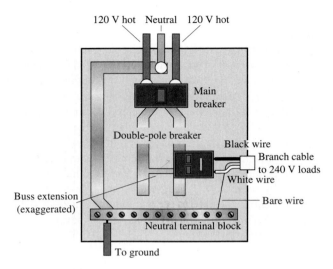

FIGURE 9–19
240-V load on one double-pole breaker.

Notice that in all three illustrations a bare wire is connected to the neutral terminal block. This is the *grounding wire*. It is connected to all exposed metal components in the wiring system, such as conduit, armored cable, junction boxes, and ceiling fixtures. Through the prong in power cords, it is also connected to the metal chassis of many loads, such as power tools.

The purpose of a grounding wire is to provide a path to ground in case of a short circuit, which causes the circuit breaker to open immediately. To understand the function of the grounding wire, we must think through several situations. The insulated white wire is the neutral wire, and the insulated black wire is the hot wire. These wires are the normal current path for all loads. Figure 9–20 shows a hand drill operating normally, drawing 3 A on these two conductors.

Now suppose the drill has a metal case, and suppose that inside the case, the hot wire comes into contact with the metal case. This can easily happen, as insulation deteriorates with age and wires are flexed with constant use. As shown in Figure 9–21, no current would flow, since there is no current path between the hot wire and neutral.

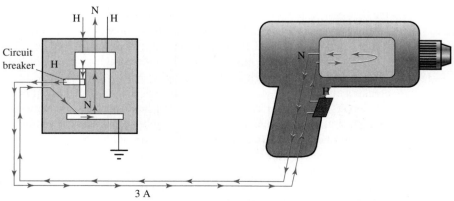

Two-conductor wiring. Current path during normal operation.

FIGURE 9–20
Normal current path.

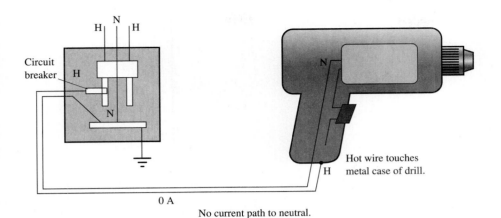

FIGURE 9–21
Shorted hot wire to metal case of drill.

Figure 9–22 illustrates the danger. If a person touches the metal case, this may create a current path to neutral through the person's body and through ground. For example, the person may be working outside and be in contact with the earth, or working inside and touch a metal object that is connected to ground through a pipe or wire. The small amount of current that would flow through the person's body would not trip the 15-A circuit breaker but would injure or kill the person.

If we add a grounding wire to the circuit, the normal operation of the drill does not change. The normal current path is still through the hot and neutral wires, as shown in Figure 9–23. However, if the hot wire shorts to the metal case, the grounding wire provides an immediate current path to neutral, as shown in Figure 9–24. With a resistance of nearly zero, the current is extremely high, which trips the circuit breaker, cutting off all current to the drill. Even if the person is touching the metal case at the moment the short occurs, no current can flow through the person because the grounding wire places the metal case at a potential of zero volts and carries the current of the short, as shown in Figure 9–25.

The ground-fault circuit interrupter *(GFCI)* is a modern improvement to wiring safety. The purpose of a ground-fault circuit interrupter is to open the hot wire if any current is returning to ground by any path other than the neutral wire.

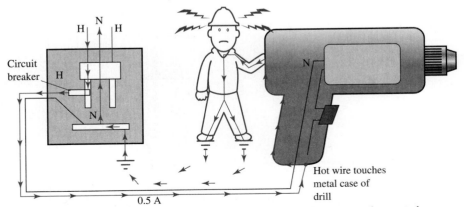

A person touching the metal case of the drill may complete the current path to neutral through ground. Low current does not trip circuit breaker. Person may be electrocuted.

FIGURE 9–22
Electrocution hazard.

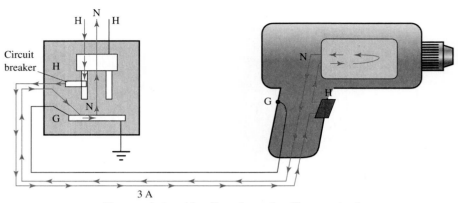

Three-conductor wiring. Normal operation. The current path
is the same as with two-conductor wiring.

FIGURE 9–23
Normal current path.

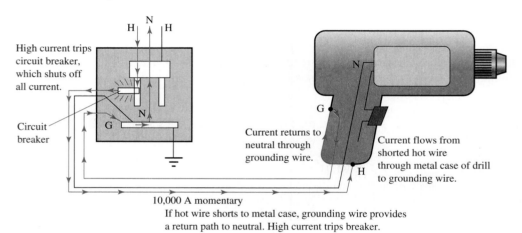

Current returns to
neutral through
grounding wire.

Current flows from
shorted hot wire
through metal case of drill
to grounding wire.

10,000 A momentary

If hot wire shorts to metal case, grounding wire provides
a return path to neutral. High current trips breaker.

FIGURE 9–24
Short circuit finds a current path through grounding wire.

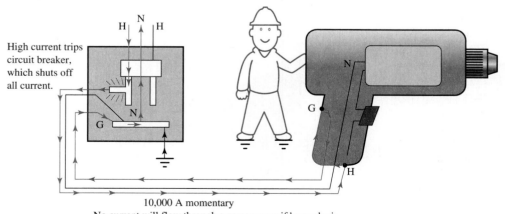

10,000 A momentary

No current will flow through a person even if he or she is
touching the drill at the moment the short occurs.

FIGURE 9–25
User is safe from shorted drill.

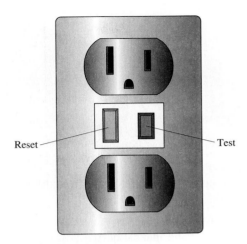

FIGURE 9–26
Ground fault circuit interrupter (GFCI)
receptacle.

Figure 9–26 shows a duplex receptacle with a built-in GFCI. The GFCI is a minia-ture circuit breaker controlled by a solid-state electronic module which, in the case of a 120-V circuit, compares the currents in the hot wire and the neutral wire. Under normal conditions, the two currents should be exactly the same, since the neutral wire is the nor-mal return path for current. However, if some of the current in the hot wire returns to ground through another path, the two currents will be different. This is considered a ground fault and is a potentially dangerous condition. If the GFCI detects a difference in the hot and neutral currents, it trips and shuts off the current.

To understand the value of a GFCI, suppose that a person is working in the yard with an electric tool connected to an outdoor extension cord. A thin film of water, perhaps from dew, rain, or sweat, has formed on the handle of the tool, along the extension cord, and down into the plug and receptacle, where the water creates a current path between the hot wire and the person's hand.

The person's body has a resistance of perhaps 1,200 Ω, and the 120-V line pro-duces a current of 0.1 A through the body and back through the earth to the copper rod connected to the neutral buss bar. If a current of 0.1 A passes through a person's chest for more than one second, the person will probably die.

Now suppose the power tool is operating normally and drawing perhaps 10 A. The branch breaker is rated at 20 A, and the total current in the hot wire is 10.1 A. The branch breaker will stay on, the tool will run, and the person may die. The grounding wire offers no protection here because there is no short between it and the hot wire.

A GFCI would save the person's life. If a GFCI detects a difference of 5 mA between the currents in the hot and neutral wires, it trips within 15 milliseconds and shuts off all current in the hot wire.

In many communities, GFCI receptacles are now required by law in kitchens and bathrooms of new buildings. Only one GFCI receptacle is needed in the room if the incoming hot and neutral wires are connected to the GFCI first, and then go to all the other receptacles. GFCI branch breakers are also available, as well as GFCI-equipped receptacles to be used with outdoor extension cords.

Plugs and Receptacles

Figure 9–27 illustrates the two most common 120-V plugs. These plugs are designed to go into the receptacle in only one way. This design is called a polarized plug. In this case,

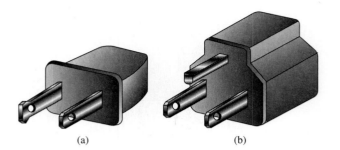

FIGURE 9–27
Polarized plugs: (a) wide neutral blade, (b) grounding prong.

the term *polarized* describes the mechanical design of the plug and receptacle and does not refer to electrical polarity (positive and negative terminals).

Many small tools and appliances are constructed with double-insulated plastic bodies to prevent the possibility of the hot wire shorting to the case. These devices will have a two-conductor plug, as shown in Figure 9–27(a). The plug is polarized by making the neutral prong wider than the hot prong. Devices with metal cases use the three-prong plug shown in Figure 9–27(b). The half-round prong is the grounding wire. This prong provides polarization, so the neutral and hot prongs are made the same width.

Many older houses still have wall outlets with two-conductor receptacles. If you must use a three-prong plug in such receptacles, use a two-to-three prong adapter, an inexpensive device available in any hardware store. The adapter should be mounted onto the receptacle using the screw on the cover plate.

Do not cut off the grounding prong on the plug. This would defeat the protection of the grounding wire and eliminate the polarization of the plug, allowing someone to insert the plug incorrectly in the receptacle. This could place the hot wire on the metal case of the appliance, a very dangerous condition.

Figure 9–28 shows the various receptacles commonly found in residential, commercial, and light industrial buildings. The purpose of the various receptacle types is to identify the voltage and current available, preventing improper connections. Notice that receptacles (e) and (f) have no grounding wire.

The t-slot arrangement in receptacles (b) and (d) causes confusion. The 120-V t-slot is on the neutral wire and simply identifies the receptacle as a 20-amp device instead of the standard 15-amp receptacle. The 240-V t-slot is on one of the hot wires and allows the same receptacle to be used with either a 15-A or 20-A plug.

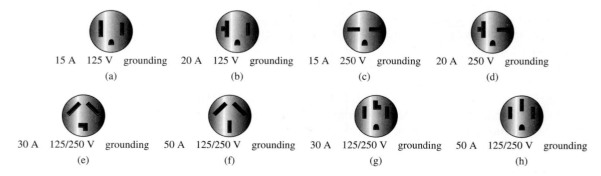

15 A   125 V   grounding   20 A   125 V   grounding   15 A   250 V   grounding   20 A   250 V   grounding
        (a)   (b)   (c)   (d)

30 A   125/250 V   grounding   50 A   125/250 V   grounding   30 A   125/250 V   grounding   50 A   125/250 V   grounding
        (e)   (f)   (g)   (h)

FIGURE 9–28
Receptacle types.

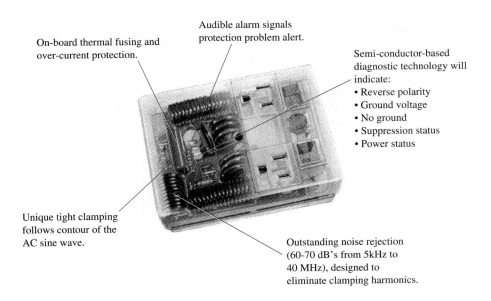

On-board thermal fusing and over-current protection.

Audible alarm signals protection problem alert.

Semi-conductor-based diagnostic technology will indicate:
• Reverse polarity
• Ground voltage
• No ground
• Suppression status
• Power status

Unique tight clamping follows contour of the AC sine wave.

Outstanding noise rejection (60-70 dB's from 5kHz to 40 MHz), designed to eliminate clamping harmonics.

FIGURE 9–29
Surge suppressor.

Surge Suppressors

A surge is a momentary increase in voltage. Surges often occur in utility company power lines for a number of reasons, including lightning strikes, transformer malfunctions, problems at generating stations, and industrial customers on the line who turn on and off large equipment such as furnaces and welders. The voltage surges may last for only a few milliseconds, but during that time the voltage on the power line entering an office or residence may reach between 6,000 V and 10,000 V, or even higher. The power switches in consumer electronic equipment are usually rated at 600 V. This means that a high-voltage surge may arc across the open contacts of the power switch and turn the machine on momentarily. During this brief moment, the high voltage on the power line produces high voltages inside the equipment, which may destroy sensitive transistors and integrated circuits.

Surge suppressors are connected between the equipment and the power line. Usually, surge suppressors plug directly into the wall outlet receptacle and then provide other receptacles for the equipment. Circuits inside the suppressor react quickly to voltage surges, limiting the maximum surge voltage to a few hundred volts, which is not enough to arc the power switch and damage the equipment. Some suppressors use the magnetic reaction of a large coil to suppress the surge. Others use solid-state components which clamp (limit) the surge voltage to safe levels.

The suppressor should be matched to the equipment. The manufacturer provides information on the types of equipment the suppressor can protect. Many suppressors also provide connections and circuits to suppress voltage surges on cable television lines and telephone lines, which may be used either for voice or data (fax machines and computer modems). Figure 9–29 illustrates a duplex surge suppressor intended to be plugged into a wall outlet.

→ *Self-Test*

12. Describe a residential wiring system.

Formula List

Formula Diagram 9–1

$$\frac{\text{Ampere-hour rating}}{6 \mid \text{Maximum continuous current}}$$

Formula 9–4

$$\text{Circular mil area} = (\text{Diameter in mils})^2$$

Formula Diagram 9–3

$$\frac{\text{Square mils}}{\text{Circular mils} \mid 0.7854}$$

● TROUBLESHOOTING: Applying Your Knowledge

Since the water heater is a 240-V appliance, it requires either one double-pole circuit breaker or two single-pole circuit breakers. You ask your neighbor what arrangement of breakers his heater has and find out that he doesn't know. He tells you he saved the old circuit breaker, and when he brings it to you, you notice it has only one handle and one screw terminal, identifying it as a single-pole breaker. You go to his load center, trace the wires, and find that the other circuit breaker is in the tripped position. It too shows signs of age. You suggest that your neighbor replace this breaker also. When he does, the water heater comes on.

The original trouble was probably that both circuit breakers had deteriorated with age and were tripping below their rated current. The old water heater probably did not need to be replaced.

CHAPTER SUMMARY: ANSWERS TO SELF-TESTS

1. A cell is a dc voltage source consisting of two electrodes immersed in an electrolyte.
2. A battery is a group of cells connected in series and/or parallel.
3. An ampere-hour rating is the continuous discharge current of a battery multiplied by a six-hour discharge cycle.
4. The ampacity of a conductor is the maximum current the conductor may carry under specific conditions according to the National Electrical Code.
5. A circular mil is the cross-sectional area of a round conductor with a diameter of 1 mil (0.001 inch).
6. The American Wire Gauge (AWG) and kilo-circular-mil (kcmil) systems measure the size of conductors. Both systems are based on the circular mil.
7. Soldering creates permanent electrical connections by melting a tin/lead alloy into the surfaces of conductors.
8. The characteristic impedance of a wire or cable is its overall opposition to current, expressed in ohms.
9. Power transmission lines carry power at high voltage to minimize power losses in the lines.
10. When voltage is applied across any insulator, a small leakage current flows through the insulator.
11. The dielectric strength of an insulator is the voltage required to produce dielectric breakdown in a layer of the material of a specific thickness.
12. A residential wiring system consists of two hot wires, a neutral wire, a grounding wire, and a variety of circuit breakers, ground-fault circuit interrupters, and surge suppressors.

CHAPTER REVIEW QUESTIONS

Determine whether the following questions are true or false.

9–1. A primary cell can be used only once.

9–2. The National Electrical Code is a federal law.

9–3. A #12 wire is smaller than a #10 wire.

9–4. Magnet wire is insulated with a thin coating of varnish.

9–5. Denatured alcohol is used as a flux solvent.

9–6. A large transformer may be filled with oil.

9–7. If a computer is properly grounded, a surge suppressor is unnecessary.

Select the response that best answers the question or completes the statement.

9–8. A battery is recharged by
   a. replacing the chemicals.
   b. replacing the electrodes.
   c. forcing current to flow backward.
   d. applying forward voltage.

9–9. A lead-acid battery is likely to be used in
   a. a portable computer.
   b. a flashlight.
   c. an automobile.
   d. a calculator.

9–10. A lithium battery is likely to be used in
   a. a portable computer.
   b. a flashlight.
   c. an automobile.
   d. a calculator.

9–11. A carbon-zinc cell is likely to be used in
   a. a portable computer.
   b. a flashlight.
   c. an automobile.
   d. a calculator.

9–12. A photoelectric cell is likely to be used in
   a. a portable computer.
   b. a flashlight.
   c. an automobile.
   d. a calculator.

9–13. Which of the following is *not* a factor in ampacity?
   a. the conductor material
   b. the way the conductor is mounted or installed
   c. the dielectric strength of the conductor
   d. the air temperature around the conductor
   e. the electrical insulation around the conductor
   f. the cross-sectional area (size) of the conductor

9–14. A conductor has a cross-sectional area of 400,000 circular mils. Its kcmil gauge is
   a. 632
   b. 200
   c. 200,000
   d. 400

Solve the following problems.

9–15. A battery has a 300 ampere-hour rating. How much continuous current can it provide?

9–16. What is the circular mil area of a wire with a diameter of 0.02 inch?

9–17. How much current will be produced if an electromagnetic wave induces 150 μV in an antenna with a characteristic impedance of 300 Ω?

9–18. How much voltage is required to produce dielectric breakdown in a sheet of glass 1/8″ thick?

The following problem is more challenging.

9–19. A transmission line has an impedance of 0.06 ohms per thousand feet. A utility company must provide 50 MW of power to a customer located 37 miles from the generating station. The utility company will allow a transmission loss of 1.5 W per kW. What transmission line voltage is required to achieve this?

GLOSSARY
------- -------

ampacity   The maximum current a conductor may carry according to the National Electrical Code.

ampere-hour rating   The continuous discharge current of a battery multiplied by a six-hour discharge cycle.

AWG   American Wire Gauge.

battery   A group of connected cells.

buss bar   A bare metal strip used as a conductor.

cable   A combination of two or more conductors.

cell   A dc voltage source consisting of two electrodes immersed in an electrolyte.

circular mil   A unit of cross-sectional area used in wire and cable gauge systems.

characteristic impedance   The overall opposition to current of a wire or cable.

dielectric   An insulator.

dielectric strength   The voltage needed to break down an insulator.

eutectic alloy   An alloy that has the lowest possible melting temperature.

flux   A liquid or paste-cleaning agent used in soldering.

GFCI   Ground-fault circuit interrupter; a safety device similar to a circuit breaker that interrupts current when the hot wire and neutral wire have different amounts of current.

grounding wire   A grounded wire that provides a momentary current path to cause a circuit breaker or fuse to open in case of a wiring fault.

hot wire   A wire that has voltage relative to ground.

kcmil   Kilo-circular-mil gauge system.

leakage current   A small current that flows through an insulator.

MCM   Same as kcmil.

neutral wire   A grounded wire used as a normal return path for current.

solder   An alloy of tin and lead used to create permanent electrical connections.

10

MAGNETISM AND ELECTRO-MAGNETISM

● SECTION OUTLINE

10–1 The Magnetic Field

10–2 Magnets

10–3 Electromagnetism

10–4 Magnetic Devices

10–5 Magnetic Quantities and Units

10–6 Induced Voltage

● TROUBLESHOOTING: Applying Your Knowledge

Your friend has a problem with her tape deck and wants your opinion. Some of her tapes sound normal when played on the deck, while others sound distorted. Someone told her she needs to demagnetize the heads. What do you think?

● OVERVIEW AND LEARNING OBJECTIVES

Electricity and magnetism are different forms of the electromagnetic force. Many areas of electronic technology are based on the relationships between electricity and magnetism. You must, therefore, have a solid understanding of the basic principles of magnetism in order to be an effective technician.

After completing this chapter, you should be able to:

1. Describe a magnetic field in terms of lines of force and magnetic poles.
2. State the laws of magnetic attraction and repulsion.
3. Describe the domain theory of ferromagnetism.
4. Describe the nature of residual magnetism and the characteristic of retentivity.
5. Describe the principles of electromagnetism.
6. Name several common electromagnetic devices.
7. Associate magnetic quantities with their corresponding electrical quantities.
8. Describe the principles of induced voltage.

SECTION 10–1: The Magnetic Field

In this section you will learn the following:

→ Magnetism is a natural force produced by the movement of charged particles.

→ The spin of electrons is the main source of magnetic forces.

→ Magnetic moment is a magnetic force with a particular direction.

→ The atoms and molecules of most materials have approximately equal numbers of electrons spinning in each direction. Thus, the total magnetic moment is nearly zero.

→ Iron atoms have more electrons spinning in one direction than in the other direction. Thus, iron atoms have a magnetic moment.

→ Materials with magnetic properties similar to iron are called ferromagnetic materials.

→ A change in an electric field produces a magnetic field.

→ The magnetic field is visualized as a number of invisible flux lines of magnetic force.

→ The shape of a magnetic field is basically circular but is often stretched and elongated.

The Magnetic Force

We have learned that charged atomic particles (electrons and protons) are the source of the electrical force. When charged particles move, they produce another natural force, the *magnetic force*. The magnetic force is a sort of invisible whirlpool in space produced by the movement of the charged particles. The magnetic force will change the motion of other charged particles that come near it.

Nearly all magnetic forces are produced by electrons. Electrons are always spinning, similar to the earth rotating on its axis. The spin of the electron produces a tiny magnetic whirlpool, as shown in Figure 10–1.

Magnetic Moment

Magnetic moment is a magnetic force with a particular direction. If two magnetic forces work in the same direction, the total magnetic force is increased. On the other hand, if two magnetic forces work in opposite directions, the total magnetic force is decreased.

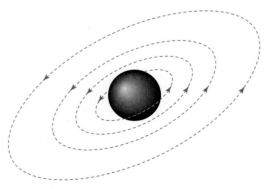

FIGURE 10–1
Magnetic moment of an electron.

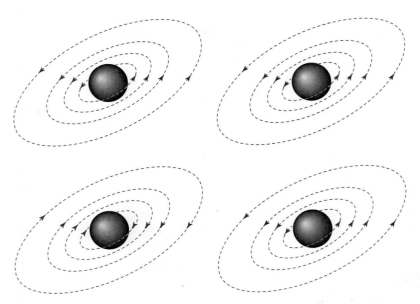

FIGURE 10–2
Electrons with (a) similar magnetic moments; (b) opposite magnetic moments.

In physics, the word *moment* means the direction of a force. We say that the spin of an electron produces a magnetic moment, which simply means that the direction the electron is spinning determines the direction of the magnetic force produced by the spin.

Electrons have only two directions of spin, called clockwise and counterclockwise spin. If two electrons spin in the same direction, they will have similar magnetic moments, and the two magnetic forces will combine into a single, stronger magnetic force. On the other hand, if two electrons spin in opposite directions, they will have opposite magnetic moments, and the two magnetic forces work against each other to produce a single magnetic force that is weaker. These principles are illustrated in Figure 10–2.

Magnetic Materials

In most atoms, approximately half the electrons have clockwise spin, and the other half have counterclockwise spin. Thus, although each electron has its own magnetic moment, the combined magnetic moment of most atoms is nearly zero, since opposite spins produce a weakened combined magnetic force.

However, there are a few materials that have more electrons spinning one way than the other, especially in the valence (outside) shell. The best example is iron, which has six electrons in its valence shell. Five of these spin clockwise and only one spins counterclockwise, as shown in Figure 10–3. Thus, the iron atom has an overall magnetic moment, a strong combined magnetic force.

Atoms of nickel and cobalt also have a magnetic moment similar to iron atoms. Therefore, these elements are called *ferromagnetic* materials (*ferrum* is the Latin word for iron). Ferromagnetic materials include compounds such as ferrite, a ceramic material widely used in electronics. Ferrite is an electrical insulator (a ceramic), whereas iron is a conductor; yet ferrite has the magnetic properties of iron.

Many molecules have enough magnetic moment to respond to external magnetic forces. A water molecule is a good example: Magnetism will cause water molecules to move. Microwave ovens operate by producing strong, vibrating magnetic waves, which

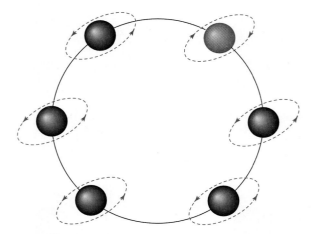

FIGURE 10–3
Valence electrons of an iron atom.

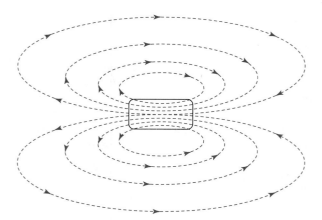

FIGURE 10–4
Magnetic flux lines in and around a bar magnet.

shake up the water molecules in food, making it hot. However, paper, plastic, and glass molecules have nearly zero magnetic moments. These atoms are not affected by the magnetic waves and remain cool in the microwave oven.

Magnetic Fields

A field is a region of space where a natural force can be felt. The space around a magnet is called a *magnetic field*. We sometimes imagine the shape of the magnetic field as similar to invisible rubber bands or balloons. The shape is basically circular but flexible and easily stretched, elongated, twisted, and distorted.

A change in an electric field produces a magnetic field. This simple statement is actually one of the most important laws in electromagnetic theory. It was discovered by the scientist Maxwell in the late nineteenth century and gave us the understanding that electricity and magnetism are actually the same force.

Each electron produces a small electric field. Because each electron spins, the electric field changes. This is what produces the magnetic field around each electron.

When electrons flow as current, a changing electric field is produced. This creates a magnetic field. Thus, whenever current flows, a magnetic field is produced.

Flux Lines

We visualize the magnetic field as made of lines of magnetic force, or *flux lines,* which pass through the magnet and the space around it, as shown in Figure 10–4. Flux lines help to give us a mental picture of the invisible magnetic forces acting in specific directions.

→ *Self-Test*

1. What is magnetism?
2. What is the main source of magnetic forces?
3. What is magnetic moment?
4. Why do most materials not have a magnetic moment?
5. Why do iron atoms have a magnetic moment?
6. What are ferromagnetic materials?
7. What produces a magnetic field?
8. How is the magnetic field visualized?
9. What is the shape of a magnetic field?

SECTION 10–2: Magnets

In this section you will learn the following:

→ A magnetic domain is a large group of ferromagnetic atoms or molecules turned in a similar direction, allowing their magnetic fields to combine. The magnetic field around a domain is elongated.

→ A magnet is a piece of ferromagnetic material whose domains produce an overall magnetic field throughout the material.

→ The north and south poles of a magnet indicate the direction those poles will point if the magnet is allowed to turn freely in the earth's magnetic field.

→ Flux lines are imagined to exit the north magnetic pole and enter the south magnetic pole.

→ When two magnetic fields have similar moments, the flux loops produce a force of physical attraction between the two magnets.

→ When two magnetic fields have opposite moments, the flux loops produce a force of physical repulsion between the two magnets.

→ Unlike magnetic poles attract. Like magnetic poles repel.

→ The flux lines of an external magnetic field are attracted by the domains in ferromagnetic material and change shape in order to pass through the material.

→ A ferromagnetic material will be magnetized by an external magnetic field. The material will stick to the magnet and act as an extension of it.

→ When a magnetizing force is applied to a ferromagnetic material and then removed, the material remains partly magnetized. This is called residual magnetism.

→ Demagnetization is the elimination of residual magnetism by applying a magnetic field of equal strength and opposite polarity, or by the random disorganizing effects of heat.

→ Retentivity is the ability of a ferromagnetic material to retain an organized magnetic field and continue to act as a magnet.

→ Magnetic saturation occurs when a piece of ferromagnetic material contains the maximum amount of flux it can absorb.

Domains

When many atoms with similar magnetic moments group together, their magnetic fields blend together into a larger, combined magnetic field, as shown in Figure 10–5. The group of atoms is called a magnetic *domain.*

Magnets

All pieces of iron and steel contain magnetic domains. However, the domains are normally scattered throughout the iron. The magnetic fields of the domains do not combine into a single magnetic field.

However, if the iron is heated, struck, or exposed to a strong external magnetic field, the atoms turn so their magnetic moments are all in the same direction. The entire piece of iron becomes a single domain with a unified magnetic field. The iron now has become a *magnet,* as shown in Figure 10–6.

Magnetic Poles

The two ends of a magnet are called the *magnetic poles.* The poles of a magnet are the points where the magnetic field emerges from the magnet. The field is strongest at the poles.

FIGURE 10–5
Magnetic domains in iron.

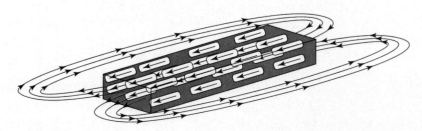

FIGURE 10–6
When domains are aligned, the iron bar becomes a magnet.

Our planet is made mostly of iron and nickel and has a strong magnetic field. The magnetic poles of the earth are near the geographical north and south poles.

If a magnet is balanced or suspended so it can move freely, its magnetic field will be affected by the earth's magnetic field, and the magnet will turn until its poles point north and south. This is the principle of the compass, which is a magnetized needle, as shown in Figure 10–7. The poles of the compass needle are labeled north and south, indicating the direction they will point.

If a compass is brought near an iron magnet, the needle will turn away from geographic north and south under the influence of the strong, nearby magnetic field. The movement of the compass needle indicates that the magnetic force acts along the flux lines in a specific direction. Therefore, we draw arrowheads on the flux lines, as shown in Figure 10–8(a). Flux-line arrowheads indicate the direction a compass needle would point if it were placed at that location. As you can see, the compass points away from the magnet's north pole and toward the magnet's south pole. Thus, the arrowheads show the flux lines leaving the north pole and entering the south pole.

FIGURE 10–7
Magnetized compass needle aligns itself with flux lines of the earth's magnetic field.

Magnetic Attraction and Repulsion

Showing flux lines with arrowheads helps us visualize the invisible magnetic field. We can use these ideas to create a set of rules to describe the interaction of magnetic fields.

In many ways, a scientific theory is like a game, with players who must obey a set of rules. In magnetic theory, the players are the flux lines. Listed next are the rules that the flux lines seem to obey.

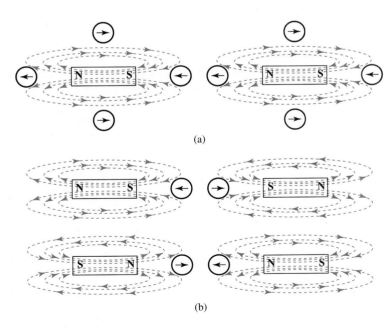

(a)

(b)

FIGURE 10–8
Arrowheads indicate direction a compass needle points at any location in the magnetic field. (a) Magnetic fields with similar moment. (b) Magnetic fields with opposite moment.

Rules of Magnetic Flux

1. Flux lines form closed loops.
2. Flux lines have moment (direction), indicated by arrowheads.
3. An increase in energy may cause flux loops to expand. Thus, energy may be stored in a magnetic field.
4. Because energy is stored in flux loops, they have a natural tendency to become smaller, releasing energy in the process.
5. Flux lines avoid touching or crossing each other.
6. If flux lines flow in similar directions, they strengthen each other.
7. If flux lines flow in opposite directions, they weaken each other.
8. Flux lines do not need a physical conductor. They pass through all materials and exist even in empty space.
9. The shape of flux loops may be irregular.

We will use these rules of flux lines to describe the well-known fact of magnetic attraction and repulsion.

If two magnets are near each other, their magnetic fields will affect each other. In Figure 10–8(a), the two magnetic fields have the same moment. This means that the flux lines flow in similar directions. You can see this by looking between the magnets. Notice that the arrowheads and compass needles are pointing in the same direction.

The magnetic fields in Figure 10–8(b) have opposite moment. Look between the magnets and you can see that the arrowheads and compass needles point in opposite directions.

You probably know that two magnets will pull together (attract) when turned one way, and push apart (repel) when turned the other way. We can now explain this by referring to whether the direction of the magnetic moments is similar or opposite.

Magnetic Attraction

In Figure 10–9(a), we see two magnetic fields with similar moment. As shown by the arrowheads, both fields circulate in a counterclockwise direction. Between the two magnets, the flux lines flow in opposite directions, making the flux weaker in these areas. However, the flux lines above and below the magnets flow in the same direction, making the flux stronger in these areas. The areas of weakness and strength cause the flux lines to break up and form a new pattern, as shown in Figure 10–9(b). The flux lines between the magnets now flow from one magnet directly into the other magnet. Above and below the magnets, the flux lines blend together. In other words, the smaller loops reform into long loops that pass through both magnets.

The same action would happen if both fields were circulating clockwise. What is important here is that the magnetic moments of the two fields are the same.

Flux loops always tend to become smaller, releasing energy in the process. The release of the energy stored in the magnetic field causes the two magnets to move together, as shown in Figure 10–9(c). Like invisible rubber bands, the flux loops hold the two magnets together.

Magnetic Repulsion

In Figure 10–10, we see two magnetic fields with opposite moment. As shown by the arrowheads, one field circulates counterclockwise and the other circulates clockwise. Between the two magnets, the flux lines flow in similar directions, making the flux stronger in these areas. However, the flux lines above and below the magnets flow in opposite directions, making the flux weaker in these areas.

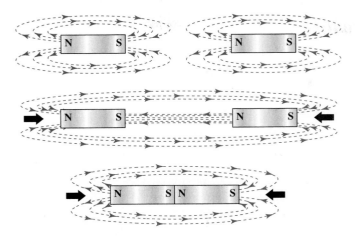

FIGURE 10–9
Flux lines of magnets with similar moments produce force of attraction between magnets.

In this situation, the two magnetic fields will remain separate. In fact, if we push the two magnets closer together, the flux lines between the magnets flatten out to avoid touching each other. The energy we apply by pushing the magnets together is stored in the magnetic fields between the magnets, producing a physical force of repulsion. When we let go of the magnets, the excess energy is released, moving the magnets apart as the fields return to their normal shape. The magnetic fields are like two rubber balloons that flatten out where they are pushed together, and then push each other apart as they return to their normal shape when we let go.

Magnetic Polarity and Physical Forces

When unlike magnetic poles (a north pole and a south pole) face each other, the magnetic moments are similar, producing a physical force of attraction between the two magnets.

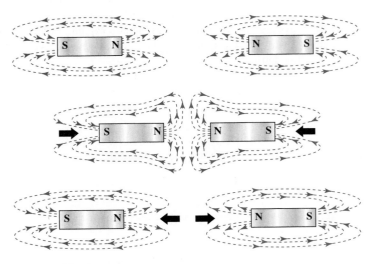

FIGURE 10–10
Flux lines of magnets with opposite moments produce force of repulsion between magnets.

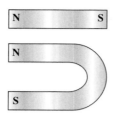

FIGURE 10–11
Bar and
horseshoe
magnets.

On the other hand, when like magnetic poles (two north poles or two south poles) face each other, the magnetic moments are opposite, producing a physical force of repulsion between the two magnets. These observations lead to the well-known laws of magnetism: Unlike magnetic poles attract, and like magnetic poles repel.

Shapes of Magnets

As shown in Figure 10–11, magnets with an elongated shape are called bar magnets. When a bar magnet is bent so its poles are opposite each other, it is called a horseshoe magnet. The magnetic flux is very strong between the poles. The horseshoe magnet is better for picking up heavier pieces of iron and steel because the horseshoe magnet touches both its poles to the metal. When the horseshoe magnet is not being used, a small steel rod is placed between the two poles. This rod is called a keeper because it keeps all the flux inside the magnet and the keeper. Nearby objects do not feel any flux and are not affected by the magnet with the keeper in place.

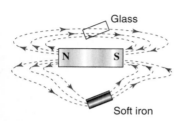

FIGURE 10–12
Effect of glass and iron on
flux lines.

Magnetization

Magnetic flux lines flow through empty space and through any material. However, they prefer to flow through ferromagnetic materials such as iron and steel. Since all magnetic fields interact and affect each other, the flux lines of an external magnetic field will bend in order to flow through a piece of iron, as shown in Figure 10–12. The external flux lines are attracted by the flux lines already within the iron. Nonmagnetic materials such as glass will not affect the external flux lines.

A strong external magnetic field will magnetize an ordinary piece of iron or steel. The ends of the piece become magnetic poles. The domains will line up to produce magnetic attraction. Thus, pieces of iron and steel are attracted by magnets and stick to them. In effect, the iron or steel becomes an extension of the magnet, as shown in Figure 10–13.

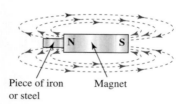

FIGURE 10–13
Iron piece acts as an
extension of the bar magnet.

Retentivity

Retentivity is the ability of a ferromagnetic material to retain a magnetic field and continue to act as a magnet. Retentivity depends on the atomic structure of the material.

Figure 10–14 shows a group of atoms arranged in a regular, orderly manner, called a crystalline structure (also called a lattice or a matrix). In general, the atoms of metals are arranged in crystalline lattices.

A strong external magnetic field will turn the atoms so they all have similar magnetic moment, magnetizing the iron. When the external magnetic field is removed, the atoms will begin to turn in random directions as they vibrate from heat energy in the material, demagnetizing the iron.

An alloy is a blend of metals. There are many different alloys of iron, including steel. Because alloys are blends of different types of atoms, their magnetic characteristics are significantly different, including retentivity, the ability to remain magnetized after the external magnetic field is removed. Alloys used to make permanent magnets are specially developed to have high retentivity. *Alnico,* an alloy of aluminum, nickel, and cobalt, is widely used as well as rare-earth magnets, whose alloys contain unusual elements such as ytterbium and tellurium.

Residual Magnetism

When ferromagnetic material has been exposed to a magnetic field, a certain amount of magnetism remains in the material because of retentivity. For example, a screwdriver

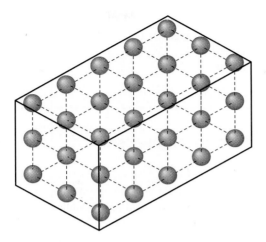

FIGURE 10–14
Atoms in a crystal lattice.

blade that has been in contact with a magnet may still be able to pick up iron filings, small nails, and so on. This leftover magnetism is called *residual magnetism.*

Demagnetization

Because of residual magnetism, many objects such as tools or the delicate recording heads of VCRs, tape decks and computer disk drives need occasional *demagnetization.* This is done by applying an external magnetic field of equal strength and opposite polarity to the residual field.

Saturation

A magnetic field will change its shape and flow into a piece of iron, attracted by the domains inside. We can imagine the piece of iron soaking up the external flux lines, similar to the way a sponge soaks up water.

Electrical machines such as motors, generators, and transformers operate because of strong magnetic fields, and all such machines contain large pieces of iron to absorb the magnetic field and hold it inside the machine where it is needed. These iron pieces are called the *core* of the machine.

However, just as a sponge can hold only a certain amount of water, an iron core can hold only a certain amount of magnetic flux. When the core holds all the flux it can, we say the core is *saturated.* If we create more flux, the core cannot hold it and the extra flux leaks into the space outside the core. This represents a waste of energy in most electrical machines and is undesirable.

→ *Self-Test*

10. What is a magnetic domain?
11. What is a magnet?
12. What is the meaning of the north and south poles of a magnet?
13. What is the relationship between flux lines and magnetic poles?
14. What happens when two magnetic fields have similar moments?
15. What happens when two magnetic fields have opposite moments?
16. Describe the interactions of like and unlike magnetic poles.

17. What is the interaction between ferromagnetic material and an external magnetic field?
18. What effect will an external magnetic field have on ferromagnetic material?
19. What is residual magnetism?
20. What is demagnetization?
21. What is retentivity?
22. What is magnetic saturation?

SECTION 10–3: Electromagnetism

In this section you will learn the following:

→ Electromagnetism is the magnetic field produced by an electric current.
→ The strength of an electromagnetic field is determined by the amount of the current and the number of turns of wire in the coil.
→ The polarity of an electromagnetic field is determined by the polarity of the current and the direction in which the wire is wound in the coil.

We know that magnetism is caused by the movement of charged particles, especially electrons. We also know that current is the movement of electrons through a conductor. Therefore, whenever current flows, a magnetic field is produced around the conductor. This important fact is called *electromagnetism*.

If the conductor is straight, the magnetic field is circular, surrounding the conductor like a shirt sleeve surrounds your arm, as shown in Figure 10–15. If the conductor is wrapped into a coil, the flux lines around the loops of the coil will break up and reform into the flux pattern shown in Figure 10–16. The coil becomes an electromagnet, with a north and south pole, and a flux pattern similar to that of a bar magnet.

Thus, we have two ways to produce a magnetic field: permanent magnets and electromagnets. Each type has its own features.

Permanent Magnets
  Require no power.
  Simple to manufacture.
  Strength of magnetic flux depends on physical size of magnet and type of alloy.
  Magnetic flux strength and polarity are constant.

Electromagnets
  Require electrical power.
  More complicated to manufacture.

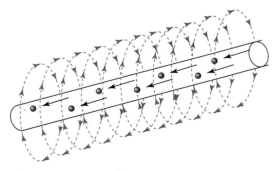

FIGURE 10–15
A circular magnetic field is produced by current in a conductor.

Strength of magnetic flux depends on amperes of current and number of turns in the coil.

Magnetic flux strength and polarity are controllable.

Permanent magnets are ideal for small, simple uses. However, electromagnets are much more useful when large amounts of magnetic energy are needed. We can create very powerful magnetic fields simply by sending more current through a coil, or adding more turns (loops of wire) to a coil. Electromagnets usually contain iron cores to concentrate the magnetic flux, but an electromagnet is always much smaller and lighter than a permanent magnet of equal strength.

Perhaps the most important advantage of the electromagnet is that the magnetic field is completely controllable, since it is caused by current, as shown in Figure 10–17. The magnetic field appears when current flows and disappears when current stops. The field becomes stronger and weaker as the current increases and decreases. The polarity of the field depends on the polarity of the current and on the direction the turns of wire are wound on the coil: clockwise or counterclockwise. Regardless of how the wire is wound, if the direction of current flow reverses, the north and south poles of the coil reverse.

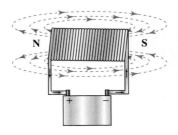

FIGURE 10–16
A coil carrying current acts as an electromagnet.

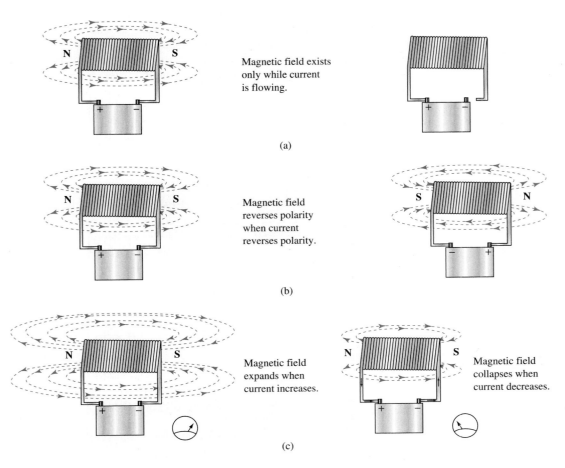

FIGURE 10–17
(a) Magnetic field exists only while current is flowing. (b) Magnetic field reverses polarity when current reverses polarity. (c) Magnetic field expands when current increases, collapses when current decreases.

→ *Self-Test*

23. What is electromagnetism?
24. What determines the strength of an electromagnetic field?
25. What determines the polarity of an electromagnetic field?

SECTION 10–4: Magnetic Devices

In this section you will learn the following:

→ Electromagnetic devices include the solenoid, relay, and contactor.
→ A plastic tape or disk coated with ferromagnetic particles is widely used for recording audio, video, and computer data.
→ An electric motor operates because of the physical forces of attraction and repulsion among electromagnetic fields inside the motor.

The simplest magnetic device is the electromagnet. Industrial electromagnets are used in handling bulk ferrous metals, such as pipe and salvage automobiles. The operator can pick up or release a load simply by turning on or off the current to the coil.

The *solenoid,* shown in Figure 10–18, uses an electromagnet to produce a short back-and-forth movement of a steel rod. Solenoids are used in a wide variety of machines. In an automobile, a solenoid mounted on top of the starter motor engages the gears of the starter motor with the gear teeth on the engine flywheel, allowing the starter motor to crank the engine. If you move the ignition switch into start position while the

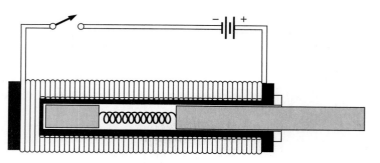

(a) Unenergized – plunger extented

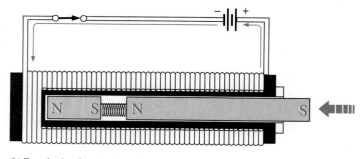

(b) Energized – plunger retracted

FIGURE 10–18
A solenoid.

engine is running, you will hear a grinding noise as the solenoid attempts to engage the gears.

Solenoids are widely used to operate valves. For example, the rocket engines of the space shuttle contain many solenoid-controlled valves. Thus, the engines can be operated automatically by computers.

CD players, VCRs, tape decks, and so on contain small solenoids, which move the various levers, pulleys, belts, and wheels into proper position under the control of a microprocessor chip. Thus, touching a single button can cause a sequence of complex mechanical movements.

A *relay* is a switch operated by an electromagnet, as shown in Figure 10–19. A movable armature is held by a spring against a contact. The armature is made of iron or steel. When the coil beneath the armature is energized with current, the magnetic field overcomes the force of the spring and pulls the armature against the other contact. Thus, the armature and the contacts function as a switch. Terminals attached to the armature and contacts allow external wiring connections. The two contacts are described as normally open (NO) or normally closed (NC), as they are held by the spring when the coil is deenergized.

Relays have several advantages over manual switches. Large relays, known as *contactors,* often carry high voltages and currents that would be unsafe to control with a manual switch. The voltage and current in the coil are often much lower than those controlled by the contactor. Thus, a low-power circuit can control a high-power circuit

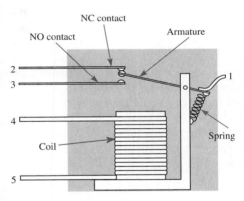

(a) Unenergized: continuity from 1 to 2

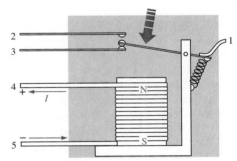

(b) Energized: continuity from 1 to 3

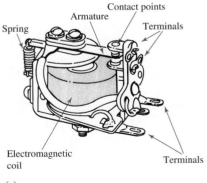

(c)

FIGURE 10–19
A relay.

through a relay. The low-power control switch can be located at a distance from the relay. A good example of this arrangement is an ordinary wall-mounted thermostat controlling a central air-conditioning unit. The switches on the thermostat are small, since they carry low currents at low voltages. Thus, they are unobtrusive and safe to mount in exposed living areas. The thermostat controls the contactor mounted on the main unit. The contactor carries the high voltage and current that the main unit requires.

Most loudspeakers operate electromagnetically, as shown in Figure 10–20. Electrical signals in the coil produce a rapidly changing magnetic field, which interacts with the field of the surrounding permanent magnet. As a result, the coil vibrates back and forth rapidly. The coil is glued to a paper diaphragm, which vibrates with the coil, producing sound waves in the air.

Magnetic principles are used in a variety of recording devices, including tapes and disks. Tapes are used for audio, video, and data recording. Disks are used primarily for data recording in computers and may be either floppy diskettes, which can be inserted and removed from the computer, or hard disks permanently mounted inside the computer. The recording material is a tough plastic coated with iron oxide particles. During the recording process, the tape or disk passes beneath a recording head, as shown in Figure 10–21. Electrical signals in the head create a magnetic field which magnetizes the tiny iron particles as they move past. The magnetic strength and polarity of the particles depend on the polarity and current of the electrical signal in the head at the moment of recording. Thus, the recording head creates a magnetic pattern on the tape that matches the electrical signals passing through the head. The particles remain magnetized, and the signal is recorded in magnetic form.

During playback, the tape or disk again passes beneath the head. The magnetic fields of the particles create small voltages in the head which are a replica of the original signal. All that is needed is a suitable amplifier to strengthen these weak signals and send them to a loudspeaker, a video screen, or the memory chips of a computer.

Electric motors operate on the principle of two magnetic fields producing physical forces of attraction and repulsion. As shown in Figure 10–22, current flowing in a conductor produces an electromagnetic field that will interact with a surrounding field to produce physical forces on the conductor. Electric motors are simply an elaboration of this basic idea, designed so the physical forces produce rotation of a shaft. We will examine motors in greater detail in a later chapter.

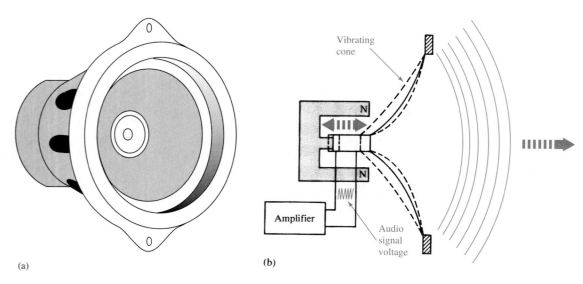

(a)   (b)

FIGURE 10–20
Details of a loudspeaker.

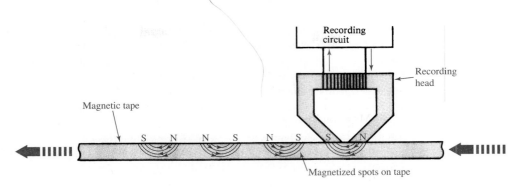

FIGURE 10–21
A magnetic record/playback head.

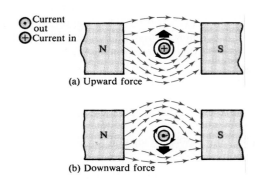

FIGURE 10–22
Principle of electric motor action.

→ *Self-Test*

26. Name several electromagnetic devices.
27. What device is widely used for recording audio, video, and computer data?
28. What causes an electric motor to operate?

SECTION 10–5: Magnetic Quantities and Units

In this section you will learn the following:

→ Flux, symbolized ϕ, is the strength of a magnetic field. The unit of flux is the weber (Wb).

→ Magnetomotive force (mmf) is the force that produces flux. The unit of mmf is the ampere-turn (At).

→ Reluctance ($\mathcal{R}$) is a measure of the opposition of a material to flux. The unit of reluctance is ampere-turns per weber, abbreviated At/Wb.

→ Flux density (β) is the flux per unit cross-sectional area. The unit of flux density is the tesla (T).

→ Field intensity (H) is the magnetomotive force per unit length. The unit of field intensity is ampere-turns per meter.

→ Permeability (μ) is a measure of the ease of establishing flux within a material. The unit of permeability is webers per ampere-turn per meter.

Before atomic theory, both electricity and magnetism were believed to be invisible weightless fluids. The quantities and units of both electricity and magnetism reflect the early theory and include three basic ideas: an invisible flow, a force which causes this flow, and differences in various materials which limit the flow. In the early nineteenth century, scientists knew that electricity and magnetism were somehow related, and so the scientists created a magnetic theory that resembled electrical theory. Although our modern understanding of electricity and magnetism is based on atomic theory, we continue to use the quantities and units of the older invisible fluid theory.

Table 10–1 summarizes the basic quantities and units of both electricity and magnetism and associates these units with the basic ideas. These quantities and units are part of the SI (International System) of measurement.

Before we begin a detailed study of the magnetic units and quantities, we should notice two things about Table 10–1.

First, there is only one set of terms for electricity, but there are two sets of terms for magnetism: one set for general magnetism, and another set for practical magnetism.

Second, in electricity we have a choice of expressing either how difficult or how easy it is for current to flow through various materials. However, in general magnetism, we express how difficult it is for flux to flow through a material, and in practical magnetism, we express how easy it is.

Flux

Flux in magnetism is similar to current in electricity. The symbol for flux is ϕ, the Greek letter phi.

TABLE 10–1

IDEA	ELECTRICITY	GENERAL MAGNETISM	PRACTICAL MAGNETISM
Flow	Current I Ampere A	Flux ϕ Weber Wb	Flux Density β Tesla T
Force	Voltage E Volt V	Magnetomotive Force mmf or F_m Ampere-turn At	Field Intensity H Ampere-turn per meter At/m
Difficulty	Resistance R Ohm Ω	Reluctance $\mathcal{R}$ Ampere-turns per weber At/Wb	None None
Ease	Conductance G Siemen S	None None	Permeability μ Webers per amp-turn-meter Wb/At-m

We imagine that a magnetic field is made of lines of flux. Thus, flux is the total strength of the magnetic field, and we may describe a stronger or weaker magnetic field as an increase or decrease in flux.

The unit of flux is the weber. We measure flux by comparing any magnetic field to a standard-strength field, such as a magnet that can pick up a certain amount of metal. We say that a standard-strength field has a flux of 1 weber, and then we compare all other magnetic fields to it.

Magnetomotive Force (mmf)

Magnetomotive force in magnetism is similar to voltage in electricity. The abbreviation for magnetomotive force is mmf.

Magnetomotive force is the force that produces flux. If current flows through a coil, the movement of the electrons will create magnetic flux. The amount of flux depends on the current and the number of turns of wire in the coil. Therefore, mmf is measured in units of ampere-turns (At).

● SKILL-BUILDER 10–2

A coil consists of ten turns of wire. A current of 2 amps flows through the coil. How much mmf is produced?

GIVEN:
   $I = 2$ A
   $N = 10$ turns

FIND:
   mmf

SOLUTION:
   mmf $= $ At
         $=$ amperes $\times$ turns
         $= 2$ A $\times 10$ turns
         $= 20$ At

Repeat the Skill-Builder using the following values. You should obtain the results that follow.

Given:

| I | 1.65 A | 490 mA | 0.023 A | 15.7 A | 908 µA |
| N | 370 t | 1560 t | 2380 t | 200 t | 110 t |

Find:

| mmf | 610.5 At | 764.4 At | 54.74 At | 3140 At | 99.88 mAt |

Reluctance

Reluctance in magnetism is similar to resistance in electricity. The abbreviation for reluctance is $\mathcal{R}$.

Reluctance is a magnetic property of materials caused by differences in atomic structure that oppose flux. Figure 10–23 shows flux flowing inside two toroids. (The word *toroid* means shaped like a doughnut.) Toroid cores are often used in magnetic devices

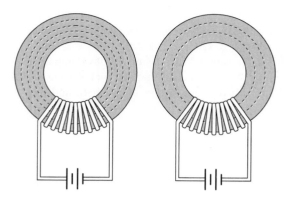

FIGURE 10–23
Flux in a toroid core.

because the shape allows all the flux lines to stay inside the object. The coils on the toroids have equal amperes and turns. Therefore, the mmf is the same for both toroids. However, the flux is not the same because the toroids are made of different materials. The second toroid, with less flux, has a higher reluctance than the first toroid.

Reluctance is measured in units of ampere-turns per weber, or At/Wb. The unit of reluctance is actually the formula for reluctance and is sometimes called the Ohm's law of magnetism.

Formula 10–1

$$\mathcal{R} = \frac{\text{mmf}}{\phi}$$

● SKILL-BUILDER 10–2

An mmf of 12 At produces a flux of 6 Wb in a toroid. What is the reluctance?

GIVEN:
  mmf = 12 At
    ϕ = 6 Wb

FIND:
  $\mathcal{R}$

SOLUTION:
  $\mathcal{R}$ = mmf ÷ ϕ
    = 12 At ÷ 6 Wb
    = 2 At/Wb

Repeat the Skill-Builder using the following values. You should obtain the results that follow.

Given:				
mmf	130.3 At	367 mAt	20.8 At	458 uAt
ϕ	73 Wb	228 mWb	63 Wb	37 µWb
Find:				
$\mathcal{R}$	1.78 At/Wb	1.61 At/Wb	0.33 At/Wb	12.38 At/Wb

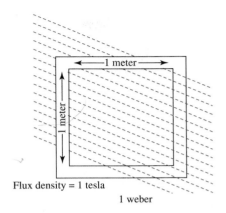

FIGURE 10–24
Flux density: number of flux lines passing through a specific cross-sectional area.

Flux Density

Originally, both current and flux were believed to be the movement of an invisible fluid. Current was later discovered to be the movement of charged particles. Flux, however, is simply a concept or visualization of the magnetic field. Nothing physical is actually flowing or moving. Therefore, flux is flexible. Flux can expand and shrink in space and flow in and out of various materials.

A magnetic field consists of a certain amount of flux. However, the flux may be spread out over a large area or all concentrated in one place. Ferromagnetic materials such as iron act like sponges to soak up and hold magnetic flux. Thus, in practical applications, we need to describe how concentrated or thinned out the flux may be. This concept is called *flux density*. The symbol for flux density is β, the Greek letter beta.

Figure 10–24 shows how we visualize flux density. We imagine an open space, like a window, 1 square meter in area. If the flux within this cross-sectional area is 1 weber, we have a flux density of 1 tesla (T). Iron becomes magnetically saturated at a flux density of approximately 2.3 teslas.

Thus, flux density equals flux in webers divided by cross-sectional area in square meters.

Formula 10–2

$$\beta = \frac{\phi}{\text{area}}$$

● *SKILL-BUILDER 10–3*

A flux of 0.2 Wb flows through an iron rod with a cross-sectional area of 0.09 square meters. What is the flux density?

GIVEN:
$\phi = 0.2 \text{ Wb}$
$A = 0.09 \text{ m}^2$

FIND:

β

SOLUTION:

β = φ ÷ A

   = 0.2 Wb ÷ 0.09 m²

   = 0.29 T

Repeat the Skill-Builder using the following values. You should obtain the results that follow.

Given:

| φ | 3.4 Wb | 15.6 mWb | 350 Wb | 0.46 Wb | 28.9 Wb |
| A | 0.37 m² | 0.016 m² | 5.37 m² | 1.39 m² | 357 m² |

Find:

| β | 9.19 T | 0.98 T | 65.18 T | 0.33 T | 0.08 T |

Field Intensity

Figure 10–25 illustrates the concept of *field intensity (H)*. If the turns of the coil are spaced closely together, all the mmf enters the rod at one point, producing a high-intensity (concentrated) magnetic field at that point. On the other hand, if the turns are spaced far apart, the mmf enters the rod along its entire length. The field strength is the same, but the intensity is low: The field is not concentrated at any one point.

Field intensity is measured in units of ampere-turns per meter (At/m). The unit of measure is actually the formula for field intensity.

Formula 10–3

$$H = \frac{\text{mmf}}{\text{length}}$$

● *SKILL-BUILDER 10–4*

A coil of 200 turns is wrapped around a rod 0.05 meters in length. A current of 25 mA flows in the coil. What is the field intensity?

GIVEN:

I = 25 mA

N = 200 turns

l = 0.05 m

FIND:

H

FIGURE 10–25
Field intensity within the iron core is higher with tightly wrapped coil, lower with loosely wrapped coil.

SOLUTION:

1.   find mmf

$$mmf = At$$
$$= 25 \text{ mA} \times 200 \text{ t}$$
$$= 5 \text{ At}$$

2.   find field intensity

$$H = mmf \div length$$
$$= 5 \text{ At} \div 0.05 \text{ m}$$
$$= 100 \text{ At/m}$$

Repeat the Skill-Builder using the following values. You should obtain the results that follow.

Given:

I	3.36 A	42.97 mA	397 μA	6.76 A	23.6 mA
N	386 t	125 t	240 t	375 t	430 t
l	0.65 m	0.15 m	0.07 m	1.75 m	0.4 m

Find:

H	1995 At/m	35.8 At/m	1.36 At/m	1449 At/m	25.4 At/m

Permeability

Permeability in magnetism is similar to conductance in electricity. The symbol for permeability is μ, the Greek letter mu.

The word *permeate* means to penetrate and spread through, like water permeating a sponge. Like reluctance, permeability is a magnetic property of materials caused by differences in atomic structure.

Permeability is measured in units of webers per ampere-turn-meter, which is actually the formula for permeability.

Formula 10–4

$$\mu = \frac{\phi}{mmf \times length}$$

$$= \frac{webers}{amperes \times turns \times meters}$$

→ Self-Test

29. What is flux and its unit?
30. What is magnetomotive force and its unit?
31. What is reluctance and its unit?
32. What is flux density and its unit?
33. What is field intensity and its unit?
34. What is permeability and its unit?

SECTION 10–6: Induced Voltage

In this section you will learn the following:

→ A change in a magnetic field creates an electric field, which induces a voltage in any conductor exposed to the magnetic field.

→ A change in either an electric or a magnetic field propagates (travels) at the speed of light.

We now see the two-way relationship between electricity and magnetism: A changing electric field produces a magnetic field, and a changing magnetic field produces an electric field. This is why scientists now consider electricity and magnetism to be two aspects of the same fundamental force, the electromagnetic force.

All conductors contain lots of free electrons. If a conductor is exposed to an electric field, the free electrons will move away from the negative pole of the electric field toward the positive pole. In fact, this is exactly what happens when you connect a wire to a battery.

When a wire is exposed to a changing magnetic field, an electric field is created inside the wire. This electric field pulls the free electrons away from one end of the wire and pushes them toward the other end of the wire, as shown in Figure 10–26. Since the wire has too many free electrons at one end and not enough free electrons at the other end, a voltage is created inside the wire. In other words, when a wire is exposed to a changing magnetic field, the wire becomes a voltage source. The ends of the wire become positive and negative terminals, just like the poles of a battery. If a load is connected to the wire, current will flow into the load.

However, as soon as the magnetic field stops changing, the electric field in the wire disappears. The free electrons spread out through the wire in a normal way, and the voltage disappears. Therefore, if we want to use the wire as a voltage source, we must find a way to keep the magnetic field always changing. A steady magnetic field will not produce voltage in the wire.

One simple way to keep the magnetic field always changing relative to a conductor is to produce physical movement between the conductor and the magnetic field, as shown in Figure 10–27. This method is used in all generators and alternators, the machines that produce electricity for our homes and automobiles.

Another way to keep the magnetic field always changing is to use a changing current to produce the magnetic field. This causes the magnetic field to become stronger and weaker. The other wire feels the changes in the magnetic field and produces a voltage, as shown in Figure 10–28. This method is used to produce voltage in transformers.

The amount of voltage induced in a conductor is determined by four factors:

1. The strength of the magnetic field (flux). A stronger flux produces a higher voltage.
2. The length of the conductor exposed to the magnetic field. If the conductor is wrapped into a coil, a higher voltage is induced in a coil with a greater number of turns.

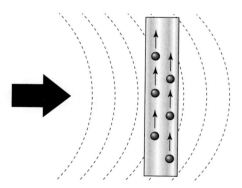

FIGURE 10–26
Induced voltage in a wire or coil.

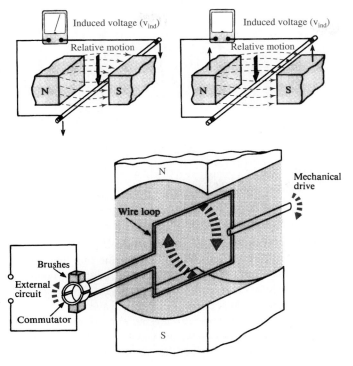

FIGURE 10–27
Relative motion between a conductor and a magnetic field.

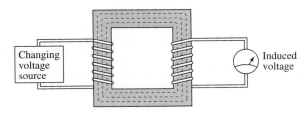

FIGURE 10–28
Transformer action.

3. The rate of change of the magnetic field. A faster relative motion between the field and the conductor produces a higher voltage.

4. The physical angle between the conductor and the flux lines. Maximum voltage is produced when the flux lines cut straight across the wire at a 90° angle. As the angle decreases, the voltage also decreases. When the flux lines move parallel to the wire at an angle of 0°, the voltage will be zero.

The polarity of *induced voltage* depends on the direction of relative motion between the magnetic field and the conductor, as shown in Figure 10–29. When the direction of relative motion reverses, the polarity of induced voltage also reverses.

Propagation of an Electromagnetic Wave

In Figure 10–30, a current is flowing in conductor *A*, establishing an electromagnetic field that extends through space to conductor *B*. If the current changes in conductor *A*, the

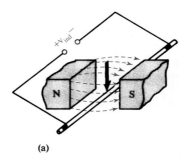 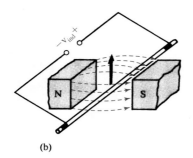

(a)   (b)

FIGURE 10–29
Polarity of induced voltage.

magnetic field will also change. However, the change takes time to propagate (spread or travel through) the magnetic field. The change in magnetic flux first occurs at the surface of conductor *A,* then travels outward through the field, reaching conductor *B* at a later time and inducing a voltage in conductor *B* when it arrives. This propagation of a change in flux throughout a magnetic field is called a magnetic wave.

A good way to visualize a magnetic wave traveling through a magnetic field is to imagine throwing a stone into a pond of water. The splash of the stone creates a wave or ripple which travels outward across the surface of the water from the place where the wave originated.

The magnetic waves in Figure 10–30 propagate in a similar manner, except that magnetic waves can propagate both away from the source and toward the source. In other words, if the magnetic field becomes stronger, flux lines expand from the source outward; but if the field becomes weaker, flux lines contract toward the source. However, in either case, the flux lines near the source are the first to change, and the flux lines at a distance from the source change at a later time. Thus, the principle of propagation is the same regardless of the nature of the flux change.

The propagation of electromagnetic waves is the basis of all forms of radio communication. In other words, radio and television waves are actually magnetic waves. In Figure 10–30, conductor *A* is a transmitting antenna, and conductor *B* is a receiving antenna.

To carry signals such as audio and video, the magnetic waves must be modified. This process is called modulation. Modulation is a fairly complex process, and you will learn more about it in later electronics courses. There are two basic forms of modulation: amplitude modulation (AM) and frequency modulation (FM). This is why commercial radio stations are classified as AM or FM, depending on which form of modulation they use.

Electromagnetic waves propagate at the speed of light, approximately 186,000 miles (300,000 kilometers) per second. Thus, there is a time delay in radio communication with distant spacecraft, for example.

In conclusion, we have presented in this chapter two very important natural facts:

1. When current flows, a magnetic field is created. This is called *electromagnetism.*
2. When a conductor feels a change in a magnetic field, voltage is created in the conductor. This is called *induced voltage.*

Almost every electrical and electronic machine operates because of these two facts of nature. That is why it is so important that you understand these two principles now.

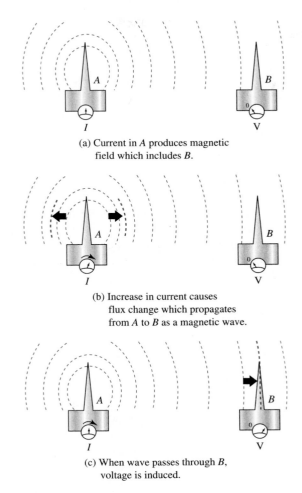

(a) Current in *A* produces magnetic
field which includes *B*.

(b) Increase in current causes
flux change which propagates
from *A* to *B* as a magnetic wave.

(c) When wave passes through *B*,
voltage is induced.

FIGURE 10–30
Propagation of an electromagnetic field.

→ *Self-Test*

35. What is the result of a change in a magnetic field?
36. What is the speed of propagation of a change in an electric or magnetic field?

Formula List

Formula 10–1

$$\mathcal{R} = \frac{\text{mmf}}{\phi}$$

Formula 10–2

$$\beta = \frac{\phi}{\text{area}}$$

Formula 10–3

$$H = \frac{\text{mmf}}{\text{length}}$$

Formula 10–4

$$\mu = \frac{\phi}{\text{mmf} \times \text{length}}$$

● TROUBLESHOOTING: Applying Your Knowledge

The record/playback heads on a tape deck need periodic demagnetization to remove residual magnetism that develops during normal use. Residual magnetism in the heads can indeed cause distortions in the audio signal. However, this distortion tends to be audible on all tapes. Your friend reports distortion on only certain tapes. While demagnetization is in general a good idea, there is no strong evidence that it will cure this particular problem. A more likely cause is some sort of magnetic or physical distortion on the individual tapes.

CHAPTER SUMMARY: ANSWERS TO SELF-TESTS

1. Magnetism is a natural force produced by the movement of charged particles.
2. The spin of electrons is the main source of magnetic forces.
3. Magnetic moment is a magnetic force with a particular direction.
4. The atoms and molecules of most materials have approximately equal numbers of electrons spinning in each direction. Thus, the total magnetic moment is nearly zero.
5. Iron atoms have more electrons spinning in one direction than in the other direction. Thus, iron atoms have a magnetic moment.
6. Materials with magnetic properties similar to iron are called ferromagnetic materials.
7. A change in an electric field produces a magnetic field.
8. The magnetic field is visualized as a number of invisible flux lines of magnetic force.
9. The shape of a magnetic field is basically circular but is often stretched and elongated.
10. A magnetic domain is a large group of ferromagnetic atoms or molecules turned in a similar direction, allowing their magnetic fields to combine. The magnetic field around a domain is elongated.
11. A magnet is a piece of ferromagnetic material whose domains produce an overall magnetic field throughout the material.
12. The north and south poles of a magnet indicate the direction those poles will point if the magnet is allowed to turn freely in the earth's magnetic field.
13. Flux lines are imagined to exit the north magnetic pole and enter the south magnetic pole.
14. When two magnetic fields have similar moments, the flux loops produce a force of physical attraction between the two magnets.

15. When two magnetic fields have opposite moments, the flux loops produce a force of physical repulsion between the two magnets.
16. Unlike magnetic poles attract. Like magnetic poles repel.
17. The flux lines of an external magnetic field are attracted by the domains in ferromagnetic material and change shape in order to pass through the material.
18. A ferromagnetic material will be magnetized by an external magnetic field. The material will stick to the magnet and act as an extension of it.
19. When a magnetizing force is applied to a ferromagnetic material and then removed, the material remains partly magnetized. This is called residual magnetism.
20. Demagnetization is the elimination of residual magnetism by applying a magnetic field of equal strength and opposite polarity, or by the random disorganizing effects of heat.
21. Retentivity is the ability of a ferromagnetic material to retain an organized magnetic field and continue to act as a magnet.
22. Magnetic saturation occurs when a piece of ferromagnetic material contains the maximum amount of flux it can absorb.
23. Electromagnetism is the magnetic field produced by an electric current.
24. The strength of an electromagnetic field is determined by the amount of the current and the number of turns of wire in the coil.
25. The polarity of an electromagnetic field is determined by the polarity of the current and the direction in which the wire is wound in the coil.
26. Electromagnetic devices include the solenoid, relay, and contactor.
27. A plastic tape or disk coated with ferromagnetic particles is widely used for recording audio, video, and computer data.
28. An electric motor operates because of the physical forces of attraction and repulsion among electromagnetic fields inside the motor.

29. Flux (φ) is the strength of a magnetic field. The unit of flux is the weber (Wb).
30. Magnetomotive force (mmf) is the force that produces flux. The unit of mmf is the ampere-turn (At).
31. Reluctance (ℛ) is a measure of the opposition of a material to flux. The unit of reluctance is ampere-turns per weber (At/Wb).
32. Flux density (β) is the flux per unit cross-sectional area. The unit of flux density is the tesla (T).
33. Field intensity (H) is the magnetomotive force per unit length. The unit of field intensity is ampere-turns per meter.
34. Permeability (μ) is a measure of the ease of establishing flux within a material. The unit of permeability is webers per ampere-turn-meter.
35. A change in a magnetic field creates an electric field, which induces a voltage in any conductor exposed to the magnetic field.
36. A change in either an electric or a magnetic field propagates (travels) at the speed of light.

CHAPTER REVIEW QUESTIONS

Determine whether the following questions are true or false.

10–1. Only certain electrons produce a magnetic field.
10–2. Ferromagnetic materials have atoms with more electrons spinning one way than the other.
10–3. A magnet is a piece of ferromagnetic material whose domains have been organized.
10–4. A magnetic pole is a point on a magnet where the flux lines enter or emerge.
10–5. Similar magnetic poles repel, and opposite poles attract.
10–6. Retentivity is the ability of a ferromagnetic material to retain an organized magnetic field.
10–7. A material with high retentivity has low residual magnetism after being magnetized.
10–8. Electromagnetism is electricity produced by magnetism.
10–9. A solenoid is seldom used in consumer electronic devices.
10–10. When a conductor is exposed to a changing magnetic field, a voltage is induced in the conductor.
10–11. An electromagnetic wave propagates at the speed of light.

Select the response that best answers the question or completes the statement.

10–12. When a ferromagnetic material is at its maximum flux density, it is described as being
 a. permeated. **c.** dominated.
 b. reluctant. **d.** saturated.
10–13. The flux pattern of an electromagnetic coil is similar to that of
 a. a bar magnet. **c.** a ferromagnetic atom.
 b. a horseshoe magnet **d.** a straight conductor.

10–14. The polarity of an electromagnetic field depends
 a. on the direction of current.
 b. on the direction of the winding.
 c. on both the direction of the current and the direction of the winding.
 d. on neither the direction of the current nor the direction of the winding.

Solve the following problems.

10–15. What is the MMF of 3 A flowing through 50 turns of wire?
10–16. What is the reluctance when 15 At produces 5 Wb?
10–17. A flux of 0.4 Wb flows through an area 0.4 meters by 0.7 meters. What is the flux density?
10–18. A mmf of 12.7 At is spread out over a rod 0.74 meters long. What is the field intensity?

GLOSSARY

contactor A large relay.
demagnetization Removal of residual magnetism.
domain A large group of ferromagnetic atoms or molecules turned in a similar direction.
electromagnetism The magnetic field produced by an electric current.
field intensity Magnetomotive force per unit length.
ferromagnetic Having the same magnetic properties as iron.
flux The strength of a magnetic field.
flux density The amount of flux per unit cross-sectional area.
flux lines Invisible lines of magnetic force.
induced voltage A voltage produced in a conductor by a change in a magnetic field.
magnet A piece of ferromagnetic material whose domains produce an overall magnetic field throughout the material.
magnetic field The space around a magnet.
magnetic force A natural force produced by the movement of charged particles.
magnetic moment Magnetic force with a particular direction.
magnetic poles The two ends of a magnet where the magnetic field emerges from the magnet.
magnetomotive force The force that produces flux.
permeability The ease of establishing flux within a material.
relay A switch operated by an electromagnet.
reluctance The opposition of a material to flux.
residual magnetism Magnetism remaining in ferromagnetic material that has been exposed to a magnetic field.
retentivity The ability of a ferromagnetic material to retain a magnetic field.
saturation The condition of a ferromagnetic core that contains as much flux as it can hold.
solenoid An electromagnet that produces a short back-and-forth movement of a steel rod.

11 ALTERNATING CURRENT

● SECTION OUTLINE

11–1   Alternators

11–2   The ac Sine Wave

11–3   ac Voltage Values

11–4   Frequency and Period

11–5   Phase

● TROUBLESHOOTING: Applying Your Knowledge

Your friend in another state is also taking a course in basic electricity and electronics, just as you are. He got excited, ran out, and bought a multimeter. Then he went home and began measuring every electrical thing he could find. He called you last night, confused and upset, believing he bought a defective multimeter. First he went out to his car and measured the battery voltage. Sure enough, the meter read about 12 V. Then he went into the house and stuck the probes into several ac wall receptacles. The meter read zero volts, yet the lights were on in his house and he knows for a fact that nothing is wrong with the receptacles. What do you think his problem might be?

● OVERVIEW AND LEARNING OBJECTIVES

In general, current flows in one of two patterns: direct or alternating. Direct current is simple. That is why we began our study of electricity with direct current circuits, powered by batteries or regulated power supplies.

We must now understand alternating current circuits for several reasons. First, all commercial electrical power is in the form of alternating current. Second, many electronic signals are alternating currents, such as the currents produced by a microphone from sound waves. Third, there are two important electrical characteristics, inductance and capacitance, that have a much greater effect in alternating current circuits than they have in direct current circuits.

After completing this chapter, you will be able to:

1. Describe the voltage values in an ac wave.
2. Describe ac time values of cycle, period, frequency, and phase.
3. Describe the relationships between voltage values, polarity, and time values in an ac wave.
4. Compare ac and dc voltage, current, and power.

SECTION 11–1: Alternators

In this section you will learn the following:

→ Mechanical alternators create ac voltage in a winding by rotating to produce relative motion between the winding and a magnetic field.

→ The voltage produced by an alternator rises from zero to a maximum value, then falls back to zero. Each time the voltage reaches zero, the polarity of the terminals reverses.

Voltage is induced in a conductor when there is relative motion between the conductor and a magnetic field. This principle is the basis of all mechanical voltage generators. The principle is illustrated in Figure 11–1. A loop of wire is mounted on a shaft and located within a magnetic field. An external mechanical force causes the shaft to rotate, producing relative motion between the loop and the magnetic field. Voltage is induced into the loop, and the two ends of the loop act as terminals of a voltage source. Each end is connected to a metal slip ring, which is insulated from the shaft. Small blocks of soft carbon make electrical contact with the slip rings by rubbing against them. The carbon blocks are called brushes and are held against the slip rings by springs which are not shown. Conductors from the brushes bring the induced voltage to an external load, where the voltage produces current.

In actual generators, the single loop of wire becomes a winding of hundreds of loops, and the slip rings and brushes may have a more complicated shape. However, the basic principle is as shown in the illustration.

In Figure 11–2, we are looking at a cross-sectional view of the loop. The shaft, slip rings, brushes, and the back of the loop are not shown. Recall from Chapter 10 that the amount of induced voltage depends on the strength of the flux, speed of relative motion, number of loops (turns) of wire, and angle between the conductors and the flux lines. In any generator, the flux, speed, and number of loops have been predetermined by the engineer who designed the generator. However, the angle between the conductors and the flux lines is constantly changing due to the rotation of the loop. When the conductors are moving past points A and C, they are moving parallel to the flux lines, not crossing them. Thus, at points A and C, the induced voltage is zero. When the conductors are moving past points B and D, they are moving straight across the flux lines. Thus, at points B and D, the induced voltage has a maximum value, depending on the other factors mentioned earlier.

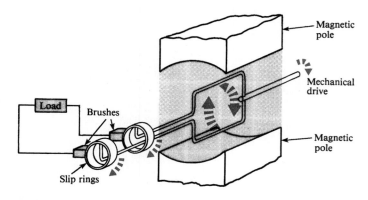

FIGURE 11–1
A basic alternator.

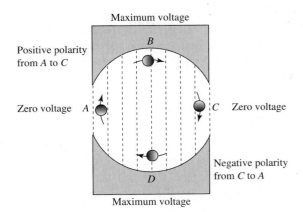

FIGURE 11–2
Rotation of a conductor through magnetic flux.

The polarity of the induced voltage depends on the direction of relative motion. A conductor moving from A to C is moving across the lines to the right, causing electrons to move into the conductor away from you as viewed in the illustration. The loop is acting as a voltage source, and electrons enter the positive terminal of a voltage source. Thus, as the conductor moves from A to C, its polarity is positive. Similarly, as the conductor moves from C to A, its direction through the flux is to the left. This reverses the polarity: The conductor is negative as it moves from C to A.

Figure 11–3 summarizes the results of rotating the loop through the flux. At point A, a conductor has an induced voltage of zero. As the conductor rotates toward point B, the induced voltage continuously increases in value with a positive polarity. At point B, the voltage reaches its maximum value. Past point B, the voltage decreases in value but the polarity remains positive. At point C, the voltage has fallen to zero again.

The pattern repeats from C to A with opposite polarity. From C to D, voltage is increasing in value with negative polarity. The voltage again reaches its maximum value at point D. From D to A, the value of the voltage is decreasing but the polarity remains negative.

In summary, at the beginning and at the halfway point in the rotation, the voltage has a value of zero. At the one-quarter and three-quarters points, the voltage reaches its maximum value. At all other points, the voltage is either increasing or decreasing in value. In other words, the voltage is continuously rising and falling, never remaining at a

Position	Voltage	Polarity
A	zero	none
	increasing	positive
B	maximum	positive
	decreasing	positive
C	zero	none
	increasing	negative
D	maximum	negative
	decreasing	negative
A	zero	none

FIGURE 11–3
Data table describing voltage and polarity as conductor moves around the magnetic field.

steady value. Moreover, the polarity reverses when the voltage reaches zero. The polarity will be positive for half the rotation and negative for the other half.

The word *alternate* means to change back and forth. Because the polarity of each terminal alternates between positive and negative, the generator is called an *alternator,* and the voltage it produces is called *alternating voltage.* Therefore, the current produced by the voltage will also alternate, with the electrons flowing one way through the load for half a rotation, then flowing the other way for the other half. Thus, the current is called *alternating current,* or simply ac.

→ *Self-Test*

1. How do mechanical alternators create ac voltage?
2. Describe the voltage and polarity produced by an alternator.

SECTION 11–2: The ac Sine Wave

In this section you will learn the following:

→ The ac voltage and polarity pattern follows an exact mathematical pattern known as a sine wave.
→ A cycle is a single repetition of the sine wave pattern.
→ Cycle time is measured in degrees. One cycle consists of 360 degrees of time.

The Graph of Alternating Voltage and Current

Figure 11–4 is a graph constructed from the data table in Figure 11–3. To show polarity as well as voltage, we locate zero volts in the middle of the graph. Thus, a voltage value with a positive polarity is located above the zero line, and a voltage value with a negative polarity is below the line.

It is important for you to understand that a voltage shown below the line is a negative polarity, not a negative value. There is no such thing as a voltage less than zero. The voltage has two maximum values, at points B and D. The voltage is increasing from A to B and also from C to D.

Like all voltage sources, an alternator has two terminals. However, the graph shows the voltage at only one terminal. In actual alternators, one terminal is usually grounded,

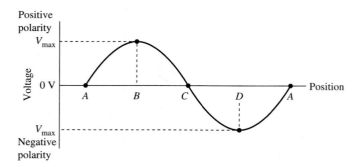

FIGURE 11–4
The graph of the data table in Figure 11–3. Because of its shape and mathematical pattern, this graph is called a sine wave.

as shown in Figure 11–5. The grounded terminal is often called the neutral terminal, and the ungrounded terminal is called the hot terminal. Thus, Figure 11–4 is a graph of the voltage on the hot terminal compared to the neutral terminal (ground).

Because we know the details of how the loop is rotating inside the alternator, we realize that the voltage and polarity of the hot terminal are caused by the position of the loop as it rotates. Thus, we have labeled the horizontal axis in the graph as indicating the position of the rotating loop.

Time

It takes time for the loop to rotate around inside the alternator. Thus, we see voltage and polarity changing with time. Therefore, it makes sense for time to be the horizontal line of the ac graph, as shown in Figure 11–6(a).

The ac pattern repeats itself as the loop continues to rotate. One repetition of the ac voltage/polarity pattern is called a *cycle*. We could measure cycle time in real time, with units of seconds, or perhaps smaller units such as milliseconds. However, we have learned through experience that it is helpful to have a different way of measuring cycle time.

A cycle is similar to a circle, and a circle can be divided into 360 units called *degrees*. In other words, we say that a circle contains 360° (degrees). In ac theory, we have borrowed the term *degree* from the language of circles and made it our unit of measuring cycle time.

In ac theory, when we speak of degrees, we are speaking of time. Thus, we say that one cycle contains 360° of time. Therefore, the horizontal axis in the graph in Figure 11–6(b) is labeled as time and marked in degrees. Point *A*, the starting time of rotation, has a time value of 0°. Point *B*, the time of maximum positive voltage, has a time value of 90° (one-quarter cycle). Point *C*, the time when voltage has fallen to zero and polarity reverses from positive to negative, has a time value of 180° (one-half cycle). Point *D*, the

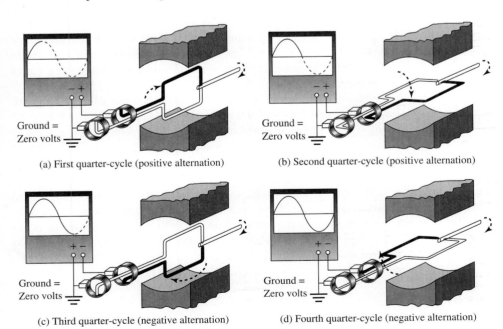

(a) First quarter-cycle (positive alternation)

Ground = Zero volts

(b) Second quarter-cycle (positive alternation)

Ground = Zero volts

(c) Third quarter-cycle (negative alternation)

Ground = Zero volts

(d) Fourth quarter-cycle (negative alternation)

Ground = Zero volts

FIGURE 11–5
The oscilloscope draws a sine wave to represent the voltage and polarity of the hot terminal compared to the neutral (grounded) terminal.

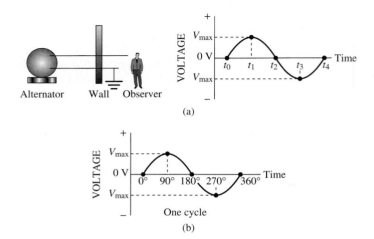

FIGURE 11–6
(a) The observer sees voltage and polarity changing with time. (b) One cycle is one repetition of the sine wave pattern. One cycle contains 360° of time.

time of maximum negative voltage, has a time value of 270° (three-quarters cycle). At the end of the cycle, the loop has returned to point *A*. The voltage has again fallen to zero, polarity reverses from negative to positive, and the next cycle begins. The time can be called either 360° (the end of one cycle) or 0° (the beginning of the next cycle).

Other ac Sources

It is possible to create alternating voltage and current without mechanical rotation. Alternating voltage and current are created by a microphone, a phonograph needle, a tape recorder or VCR playback head, and several other electronic devices. These ac voltages and currents are called signals because they represent information: sounds and pictures that are important or interesting to us.

Of course, our electronic equipment may also pick up or produce unwanted signals, which we call noise. However, both signals and noise often have the form of ac waves.

Alternating voltage and current can also be created by using a special electronic circuit called an oscillator. Oscillators are widely used in radio and television circuits. Oscillators are also used to create signals in electronic musical instruments such as a synthesizer.

Thus, audio, video, and radio electronics contain numerous ac signal voltages and currents. To troubleshoot and repair these circuits, technicians use a signal generator, which is a piece of test equipment that generates small ac voltages and currents that act as substitute signals. You will use signal generators in your experimental circuits as you progress in your studies.

The Sine Wave

The ac graph shown in Figure 11–6 follows an exact, well-known mathematical pattern called a sine function. We will learn more about this in the section on trigonometry. The graph has the visual appearance of a wave and thus the ac graph is commonly called an ac *sine wave*. We say that ac voltages and currents follow a sinusoidal pattern.

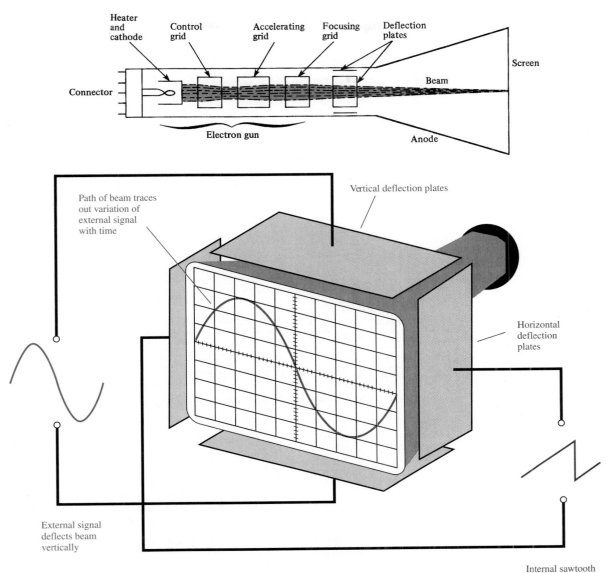

FIGURE 11–7
An oscilloscope uses an electron beam and phosphor screen to draw a graph of voltage.

The Oscilloscope

The *oscilloscope* is an important electronic test instrument designed to draw a graph of voltage versus time, using a cathode ray tube (CRT). The electron beam is deflected by two sets of plates, as shown in Figure 11–7. The voltage on the horizontal plates varies with time and moves the beam from left to right. The voltage on the vertical plates varies with the voltage detected on the test probes and moves the beam up or down. Just as you use a pencil to draw a graph on paper, the oscilloscope uses the electron beam to draw a graph on the phosphor coating on the inner surface of the glass screen. Thus, the oscilloscope draws a graph of the voltage on the test probe versus time.

Because the oscilloscope operates at electronic speed, it draws the graph in real time, as the voltage is changing.

Basic oscilloscopes work best when the test voltage is a repeating pattern, such as a sine wave. The oscilloscope redraws the graph over and over on the screen and is synchronized to draw the pattern at the same location on the screen each time. The phosphor dots on the screen thus glow continuously, and the pattern appears to be stationary on the screen. The screen of the oscilloscope has a grid (crisscross lines) etched on its surface to aid in reading and interpreting the graph. The controls on the oscilloscope change the scaling of the grid, allowing the screen to display larger or smaller ranges of both voltage and time. Other controls affect the way the scope displays the wave pattern on the screen. More advanced oscilloscopes use digital circuits including microprocessors and memory chips to capture nonrepeating patterns and events (such as a single voltage spike when a switch closes) and store them in memory or record them on a diskette for later examination.

→ *Self-Test*

- - - - - - -

3. What is the name of the ac voltage and polarity pattern?
4. What is a cycle?
5. What is the unit of cycle time?

- -

SECTION 11–3: ac Voltage Values

In this section you will learn the following:

→ The ac peak value is the highest voltage on the sine wave.

→ The ac rms value is the value of the dc voltage that produces an equivalent power in an identical load.

→ The ac average value equals half the sum of the positive peak and the negative peak.

→ The average value of an ac cycle is zero because the two ac half-cycles are identical except for polarity.

→ When an ac signal voltage rides on a dc reference-level voltage, the ac average value equals the dc reference level.

→ An ac voltmeter reads rms value. A dc voltmeter reads average value.

→ The ac peak-to-peak value equals the positive peak value minus the negative peak value. Normally, the ac peak-to-peak value is twice the peak value.

→ The peak value is an instantaneous value. All other ac values are not instantaneous and are based on the entire ac cycle.

→ The true (rms) power of ac voltage and current is half the peak power.

We have learned that power is the rate at which energy is released and, therefore, the rate at which work is done. Heat is a common form of energy, and we may compare levels of power in two different electrical circuits with resistive loads by comparing the amounts of heat dissipated (released) in the loads. If the loads are identical, then equal power levels in the two circuits would produce equal rates of heat dissipation in the loads, which would result in equal load temperatures.

As shown in Figure 11–8, the temperature of a dc resistive load remains at a constant value because dc voltage and current are constant values and, therefore, dc power is also a constant value. Electrical energy is converted into heat at a constant rate, producing a steady load temperature.

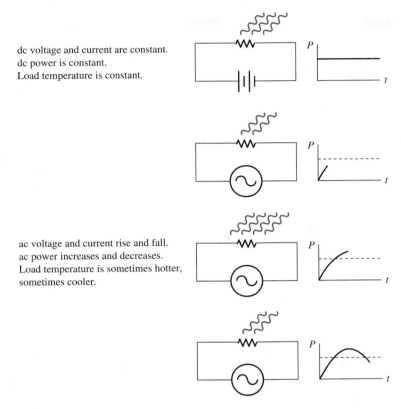

dc voltage and current are constant.
dc power is constant.
Load temperature is constant.

ac voltage and current rise and fall.
ac power increases and decreases.
Load temperature is sometimes hotter,
sometimes cooler.

FIGURE 11–8
Comparison of dc and ac power.

Practical ac voltages change rapidly and thus power is delivered to the load in pulses, similar to the firing of pistons in an automobile engine. Like the automobile engine, the load reaches a steady temperature, as the cycle of heating and cooling reaches balance.

The highest voltage on the ac sine wave is called the *peak* voltage. Scientists have found that the temperature of the load will be equal to the temperature of a similar load connected to a dc source with a voltage equal to 0.707 of the ac peak voltage, as shown in Figure 11–9. In other words, an ac generator that produces a peak voltage of 100 V produces exactly the same power as a dc battery of 70.7 V.

For any sinusoidal ac voltage, the value of the dc equivalent voltage is called the *rms* (root-mean-square) voltage and is equal to 0.707 of the ac peak voltage. Thus, when we say that an ac sine wave with a peak value of 100 V has an rms value of 70.7 volts, we mean that the ac sine wave with a 100-V peak produces exactly the same power as a steady 70.7 volts dc. The expression *root-mean-square* refers to the mathematical procedure used to determine the sine wave's dc equivalent value.

The rms value is the ac value most widely used. All ac meters are designed to measure rms values, and when people speak of ac voltages such as 120 V or 240 V, they mean the rms value. The relationship between the ac peak and ac rms values is given in Formula Diagram 11–1.

Formula Diagram 11–1

$$\frac{V_{\text{rms}}}{0.707 \mid V_{\text{peak}}}$$

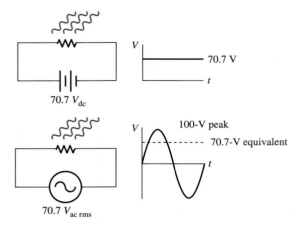

FIGURE 11–9
rms voltage.

● *SKILL-BUILDER 11–1*
- -

An ac voltmeter measures 120 V in a wall outlet. What is the peak value of the voltage?

GIVEN:
  V_{rms} = 120 V

FIND:
  V_{peak}

STRATEGY:
  Use Formula Diagram 11–1 to find V_{peak}.

SOLUTION:

V_{peak} = V_{rms} ÷ 0.707	determine correct formula.
= 120 V ÷ 0.707	substitute values
= 169.7 V	calculate

Repeat the Skill-Builder using the following values. You should obtain the results that follow.

Given:					
V_{rms}	240 V	345 kV	277 V	15 μV	38 mV
Find:					
V_{peak}	339 V	488 kV	392 V	21.2 μV	53.7 mV

- -

All insulators and devices in an ac circuit must withstand the peak voltage, which we see is about 41% greater than the rms value. In low-voltage circuits, the difference between rms and peak is usually not enough to matter. However, in high-voltage power transmission lines, the difference is large and important. For example, a 345-kV transmission line reaches a peak value of about 488 kV. If the line were designed to insulate against only 345 kV, the extra 143 kV would probably cause arcing and line failure.

Average Voltage

The *average* voltage is the dc measured value of a changing voltage or current. In other words, the average value is what a multimeter will read when set on a dc scale.

Finding any average value is a mathematical process of adding a number of values, then dividing the total by the number of samples. In the case of voltage, positive voltages are given positive numerical values, and negative voltages are given negative values.

Because the ac sine wave consists of two half-cycles which are identical except for polarity, the average value of an ac sine wave is zero volts. This is why we use the root-mean-square method of obtaining the equivalent dc voltage. The rms procedure eliminates the positive and negative signs and produces a nonzero result.

As you progress in your study of electronics, you will learn various formulas for average voltages. Always remember that any formula for average voltage applies only to one particular waveform. In other words, if the waveform changes, the formula for average voltage also changes.

Using Multimeters to Measure ac Voltage

Multimeters can be set to measure either dc or ac voltage. Because of the way the meter is designed, the ac voltage reading is the rms value, and the dc voltage reading is the average value. For example, in Figure 11–10, an alternator produces a voltage with a peak value of 169 V. A multimeter set for ac voltage reads the alternator output as 120 V, which is the rms value (169 V × 0.707 = 120 V). The same multimeter set for dc voltage reads 0 V, which is the average value of any standard sine wave.

Reading ac Voltage with an Oscilloscope

The graph of any voltage is called the waveform of the voltage. An oscilloscope draws the waveform of the voltage on its probes, as shown in Figure 11–11. Thus, an oscilloscope allows us to measure peak voltage.

The grid is the set of lines drawn on the screen of the oscilloscope. The distance between two lines on the grid is called a division. The oscilloscope has two control knobs, which set the value of the divisions. If you place your finger on the grid and move your finger from left to right, each grid line you cross is one horizontal division. The

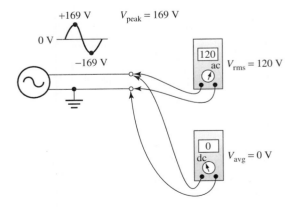

FIGURE 11–10
The ac voltmeters read rms values; dc voltmeters read average values.

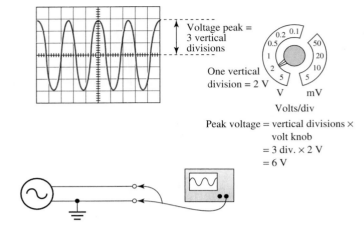

FIGURE 11–11
Oscilloscopes display peak values.

time-division knob sets the value of each horizontal division on the grid. If you move your finger up and down (vertically), each grid line you cross is one vertical division. The volt-division knob sets the value of each vertical division.

The waveform shown in Figure 11–11 has three vertical divisions (three screen lines) between the zero-volt line and the peak of the wave. The volt-division knob is set so that each vertical division has a value of 2 V. Therefore, the peak value of the wave is 3 divisions multiplied by 2 V, which is 6 V. An ac voltmeter connected to the same source would read 4.24 V_{rms} (6 V × 0.707 = 4.24 V).

● **SKILL-BUILDER 11–2**

- -

Figure 11–12 shows five examples of oscilloscope waveforms and voltage-division knob settings. For each of these, determine the peak and rms voltage values. You should obtain the following results.

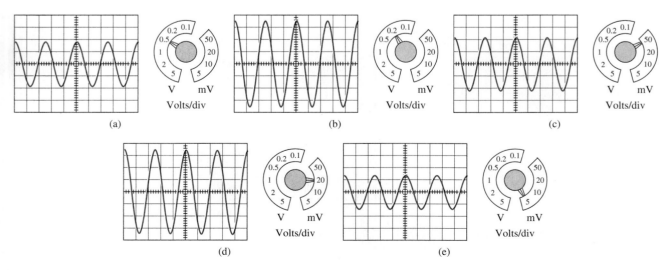

(a)   (b)   (c)

(d)   (e)

FIGURE 11–12

vertical divisions	volts per division	peak voltage	rms voltage
a. 1.8	0.5	0.9 V	0.636 V
b. 3.5	0.2	0.7 V	0.495 V
c. 2.2	50 mV	110 mV	77.78 mV
d. 3.5	20 mV	70 mV	49.5 mV
e. 1.4	5 mV	7 mV	4.95 mV

Peak-to-Peak Values

In audio, video, radio, and television circuits, the voltage is often a combination of dc and ac. The dc voltage is produced by the machine's power supply and is necessary to operate the tubes, transistors, and integrated circuits which control current in the machine's circuits. The smaller ac voltage represents the audio and video signals and any radio carrier waves necessary for transmission.

Figure 11–13 shows an ac signal riding on a dc reference level. Notice that the ac voltage wave is not centered on zero volts and does not change polarity. In this case, we measure the *peak-to-peak* value of the ac waveform.

The peak-to-peak value is measured from the lower peak to the upper peak. In Figure 11–13, the peak-to-peak value is 6 V (14 V − 8 V).

The peak-to-peak value is always twice the peak value, as shown in Formula Diagram 11–2.

Formula Diagram 11–2

$$\frac{V_{\text{peak-to-peak}}}{2 \mid V_{\text{peak}}}$$

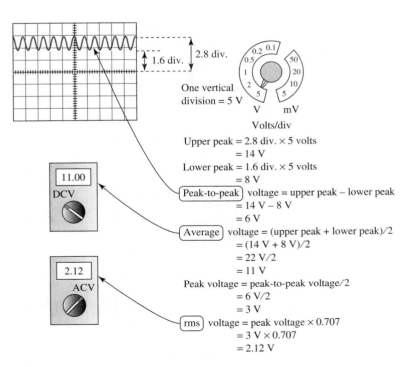

Upper peak = 2.8 div. × 5 volts
          = 14 V

Lower peak = 1.6 div. × 5 volts
          = 8 V

Peak-to-peak voltage = upper peak − lower peak
          = 14 V − 8 V
          = 6 V

Average voltage = (upper peak + lower peak)/2
          = (14 V + 8 V)/2
          = 22 V/2
          = 11 V

Peak voltage = peak-to-peak voltage/2
          = 6 V/2
          = 3 V

rms voltage = peak voltage × 0.707
          = 3 V × 0.707
          = 2.12 V

FIGURE 11–13
Relationship of peak-to-peak, average, and rms values.

Thus, the peak value of the waveform in Figure 11–13 is 3 V (6 V ÷ 2). An ac voltmeter connected to this source would ignore the dc voltage component and read the rms value of the ac component, which in this case is 2.12 V (3 V_{peak} × 0.707). On the other hand, a dc voltmeter would ignore the ac voltage component and read the dc component. In this case, the dc component is the voltage halfway between the upper and lower peaks, which is 11 V. We have learned that a dc voltmeter always reads the average value of any waveform. Therefore, the average value of an ac sine wave is the value halfway between the upper and lower peaks. This is consistent with the earlier statement that the average value of a standard sine wave is zero volts. A standard ac sine wave does not ride on a dc reference level, and the voltage halfway between the two peaks is zero volts.

What Figure 11–13 shows is an ac signal of 2.12 V_{rms} (which is also 3 V_{peak} and 6 $V_{peak-to-peak}$) riding on a dc reference level of 11 V. The 3-V negative peak of the signal reduces the reference level to 8 V (11 V − 3 V), and the 3-V positive peak increases the reference level to 14 V (11 V + 3 V).

When you work on actual audio and video circuits, you must understand clearly the relationships between ac signal levels and dc reference levels. Figure 11–13 is important to your future advanced studies. Make sure you understand it thoroughly.

● SKILL-BUILDER 11–3

An oscilloscope screen displays a sinusoidal voltage with an upper peak of 15.7 V and a lower peak of 10.3 V. Find the dc reference level and the rms value of the ac component.

GIVEN:
$V_{max} = 15.7$ V
$V_{min} = 10.3$ V

FIND:
1. V_{avg} dc reference level = V_{avg}
2. V_{rms}

SOLUTION:

V_{avg} = $(V_{max} + V_{min}) \div 2$ formula for average
     = (15.7 V + 10.3 V) ÷ 2   value of a sine wave
     = 26 V ÷ 2
     = 13 V

V_{p-p} = $V_{max} - V_{min}$ 1. Find peak-to-peak
     = 15.7 V − 10.3 V   value.
     = 5.4 V

V_{peak} = $V_{p-p} \div 2$ 2. Find peak value.
     = 5.4 V ÷ 2
     = 2.7 V

V_{rms} = $V_{peak} \times 0.707$ 3. Find rms value.
     = 2.7 V × 0.707
     = 1.91 V

Repeat the Skill-Builder using the following values. You should obtain the results that follow.

Given:

V_{max}	6.8 V	28.7 V	3.64 V	12.3 V	1.015 V
V_{min}	3.2 V	22.9 V	3.36 V	11.6 V	1.007 V

Find:

V_{avg}	5.0 V	25.8 V	3.5 V	11.95 V	1.011 V
V_{p-p}	3.6 V	5.8 V	280 mV	700 mV	8 mV
V_{peak}	1.8 V	2.9 V	140 mV	350 mV	4 mV
V_{rms}	1.27 V	2.05 V	99 mV	247 mV	2.83 mV

Instantaneous and Noninstantaneous Values

An *instantaneous* value is a value of voltage, current, or power at an exact moment (instant) in time. The sine wave is a graph of all the instantaneous values.

The peak value is an instantaneous value, the highest instantaneous value on the wave. All other values (rms, average, peak-to-peak) are not instantaneous. They are based on the entire ac cycle.

True Power Is Half of Peak Power

The rms power is often called *true power* to emphasize that it is the dc equivalent value.

True ac power is half of peak ac power. Suppose that an ac wave has a peak voltage of 100 V and produces a peak current of 20 A. The peak power then is:

$$
\begin{aligned}
P_{peak} &= V_{peak} \times I_{peak} \\
&= 100 \text{ V} \times 20 \text{ A} \\
&= 2,000 \text{ W}
\end{aligned}
$$

We have learned that the dc equivalent of ac voltage or current is the ac peak value multiplied by 0.707. Therefore,

$$
\begin{aligned}
V_{rms} &= V_{peak} \times 0.707 \\
&= 100 \text{ V} \times 0.707 \\
&= 70.7 \text{ V} \\
I_{rms} &= I_{peak} \times 0.707 \\
&= 20 \text{ A} \times 0.707 \\
&= 14.14 \text{ A}
\end{aligned}
$$

The rms power is the product of rms and voltage and current:

$$
\begin{aligned}
P_{rms} &= V_{rms} \times I_{rms} \\
&= 70.7 \text{ V} \times 14.14 \text{ A} \\
&= 1,000 \text{ W}
\end{aligned}
$$

Notice that the rms power is exactly half the peak power. This is true because 0.707×0.707 equals 0.5. Thus, when we multiply both peak voltage and peak current by 0.707, in effect we are multiplying peak power by 0.5.

→ Self-Test

6. What is peak value?
7. What is rms value?
8. What is average value?
9. What is the average value of an ac cycle?
10. What is the average value of an ac signal riding on a dc reference level?
11. What meter reads rms value, and what meter reads average value?
12. What is the peak-to-peak value?
13. Which ac values are instantaneous, and which are not?
14. What is the relationship between true power and peak power?

SECTION 11–4: Frequency and Period

In this section you will learn the following:

→ Frequency *(f)* is the number of cycles per second of a repeating waveform. The unit of frequency is hertz (Hz) or cycles per second (cps).

→ The North American standard frequency for commercial ac power is 60 Hz. The European standard is 50 Hz.

→ Audio frequencies are below 15 kHz. Radio frequencies are above 15 kHz.

→ The period of a waveform *(T)* is the time of a single cycle. The unit of period is the second.

→ Frequency and period are mutual reciprocals.

→ To measure period with an oscilloscope, count the number of horizontal divisions in one cycle of the wave, and multiply by the setting of the time-division knob.

→ To measure frequency with an oscilloscope, first measure period, and then reciprocate the value.

→ Wavelength (λ) is the distance between electromagnetic waves traveling in space. The unit of wavelength is the meter.

→ Wavelength equals the speed of light multiplied by the period, or the speed of light divided by the frequency.

→ A band is a specific group of frequencies.

Frequency

The number of cycles produced in one second of time by a repeating waveform is called the *frequency* of the wave, abbreviated *f*. For many years, frequency was measured in cycles per seconds (cps). The modern International System of measurement measures frequency in *hertz* (Hz). One hertz is the same as one cycle per second. Thus, we may say that a sine wave has a frequency of either 60 cps or 60 Hz, which means 60 repetitions of the sine wave pattern in a time of one second. Frequency is illustrated in Figure 11–14.

Electrical utility companies in North America produce ac power at a standard frequency of 60 Hz. In Europe, the standard is 50 Hz. Certain special alternators, such as those used in aircraft, may produce ac voltages at higher frequencies, such as 440 Hz.

The concept of frequency may be applied to any waveform, including sound waves. For example, the keys on a piano produce sound waves at frequencies ranging from 55

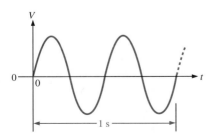

(a) Lower frequency: fewer cycles per second

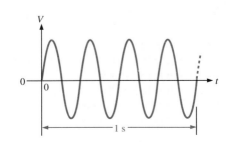

(b) Higher frequency: more cycles per second

FIGURE 11–14
The frequency is the number of cycles per second.

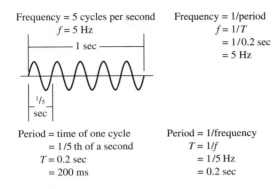

Frequency = 5 cycles per second
f = 5 Hz

Frequency = 1/period
$f = 1/T$
   $= 1/0.2$ sec
   $= 5$ Hz

Period = time of one cycle
   = 1/5 th of a second
$T = 0.2$ sec
   = 200 ms

Period = 1/frequency
$T = 1/f$
   = 1/5 Hz
   = 0.2 sec

FIGURE 11–15
The relationship of period and frequency.

Hz to about 6 kHz. The average adult can hear sound waves at frequencies between 20 Hz and 15 kHz.

When sound waves are converted by a microphone into electrical signals, the electrical waves have the same frequencies as the original sound waves. This is called the audio frequency range of ac waves. An audio-frequency (AF) generator is an electronic machine which produces low-voltage ac waves at audio frequencies. These ac waves are used to test audio circuits. The test waves are called signals because they are substitutes for actual signals produced by sound waves. Therefore, the audio generator may also be called a signal generator.

The magnetic fields produced by ac electrical signals vibrate at the same frequency as the signals. At frequencies starting around 15 kHz, the vibrating magnetic field begins to produce magnetic waves which travel through space. These magnetic waves are called radio waves and have the same frequency as the ac electrical waves that produced them. Thus, the radio-frequency (RF) range begins where the audio-frequency range ends. A radio-frequency (RF) generator is an electronic machine which produces low-voltage ac waves at radio frequencies. The radio-frequency range is much wider than the audio range, extending into the range of heat waves.

Period

The time of a single cycle is called the *period,* abbreviated *T.* The period is the reciprocal of the frequency. The period is measured in actual time, as shown in Figure 11–15. If an ac generator produces 60 cycles per second, then the time required to generate each cycle is 1/60th of a second, or 16.67 milliseconds (ms). The period of a 50-Hz wave is 1/50th of a second, or 20 ms.

Remember that if you reciprocate a number larger than 1, you get a fraction. If you then reciprocate the fraction, you get the original number. Therefore, since the period is the reciprocal of the frequency, the frequency is also the reciprocal of the period. For example, the reciprocal of 16.67 ms is 1 / 16.67 ms, which equals 60 Hz.

The mathematical relationship between frequency and period is given by Formula Diagram 11–3.

Formula Diagram 11–3

$$\frac{1}{f \mid T}$$

● *SKILL-BUILDER 11–4*
- -

Find the period of a 1-kHz sine wave.

GIVEN:
 $f = 1$ kHz

FIND:
 T

SOLUTION:

$T = \dfrac{1}{f}$ determine correct formula

 $= 1/1$ kHz   substitute values

 $= 1$ ms   calculate

      Repeat the Skill-Builder using the following values. You should obtain the results that follow.

Given:					
f	200 Hz	150 kHz	565 kHz	100 MHz	3 GHz
Find:					
T	5 ms	6.67 µs	1.77 µs	10 ns	333 ps

- -

Calculating Period and Frequency from the Sine Wave Graph

Figure 11–16 shows a sine wave displayed on the screen of an oscilloscope. As mentioned earlier, the distance between two grid lines is called a division, and the value of each division along the horizontal time line is set by the volt-division knob. By counting the divisions in one cycle and multiplying by the value of each division (the knob setting), we can find the period. Then we can reciprocate the period to find the frequency.

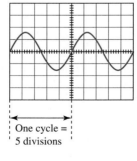
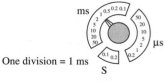

One division = 1 ms

Time/div

Period = horizontal divisions × time knob
= 5 div. × 1 ms
$T = 5$ ms
frequency = 1/period
= 1/5 ms
$f = 200$ Hz

One cycle = 5 divisions

FIGURE 11–16
An oscilloscope displays period. Frequency is calculated from period.

The 0° cycle time is called the positive *transition* of the wave because the wave is changing (in transition) to a positive polarity. The 90° cycle time is the positive peak, the 180° cycle time is the negative transition, and the 270° cycle time is the negative peak. The 270° cycle time is also called the −90° time.

Note carefully in Figure 11–16 that there are five time-division lines between positive transitions. The time-division line at the first positive transition is not counted.

● SKILL-BUILDER 11–5

Find the period and frequency of the wave in Figure 11–16.

GIVEN:
Graph of the waveform
Setting of the time-division knob

FIND:
Period of the wave
Frequency of the wave

STRATEGY:
1. Read the graph and count the number of time divisions in one cycle of the wave.
2. Multiply the number of time divisions in one cycle by the value of each time division. The result is the period of the wave.
3. Reciprocate the period to find the frequency.

SOLUTION:
1. By reading the graph, we count 5 time divisions between the first positive transition and the next positive transition.

   Positive transition to positive transition is one cycle.

2. $T = 5 \text{ td} \times 1 \text{ ms}$
   $= 5 \text{ ms}$

   Period equals number of time divisions between positive transitions multiplied by value of each time division.

3. $f = \dfrac{1}{T}$

   Frequency is the reciprocal of period.

   $= \dfrac{1}{5 \text{ ms}}$

   Substitute values.

   $= 200 \text{ Hz}$

   Calculate.

Repeat the Skill-Builder using the waveforms shown in Figure 11–17. You should obtain the following results.

Given:

Number of time divisions				
6.8	4.5	5.6	7.2	4.8

Value of each time division				
100 µs	50 µs	10 µs	5 µs	5 µs

Find:

T	680 µs	225 µs	56 µs	36 µs	24 µs
f	1.47 kHz	4.44 kHz	17.86 kHz	27.78 kHz	41.67 kHz

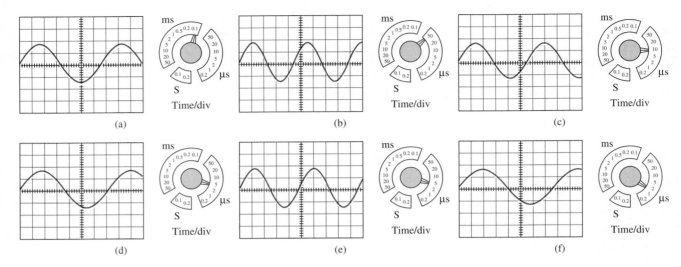

FIGURE 11–17

Wavelength

Wavelength is the distance between electromagnetic waves traveling in space. The symbol for wavelength is the Greek letter lambda (λ).

In Figure 11–18, an ac current in a transmitting antenna has risen to a positive peak. The magnetic energy around the antenna also reaches a peak and begins traveling away from the antenna as the crest of a magnetic wave. One cycle later, the current again has reached a positive peak and another wave of magnetic energy leaves the antenna. The question is this: How far from the antenna has the first magnetic wave traveled by the time the second magnetic wave is released?

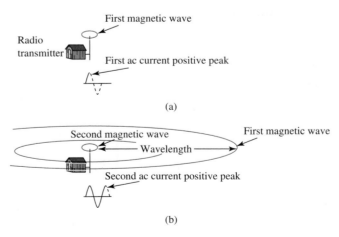

FIGURE 11–18
Wavelength is the distance between successive
electromagnetic waves.

Distance equals speed multiplied by time. Therefore, we can calculate the distance that the first magnetic wave has traveled by multiplying the speed of the wave and the time between waves. The speed of the magnetic wave is the speed of light: 3×10^8 meters (300,00 kilometers) per second. The time between magnetic waves is the period of the ac current producing the waves. Therefore, the distance the wave has traveled equals the speed of light *(c)* multiplied by the period *(T)*.

Since the period is the reciprocal of the frequency, another formula for wavelength is the speed of light divided by the frequency. Both formulas are given in Formula 11–4.

Formula 11–4

$$\lambda = T \times c$$

$$= \frac{c}{f}$$

Notice in Figure 11–18 that a magnetic wave is generated by the negative peak of ac current, and this wave is shown in the illustration. However, by definition, wavelength is the distance traveled by the wave in one complete cycle of the current. Therefore, this middle wave is not counted.

● *SKILL-BUILDER 11–6*

The main frequency of a radio transmitter is called the carrier frequency. What is the wavelength of the carrier wave of an AM radio station transmitting at a frequency of 740 kHz?

GIVEN:
   $f = 740$ kHz

FIND:
   λ

SOLUTION:

$\lambda = \dfrac{c}{f}$ determine correct formula

$= \dfrac{3 \times 10^8 \text{ m/sec}}{740 \text{ kHz}}$ substitute values

$= 405$ meters   calculate

Repeat the Skill-Builder using the following values. You should obtain the results that follow.

Frequency	Wavelength
10 MHz (amateur shortwave radio)	30 meters
57 MHz (television channel 2)	5.3 meters
88 MHz (FM radio station)	3.4 meters
108 MHz (FM radio station)	2.78 meters
213 MHz (television channel 13)	1.4 meters
3 GHz (radar, microwave oven)	0.1 meter

MATH TIP

The correct way to enter 3×10^8 into your calculator is 3 EXP 8. Do *not* key in "× 10."

Bands

A specific group of frequencies is called a *band*. Bands may be classified either by frequency or by wavelength.

Band Name	Frequency	Wavelength
VLF Very low frequency	< 30 kHz	> 10 km
LF Low frequency	30–300 kHz	10 km–1 km
MF Medium frequency	300–3,000 kHz	1,000–100 m
HF High frequency	3–30 MHz	100–10 m
VHF Very high frequency	30–300 MHz	10–1 m
UHF Ultra high frequency	0.3–3 GHz	100–10 cm
SHF Super high frequency	3–30 GHz	10–1 cm
EHF Extra high frequency	30–300 GHz	1–0.1 cm

AM radio stations operate in the MF band between 535 and 1,650 kHz. FM radio stations operate in the VHF band between 88 and 108 MHz. Television stations are classified as either VHF or UHF, according to the band within which they operate. Channels 2 to 13 are VHF, and channels 14 to 83 are UHF. The HF and VHF bands are also used for mobile communication such as aircraft, police, fire, ambulance, taxis, and so on. The HF band is called the shortwave band and is widely used by amateur radio operators. Frequencies in the UHF band and higher are called microwave frequencies. Radar and spacecraft communications use these frequencies, as well as the familiar microwave oven, which operates at about 3 GHz, or 3 billion waves per second.

→ *Self-Test*

15. What is frequency and its unit?
16. What are the standard frequencies for commercial ac power?
17. What are audio and radio frequencies?
18. What is period and its unit?
19. What is the relationship between period and frequency?
20. What is the procedure for measuring period with an oscilloscope?
21. What is the procedure for measuring frequency with an oscilloscope?
22. What is wavelength and its unit?
23. What are the relationships between wavelength, the speed of light, period, and frequency?
24. What is a band?

SECTION 11–5: Phase

In this section you will learn the following:

→ Phase is the timing relationship between ac waves.

→ Two waves are in phase if similar transitions occur at the same time. Otherwise, the waves are out of phase.

→ Phase angle (θ) is the timing difference between two waves. The unit of phase angle is the degree.

→ If two waves are out of phase, the wave that is farther ahead in its cycle is the leading wave. The wave that is farther behind in its cycle is the lagging wave.

→ As viewed on an oscilloscope, the wave that transitions (crosses zero volts) farther to the left is leading, and the wave that transitions farther to the right is lagging.

Phase is the timing relationship between waves. If two waves are in time with each other, rising and falling together, reaching their peaks and polarity reversals at exactly the

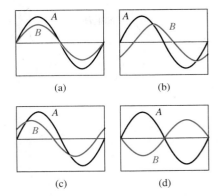

FIGURE 11–19
(a) *A* and *B* in phase. (b) *A* leads *B*.
(c) *A* lags *B*. (d) *A* and *B* out of
phase 180°.

same time, we say the waves are in phase. Otherwise, if the waves do not rise and fall together, we say they are out of phase.

When two sine waves are drawn on the same graph, as shown in Figure 11–19, it is easy to recognize whether the waves are in or out of phase. In Figure 11–19(a), waves *A* and *B* are in phase. They cross the zero-volt line at the same location on the graph, going in the same direction. Crossing at the same location on the graph means the waves cross zero volts at the same time. We would say that the waves have similar transitions (polarity reversals) at the same time.

In Figure 11–19(b) and (c), the waves cross zero at different locations on the graph, which means different times. Therefore, the waves are out of phase. In (b), wave *A* crosses before wave *B*. In (c), wave *B* crosses first.

If the waves happen to be exactly half a cycle out of phase, they will cross zero at the same location on the graph, but going in different directions, as shown in Figure 11–19(d). The transitions occur at the same time, but the polarity reversals are opposite, not similar. Wave *A* is going from positive to negative, and wave *B* from negative to positive.

The graphs in Figure 11–19 are important to you. As an electronics technician, you will do a lot of work with oscilloscopes, and you must learn to recognize phase relationships quickly. What you see in Figure 11–19 is exactly what you would see on the screen of an oscilloscope. Make sure you understand and remember how to interpret phase relationships.

Phase Angle

When two waves are out of phase, we describe this by speaking of the *phase angle* between the waves. The phase angle is the difference in cycle time between the two waves and is measured in degrees.

In Figure 11–20(a), at time t_1, wave *A* is at 90° cycle time, and wave *B* is at 180°. The difference in cycle time is 90° (180° − 90°). Therefore, the phase angle between the waves is 90°.

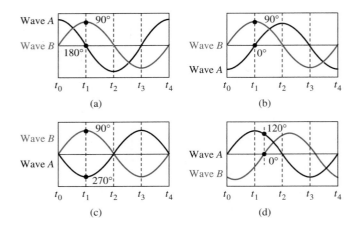

FIGURE 11–20
Standard phase angles.

In Figure 11–20(b), at t_1, A is at 0° and B is at 90°. The difference is still 90° (90° − 0°) and, thus, this phase angle is also 90°.

In Figure 11–20(c), at t_1, A is at 270° and B is at 90°. The difference is 180° (270° − 90°) and, thus, the phase angle is 180°.

In Figure 11–20(d), A is at 120° when B is at 0°. The difference is 120° (120° − 0°) and, thus, the phase angle is 120°.

A change in phase angle from one value to another is called a phase shift. For example, if two waves have a phase angle of 30°, and the phase angle changes to 45°, the phase shift is 15°, as shown in Figure 11–21.

At any instant in time, we may have a particular phase angle between two waves. If the two waves have exactly the same frequency, then the phase angle will remain constant. If the waves have different frequencies, the phase angle will shift continuously.

To understand this, let's make a comparison. Suppose two athletes are running laps around a circular track, and one runner is ahead of the other. If both are running at the same speed, the distance between them will stay the same. This means there will be a certain interval of time between them, which also will not change. For example, one runner may keep a 3-second lead over the other throughout the race.

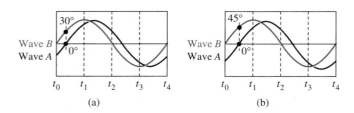

FIGURE 11–21
15° phase shift: A leads B by 30° at (a), increasing to 45° at (b).

On the other hand, if the athletes run at different speeds, the distance between them will change continuously. As the faster runner pulls ahead, the distance between the runners will increase until the faster runner is half a lap ahead. Then the distance will decrease as the faster runner overtakes and passes the slower. Since the distance between the runners is continuously changing, so is the time interval between them.

Applying this comparison to our two electrical waves, if two waves have the same frequency, this is like the two athletes running at the same speed. The phase angle is the time interval between the waves, and this will not change when both have the same frequency. However, if the waves have different frequencies, the wave with the higher frequency completes its cycle more quickly than the wave with the lower frequency. Like the faster runner, the higher-frequency wave gets ahead of the other, then catches up with it. Thus, the phase angle between the waves is continuously changing, increasing to 180°, then decreasing to 0°.

Synchronization

Oscilloscopes have control knobs that *synchronize* the scope with the waves on the screen. This makes the waves appear to stand still. Synchronizing the scope means adjusting the controls until the timing of the scope matches the timing of the waves. If the waves seem to move left or right and won't stand still, the problem is either that the synchronization sensitivity control is set too low, or else there is a difference in frequency between the wave used for synchronization and the wave appearing on the screen.

Constant Phase Angle: Lead and Lag

If two waves are out of phase, one wave will be farther along in its cycle than the other by an amount equal to the phase angle. For example, Figure 11–22 shows two waves, *A* and *B*, that have a phase angle of 20°. When wave *A* is farther along at 90° on its cycle, wave *B* is 20° behind at 70° on its own cycle. If *A* and *B* have the same frequency, the phase angle remains constant. Therefore, at a later time, when *A* is at 160° on its cycle, *B* is still 20° behind at 140° on its own cycle.

We use the words *lead* and *lag* to describe the phase relationship of two waves. The *leading* wave is the one that is farther ahead in its cycle and, thus, the *lagging* wave is the one that is farther behind. In Figure 11–22, we can describe the phase rela-

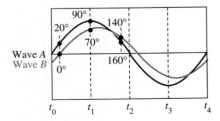

FIGURE 11–22
Wave *A* leads wave *B*, or wave *B* lags wave *A*.

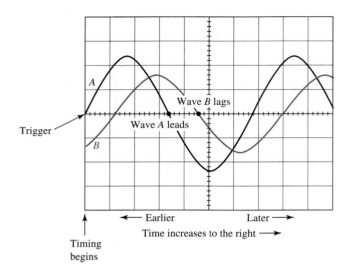

FIGURE 11–23
Leading events are seen on the left of the screen;
lagging events on the right.

tionship of *A* and *B* in two ways: *A* leads *B,* or *B* lags *A.* Either way, the phase angle
is 20°.

When you see two waves on an oscilloscope screen, you must be able to say which
is leading and lagging. Figure 11–23 makes this clear. Since the time scale on the graph
begins on the left and increases to the right, a point on the time line farther to the left represents an event that happened earlier in time. Similarly, a point on the time line farther to
the right represents an event that happened later in time.

In this drawing, we are examining the negative transitions of the waves rather than
the positive transitions. We do this for two reasons. First, there is no natural reason to prefer the positive transition. All that matters is that we are comparing identical cycle times

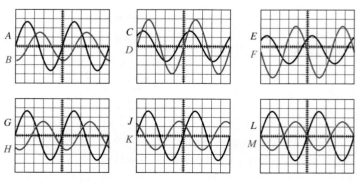

FIGURE 11–24

on the two waves. Second, when an oscilloscope is set to trigger (begin its timing) on positive transitions, it is often easier to determine phase relationships by looking at the negative transitions.

Wave A's negative transition is farther to the left on the time line, and wave B's negative transition is farther to the right. Since the time line increases from left to right, wave A's negative transition happened earlier or sooner than wave B's. Therefore, wave A is leading (transition more to the left on the time line), and wave B is lagging (transition more to the right).

● **SKILL-BUILDER 11–7** -

Refer to Figure 11–24. For each group of waves, determine which is leading and which is lagging.

You should obtain the results that follow.

A leads B	or	B lags A
C leads D	or	D lags C
E leads F	or	F lags E
G leads H	or	H lags G
J leads K	or	K lags J
L leads M	or	M lags L

Relationship Between Period and Phase Angle

We have just learned that the horizontal distance between similar transitions of two waves is the phase angle (timing difference) between the waves. Since oscilloscope screens are calibrated (marked) in real time, the phase angle is displayed in real time. We need to know how to measure the phase angle from the screen in real time (seconds) and convert the measurement into cycle time (degrees).

Since all waves contain 360° in one cycle, we can think of the phase angle as a fraction of 360°. Thus, we will follow these steps to calculate the phase angle:

1. Measure the period in real time.
2. Measure the phase angle in real time.
3. Express the phase angle as a fraction of the period: phase angle divided by period.
4. Multiply this fraction by 360° to calculate phase angle in degrees.

The process can be written as a single formula and is given in Formula 11–5.

Formula 11–5

$$\text{Phase angle (deg)} = \frac{\text{phase angle (real time)}}{\text{period (real time)}} \times 360$$

● **SKILL-BUILDER 11–8** -

Referring to Figure 11–25(a) and (b), calculate the phase angle in degrees.

1. In Figure 11–25(a), each time division is 10 μs. The period of wave A (distance between similar transitions) is 6.4 time divisions. Therefore, the period is 6.4 div × 10 μs = 64 μs. The period of wave B is the same. The two waves have the same frequency (1 / 64 μs = 15.625 kHz), and the phase angle is constant.

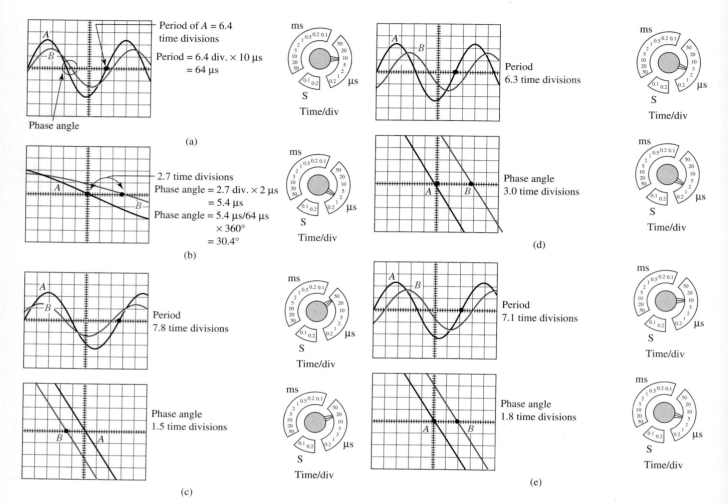

Period of A = 6.4 time divisions

Period = 6.4 div. × 10 μs
= 64 μs

Phase angle

(a)

2.7 time divisions

Phase angle = 2.7 div. × 2 μs
= 5.4 μs

Phase angle = 5.4 μs/64 μs
× 360°
= 30.4°

(b)

Period
7.8 time divisions

Phase angle
1.5 time divisions

(c)

Period
6.3 time divisions

Phase angle
3.0 time divisions

(d)

Period
7.1 time divisions

Phase angle
1.8 time divisions

(e)

FIGURE 11–25
Steps in calculating phase angle from oscilloscope displays.

2. In Figure 11–25(b), each time division is 2 μs. We have increased the setting of the time-division knob to read the phase angle more closely. In effect, we have magnified the view. The distance between similar transitions of waves A and B is 2.7 time divisions. Therefore, the phase angle between the waves is 2.7 div × 2 μs = 5.4 μs.

3. We express the phase angle as a fraction of the period:

$$\frac{\text{phase angle}}{\text{period}}$$

$$\frac{5.4 \ \mu s}{64 \ \mu s} = 0.084375$$

4. We multiply this fraction by 360° to calculate the phase angle in degrees:

$$0.084375 \times 360° = 30.4°$$

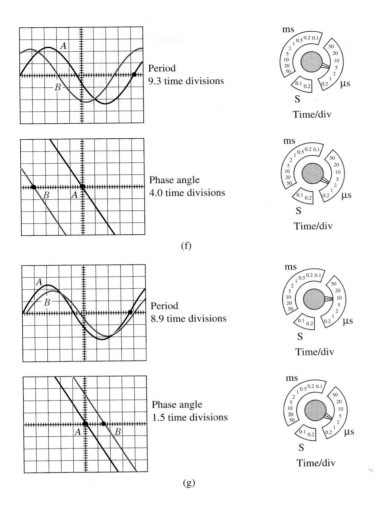

Period
9.3 time divisions

ms

S
Time/div

Phase angle
4.0 time divisions

ms

S
Time/div

(f)

Period
8.9 time divisions

ms

S
Time/div

Phase angle
1.5 time divisions

ms

S
Time/div

(g)

FIGURE 11–25 *(CONTINUED)*

Using Formula 11–5,

$$\text{Phase angle (deg)} = \frac{\text{phase angle (real time)}}{\text{period (real time)}} \times 360$$

$$= \frac{5.4 \ \mu s}{64 \ \mu s} \times 360$$

$$= 30.4°$$

Repeat the Skill-Builder using the following values. You should obtain the results that follow.

Given:

Figure	(c)	(d)	(e)	(f)	(g)
T	390 μs	31.5 μs	71 μs	46.5 μs	89 μs

Find:

θ (real)	30 μs	6 μs	9 μs	8 μs	3 μs
θ (deg)	27.7°	68.6°	45.6°	61.9°	12.1°

→ *Self-Test*

25. What is phase?
26. When are waves in or out of phase?
27. What is phase angle and its unit?
28. If two waves are out phase, which is leading and which is lagging?
29. If two waves out of phase are viewed on an oscilloscope, which is leading and which is lagging?

Formula List

Formula Diagram 11–1

$$\frac{V_{rms}}{0.707 \mid V_{peak}}$$

Formula Diagram 11–2

$$\frac{V_{peak\text{-}to\text{-}peak}}{2 \mid V_{peak}}$$

Formula Diagram 11–3

$$\frac{1}{f \mid T}$$

Formula 11–4

$$\lambda = T \times c$$

$$\lambda = \frac{c}{f}$$

Formula 11–5

$$\text{Phase angle (deg)} = \frac{\text{phase angle (real time)}}{\text{period (real time)}} \times 360$$

● TROUBLESHOOTING: Applying Your Knowledge

Your friend's multimeter measured the 12-V dc auto battery successfully but measured the ac wall receptacles as zero volts. Most likely your friend did not set the meter to measure ac voltage. In other words, the meter is still set to measure dc voltage. A meter set for dc voltage will measure an ac average voltage value, and for a full sine wave, the average value is exactly zero. Therefore, the meter is probably functioning correctly. Switching the meter to the ac voltage position will allow it to measure the ac rms value of about 120 V.

CHAPTER SUMMARY:
ANSWERS TO SELF-TESTS

1. Mechanical alternators create ac voltage in a winding by rotating to produce relative motion between the winding and a magnetic field.
2. The voltage produced by an alternator rises from zero to a maximum value, then falls back to zero. Each time the voltage reaches zero, the polarity of the terminals reverses.
3. The ac voltage and polarity pattern follows an exact mathematical pattern known as a sine wave.
4. A cycle is a single repetition of the sine wave pattern.
5. Cycle time is measured in degrees. One cycle consists of 360 degrees of time.
6. The ac peak value is the highest voltage on the sine wave.
7. The ac rms value is the value of the dc voltage that produces an equivalent power in an identical load.
8. The ac average value equals half the sum of the positive peak and the negative peak.
9. The average value of an ac cycle is zero because the two ac half-cycles are identical except for polarity.
10. When an ac signal voltage rides on a dc reference-level voltage, the ac average value equals the dc reference level.
11. An ac voltmeter reads rms value. A dc voltmeter reads average value.
12. The ac peak-to-peak value equals the positive peak value minus the negative peak value. Normally, the ac peak-to-peak value is twice the peak value.
13. The peak value is an instantaneous value. All other ac values are not instantaneous and are based on the entire ac cycle.
14. The true (rms) power of ac voltage and current is half the peak power.
15. Frequency (*f*) is the number of cycles per second of a repeating waveform. The unit of frequency is the hertz (Hz) or cycles per second (cps).
16. The North American standard frequency for commercial ac power is 60 Hz. The European standard is 50 Hz.
17. Audio frequencies are below 15 kHz. Radio frequencies are above 15 kHz.
18. The period (*T*) of a waveform is the time of a single cycle. The unit of period is the second.
19. Frequency and period are reciprocals of each other.
20. To measure period with an oscilloscope, count the number of horizontal divisions in one cycle of the wave, and multiply by the setting of the time-division knob.
21. To measure frequency with an oscilloscope, first measure period, and then reciprocate the value.
22. Wavelength (λ) is the distance between electromagnetic waves traveling in space. The unit of wavelength is the meter.
23. Wavelength equals the speed of light multiplied by the period, or the speed of light divided by the frequency.
24. A band is a specific group of frequencies.
25. Phase is the timing relationship between ac waves.
26. Two waves are in phase if similar transitions occur at the same time. Otherwise, the waves are out of phase.
27. Phase angle (θ) is the timing difference between two waves. The unit of phase angle is the degree.
28. If two waves are out of phase, the wave that is farther ahead in its cycle is the leading wave. The wave that is farther behind in its cycle is the lagging wave.
29. As viewed on an oscilloscope, the wave that transitions (crosses zero volts) farther to the left is leading, and the wave that transitions farther to the right is lagging.

CHAPTER REVIEW QUESTIONS

Determine whether the following questions are true or false.

11–1. The ac voltage reverses polarity when it reaches its peak value.
11–2. The frequency cycle contains 360°.
11–3. The ac waves can be produced only by alternators.
11–4. An oscilloscope draws graphs of voltage waveforms in real time (as they occur).
11–5. True power is twice peak power.
11–6. The standard commercial power frequency in North America is 120 Hz.
11–7. One cycle equals 1 hertz per second.
11–8. Waves *A* and *B* have the same frequency. When wave *B* is at its negative transition, wave *A* is at its positive peak. Therefore, wave *B* leads wave *A* by 90°.

11–9. Waves *A* and *B* have the same frequency. When wave *B* is at its negative transition, wave *A* is at its negative peak. Therefore, wave *B* lags wave *A* by 90°.

Select the response that best answers the question or completes the statement.

11–10. Which of the following is *not* a factor in determining the amount of induced voltage in an alternator?
   a. strength of the flux
   b. speed of relative motion
   c. direction of relative motion
   d. number of loops (turns) of wire
   e. angle between the conductors and the flux lines

11–11. Which of the following is the chief factor in determining the polarity of induced voltage in an alternator?
   a. strength of the flux
   b. speed of relative motion
   c. direction of relative motion
   d. number of loops (turns) of wire
   e. angle between the conductors and the flux lines

11–12. One repetition of a waveform is called
   a. period.
   b. cycle.
   c. hertz.
   d. frequency.

11–13. The time required for one repetition of a waveform is called
   a. period.
   b. cycle.
   c. hertz.
   d. frequency.

11–14. The number of repetitions of a waveform per second is called
   a. period.
   b. cycle.
   c. hertz.
   d. frequency.

11–15. The _____ value is the highest voltage on the sine wave.
   a. peak
   b. rms
   c. average
   d. peak-to-peak

11–16. The _____ value is the value of the dc voltage that produces an equivalent power in an identical load.
   a. peak
   b. rms
   c. average
   d. peak-to-peak

11–17. The _____ value of a sine wave is zero.
   a. peak
   b. rms
   c. average
   d. peak-to-peak

11–18. The _____ value is normally twice the peak value.

   a. peak
   b. rms
   c. average
   d. peak-to-peak

11–19. Wavelength equals
   a. period times frequency.
   b. period times the speed of light.
   c. period divided by frequency.
   d. speed of light divided by frequency.
   e. a. and d. above.
   f. b. and c. above.
   g. b. and d. above.

Solve the following problems.

11–20. An ac voltmeter measures 208 V. What is the peak voltage?

11–21. An ac waveform on an oscilloscope screen extends 3.2 vertical divisions above the zero-volt reference level. The volt-division control is set at 0.5. What is the peak voltage and the rms voltage?

11–22. An oscilloscope screen displays a sinusoidal voltage with an upper peak of 12.4 V and a lower peak of 9.7 V. Find the dc reference level (V_{avg}) and the rms value of the ac component.

11–23. A 6.81 V_{rms} ac voltage rides on a 21.7-V dc reference level. What is the maximum and minimum value of the waveform?

11–24. Find the period of a 6.3-kHz sine wave.

11–25. An oscilloscope displays a sine wave whose positive transitions are 6.8 horizontal divisions apart. The time-division control is set at 5 μs. What is the period and frequency of the wave?

11–26. What is the wavelength of a police radio transmitting at 275 MHz?

11–27. An ac voltage has a peak value of 170 V. What is the instantaneous value at 30°?

The following problems are more challenging.

11–28. An oscilloscope indicates a peak voltage of 2.38 V across a 600-Ω load. What is the true power dissipated in the load?

11–29. An ac voltage *A* has a peak value of 13.7 V, and an ac voltage *B* has a peak value of 21.8 V. Voltage *A* leads voltage *B* by 30°. What is the instantaneous value of both waves when voltage *A* is at 70° cycle time?

11–30. An oscilloscope displays two waves of identical frequency. The time-division control is set on 2 μs. The distance between successive positive transitions of wave *A* is 7.2 divisions. Wave *B* has a positive transition 0.8 divisions to the left of a positive transition of wave *A*. What is the phase angle in degrees between the waves, and which one is lagging?

11–31. An amateur radio operator tells you he can receive signals between 1 meter and 5 meters. What are the corresponding frequencies?

GLOSSARY

alternating current Current that flows alternately back and forth.

alternating voltage A voltage that follows a sine wave pattern, rising and falling between zero and a peak value, and reversing polarity each time it reaches zero volts.

alternator A mechanical generator that produces ac voltage.

average The dc measured value of a changing voltage or current. The average value of an ac sine wave is zero.

band A group of frequencies.

cycle One repetition of the ac sine wave pattern.

degree A unit of cycle time. One cycle consists of 360°.

frequency The number of cycles per second of an ac wave.

hertz (Hz) The unit of frequency; cycles per second.

instantaneous The value of voltage or current at a specific instant of time.

lagging Behind in time or phase.

leading Ahead in time or phase.

oscilloscope An electronic test instrument that uses a cathode ray tube to draw a graph of voltage versus time.

peak The highest value of voltage or current on an ac sine wave.

peak-to-peak The difference between the maximum and minimum values of an ac wave.

period The time of a single ac cycle.

phase The timing relationship between ac waves.

phase angle The timing difference between two ac waves, expressed in degrees.

rms The dc equivalent value of an ac voltage or current, equal to 0.707 times the peak value.

sine wave A mathematical pattern; the graph of alternating voltage and current.

synchronization A timing feature of an oscilloscope that allows waves on the screen to appear to stand still.

transition The point on a sine wave where the value is zero; the 0° and 180° points.

true power The rms (dc equivalent) power of an ac voltage and current, equal to half the peak power.

wavelength The distance between electromagnetic waves.

CHAPTER

12

INDUCTANCE

● SECTION OUTLINE

12–1   Inductance
12–2   Inductors
12–3   Inductive Opposition to dc Current: Transient Time
12–4   Inductive Opposition to ac Current: Reactance

● TROUBLESHOOTING: Applying Your Knowledge

Your cousin is taking vocational training to become a welder and is presently learning to use an electric arc welder. He bought one to use at home and is having trouble with it. For some reason, it won't produce enough heat to weld properly. He suspects an electrical problem and asks you to come over. As you watch him using the welder, you notice that he keeps the welding cables coiled up while welding. His welder is the type that uses ac current to produce the arc. What do you think the trouble might be?

● OVERVIEW AND LEARNING OBJECTIVES

There are three electrical characteristics that affect current: resistance, inductance, and capacitance. In this chapter we will learn about the second characteristic, inductance.

Many important devices, such as electromagnets, transformers, motors, generators, and so on, either operate because of inductance or else have a significant effect on current because of their inductance. Thus, understanding the basic principle of inductance is essential to understanding many of these and other electrical devices.

After completing this chapter, you will be able to:

1. Define inductance and describe the conditions likely to produce a significant amount of inductance.
2. Describe the physical properties of a device that cause inductance.
3. Explain the concept of transient time in an inductive circuit.
4. Describe the effects of inductance in a dc circuit.
5. Explain the concept of reactance in an inductive circuit.

SECTION 12–1: Inductance

In this section you will learn the following:

→ When current changes in a conductor, the corresponding change in magnetic flux induces a voltage within the conductor.

→ The polarity of a self-induced voltage opposes the change in current that produces the voltage.

→ Inductance *(L)* is the ability of a conductor to create a voltage within itself due to a change in its current.

We have learned about electromagnetism and induction. These two natural principles are so closely related that they can occur at the same time in a single electrical device.

In Figure 12–1(a), a current flowing in a conductor has created a magnetic flux around the conductor. This is the principle of electromagnetism. In Figure 12–1(b), the current has increased, which causes the magnetic flux to increase. Since flux lines cannot touch each other, the original flux lines have to move farther out to make room for the new flux lines. In other words, the increase in current causes the magnetic field to expand.

Where do the flux lines come from? The flux lines appear to originate inside the conductor. They are then pushed outside the conductor by new flux lines that form as the current increases. In other words, the increase in current causes flux lines to move through the conductor, from inside to outside.

This movement of the flux lines within the conductor causes induction to occur in the conductor. Remember that induction is the creation of a voltage inside a conductor when there is relative motion between a conductor and magnetic flux lines.

To visualize this process more clearly, look at Figure 12–2. In (a), V_S, a dc voltage source, produces a steady current in a coil, which produces a steady magnetic flux around the coil. In (b), V_S increases in value, causing an increase in current. This in turn causes an increase in flux, and the magnetic field expands, moving outward from the coil. The outward movement of the flux induces a second voltage, V_L, in the coil itself.

Figure 12–2 is important to your understanding of the next several chapters. When the current is steady, there is only one voltage in the circuit, the source voltage V_S which is producing the current. However, when the current changes, a second voltage, V_L, is created in the circuit. V_L is a voltage inside the coil, created by induction as the coil feels its own magnetic flux changing, due to the change in current. V_L exists only during the time the current is changing. As soon as the current reaches a new steady value, V_L disappears because induction ceases when the magnetic flux stops moving.

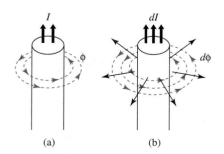

(a)   (b)

FIGURE 12–1
When current increases, magnetic flux increases and expands outward.

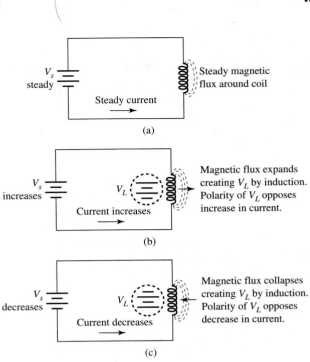

FIGURE 12–2
Changes in current produce V_L. Polarity of V_L opposes change in current.

As shown in Figure 12–2(b), the polarity of V_L is the opposite of V_S while the current is increasing. As a general rule, whenever a natural force tries to cause a change, another natural force opposes the change. In this case, the natural opposition to change is in the form of a second voltage, V_L, whose polarity opposes the effort of V_S to increase the current.

In Figure 12–2(c), V_S is decreasing to a lower value, causing a decrease (change) in current. Now the magnetic flux is collapsing (moving back into the coil). Once again, the movement of flux within the coil induces (creates) V_L, which opposes the change in current. Since the current is now decreasing, V_L opposes the change by assuming the same polarity as V_S.

We may ask how strong V_L will be, and how much it will affect the current. The value and effect of V_L depend on several factors, discussed in the following sections. The important point here is that two different coils may produce different amounts of V_L when exposed to equal changes in current. Thus, we may speak of the ability of a coil to produce V_L when the coil's current changes. This ability is called self-inductance, or simply *inductance*. The symbol for inductance is *L*.

Inductance is one of the three basic electrical characteristics, along with resistance and capacitance. All conductors possess a certain amount of inductance, but coils and windings possess more of it than straight conductors.

→ *Self-Test*

1. What happens when current changes in a conductor?
2. What is the relationship between the polarity of a self-induced voltage and the change in current that produces the voltage?
3. What is inductance?

SECTION 12–2: Inductors

In this section you will learn the following:

→ An inductance of 1 henry is the ability to produce 1 volt when the current changes at the rate of 1 ampere per second.
→ The four physical factors of inductance are the number of turns of wire, the cross-sectional area, the length of the core, and the permeability of the core.
→ The total inductance of inductors in series is the sum of the inductance of each inductor.
→ The total inductance of inductors in parallel is the reciprocal of the sum of the reciprocals of the inductance of each inductor.

An *inductor* is a device specifically designed to produce a substantial amount of inductance. Inductors are usually coils of wire wrapped around a solid core of ferromagnetic material. The schematic symbol for an inductor is shown in Figure 12–3.

Unit of Inductance

The unit of inductance is the *henry,* symbolized H. An inductance of 1 henry is the ability to produce 1 volt when the current changes at the rate of 1 ampere per second.

To understand this definition, suppose the current in an inductor is changing so that at the end of each second of time, the current has either increased or decreased by exactly 1 ampere. If this rate of change of current produces a self-induced voltage of 1 volt in the inductor, then the inductor has an inductance value of 1 henry.

Notice that the rate of change of the current is what matters, and not the actual value of the current. In other words, a current that changes from 2 A to 3 A in 1 second has a rate of change of 1 ampere per second, and so does a current that changes from 250 A to 251 A in 1 second. Either of these would produce 1 volt in a 1-henry inductor.

Types of Inductors

Figure 12–4 illustrates several types of inductors. Figure 12–4(a) shows examples of larger inductors used in radios and televisions. The inductor consists of a coil of fine wire wrapped around a core. This type of inductor is called a choke. Figure 12–4(b) shows miniature inductors encapsulated in plastic cases.

Inductors are classified according to the type of core material as air, iron, or ferrite core. These types may be either fixed value or variable. Figure 12–5 shows the schematic symbols for the various types.

Ferrite is a ceramic material with fine particles of iron mixed in. The iron particles give ferrite its ferromagnetic properties, and the ceramic material makes it an electrical insulator. In high-frequency circuits, a ferrite bead may be threaded onto a straight section of wire to act as an inductor with a small value of *L*.

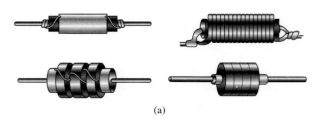

(a)

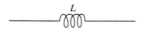

FIGURE 12–3
Symbol for inductor.

FIGURE 12–4
Typical inductors.

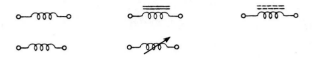

FIGURE 12–5
Symbols indicate inductor core type and whether
inductor is fixed or variable.

As shown in Figure 12–5, a diagonal arrow added to the schematic symbol indicates the inductor is a variable type. Variable inductors have threaded cores which can be screwed in and out of the winding. This changes the amount of the core exposed to the magnetic field, which changes the inductance. Variable inductors are often used in tuning circuits in radios and televisions. Some have a slot for a screwdriver blade, and others have a recessed hexagonal socket for an Allen wrench.

Inductors must be adjusted with plastic tools. If a steel tool is used, the magnetic field will flow through the tool as well as through the core. Thus, the value of L will be affected unpredictably by the contact of the tool. Special plastic tools called alignment tools are used for adjusting inductors.

Physical Characteristics of an Inductor

As shown in Figure 12–6, there are four physical characteristics of an inductor that significantly affect its ability to create a self-induced voltage:

1. N, the number of turns of wire.
2. A, the cross-sectional area in meters squared.
3. l, the length of the core in meters.
4. μ, the permeability of the core.

These four factors are expressed in Formula 12–1.

Formula 12–1

$$L = \frac{N^2 \mu A}{l}$$

● SKILL-BUILDER 12–1

An inductor consists of 400 turns of wire wrapped on a core with a cross-sectional area of 7.1256×10^{-5} square meters and a length of 2.225×10^{-2} meters. The core has a permeability of 3.7×10^{-3}. What is the inductance?

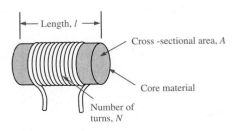

FIGURE 12–6
Physical factors that determine
inductance.

GIVEN:
    $N = 400$
    $A = 7.1256 \times 10^{-5}$ square meters
    $l = 2.225 \times 10^{-2}$ meters
    $\mu = 3.7 \times 10^{-3}$

FIND:
    L

SOLUTION:

$$L = \frac{N^2 \mu A}{l}$$ Formula 12–1

$$\begin{aligned} = \ & 400^2 \times 3.7 \times 10^{-3} \\ & \times 7.1256 \times 10^{-5} \\ & \div 2.225 \times 10^{-2} \end{aligned}$$ substitute values

$$= 1.895 \text{ H}$$

keystrokes:
400 [x^2]
[×] 3.7 [EXP] 3 [±]
[×] 7.1256 [EXP] 5 [±]
[÷] 2.225 [EXP] 2 [±]
[=]

- -

Inductors in Series

When inductors are placed in series, the total inductance is the sum of each component's inductance, as shown in Figure 12–7 and stated in Formula 12–2. Notice that although inductance and resistance are entirely different electrical characteristics, the formulas for total values in series are similar.

Formula 12–2

$$L_T = L_1 + L_2 + L_3 + \ldots + L_n$$

● SKILL-BUILDER 12–2
- -

In Figure 12–8, three inductors are connected in series. What is the total inductance if $L_1 = 1.2$ mH, $L_2 = 600$ μH, and $L_3 = 0.8$ mH?

GIVEN:
    $L_1 = 1.2$ Mh
    $L_2 = 600$ μH
    $L_3 = 0.8$ mH

FIND:
    L_T

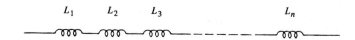

L_1 L_2 L_3 L_n

FIGURE 12–7
Total series inductance is the sum of component values.

SOLUTION:

$$L_T = L_1 + L_2 + L_3$$
$$= 1.2 \text{ mH} + 600 \text{ μH} + 0.8 \text{ mH}$$

Formula 12–2
keystrokes:
1.2 [EXP] 3 [±] [+]
600 [EXP] 6 [±] [+]
0.8 [EXP] 3 [±] [=]

$$= 2.6 \text{ mH}$$

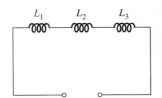

FIGURE 12–8
Circuit for Skill-Builder
12–3.

Repeat the Skill-Builder using the following values. You should obtain the results that follow.

Given:

L_1	300 μH	33,000 μH	1.87 H	780 μH	7.85 H
L_2	1,500 μH	27,000 μH	0.57 H	3.5 mH	130,000 μH
L_3	0.45 mH	46,000 μH	350 mH	2,200 μH	1,700 mH

Find:

L_T	2.25 mH	106 mH	2.79 H	6.48 mH	9.68 H

Inductors in Parallel

When inductors are placed in parallel, the total inductance is the reciprocal of the sum of the reciprocal of each component's inductance, as shown in Figure 12–9 and stated in Formula 12–3. Notice that although inductance and resistance are entirely different electrical characteristics, the formulas for total values in parallel are similar.

Formula 12–3

$$L_T = \cfrac{1}{\cfrac{1}{L_1} + \cfrac{1}{L_2} + \cfrac{1}{L_3} + \ldots + \cfrac{1}{L_n}}$$

● *SKILL-BUILDER 12–3*

In Figure 12–10, three inductors are connected in parallel. What is the total inductance if $L_1 = 1.2$ mH, $L_2 = 600$ μH, and $L_3 = 0.8$ mH?

GIVEN:

$L_1 = 1.2$ Mh
$L_2 = 600$ μH
$L_3 = 0.8$ mH

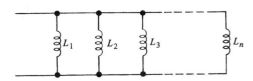

FIGURE 12–9
Total parallel inductance is the reciprocal of the sum of branch reciprocals.

CHAPTER 12

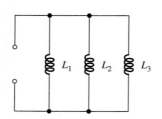

FIGURE 12–10
Circuit for Skill-Builder
12–4.

FIND:
 L_T

SOLUTION:

$$\frac{1}{L_1} = \frac{1}{1.2 \text{ mH}}$$

   $= 833.33$ reciprocal of L_1

$$\frac{1}{L_2} = \frac{1}{600 \text{ μH}}$$

   $= 1,666.67$ reciprocal of L_2

$$\frac{1}{L_3} = \frac{1}{0.8 \text{ mH}}$$

   $= 1,250$ reciprocal of L_3

$$\begin{array}{r} 833.33 \\ + 1,666.67 \\ + 1,250 \\ \hline 3,750 \end{array}$$ sum of reciprocals

$$L_T = \frac{1}{3,750}$$ reciprocate the sum

  $= 266.67 \text{ μH}$ keystrokes:
              1.2 [EXP] 3 [±] [1/x]
              [+]
              600 [EXP] 6 [±] [1/x]
              [+]
              0.8 [EXP] 3 [±] [1/x]
              [=]
              [1/x]

    Repeat the Skill-Builder using the following values. You should obtain the results that follow.

Given:

L_1	300 μH	33,000 μH	1.87 H	780 μH	7.85 H
L_2	1,500 μH	27,000 μH	0.57 H	3.5 mH	130,000 μH
L_3	0.45 mH	46,000 μH	350 mH	2,200 μH	1,700 mH

Find:

L_T	250 μH	11.23 μH	194 μH	494 μH	73 μH

→ *Self-Test*

4. What is an inductance of 1 henry?
5. What are the four physical factors of inductance?
6. What is the total inductance of inductors in series?
7. What is the total inductance of inductors in parallel?

SECTION 12–3: Inductive Opposition to dc Current: Transient Time

In this section you will learn the following:

→ The self-induced voltage in a coil equals the inductance multiplied by the rate of change of current. The polarity of the self-induced voltage opposes the change in current.

→ The transient time of a circuit is the time required for output voltage to reach a steady-state value following a change in input voltage.

→ One time constant is the time required for the output to change an amount equal to approximately 63% of the difference between the output's present value and its final value.

→ The transient time of any circuit is approximately five time constants.

CEMF or Back Voltage

The self-induced voltage in a coil equals the inductance multiplied by the rate of change of current. The polarity of the self-induced voltage opposes the change in current. Because the self-induced voltage V_L opposes changes in current, it is often called counter-electromotive force *(CEMF)*. It is also called *back voltage.*

Inductor Kick

When dc current is first applied to a coil, the sudden change in current produces a voltage equal to the source voltage and opposite in polarity. At first you might think that no current could flow. However, there is a important detail to remember here. The back voltage V_L is produced by the change in current, not by the current itself. If V_L could actually stop the current from changing, then V_L would disappear, and the current would go ahead and change! What actually happens is an automatic balance of forces, where V_L at first is equal to V_S, then gradually falls to zero as the current gradually rises to its normal value. This kind of situation is too complicated to be calculated using ordinary algebra, and a higher form of mathematics, calculus, is used to obtain exact results.

The point here is that when the switch closes, V_L is momentarily equal to V_S, then falls off to zero.

However, quite a different situation arises when the switch opens again and the current suddenly drops toward zero. V_S is no longer connected to the coil, and the rapid collapse of the magnetic field may produce voltages many times larger than V_S. This phenomenon is known as *inductive kick* and is used in conventional automobile ignition systems to produce the 20-kV voltages necessary for the spark plugs from the low 12 V available from the battery.

Machines such as electric arc welders and large motors contain wire windings which produce the magnetic fields essential for the machine's operation. However, when the machine is switched off, these windings may act as strong inductors and produce high voltages. This inductive kick can cause arcing and burning of the switch contacts and serious injury to people who contact these brief but large voltages.

Transients

A *transient* is a change in input voltage that lasts for a short time. A high transient voltage is often nicknamed a voltage *spike* because of the way it looks on an oscilloscope screen.

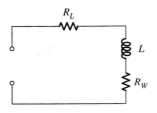

FIGURE 12–11
Winding resistance is
included with load
resistance to calculate
transient time.

Transients cause current to change rapidly. Since inductors oppose a change in current, inductors are an effective way to prevent unwanted changes in current due to transients.

Transients often occur on commercial power lines due to lightning strikes, industrial machines such as arc welders, and so on. Transients can damage electronic equipment even when it is turned off. A device called a surge suppressor is used to protect computers, audio and video equipment, and so on from high transient voltage. Surge suppressors reduce transients with circuits that include large inductors.

Transients also occur in signal lines, producing noise in radio and telephone communication. These transients can be caused by thunderstorms, high-voltage power lines and transformers, spark plug wires in engines, industrial machinery, and so on. This type of noise is generally called electromagnetic interference *(EMI)* or sometimes radio frequency interference *(RFI)*. Circuits called filters use inductors to eliminate the noise produced by these transients.

dc Effects: Time Constants and Transient Time

Inductors cause a time delay in changes in current. The exact amount of time depends on the value of the inductance and the resistance in the circuit and is given by Formula 12–4. This amount of time is called the *transient time* of the circuit, since any transient which lasts for less time than this will be more or less suppressed by the inductor.

Formula 12–4: Transient Time

$$t_{\text{transient}} = 5 \times \frac{L}{R_T}$$

An inductor made of a wire coil has a measurable amount of resistance in the wire. This is called winding resistance (R_W) and must be included in the calculation for transient time. Thus, in Figure 12–11, R_W, the winding resistance of the inductor, is shown as a separate symbol to distinguish it from R_L, the normal load resistance.

In Figure 12–12(a), $L = 250$ mH, $R_W = 7\ \Omega$, and $R_L = 100\ \Omega$. Then $R_T = 107\ \Omega$ and the transient time, $t_{\text{transient}}$, is 5×250 mH $\div\ 107\ \Omega = 11.68$ ms. This means that if the voltage changes suddenly to another value, there will be a time delay of 11.68 ms before the current reaches its new value.

In Figure 12–12(a), the initial voltage is 12 V, producing a current of 112 mA. In Figure 12–12(b), the voltage suddenly changes to 18 V. However, the current remains temporarily at 112 mA. The reason is the induced voltage of 6 V in the coil, acting against the change.

In Figure 12–12(c), the transient time of 11.68 ms has passed. Now the induced voltage in the coil has dropped to zero and the current has reached its new value of 168 mA.

● *SKILL-BUILDER 12–4*
- -

In Figure 12–12, what is the transient time when $R_L = 300\ \Omega$, $L = 400$ mH, and $R_W = 25\ \Omega$?

GIVEN:
  $R_L = 300\ \Omega$
   $L = 400$ mH
  $R_W = 25\ \Omega$

FIND:
  $t_{\text{transient}}$

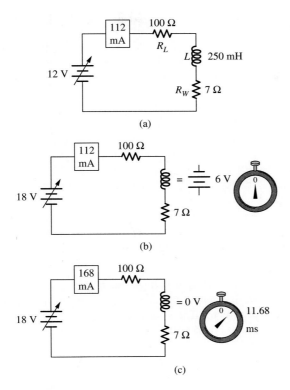

FIGURE 12–12
An example of transient time.

STRATEGY:
Find R_T by adding R_L and R_W. Then use Formula 12–4.

SOLUTION:
$$R_T = R_L + R_W$$
$$= 300\ \Omega + 25\ \Omega$$
$$= 325\ \Omega$$

$$t_{transient} = 5 \times \frac{L}{R_T}$$

$$= 5 \times \frac{300\ mH}{325\ \Omega}$$

$$= 4.62\ ms$$

Repeat the Skill-Builder using the following values. You should obtain the results that follow.

Given:

R_L	500 Ω	75 Ω	120 Ω	680 Ω	750 Ω
L	1,500 mH	125 mH	50 mH	250 mH	1,900 mH
R_W	60 Ω	4 Ω	1 Ω	8 Ω	85 Ω

Find:

R_T	560 Ω	79 Ω	121 Ω	688 Ω	835 Ω
t_{trans}	13.39 ms	7.91 ms	2.07 ms	1.82 ms	11.38 ms

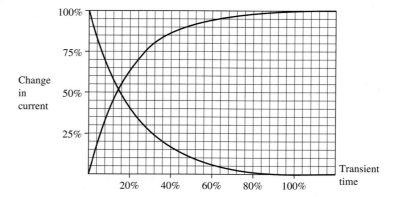

FIGURE 12–13
Universal exponential curves. Current changes at an exponential rate during the transient time.

The value of the initial and final voltage has nothing to do with how long it takes for the current to change. It doesn't matter if the change in source voltage is 1 volt or 1,000 volts. It also doesn't matter if the voltage change is an increase or a decrease. The same amount of time will pass before the current changes from its initial value to its final value.

The change in current always follows a specific pattern. Figure 12–13 is a graph of the pattern, which is called a universal exponential curve.

Increase in Current: Expanding Magnetic Field

In our example (Figure 12–11), the voltage suddenly changed from 12 V to 18 V, causing the current to increase. The expanding magnetic field induced a voltage in the coil which delayed this change. This time delay in the current change *(dI)* is the transient time *(dt)*. The symbol *d* is often used to indicate the change in current. Follow the graph shown in Figure 12–14(a).

We calculated that *dI* was 56 mA (168 mA − 112 mA) during a *dt* of 11.68 ms. To see more closely what was happening to the current during the transient time, refer to Figure 12–14. The curves show that most of *dI* occurs during the first 40% of *dt,* with the remainder of *dI* tapering off during the last 60% of *dt.* Applying the values of Figure 12–14 to our example, we get the following breakdown:

% of dt	value of dt	% of dI	value of dI	value of I
0%	0.00 ms	0%	0.00 mA	112.00 mA
20%	2.34 ms	63%	35.28 mA	147.28 mA
40%	4.67 ms	86%	48.16 mA	160.16 mA
60%	7.01 ms	95%	53.20 mA	165.20 mA
80%	9.34 ms	98%	54.88 mA	166.88 mA
100%	11.68 ms	100%	56.00 mA	168.00 mA

Decrease in Current: Collapsing Magnetic Field

If the voltage in our example were to suddenly drop from 18 V back to 12 V, the current would change from 168 mA back to 112 mA. Thus, *dI* would be the same as before: 56 mA. Now the magnetic field is collapsing, inducing a voltage in the coil, which tries to keep the current from decreasing during the transient time. The change in current now

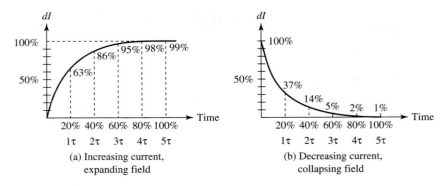

FIGURE 12–14
Exponential charge and discharge curves. Current changes approximately 63% during each 20% time interval (each time constant).

follows the graph shown in Figure 12–14(b). Applying this graph, we obtain the following table for the decrease in current:

% of dt	value of dt	% of dI remaining	value of dI remaining	value of I
0%	0.00 ms	0%	56.00 mA	168.00 mA
20%	2.34 ms	37%	20.72 mA	132.72 mA
40%	4.67 ms	14%	7.84 mA	119.84 mA
60%	7.01 ms	5%	2.80 mA	114.80 mA
80%	9.34 ms	2%	1.12 mA	113.12 mA
100%	11.68 ms	0%	0.00 mA	112.00 mA

The relationship between the growth (increase) curve and the decay (decrease) curve is complimentary. For example, on the growth curve, after 20% of *dt*, 63% of the change in current has occurred. Therefore, 37% of the change still remains. Likewise, on the decay curve, after the same amount of time, 63% of the change in current has occurred, and 37% remains.

Exponential and Logarithmic Curves

In mathematics, exponents can be converted into another kind of number called a *logarithm*. The concept of logarithms was developed in the seventeenth century by John Napier, a Scottish mathematician. Logarithms were used to speed up calculations involving multiplication, division, and exponents. In developing the system of logarithms, lengthy tables of numbers had to be calculated by hand. Napier discovered that the number 2.71828 was useful to him in speeding up his calculations. Therefore, this number is called the base of natural logarithms. The number is symbolized by *e*. In other words, $e = 2.71828$.

Later scientific studies in physics, chemistry, and biology revealed a great many natural phenomena that change at an exponential rate. These phenomena generally involve the natural process of increase or decrease, particularly processes of growth and decay. Scientists gathered data for these processes and graphed the data. When graphed with ordinary horizontal and vertical scales, the graphs appeared as curved lines, similar to Figure 12–13. However, when graphed with scales based on natural logarithms, the graphs became straight lines.

Figure 12–13 is called a universal exponential curve. Exponential curves are significant because they lead to formulas for the process of change. Such formulas often include the number *e* as an exponent.

The reciprocal of *e* is 1/2.71828, which equals 0.36788, or 36.7%. This rounds off to 37%. The curves in Figure 12–14 are based on this value of 37%. During each of the 20% intervals of *dt,* the amount of *dI* that still remains changes 63%, leaving 37% of the *dI* still to change:

A % of transient time	B % of change completed at beginning of this interval	C % of change remaining	D 63% of change remaining	E = B + D % of change completed at end of this interval
0%	0%	100%	63%	63%
20%	63%	37%	23%	86%
40%	86%	14%	9%	95%
60%	95%	5%	3%	98%
100%	98%	2%	1.2%	99.2%

As the table shows, the change in current doesn't quite reach 100% during the transient time. For practical purposes, we ignore this slight difference and say that the change is 100% during the transient time.

The Time Constant, Tau

Because the percentage of change is a constant 63% during each of the 20% time intervals, we call each of the intervals a *time constant*. The symbol for a time constant is the Greek letter tau (τ). Thus, when we say one time constant or one tau, or when we write 1 τ, we are referring to an interval of time equal to 20% of the transient time. Table 12–1 summarizes the relationships between the number of time constants that have passed and the percentage of change in current that has been completed.

Since Formula 12–4 is the formula for the entire transient time, and since the transient time equals 5 τ, we can divide by 5 and get the formula for one time constant (one tau), Formula 12–5.

Formula 12–5

$$\tau = \frac{L}{R}$$

TABLE 12–1

TIME CONSTANT	% OF CHANGE IN VALUE COMPLETED	% OF CHANGE IN VALUE REMAINING
1 τ	63%	37%
2 τ	86%	14%
3 τ	95%	5%
4 τ	98%	2%
5 τ	100%	0%

● SKILL-BUILDER 12–5

The voltage applied to an inductor suddenly increases, causing the magnetic field to expand and inducing an opposing voltage in the inductor. The transient time is 50 ms. What percentage of the current increase has occurred after 20 ms?

GIVEN:

$t_{transient} = 50$ ms
$t_{interval} = 20$ ms

FIND:

% dI

STRATEGY:

1. Find the value of one time constant ($1\ \tau$).
2. Find the number of time constants (number of τ) in the time interval.
3. Refer to Table 12–1 to determine the appropriate percentage for the number of time constants.

SOLUTION:

1. $\tau = t_{transient} \div 5$ Real-time value of $1\ \tau$ equals
    $= 50$ ms $\div 5$ total transient time divided by 5.
    $= 10$ ms

2. $N_\tau = t_{interval} \div \tau$ Value of interval divided by
     value of $1\ \tau$ equals number of τ
     in the interval.

    $= 20$ ms $\div 10$ ms
    $= 2\ \tau$ Interval contains 2 time constants.

3. % dI = 86%   Table 12–1 indicates that the
     current change is 86% complete
     after an interval of 2 time constants.

Repeat the Skill-Builder using the following values. You should obtain the results that follow.

Given:

t_{trans}	8.0 ms	600 ms	1.2 ms	75 μs
t_{int}	1.6 ms	480 ms	720 μs	30 μs

Find:

τ	1.6 ms	120 ms	240 μs	15 μs
N_τ	1	4	3	2
% dI	63%	98%	95%	86%

→ Self-Test

8. What is the transient time of a circuit?
9. What is a time constant?
10. What is the relationship between transient time and time constants?

SECTION 12–4: Inductive Opposition to ac Current: Reactance

In this section you will learn the following:

→ Inductors oppose changes in ac current constantly because ac current is constantly changing.

→ Inductive reactance, symbolized X_L and measured in ohms, is the opposition of an inductor to ac current due to the continuous self-induced voltage in the inductor.

→ The inductance of the coil and the frequency of the ac current determine the value of the inductive reactance.

→ Inductive reactance is low at low frequencies and high at high frequencies.

→ When ac current flows in a coil, the self-induced voltage V_L leads the current by 90°.

→ The quality factor Q of a coil is the ratio of reactance to resistance at a specific frequency.

When current changes in a coil, the value of the self-induced voltage depends on the inductance of the coil and on the rate of change of the current. Changes in dc current are generally caused by a switching action in the circuit. This means that changes in dc current are sudden and sharp, producing a self-induced voltage (V_L) that is momentarily equal to or greater than the source voltage. V_L then falls rapidly, reaching zero at the end of the transient time.

However, changes in ac current are smooth and continuous because ac current is always changing, following the sine wave pattern. Therefore, when ac current flows through a coil, the self-induced voltage is constantly present and has the same sine wave pattern as the current. Thus, an inductor in an ac circuit opposes current all the time because of the inductor's back voltage V_L.

Inductive Reactance

The opposition to ac current due to the inductance of a coil is called *inductive reactance* and is measured in ohms. The symbol for inductive reactance is X_L. You will find X_L only in ac circuits. Inductors in dc circuits do not produce inductive reactance.

Resistance is the only kind of opposition to current that we have studied so far. Resistance is caused by differences in the atomic structures of atoms and molecules. We think of resistance as an opposition to current, measured in ohms. You will find resistance in both dc and ac circuits.

V_L, the back voltage of an inductor, is another kind of opposition to current, which is entirely different from resistance. Although V_L is actually a voltage within the coil, we have learned through experience that it is better to measure all kinds of opposition to current in ohms. This allows us to use our regular Ohm's Law and Watt's Law formulas.

Remember that in dc circuits you will have only one kind of opposition to current: ohms of resistance. However, in ac circuits that contain inductors, you will have two kinds of opposition to current: ohms of resistance and ohms of reactance.

The Formula for Inductive Reactance

The strength of the back voltage V_L depends both on the inductance of the coil and the frequency of the current. Larger coils can produce a stronger V_L, and any coil will produce more V_L at higher frequencies when the magnetic field is changing more rapidly. In other words, coils have less reactance (less opposition to ac current) at low frequencies and more reactance at high frequencies. Formula 12–6 is used to calculate ohms of inductive reactance. As the formula shows, X_L depends on inductance *(L)* and frequency *(f)*. The constant number π (3.1416) is necessary to produce the correct value of ohms.

Formula 12–6

$$X_L = 2\pi fL$$

● **SKILL-BUILDER 12–6**

A coil has an inductance of 50 mH. What is its opposition to an ac current at 25 kHz?

GIVEN:
  $L = 50$ mH
  $f = 25$ kHz

FIND:
  X_L

SOLUTION:
  $X_L = 2\pi fL$
    $= 2 \times 3.1416 \times 25$ kHz $\times 50$ mH
    $= 7,854 \ \Omega$

  Repeat the Skill-Builder using the following values. You should obtain the results that follow.

Given:

f	200 Hz	16 kHz	125 kHz	75 MHz	400 MHz
L	150 mH	8 mH	300 μH	6 μH	0.8 μH

Find:

X_L	188 Ω	804 Ω	236 Ω	2,827 Ω	2,010 Ω

Remember that values of X_L are given in ohms and can be used in the Ohm's Law and Watt's Law formula in place of *R*. In the next exercise, we will use the formulas to see how the same coil has less opposition to current at low frequencies and more opposition at high frequencies.

● **SKILL-BUILDER 12–7**

What is the current in the circuit in Figure 12–15 at frequencies of 10 kHz, 25 kHz, 50 kHz, and 75 kHz? (*Note:* For now, ignore winding resistance R_W and use only X_L to calculate current.)

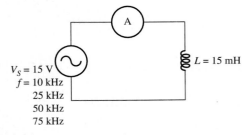

FIGURE 12–15
Circuit for Skill-Builder 12–11.

GIVEN:
$V_S = 15\ V_{rms}$
$L = 15\ mH$

FIND:
I_{10kHz}
I_{25kHz}
I_{50kHz}
I_{75kHz}

STRATEGY:
For each frequency, calculate X_L. Use X_L in Ohm's Law to find current.

SOLUTION:
1. at 10 kHz:
$$
\begin{aligned}
X_L &= 2\pi fL \\
&= 2 \times 3.1416 \times 10\ kHz \times 15\ mH \\
&= 942\ \Omega \\
I &= 15\ V \div 942\ \Omega \\
&= 15.92\ mA
\end{aligned}
$$

2. at 25 kHz:
$$
\begin{aligned}
X_L &= 2\pi fL \\
&= 2 \times 3.1416 \times 25\ kHz \times 15\ mH \\
&= 2{,}356\ \Omega \\
I &= 15\ V \div 2{,}356\ \Omega \\
&= 6.37\ mA
\end{aligned}
$$

3. at 50 kHz:
$$
\begin{aligned}
X_L &= 2\pi fL \\
&= 2 \times 3.1416 \times 50\ kHz \times 15\ mH \\
&= 4{,}712\ \Omega \\
I &= 15\ V \div 4{,}712\ \Omega \\
&= 3.18\ mA
\end{aligned}
$$

4. at 75 kHz:
$$
\begin{aligned}
X_L &= 2\pi fL \\
&= 2 \times 3.1416 \times 75\ kHz \times 15\ mH \\
&= 7{,}069\ \Omega \\
I &= 15\ V \div 7{,}069\ \Omega \\
&= 2.12\ mA
\end{aligned}
$$

Repeat the Skill-Builder using the following values. You should obtain the results that follow.

Given:

V_S	15 V	15 V	15 V	15 V
L	10 mH	25 mH	50 mH	75 mH

Find:

X_L @ 10 kHz	628 Ω	1,571 Ω	3,142 Ω	4,712 Ω
I @ 10 kHz	23.87 mA	9.55 mA	4.77 mA	3.18 mA
X_L @ 25 kHz	1,571 Ω	3,927 Ω	7,854 Ω	11,781 Ω
I @ 25 kHz	9.55 mA	3.82 mA	1.91 mA	1.27 mA
X_L @ 50 kHz	3,142 Ω	7,854 Ω	15.71 kΩ	23.56 kΩ
I @ 50 kHz	4.77 mA	1.91 mA	0.95 mA	0.64 mA
X_L @ 75 kHz	4,712 Ω	11.78 kΩ	23.56 kΩ	35.34 kΩ
I @ 75 kHz	3.18 mA	1.27 mA	0.64 mA	0.42 mA

90° Phase Shift Between V_L and I_L

When ac current flows in a coil, the self-induced voltage V_L leads the current by 90°. This happens because the amount of V_L depends on the rate of change of the current.

All ac waves follow the same sinusoidal pattern with maximum rate of change occurring near 0° and 180°, and zero rate of change occurring at 90° and 270°.

Figure 12–16(a) shows the phase relationship between I_L and V_L. Near 0° and 180°, the current is low but is changing rapidly. Since the current is changing rapidly, V_L will be at maximum value, as shown.

Near 90° and 270°, the current is high but is changing slowly. Therefore, at these times V_L will be at zero, as shown.

In other words, when I_L is at 0° in its cycle, V_L is at 90° in its cycle. Since the difference is 90°, we can see that the phase angle between I_L and V_L is 90°. V_L is farther along in its cycle than I_L, and so we say that V_L leads I_L by 90°.

Phasor Diagrams

Phasor diagrams are used to visualize and to calculate ac quantities (voltage, current, and so on) that are out of phase. These diagrams consist of a group of *phasors,* which look like arrows pointing in different directions. The length of the arrow represents the magnitude (amount) of the quantity, and the angle between the arrows represents the phase angle (the timing difference) between the quantities.

Phasor diagrams are something like the hands of a clock: they tell time by the direction in which they point. Figure 12–17 will help you understand a phasor diagram. A phasor that points to the right is at 0° in its cycle. We imagine that the phasor then rotates, as indicated by the curved arrows. Thus, when the phasor points up, it is at 90° in its cycle. Similarly, pointing to the left indicates 180° in the cycle, and pointing down indicates 270°.

Figure 12–16(b) uses a phasor diagram to show the relationship between V_L and I. Notice that the phasor for I points to the right. This means that the current is at 0° in its cycle. At the same time, the phasor for V_L points up, which means that V_L is at 90° in its cycle. Thus, the diagram shows that V_L leads I by 90°.

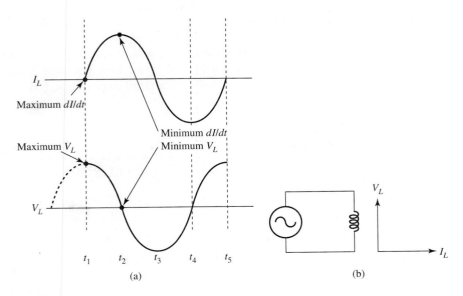

(a)   (b)

FIGURE 12–16

(a) Phase relationship between I_L and V_L. (b) Phasor diagram showing V_L leading I_L by 90°.

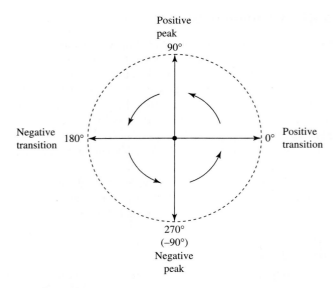

FIGURE 12–17

The Quality (*Q* Factor) of a Coil

The *quality factor Q* of a coil is the ratio of reactance to resistance at a specific frequency, calculated by Formula 12–7.

Formula 12–7

$$Q = \frac{X_L \text{ at a specific frequency}}{R_W}$$

We have learned from experience that a good coil should have a high *Q* factor, which means that X_L should be many times larger than R_W. A high-*Q* coil performs better in tuning and filtering circuits.

When two numbers have the same unit, the ratio of those numbers will not have a unit. This is the case with the *Q* factor. Because X_L and R_W are both measured in ohms, the *Q* factor has no unit. It is a dimensionless number that helps us determine how a coil will perform in a circuit.

● *SKILL-BUILDER 12–8*

A coil has an inductance of 7 mH and a winding resistance of 12.3 Ω. Find the *Q* of the coil at 5 kHz, 30 kHz, and 50 kHz.

GIVEN:
$L = 7$ mH
$R_W = 12.3\ \Omega$
$f_1 = 5$ kHz
$f_2 = 30$ kHz
$f_3 = 50$ kHz

FIND:
Q at f_1, f_2, and f_3

SOLUTION:
1. at $f = 5$ kHz:

$$X_L = 2\pi f L$$
$$= 2 \times 3.1416 \times 5 \text{ kHz} \times 7 \text{ mH}$$
$$= 220 \ \Omega$$

$$Q = \frac{X_L}{R_W}$$

$$= \frac{220 \ \Omega}{12.3 \ \Omega}$$

$$= 18$$

Q is a dimensionless number; it has no units.

2. at $f = 30$ kHz:

$$X_L = 2\pi f L$$
$$= 2 \times 3.1416 \times 30 \text{ kHz} \times 7 \text{ mH}$$
$$= 1{,}320 \ \Omega$$

$$Q = \frac{X_L}{R_W}$$

$$= \frac{1{,}320 \ \Omega}{12.3 \ \Omega}$$

$$= 107$$

3. at $f = 50$ kHz:

$$X_L = 2\pi f L$$
$$= 2 \times 3.1416 \times 50 \text{ kHz} \times 15 \text{ mH}$$
$$= 2{,}199 \ \Omega$$

$$Q = \frac{X_L}{R_W}$$

$$= \frac{2{,}199 \ \Omega}{12.3 \ \Omega}$$

$$= 179$$

Repeat the Skill-Builder using the following values. You should obtain the results that follow.

Given:

L	800 μH	1.2 mH	35 μH	300 μH	7 μH
R_W	5.8 Ω	6.4 Ω	2.4 Ω	4.7 Ω	0.8 Ω

Find:

f_1	50 kHz	40 kHz	125 kHz	50 kHz	150 kHz
X_L	251 Ω	302 Ω	27.5 Ω	94 Ω	6.6 Ω
Q	43	47	11	20	8
f_2	125 kHz	75 kHz	300 kHz	175 kHz	800 kHz
X_L	628 Ω	566 Ω	66 Ω	330 Ω	35 Ω
Q	108	88	27	70	44
f_3	270 kHz	125 kHz	1.5 MHz	350 kHz	2.5 MHz
X_L	1,357 Ω	943 Ω	330 Ω	660 Ω	110 Ω
Q	234	147	137	140	137

A Coil Is a dc Short and an ac Open

Technicians often say that a coil is a dc short and an ac open. What does this mean?

Dc current does not produce back voltage in a coil. The only opposition to current is the resistance of the wire, which is nearly zero. Therefore, the coil acts like a short to dc current: a piece of wire with almost zero resistance.

Ac current produces inductive reactance due to back voltage. As the frequency increases, so does the inductive reactance. At some frequency, the reactance is so high that the current is nearly zero. Therefore, the coil acts like an open to ac current.

→ *Self-Test*

11. When do inductors oppose a change in ac current?
12. What is inductive reactance and its unit?
13. What determines inductive reactance?
14. What is the relationship between inductive reactance and frequency?
15. What is the phase angle between self-induced voltage and ac current in a coil?
16. What is the quality factor of a coil?

Formula List

Formula 12–1

$$L = \frac{N^2 \mu A}{l}$$

Formula 12–2

$$L_T = L_1 + L_2 + L_3 + \ldots + L_n$$

Formula 12–3

$$L_T = \frac{1}{\dfrac{1}{L_1} + \dfrac{1}{L_2} + \dfrac{1}{L_3} + \ldots + \dfrac{1}{L_n}}$$

Formula 12–4

$$t_{\text{transient}} = 5 \times \frac{L}{R_T}$$

Formula 12–5

$$\tau = \frac{L}{R}$$

Formula 12–6

$$X_L = 2\pi f L$$

Formula 12–7

$$Q = \frac{X_L \text{ at a specific frequency}}{R_W}$$

● TROUBLESHOOTING: Applying Your Knowledge

The problem with the arc welder is that the coiled cables are acting as an inductor. The large ac electrode current is producing a significant back voltage in the cables, which is reducing current. Thus, the arc does not produce the proper amount of heat. You suggest to your cousin that he uncoil the cables and lay them out on the floor while welding. This eliminates the inductive reactance and allows the current to increase. Sure enough, the arc produces more heat and welds properly.

CHAPTER SUMMARY: ANSWERS TO SELF-TESTS

1. When current changes in a conductor, the corresponding change in magnetic flux induces a voltage within the conductor.

2. The polarity of a self-induced voltage opposes the change in current that produces the voltage.

3. Inductance *(L)* is the ability of a conductor to create a voltage within itself due to a change in its current.

4. An inductance of 1 henry is the ability to produce 1 volt when the current changes at the rate of 1 ampere per second.

5. The four physical factors of inductance are the number of turns of wire, the cross-sectional area, the length of the core, and the permeability of the core.

6. The total inductance of inductors in series is the sum of the inductance of each inductor.

7. The total inductance of inductors in parallel is the reciprocal of the sum of the reciprocals of the inductance of each inductor.

8. The transient time of a circuit is the time required for output voltage to reach a steady-state value following a change in input voltage.

9. One time constant is the time required for the output to change an amount equal to approximately 63% of the difference between the output's present value and its final value.

10. The transient time of any circuit is approximately five time constants.

11. Inductors oppose changes in ac current constantly because ac current is constantly changing.

12. Inductive reactance, symbolized X_L and measured in ohms, is the opposition of an inductor to ac current due to the continuous self-induced voltage in the inductor.

13. The inductance of the coil and the frequency of the ac current determine the value of the inductive reactance.

14. Inductive reactance is low at low frequencies and high at high frequencies.

15. When ac current flows in a coil, the self-induced voltage V_L leads the current by 90°.

16. The quality factor Q of a coil is the ratio of reactance to resistance at a specified frequency.

CHAPTER REVIEW QUESTIONS

Determine whether the following questions are true or false.

12–1. The polarity of a self-induced voltage aids the change in current.

12–2. Induced voltage exists only during the time when current is changing.

12–3. The unit of inductance is the ohm.

12–4. Inductors are classified according to the type of core material.

12–5. A ferrite bead is used as an inductor in low-frequency circuits.

12–6. The transient time is approximately five time constants.

12–7. Inductors oppose changes in dc current for five transient times.

12–8. Inductive reactance is high at low frequencies.

12–9. The quality factor Q of a coil is the ratio of reactance to resistance at a specified frequency.

Select the response that best answers the question or completes the statement.

12–10. Which of the following is *not* a factor in inductance?

a. the length of the core
b. the number of turns of wire
c. the resistance of the core
d. the number of turns of wire
e. the permeability of the core

12–11. When inductors are placed in series, the total inductance
　a. does not change.
　b. is the reciprocal of the sum of the reciprocals of each inductance value.
　c. is the sum of the inductance values.
　d. is the square root of resistance squared plus inductance squared.

12–12. When inductors are placed in parallel, the total inductance
　a. does not change.
　b. is the reciprocal of the sum of the reciprocals of each inductance value.
　c. is the sum of the inductance values.
　d. is the square root of resistance squared plus inductance squared.

12–13. Back voltage is
　a. the opposite of front voltage.
　b. self-induced voltage.
　c. produced by reactance.
　d. found only in dc circuits.

12–14. A transient is
　a. a voltage that is unusually high for the circuit and that lasts for a short time.
　b. a voltage that is unusually low for the circuit and that lasts for a short time.
　c. a voltage that is unusually high for the circuit and that lasts for a long time.
　d. a voltage that is unusually low for the circuit and that lasts for a long time.

12–15. The time required for a circuit's output to reach a steady value following a change in input values is called
　a. surge time.　　c. lead time.
　b. reactance time.　　d. transient time.

12–16. Which of the following statements is false?
　a. An increasing current in a coil reaches 63% of its final value in one time constant.
　b. A decreasing current in a coil reaches 37% of its original value in one time constant.
　c. An increasing current in a coil reaches 37% of its final value in one time constant.
　d. A decreasing current in a coil loses 63% of its original value in one time constant.

12–17. The symbol for time constant is the Greek letter
　a. alpha.　　c. tau.
　b. omega.　　d. lambda.

12–18. The ac phase angle between V_L and I_L is
　a. 90°.　　c. zero.
　b. 63°.　　d. 2π.

12–19. Q is
　a. the product of resistance and reactance.
　b. the difference between resistance and reactance.
　c. the phasor sum of resistance and reactance.
　d. the ratio of reactance to resistance.

Solve the following problems.

12–20. An inductor is placed in a circuit where the current changes from 780 mA to 940 mA in 63 ms. The self-induced voltage in the inductor is 2.7 V. What is the inductance value?

12–21. What is the total inductance of three inductors in series if $L_1 = 3.7$ mH, $L_2 = 450$ μH, and $L_3 = 0.4$ mH?

12–22. What is the total inductance of the three inductors in the previous problem if they are placed in parallel?

12–23. What is the transient time when $R_L = 600$ Ω, $L = 200$ mH, and $R_W = 15$ Ω?

12–24. The voltage applied to an inductor suddenly increases, causing the magnetic field to expand and inducing an opposing voltage in the inductor. The transient time is 40 ms. What percentage of the current increase has occurred after 8 ms?

12–25. The dc current in an inductor changes from 15 mA to 75 mA in a time of 12 ms. What is the current at the end of each time constant? How much real time has elapsed at the end of each time constant?

12–26. A coil has an inductance of 30 mH. What is its opposition to an ac current at 45 kHz?

12–27. A 2.7-mH coil is connected to a 4.3-V dc source. The current is 678 mA. What will the current be if the coil is connected to a 3.6-V ac source at a frequency of 2.5 kHz?

12–28. A coil has an inductance of 37 mH and a winding resistance of 4.3 Ω. Find the Q of the coil at 1,500 Hz.

GLOSSARY

back voltage   The self-induced voltage in an inductor caused by a change in inductor current.

CEMF   Counterelectromotive force; same as back voltage.

EMI   Electromagnetic interference.

ferrite   A ceramic material with ferromagnetic properties.

henry   The unit of inductance.

inductance   The ability of a conductor to create a voltage within itself due to a change in its current.

inductive kick   A brief, high dc voltage produced in an inductor when a switch opens or closes.

inductive reactance   Opposition in ohms to ac current due to inductance.

inductor   A device specifically designed to produce a substantial amount of inductance; a coil.

phasor   An arrow that represents the magnitude and phase angle of an ac quantity.

phasor diagram   A diagram consisting of a group of phasors showing the relationships among several ac quantities.

quality factor   The ratio of reactance to resistance in a coil at a specific frequency.

RFI   Radio frequency interference.

spike   A large transient voltage.

time constant   The time required for the output to change an amount equal to approximately 63% of the difference between the output's present value and its final value.

transient   A voltage that is unusually high for the circuit and that lasts for a short time.

transient time   The time required for output voltage to reach a steady-state value following a change in input voltage; five time constants.

CAPACITANCE

● SECTION OUTLINE

13–1   Capacitance
13–2   Capacitors
13–3   Capacitors in Series and Parallel
13–4   Effect of Capacitance on Current
13–5   Capacitor Applications
13–6   Testing and Troubleshooting Capacitors

● TROUBLESHOOTING: Applying Your Knowledge

Your fellow student has bought a surplus power supply from an electronics store and tested it with an oscilloscope. He observes a periodic waveform of the output voltage rising from zero to +20 V, then falling to zero. The waveform has a frequency of 120 Hz. He knows the supply is supposed to produce a steady +20 V dc. What do you think might be wrong?

● OVERVIEW AND LEARNING OBJECTIVES

Capacitance, the ability to store charge, is the third major electrical characteristic, along with resistance and inductance. Capacitors are among the most useful electronic devices and have many different applications in electronic circuits and electrical machines.

After completing this chapter, you will be able to:

1. Define capacitance and its unit of measurement.
2. Describe the basic construction of a capacitor and state the relationship between its physical factors and the value of its capacitance.
3. Calculate the total capacitance of capacitors in series and in parallel.
4. Interpret the imprinted figures on a capacitor to determine its capacitance.
5. Name several types of capacitors in common use.
6. Describe the hazards of large capacitors and the precautions to be followed in working with them.
7. Name a variety of capacitor applications.
8. Evaluate the effects of capacitance on dc current.
9. Evaluate the effects of capacitance on ac current.
10. Describe typical capacitor faults and proper testing methods.

SECTION 13–1: Capacitance

In this section you will learn the following:

→ Capacitance *(C)* is the ability to store charge.

Capacitance

Capacitance *(C)* is the ability to store charge. Capacitance is one of the three basic electrical characteristics, along with resistance and inductance.

Remember that the modern meaning of charge is either free electrons or ions (atoms with unequal numbers of protons and electrons). The symbol of charge is Q, and the unit of charge is the coulomb. A coulomb is a unit of volume: 1 coulomb equals 6.25×10^{18} electrons. In other words, just as a gallon is a certain amount of liquid, a coulomb is a certain amount of charge (free electrons).

Storing charge means maintaining an imbalance of charge. Remember that normal atoms contain equal numbers of electrons (negative charge) and protons (positive charge). Having equal amounts of the two kinds of charge means that in normal materials, the charges are in balance, leaving the material in a neutral or uncharged condition.

However, as we have learned, it is possible to remove some electrons from the outside (valence) shell of an atom, leaving a hole (vacancy) in the shell. The free electrons naturally have negative charge, and the holes act like they have positive charge.

As long as the free electrons remain in the same material as the atoms they came from, the overall charge of the material is still zero (neutral), since the overall count of electrons and protons is still the same. However, as shown in Figure 13–1, if free electrons are removed from one object and forced onto another object, both objects become charged, one positively and the other negatively. If the objects remain in this charged (imbalanced) condition, we say that charge has been stored in the objects.

Notice that we think of charge being stored in both objects, not just the one that got the electrons. The holes create a positive charge in object *A* whose electrical force is just as strong as the electrical force created by the negative charge of the extra electrons in object *B*. Therefore, both objects have charge stored in them.

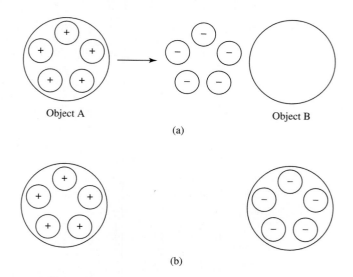

FIGURE 13–1
(a) When charge is transferred from *A* to *B*, both objects become charged. (b) *A* and *B* store equal amounts of charge.

The Basic Capacitor

A device designed to store charge is called a *capacitor.* As shown in Figure 13–2(a), a capacitor consists of two conductive surfaces separated by an insulator. The conductive surfaces are called *plates* and are usually made of thin aluminum or foil. The insulator material between the plates is called the *dielectric.* The schematic symbol for a capacitor is shown in Figure 13–2(b).

Figure 13–3 illustrates how a capacitor can store charge. In Figure 13–3(a), we see a battery with its terminals packed with charge. The chemical reaction inside the battery has taken free electrons from one terminal and packed them onto the other. The terminal that lost the electrons has positive charge. The other terminal that gained the free electrons has negative charge.

The voltage between the battery terminals represents a natural balance of forces. When the potential energy of the charged terminals equals the potential energy of the chemical reaction, the forces are in balance and the electron movement stops.

In Figure 13–3(a), two lengths of wire are near but not connected to the terminals. The wires have a normal number of free electrons and holes. Therefore, we would say the wires are uncharged. However, in Figure 13–3(b), the wires are connected to the terminals. The wires become extensions of the terminals, increasing the amount of metal affected by the chemical reaction. This allows the chemical reaction to remove a few more electrons from the positive terminal and force them onto the negative terminal. Thus, a current (movement of electrons) occurs for a brief time until the wires are just as packed with charge as the terminals. Then the balance of forces stops any further charge movement.

In Figure 13–3(c), a large flat metal plate has been connected to each wire. Because more metal has been added, the chemical reaction can move more charges for a brief time until the plates have become just as packed with charge as the terminals. Then, once more, the balance of forces stops any further charge movement.

The gap between the plates in Figure 13–3(c) prevents electrons from moving from the negative plate to the positive. Thus, the gap acts like an insulator. The gap is the dielectric, the region or space between the plates that prevents charge from moving from one plate to the other.

The most important feature of a capacitor is shown in Figure 13–3(d). The gap has been filled with a solid dielectric material that allows a strong electric field between the plates. To appreciate the effect, we must first review some basic facts about an electric field.

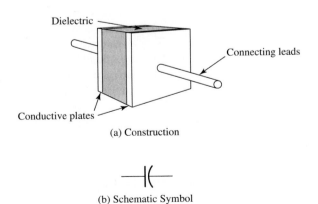

(a) Construction

(b) Schematic Symbol

FIGURE 13–2
The basic capacitor: (a) construction, (b) schematic symbol.

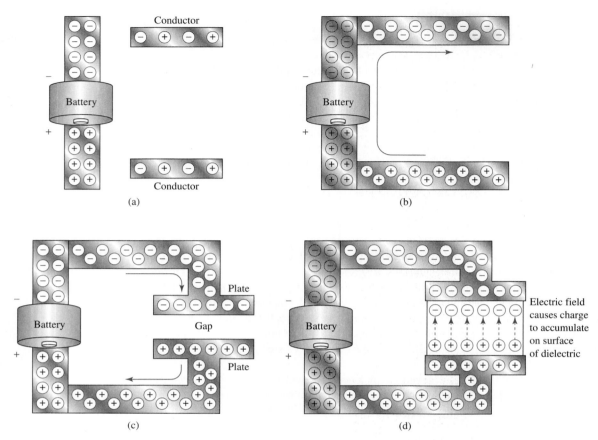

FIGURE 13–3
How a capacitor stores charge.

An electric field is a region of space between two regions of charge. The electrical forces of attraction or repulsion pass through this space and are symbolized as lines from the area of positive charge to the area of negative charge.

For any given plate voltage, the intensity of the electric field within the dielectric depends on the atomic structure of the material. In Figure 13–3(d), the dielectric material allows a strong electric field between the charged plates. This allows the charges on the two plates to feel each other across the gap. The electric field in the dielectric pulls the charges off the plates and onto the surface of the dielectric.

This is what we mean when we say that capacitance is the ability to store charge. One surface of the dielectric has an excess of positive charges, and the other surface has an excess of negative charges. The charges are held on the surfaces by an electric field due to the voltage on the plates. Because the dielectric is an insulator, the charges cannot move through the dielectric. They are trapped or stuck on the surfaces, and so we say they are stored there.

Describing the Amount of Charge That Is Stored

Benjamin Franklin made two important discoveries about electrical charge: electrostatic induction and conservation of charge. Electrostatic induction means that an uncharged object will become charged when a charged object is brought near. In Figure 13–4, a negatively charged rod is brought near an uncharged rod. The electric field induces the free

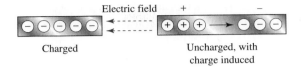

FIGURE 13–4
A charged object creates an electric field that
induces charge in an uncharged object.

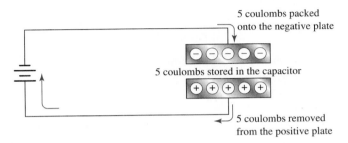

FIGURE 13–5
Describing the amount of charge stored in a capacitor.

electrons in the uncharged rod to move to the other end. Thus, although the number of free electrons and holes remains in balance, the ends of the rod now appear to be charged. When the charged rod is taken away, the free electrons in the uncharged rod spread throughout the rod in a normal way, and the rod no longer seems to be charged.

The principle of conservation of charge says that the total amount of charge in a group of objects remains constant as charge is transferred from one object to another. Conservation of charge is illustrated in Figure 13–1. If five units of charge leave object *A* and enter object *B,* the total amount of charge in the two objects does not change. The only change is in the distribution of the charge.

This is important in describing the amount of charge in a capacitor. In Figure 13–5, the battery has removed 5 coulombs of electrons from the positive plate and packed that same 5 coulombs onto the negative plate. The total number of electrons and holes remains the same (conservation of charge); but because the distribution of charge is out of balance by 5 coulombs, we say the capacitor has stored 5 coulombs of charge. Although each plate has 5 coulombs of charge stored on it, it would be wrong to think that the total charge stored is 10 coulombs. The correct thinking is +5 coulombs and −5 coulombs equals zero: although the charges are equal in number, they are opposite in polarity, so the total amount of charge has not changed.

→ *Self-Test*

1. What is capacitance?

SECTION 13–2: Capacitors

In this section you will learn the following:

→ The unit of capacitance is the farad (F). A capacitance of 1 farad is the ability to store 1 coulomb of charge at a potential of 1 volt.

→ An increase in plate area causes an increase in capacitance.

➜ An increase in dielectric thickness causes a decrease in capacitance.

➜ An increase in value of the dielectric constant causes an increase in capacitance.

Units of Capacitance

Formula Diagram 13–1 defines capacitance. This formula can be expressed as a T-diagram, which will be helpful to us.

Formula Diagram 13–1

$$\frac{Q \text{ coulombs}}{C \text{ farads} \mid V \text{ volts}}$$

Notice that C is the symbol for both coulomb and capacitance. Remember that capacitance is a quantity, and its unit is the *farad*. Charge is also a quantity, and its unit is the coulomb. In an equation, a unit always appears after a number; but a quantity does not. Thus, the equation $Q = 1$ C means a charge *(Q)* of 1 coulomb (C). However, the equation $C = Q / V$ means capacitance *(C)* equals charge *(Q)* divided by voltage *(V)*.

To understand Formula Diagram 13–1, let's make a comparison with something more familiar. The relationship between a voltage source and a capacitor is similar to the relationship between an air compressor and an air-storage tank. An air compressor creates air pressure, and an air tank stores a large amount of compressed air. Similarly, a voltage source creates electrical pressure by giving energy to charge (electrons), and a capacitor stores a large amount of charge under pressure (voltage). The relationship is illustrated in Figure 13–6.

Air pressure is measured in pounds per square inch (psi). The air compressor can produce only a certain amount of pressure. Suppose that the compressor in Figure 13–6 can produce a maximum air pressure of 100 psi. It makes sense, then, that the air pressure in the tank could be less than 100 psi, but never more. Also, if the tank is empty, the compressor has to run for a certain amount of time to fill the tank. During this time, the pressure in the tank builds up until it equals the pressure created by the compressor.

In a similar way, a discharged capacitor is like an empty tank. If the capacitor is connected to a voltage source, charge flows under pressure (voltage) from the source to the capacitor. As the capacitor charges up, the voltage between its plates increases. After a

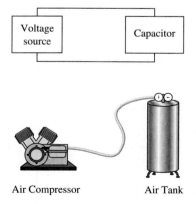

FIGURE 13–6
A voltage source connected to a capacitor is similar to an air compressor connected to an air tank.

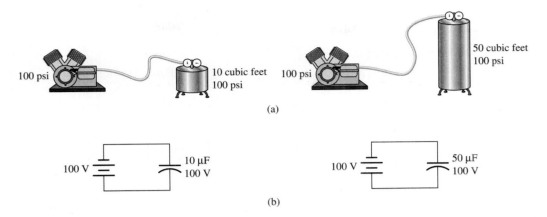

FIGURE 13–7
Capacitors of different size are similar to air tanks of different size.

certain amount of time, the charge stored in the capacitor gives the plates a voltage equal to the source voltage. This creates a balance of forces (equal pressures), and no more charge will flow into the capacitor from the source. In other words, the voltage across the capacitor will equal the voltage applied to the capacitor if the capacitor has enough time to fill up with charge.

Figure 13–7(a) takes the comparison a step further. The same compressor is shown connected to tanks of different size. Given enough time, the compressor will pressurize both tanks to 100 psi. However, the larger tank holds five times as much air at the same pressure. If you have ever worked with compressed air, you know that the larger tank can do a lot more work than the smaller tank because of the increased volume of air stored in it.

In a similar way, Figure 13–7(b) shows the same voltage source connected to capacitors of different size (farads). Given enough time, the battery will charge both capacitors to 100 volts. However, the larger capacitor will have five times as many electrons (charge) pumped into it compared to the smaller capacitor. Like the larger air tank, the larger capacitor can do more work because it contains more energy: a larger volume of charge under pressure (voltage).

This is the meaning of Formula Diagram 13–1. The formula compares the amount of charge (electrons) stored in a capacitor with the amount of voltage (pressure) needed to pump the electrons into the capacitor. Thus, capacitance in farads is a ratio (charge divided by voltage) that indicates several things: the amount of energy that can be stored in the capacitor, the amount of current the capacitor can produce when it discharges, and the amount of time the capacitor needs to charge and discharge.

In other words, if 1 volt of pressure is enough to fill a capacitor with 1 coulomb of charge, then the capacitor is rated at 1 farad. It turns out that 1 farad is a lot of capacitance. Even the largest capacitors used in electronics are much less than 1 farad.

● *SKILL-BUILDER 13–1*

PART 1:

A 33-μF capacitor is charged to 24 V. How many coulombs of charge are stored in it?

GIVEN:
$C = 33$ μF
$V = 24$ V

FIND:
   Q

SOLUTION:
   Q = C × V   Formula Diagram 13–1
      = 33 μF × 24 V
      = 792 μC   μC means microcoulombs

PART 2:

   A 10,000-μF capacitor contains 0.2 coulombs of charge. What is the voltage across its plates?

GIVEN:
   C = 10,000 μF
   Q = 0.2 C

FIND:
   V

SOLUTION:
   V = Q ÷ C   Formula Diagram 13–1
      = 0.2 C ÷ 10,000 μF
      = 20 V

   Develop your understanding of Formula Diagram 13–1 by completing the blanks in the following data table. The correct values are given in parentheses at the end of each row.

Capacitance	Charge	Voltage	
68,000 μF	_____ C	35 V	(2.38 C)
_____ μF	0.005 C	5.2 V	(961 μF)
220 μF	0.002 C	_____ V	(9.09 V)
470 μF	_____ C	27 V	(0.0127 C)
_____ μF	0.014 C	125 V	(112 μF)
6,800 μF	0.007 C	_____ V	(1.03 V)

Physical Characteristics of a Capacitor

Capacitance depends on three physical characteristics:

   1. The size of the plates (plate area).
   2. The size of the gap between the plates (dielectric thickness).
   3. The material used for the dielectric (dielectric constant).

   All of these three factors affect the intensity of the electric field between the plates, which is the primary cause of capacitance.

Plate Area

Figure 13–8 illustrates plate area. A larger plate area means there is a larger dielectric surface, which can hold more charge. Thus, a larger plate area produces a larger capacitance (more farads). Notice the difference in Figure 13–8 between physical plate area and effective plate area. The effective plate area is the amount of overlap between the plates.

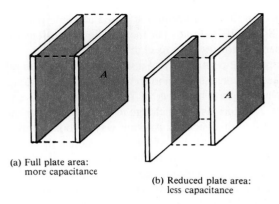

(a) Full plate area:
   more capacitance

(b) Reduced plate area:
   less capacitance

FIGURE 13–8
Capacitance is directly proportional to
effective plate area.

Dielectric Thickness

Figure 13–9 illustrates dielectric thickness, the gap or distance between the plates. When
the dielectric is thin, the electric field is more intense, which pulls more charge onto the
dielectric surface and increases the capacitance. In other words, a smaller thickness or
gap creates a larger capacitance.

Dielectric Constant

The intensity of the electric field partly depends on the atomic structure of the dielectric.
Some materials allow the electric field to pass through them more easily. This happens
because the electric field in the dielectric causes the atoms either to move slightly off center or else to twist or rotate. This distortion in the atomic structure of the dielectric affects
the intensity of the electric field and thus affects the capacitance.

The *dielectric constant* is a value that represents the ability of a dielectric material
to support an electric field. The symbol for the dielectric constant is either the Greek letter ε (epsilon) or the Latin letter *k*.

The concept of the dielectric constant is based on the fact that a vacuum is a perfect
dielectric, since a vacuum contains no material whatsoever. Thus, in a vacuum, there are
no atoms to move or twist, and no charges to accumulate near the plates. Therefore, since
a vacuum is a perfect dielectric, it does nothing to increase capacitance.

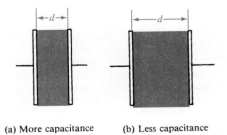

(a) More capacitance   (b) Less capacitance

FIGURE 13–9
Capacitance is inversely proportional
to dielectric thickness.

The dielectric constant of a vacuum is called absolute permittivity, symbolized ε_0. The value of ε_0 is 8.85×10^{-12} farads per meter. The dimensions of ε_0, farads per meter, are arbitrarily chosen to allow ε_0 to be used in practical formulas.

We then compare the permittivity of a vacuum to that of all other dielectric materials. This is the dielectric constant, symbolized ε_r. A vacuum is assigned a dielectric constant value of 1. Thus, if glass has an ε_r of 7.5, this means that a capacitor with a glass dielectric will allow 7.5 times as much charge to accumulate near the plates as it would if the glass were replaced with a vacuum. Another way to look at this is to say that the electric field between two plates will be 7.5 times as intense through glass as it would be through a vacuum. Figure 13–10 illustrates this idea.

Applied to capacitors, the dielectric constant indicates the effect the dielectric material will have on the capacitance. If two capacitors have equal plate area and dielectric thickness but different dielectric constants, the one with the higher dielectric constant value will have higher capacitance.

Physical Formula for Capacitance

Formula 13–2 relates the three physical factors that affect capacitance.

Formula 13–2

$$C = \frac{A\,\varepsilon_r\,\varepsilon_0}{d}$$

C = capacitance in farads
A = effective plate area in square meters
ε_r = dielectric constant at a specific frequency and temperature
ε_0 = absolute permittivity of a vacuum (8.85×10^{-12} farads per meter)
d = distance between the plates (thickness of the dielectric) in meters

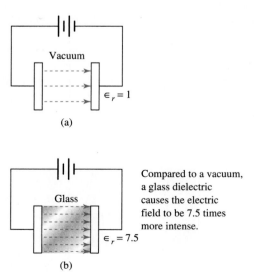

FIGURE 13–10
Capacitance is directly proportional to dielectric constant.

This formula allows us to see how large 1 farad really is. Suppose two aluminum plates were separated by an air gap of 1 millimeter (about 0.04 inches). To have a capacitance of 1 farad, the plates would need a surface area of 113 million square meters. If the plates were square, each side would measure 10.63 kilometers, or about 6.4 miles!

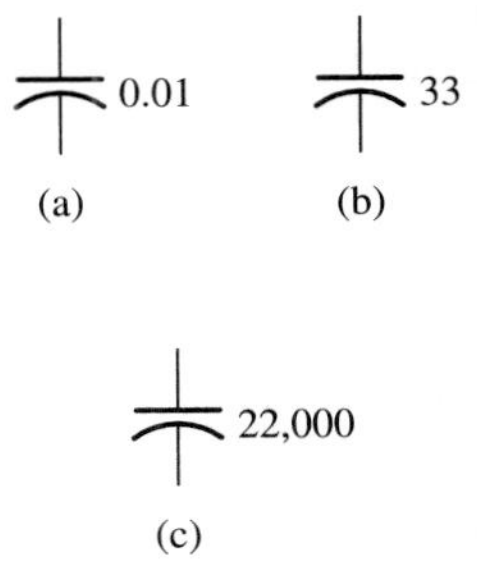

Determining Capacitance

As we have seen, the farad is a large unit. Most capacitors used in electronics are rated either in microfarads (µF) or picofarads (pF). Manufacturers do not rate capacitors either in millifarads (mF) or nanofarads (nF). Therefore, you will see capacitors with values like 33,000 µF, which is really 33 mF; or 0.047 µF, which is really 47 nF.

In schematic diagrams, capacitor values may appear as shown in Figure 13–11. In such cases where a number without a unit is written near a capacitor, you may assume the unit is microfarads (µF).

Many capacitors are physically large enough to allow the capacitance value to be printed plainly on the case or body. Smaller capacitors may have a three-digit code printed on them, as shown in Figure 13–12. The first two digits will be 10, 22, 33, 47, 56, 68, or 82. The last digit will be 0, 1, 2, 3, or 4. These small capacitors are always rated in picofarads (pF). To find the rating, write down the first two digits and add the number of zeros indicated by the third digit.

● **SKILL-BUILDER 13–2**

A small capacitor has the digits "103" printed on it. What is its capacitance?

SOLUTION:

10	Write down the first two digits.
10 <u>000</u> 3	Write down the number of zeros indicated by the third digit.
10,000 pF	These small capacitors are rated in picofarads.

Repeat the Skill-Builder using the following values. You should obtain the results that follow.

Printed code	222	470	563
First two digits	22	47	56
Number of zeros	2	0	3
Value	2,200 pF	47 pF	56,000 pF

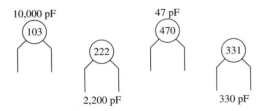

FIGURE 13–12
Small capacitors with values indicated by an imprinted numeric code which requires interpretation.

FIGURE 13–11
Typical schematic indications of capacitance values in µF.

(a) 0.01 (b) 33

(c) 22,000

Capacitor Color Code

Small rectangular and tubular capacitors may have colored dots or stripes, similar to the examples in Figure 13–13. The two dots or stripes labeled as digits represent the first two significant digits of the capacitance, and the multiplier dot or stripe represents the number of zeros after the significant digits. The unit for these small capacitors is picofarads (pF).

Many DMMs (digital multimeters) are able to measure capacitance. Usually, the meter reads the capacitance in nanofarads (nF). Thus, the technician must become familiar and comfortable with a situation where capacitor values may be printed on the capacitor or on the schematic diagram in units of either microfarads or picofarads, but meters will measure these same capacitors in nanofarads. For example, in Figure 13–14, a schematic diagram indicates a capacitance of 0.1 µF. The actual capacitor in the circuit is coded 104, which works out to 100,000 pF. When this capacitor is tested in a DMM, the reading is 100 nF. All these values are the same, written with different SI prefixes.

Tolerance

The tolerance of a capacitor depends on its use in a circuit. Most capacitance values are not critical, and it is common to find capacitors with tolerances as high as 40%. Capacitors used in high-frequency tuning and filtering circuits have typical tolerances of 2% to 5%.

dc Working Voltage (Breakdown Voltage)

The dielectric is a thin insulator, which can easily be cracked or burned through by an excessive voltage on the plates. Thus, all capacitors are rated according to the maximum sustained plate voltage they can withstand. This voltage limit is called the dc working voltage or *breakdown voltage* of the capacitor. These ratings are conservative. Most

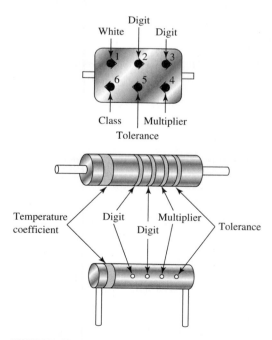

FIGURE 13–13
Capacitor color code.

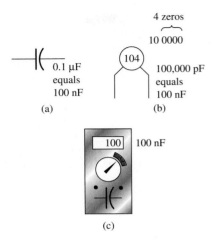

FIGURE 13–14
Comparisons of capacitor values.

capacitors can withstand a voltage at least four times greater than the WDCV (working dc voltage) for a brief time. However, if you exceed the WDCV, you void the manufacturer's warranty. When replacing a capacitor, you must know both the capacitance and the WDCV. The replacement capacitor should have the same capacitance as the original, and a WDCV equal to or greater than the original.

The WDCV is directly related to a property of the dielectric called dielectric strength, the ability of the dielectric to withstand the electric field. Dielectric strength is different from dielectric constant. Figure 13–15 lists both dielectric strength and dielectric constant values for similar materials. The dielectric strength shown is the amount of voltage necessary to break down a 1-mil (0.001-inch) layer of the material. Thus, at 80 volts per mil for air, it would take 80,000 volts to create a spark across 1 inch of air.

A high voltage literally cracks an insulator by producing an electric field so strong that the atoms are twisted and torn completely out of their normal position in the solid material, producing cracks. An electric arc (a spark) passes through the cracks, heating the material enough to melt it and destroy it even more.

dc Leakage Resistance

All dielectric materials contain some charge carriers and, thus, the electric field between the plates causes a small current to flow through the dielectric. This is called leakage cur-

Material	Dielectric Strength (volts/mil)	Material	Typical ϵ_r Values
Air	80	Air (vacuum)	1.0
Oil	375	Teflon®	2.0
Ceramic	1000	Paper (paraffined)	2.5
Paper (paraffined)	1200	Oil	4.0
Teflon®	1500	Mica	5.0
Mica	1500	Glass	7.5
Glass	2000	Ceramic	1200

FIGURE 13–15
Tables of dielectric strength and relative permittivity.

rent. Although it is unwanted, a certain amount of it can be tolerated in most cases. Thus, the dc leakage resistance (R_1) is an important capacitor rating, since it allows us to calculate the leakage current at various dc voltages. Capacitors with paper, mica, and ceramic dielectrics have a dc leakage resistance of over 100 MΩ. Electrolytic capacitors have a thin dielectric and a lower dc leakage resistance around 0.5 MΩ.

Types of Capacitors

Capacitors are generally described either by construction features or by the type of dielectric material.

Construction Features

In Figure 13–16, many capacitors are described according to the shape of the body: tubular, disc, or chip. The method of applying the outer insulation may be molded, dipped, solid, or hermetically sealed. The method of mounting may be *axial* or *radial* lead, snap-mount, DIP (double in-line pin), or surface mount.

Figure 13–17 shows axial-lead and radial-lead configurations. If a radial-lead capacitor is also a polarized capacitor, the shorter lead is the negative lead.

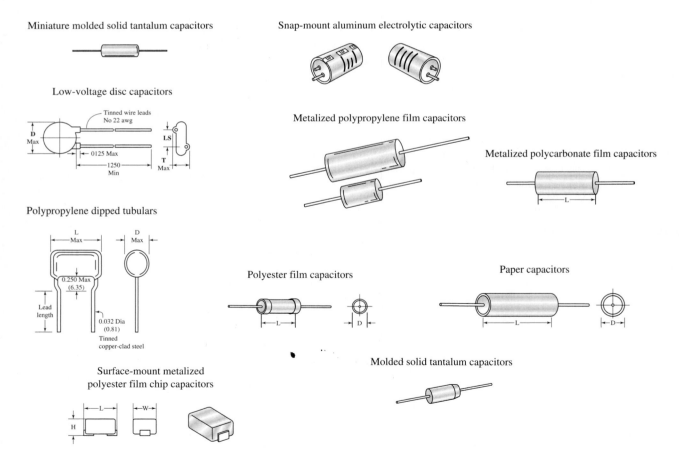

FIGURE 13–16
Various capacitors.

Dip monolythic ceramic capacitors

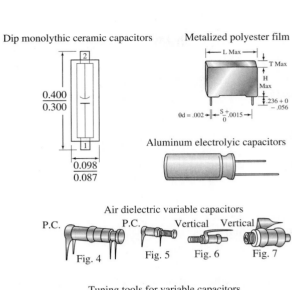

$\dfrac{0.400}{0.300}$

$\dfrac{0.098}{0.087}$

Metalized polyester film

L Max

T Max

H Max

.236 + 0
     – .056

θd = .002 $\dfrac{S}{0}$ + .0015

ac motor-start capacitors

Variable ceramic capacitors

Aluminum electrolyic capacitors

Air dielectric variable capacitors

P.C.   P.C.   Vertical   Vertical

Fig. 4   Fig. 5   Fig. 6   Fig. 7

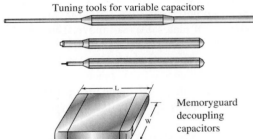

Ceramic trimmer capacitors

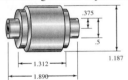

Tuning tools for variable capacitors

Memoryguard
decoupling
capacitors

L

W

Transmitting ceramic capacitors

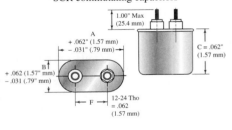

.375

.5

1.312

1.890

1.187

Mylar paper filter capacitors

SCR commutating capacitors

1.00" Max
(25.4 mm)

A
+ .062" (1.57 mm)
– .031" (.79 mm)

B
+ .062 (1.57" mm)
– .031 (.79" mm)

C = .062"
(1.57 mm)

F

12-24 Tho
= .062
(1.57 mm)

dc power packs

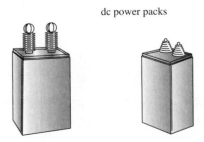

FIGURE 13–16 *(Continued)*

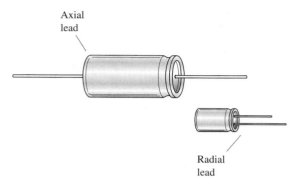

FIGURE 13–17
Axial and radial leads.

Capacitors may be variable, which means the value of capacitance can be changed or adjusted. Figure 13–18 is a common variable capacitor. One set of plates moves when the shaft is turned. The other set of plates does not move. The sets of plates are separated by an air gap, which is the dielectric. The value of the capacitance varies according to the amount of overlapping plate area. When the movable plates are completely inside the air gap, the overlap is maximum and, thus, the capacitance is also maximum. As the movable plates are rotated out of the air gap, the overlapping area decreases, which also decreases the capacitance. This type of variable capacitor is widely used as a tuning control in radios.

Variable ceramic capacitors, shown in Figure 13–16, require special plastic tools to turn or screw the plates or the dielectric. These small ceramic capacitors may be called trimmers, referring to their application.

An ordinary nonadjustable capacitor is called a fixed capacitor.

Dielectric Material

Capacitors are also described by the dielectric material: oil, paper, ceramic, mica, air, or a variety of plastic films such as polyester, polystyrene, polypropylene, and polycarbonate. Polyester film is sometimes called mylar, a brand name.

Mica is a rocklike mineral with a crystalline structure. Mica is easy to cut into thin slices and is a good insulator, both for charge and heat. Figure 13–19 shows the details of construction of a typical mica capacitor.

FIGURE 13–18
A typical variable-air capacitor.

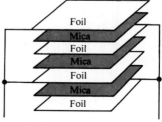

(a) Stacked layer arrangement

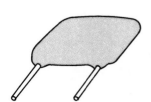

(b) Layers are pressed together and encapsulated

FIGURE 13–19
Details of a mica capacitor.

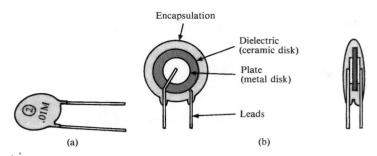

FIGURE 13–20
Details of a ceramic-disk capacitor.

Ceramic capacitors are widely used on printed circuit boards, in both disc and capsule-body styles. Details of ceramic capacitor construction are shown in Figure 13–20 and Figure 13–21.

Paper and plastic film capacitors are often constructed in tubular form, as shown in Figure 13–22. Strips of aluminum foil serve as the plates. The foil strips are offset slightly to allow attachment of the leads. Thus, tubular capacitors generally have axial leads. Paper dielectrics are often soaked in oil or paraffin (wax) to improve their function. Metalized film capacitors have plates made of thin layers of metal deposited on the plastic film instead of strips of aluminum foil.

One advantage of metalized film capacitors is their ability to survive voltage overloads. If the breakdown voltage of any capacitor is greatly exceeded, a spark will punch a hole through the dielectric. In most capacitors, the plates will remain in contact with each other through the hole, permanently shorting out the capacitor. However, in the case of metalized film capacitors, the heat of the spark vaporizes the metal deposits near the edge

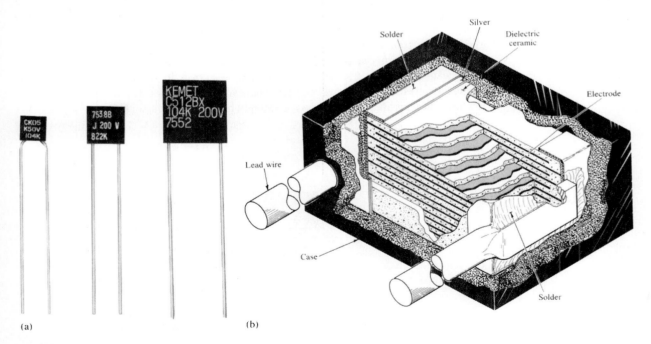

FIGURE 13–21
Details of an encapsulated ceramic capacitor.

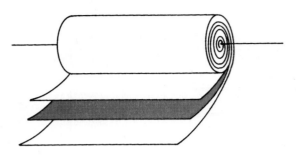

FIGURE 13–22
Details of a tubular paper or plastic capacitor.

of the hole in the film. Because the metal layer is burned back from the edge of the hole, often the capacitor is not permanently shorted out and may continue to function properly after the voltage overload passes.

Oil is an excellent high-voltage insulator and is used as the dielectric in power capacitors that accumulate large amounts of energy for industrial machines, such as certain types of welding equipment. Oil is also an insulator in power transformers used by utility companies.

Electrolytic Capacitors

Electrolytic capacitors have several unique characteristics. An electrolyte is any liquid or paste that is conductive. As shown in Figure 13–23, an electrolytic capacitor consists of two foils of either aluminum or titanium, separated by a layer of paper or gauze soaked with an electrolytic solution of borax, phosphate, or carbonate. A chemical reaction between the electrolyte and the metal foil forms a thin layer of oxide (corrosion) on the surface of one foil. This layer of oxide is the dielectric. The conductive electrolyte and one foil together form one plate; the other foil is the other plate.

Because the dielectric layer is so thin, the electrolytic capacitor has high capacitance for its physical size. However, there are several drawbacks. The thin oxide layer has a low breakdown voltage and allows a relatively high leakage current.

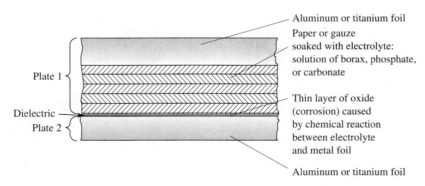

FIGURE 13–23
Details of electrolyte and oxide layer in an electrolytic capacitor.

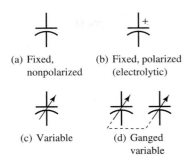

(a) Fixed,
nonpolarized

(b) Fixed, polarized
(electrolytic)

(c) Variable

(d) Ganged
variable

FIGURE 13–24
Schematic symbols for
capacitors.

The dielectric properties of the oxide layer vary according to the metal used for the plates. For example, electrolytic capacitors made with titanium plates have less leakage current and a higher breakdown voltage than ordinary electrolytic capacitors made with aluminum plates because of differences in the chemistry of the oxide layer.

Because the dielectric is formed by a chemical reaction which deteriorates over time, electrolytic capacitors have a limited shelf life. An old capacitor may not perform properly when it is installed. Capacitors in a circuit also wear out with age due to chemical deterioration. The capacitor becomes leaky, which means an increase in leakage current. Sometimes they become physically leaky as well: The can (the aluminum outer shell or body) may swell and electrolyte may start to ooze out.

Many electrolytic capacitors are polarized, making them suitable only for dc circuits. Technicians must be careful to install electrolytic capacitors correctly. Reversing the polarity can reverse the chemical reaction and dissolve the oxide layer. Without the dielectric, the capacitor is a short in the circuit. Usually this will allow a high current through the capacitor, heating the electrolyte to the boiling point. The capacitor may explode, sending hot, corrosive paste everywhere. You can be severely burned or blinded by an exploding electrolytic capacitor. Always wear eye protection when working with electrolytic capacitors, make certain you have installed the capacitor correctly, stand back when applying power, and wait a few minutes before deciding that everything is all right.

Schematic Symbols

Figure 13–24 shows several additional schematic symbols for capacitors. Note the symbol for a ganged capacitor, which is several separate variable capacitors mounted on a single shaft so they all adjust together.

→ *Self-Test*

2. What is the unit of capacitance?
3. What is the relationship between capacitance and plate area?
4. What is the relationship between capacitance and dielectric thickness?
5. What is the relationship between capacitance and the value of the dielectric constant?

SECTION 13–3: Capacitors in Series and Parallel

In this section you will learn the following:

→ When capacitors are charged in series, each plate stores an identical amount of charge.
→ When capacitors are connected in series, the total capacitance is the reciprocal of the sum of the reciprocals of each capacitor.
→ When capacitors are charged in series, plate voltage is inversely proportional to capacitance for each capacitor.
→ When capacitors are connected in parallel, the total capacitance equals the sum of the capacitance of each branch.

Capacitors in Series

In Figure 13–25(a), if 1 coulomb of negative charge is forced onto plate 1, then 1 coulomb of negative charge is forced off plate 2, leaving it with 1 coulomb of positive charge. Therefore, the two plates have identical amounts of charge stored on them.

This situation remains the same if two capacitors are charged in series, as shown in Figure 13–25(b). Although plates 2 and 3 are joined together as a single piece of

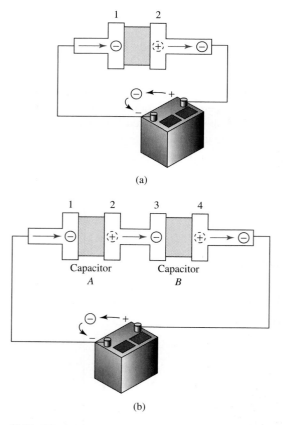

(a)

(b)

FIGURE 13–25
Capacitors in series.

metal, negative charges (free electrons) are pushed to one end (plate 3), leaving the other end (plate 2) with positive charge. Thus, if plate 1 has a charge of 1 coulomb, then so do plates 2, 3, and 4; and so does capacitor *A* and *B;* and so does the combination of *A* and *B!*

Charge is stored on the plates of the capacitors because charge has moved or shifted in the circuit, from the voltage source's negative terminal toward the positive terminal. No matter where you look, you find the same amount of charge stored because the shift or movement has been the same. After all, in a series circuit, current has the same value at all points. We can think of the charge stored in a series of capacitors as a current that has stopped moving and is trapped on the plates. Its value is still the same at all locations.

Equivalent Value of Capacitors in Series

In Figure 13–26(a), a capacitor has 10 C of charge stored on its plates at a potential of 10 V. Therefore, its capacitance is 1 F, according to Formula Diagram 13–1.

Now in Figure 13–26(b), we imagine that another plate has been inserted into the dielectric so that it has a potential of 2 V with respect to the top plate, and 8 V with respect to the bottom plate. If we now separate the dielectric at this point and connect the two pieces as shown in Figure 13–26(c), we have two capacitors in series. Since the charge is the same on all plates, both the upper and lower capacitors have 10 C stored on their plates. According to Formula Diagram 13–1, the upper capacitor, with 10 C and 2 V, has a capacitance of 5 F; and the lower capacitor, with 10 C and 8 V, has a capacitance of 1.25 F.

If we began by taking a 5-F capacitor and placing it in series with a 1.25-F capacitor, we would have the same result: a total capacitance of 1 F. From this experiment, we see that when capacitors are connected in series, the total capacitance is reduced and is smaller than the capacitance of the smallest single capacitor.

Connecting capacitors in series has the effect of increasing the dielectric thickness or gap between the plates, which decreases the total capacitance. This also increases the total breakdown voltage rating. In other words, the breakdown voltage of a series of capacitors is the sum of the breakdown voltage of each capacitor.

The total capacitance of a number of capacitors in series is calculated using the reciprocal method we learned for resistors in parallel.

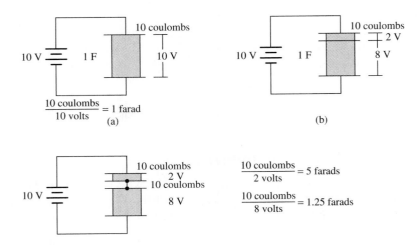

FIGURE 13–26
Capacitance decreases when capacitors are connected in series.

Formula 13–3

Total capacitance in series:

$$C_T = \frac{1}{\dfrac{1}{C_1} + \dfrac{1}{C_2} + \ldots + \dfrac{1}{C_n}}$$

● *SKILL-BUILDER 13–3*

Three capacitors are connected in series: $C_1 = 33\ \mu F$, $C_2 = 47\ \mu F$, and $C_3 = 22\ \mu F$. What is the total capacitance?

GIVEN:
   $C_1 = 33\ \mu F$
   $C_2 = 47\ \mu F$
   $C_3 = 22\ \mu F$

FIND:
   C_T

SOLUTION:
   $C_T = 1/(1/C_1 + 1/C_2 + 1/C_3)$
      $= 1/(1/33\ \mu F + 1/47\ \mu F + 1/22\ \mu F)$
      $= 10.31\ \mu F$

keystrokes:

33 [1/x] [+] 47 [1/x] +
22 [1/x] [=] [1/x]

Tip: if all values have the same SI prefix, it is not necessary to include the prefix in the keystrokes, as long as you give the same prefix to the answer. In this case, all values had the prefix of micro (μ).

Repeat the Skill-Builder using the following values. You should obtain the results that follow.

Given:

C_1	0.47 μF	100 pF	0.22 μF	470 pF	12 nF
C_2	0.68 μF	68 pF	1 μF	0.001 μF	820 pF
C_3	0.1 μF	330 pF	0.47 μF	680 pF	0.047 μF

Find:

C_T	0.0735 μF	36 pF	0.13 μF	217 pF	755 pF

Series Capacitor Voltage Divider

If we look at the way voltage is divided in Figure 13–26, we see that the smallest capacitor (1.25 F) has the largest voltage (8 V) across its plates. Capacitors in series act as a special kind of voltage divider, according to Formula 13–4.

Formula 13–4

$$V_x = \frac{C_T}{C_x} \times V_T$$

● *SKILL-BUILDER 13–4*

A 16-V source is connected to three capacitors in series. The values of the capacitors are $C_1 = 470$ μF, $C_2 = 560$ μF, and $C_3 = 680$ μF. What is the voltage potential across the plates of each capacitor?

GIVEN:
$V_S = 16$ V
$C_1 = 470$ μF
$C_2 = 560$ μF
$C_3 = 680$ μF

FIND:
V_1
V_2
V_3

SOLUTION:
$C_T = 1 / (1/C_1 + 1/C_2 + 1/C_3)$
  $= 185.7$ μF   Use reciprocal method

$V_1 = C_T / C_1 \times V_T$ Formula 13–4
  $= 185.7$ μF $/ 470$ μF $\times 16$ V
  $= 6.32$ V

$V_2 = C_T / C_2 \times V_T$
  $= 185.7$ μF $/ 560$ μF $\times 16$ V
  $= 5.31$ V

$V_3 = C_T / C_3 \times V_T$
  $= 185.7$ μF $/ 680$ μF $\times 16$ V
  $= 4.37$ V

Repeat the Skill-Builder using the following values. You should obtain the results that follow.

Given:

V_T	12 V	28 V	5.6 V	30 V	5 V
C_1	0.033 μF	220 pF	47 μF	100 pF	0.010 μF
C_2	0.100 μF	100 pF	10 μF	56 pF	0.022 μF
C_3	0.022 μF	680 pF	68 μF	22 pF	0.047 μF

Find:

C_T	11.66 μF	62.4 pF	7.35 μF	13.64 pF	0.006 μF
V_1	4.24 V	7.95 V	0.88 V	4.09 V	3.00 V
V_2	1.40 V	17.48 V	4.12 V	7.31 V	1.36 V
V_3	6.36 V	2.57 V	0.61 V	18.60 V	0.64 V

Capacitors in Parallel

Suppose we take the same two capacitors in Figure 13–26 and connect them in parallel to the same voltage source, as shown in Figure 13–27. Now each capacitor has its own current path to the 10-V voltage source and can charge independently of the other capacitor. Therefore, C_1, with a capacitance of 5 F, stores a charge of 50 C. Likewise, C_2, with a capacitance of 1.25 F, stores 12.5 C. The total charge stored in the circuit, Q_T, is clearly

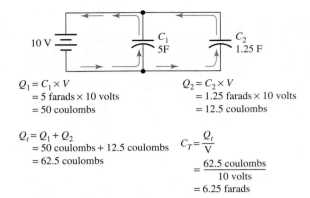

$Q_1 = C_1 \times V$
      $= 5$ farads $\times 10$ volts
      $= 50$ coulombs

$Q_2 = C_2 \times V$
      $= 1.25$ farads $\times 10$ volts
      $= 12.5$ coulombs

$Q_t = Q_1 + Q_2$
      $= 50$ coulombs $+ 12.5$ coulombs
      $= 62.5$ coulombs

$C_T = \dfrac{Q_t}{V}$

   $= \dfrac{62.5 \text{ coulombs}}{10 \text{ volts}}$

   $= 6.25$ farads

FIGURE 13–27
Capacitance increases when capacitors are
connected in parallel.

the sum of Q_1 and Q_2. Thus, Q_T is 62.5 C (50 C + 12.5 C). Therefore, the total capacitance of the circuit, C_T, equals the total stored charge divided by the voltage: $Q_T \div V_S$ equals 62.5 C ÷ 10 V, which results in 6.25 F.

From this example, we can see that when capacitors are connected in parallel, the total capacitance equals the sum of the capacitance of each branch.

Formula 13–5

Total capacitance in parallel:

$$C_T = C_1 + C_2 + \ldots + C_n$$

● SKILL-BUILDER 13–5

What is the total capacitance of three capacitors in parallel when $C_1 = 33$ μF, $C_2 = 47$ μF, and $C_3 = 22$ μF?

GIVEN:
  $C_1 = 33$ μF
  $C_2 = 47$ μF
  $C_3 = 22$ μF

FIND:
  C_T

SOLUTION:
  $C_T = C_1 + C_2 + C_3$ Formula 13–5
     $= 33$ μF $+ 47$ μF $+ 22$ μF
     $= 102$ μF

Repeat the Skill-Builder using the following values. You should obtain the results that follow.

Given:

C_1	0.47 μF	100 pF	0.22 μF	470 pF	12 nF
C_2	0.68 μF	68 pF	1 μF	0.001 μF	820 pF
C_3	0.1 μF	330 pF	0.47 μF	680 pF	0.047 μF

Find:

| C_T | 1.25 µF | 498 pF | 1.69 µF | 2.15 nF | 59.82 nF |

Capacitors in parallel amount to an increase in plate area, which increases total capacitance. However, the breakdown voltage rating of the circuit is limited to the smallest breakdown voltage rating of any branch.

→ Self-Test

6. When capacitors are charged in series, how much charge is stored on each plate?
7. What is the total capacitance of capacitors in series?
8. What is the relationship between capacitance and plate voltage when capacitors are charged in series?
9. What is the total capacitance of capacitors connected in parallel?

SECTION 13–4: Effect of Capacitance on Current

In this section you will learn the following:

→ The amount of time required for a capacitor to charge and discharge depends on the value of capacitance and the value of series resistance.

→ A capacitor requires five time constants either to charge or discharge completely from any voltage to any other voltage. Therefore, 5 τ (five time constants) is the transient time of a capacitor.

→ Capacitive reactance, symbolized X_C and measured in ohms, is the opposition of a capacitor to ac current due to the variations in the capacitor's plate voltage as it charges and discharges.

→ When ac current is applied to a capacitor, the capacitor's plate voltage V_C lags the current I by 90°.

The Charge and Discharge Cycle

Figure 13–28(a) shows a capacitor charging through a resistor. The resistor represents any resistance in the circuit, including the internal resistance of the capacitor and the voltage source. The circuit explains how Kirchhoff's Voltage Law is maintained during the charge cycle.

Kirchhoff's Voltage Law requires that at all times the sum of the voltage drops and rises in a closed loop equals zero. At the moment the switch closes, the capacitor is uncharged and the voltage potential across its plates is zero. The inrush of current through the resistor produces a voltage drop equal to the source voltage, fulfilling Kirchhoff's Voltage Law.

As time passes, the inflow of charge onto the capacitor's plates increases the voltage potential between the plates. Notice that the polarity of the capacitor with respect to the direction of current flow is the same as the polarity of the resistor, and opposite the polarity of the battery. This allows us to treat the voltage potential across the capacitor as if it were a voltage drop.

The increasing voltage across the capacitor opposes the battery voltage and, thus, decreases current. By Ohm's Law, the voltage drop across the resistor depends on current ($E = I \times R$). Therefore, the resistor's voltage drop decreases as the capacitor's voltage potential increases; yet at all times $V_R + V_C$ equals V_S, fulfilling Kirchhoff's Voltage Law.

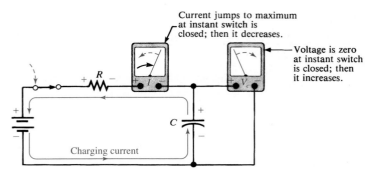

(a) Charging: Capacitor voltage increases
as the current and resistor voltage decrease.

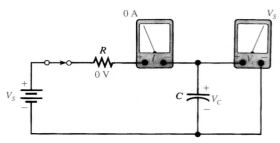

(b) Fully charged: Capacitor voltage equals
source voltage. The current is zero.

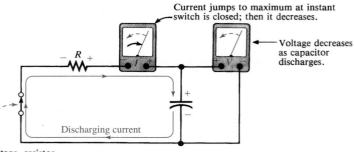

(c) Discharging: Capacitor voltage, resistor
voltage, and the current decrease from
initial maximums. Note that the discharge
current is opposite to the charge current.

FIGURE 13–28

Capacitor charging and discharging through resistance.

In time, the capacitor becomes fully charged, as shown in Figure 13–28(b). Its plate potential equals the voltage source. All current stops, and the resistor has no voltage drop. Since the polarity of the capacitor still opposes the polarity of the battery, $V_S - V_C$ equals zero, again fulfilling Kirchhoff's Voltage Law.

If the capacitor were physically disconnected from the circuit while charged, the charge would remain trapped on the plates and the capacitor would still have all its voltage. This characteristic is one reason why you must always treat large capacitors with caution and respect. Even with the power turned off, a large capacitor may remain charged to a high voltage.

In Figure 13–28(c), the battery has been replaced with a closed switch, creating a current path between the capacitor's plates and allowing the capacitor to discharge through the resistor. Now the capacitor is acting as the voltage source in the circuit, producing a current through the resistor, which in turn produces a voltage drop across the resistor. As the capacitor discharges, its plate voltage decreases. This decreases the current through the resistor and thus decreases the voltage drop across the resistor. However, at all times during discharge V_R equals V_C, fulfilling Kirchhoff's Voltage Law until the capacitor is completely discharged to zero volts.

It is important to have some resistance in the circuit when a capacitor is discharging to limit the amount of current. The maximum current will occur at the moment discharge begins. This amount is equal to $V_C \div R$ (Ohm's Law).

A larger resistance will limit the discharge current to a smaller amount. As a result, the discharge will take more time because the charge is flowing at a reduced volume. Remember that current equals charge divided by time: 1 amp is a flow of 1 coulomb per second. The capacitor contains a certain number of coulombs of charge. If the resistor is limiting current (fewer coulombs per second), more time is needed to let all the coulombs out of the capacitor.

dc Effects: Time Constant

A capacitor requires five time constants either to charge or discharge completely from any voltage to any other voltage. Therefore, $5\,\tau$ (five time constants) is the transient time of a capacitor.

The charge and discharge cycle of a capacitor follows the same time pattern as the expansion and collapse cycle of an inductor's magnetic field. Thus, we use the same universal exponential curves (graphs of voltage or current versus time) for both inductors and capacitors. These curves are shown in Figure 13–29 and Figure 13–30.

For capacitors, the value of one time constant (1 tau or τ) equals the capacitance multiplied by the series resistance.

Formula 13–6

$$\tau = RC$$

As shown in Figure 13–29, during the charge cycle, the capacitor reaches 63% of final voltage in a time of 1 tau, 86% in 2 tau, 95% in 3 tau, 98% in 4 tau, and 99+% (practically 100%) in 5 tau.

The discharge curve is similar. In 1 tau the capacitor's voltage drops to 37% of its original value (63% loss), in 2 tau to 14% of original value (86% loss), in 3 tau to 5% of original value (95% loss), in 4 tau to 2% of original value (98% loss), and in 5 tau to less than 1% (practically 100% loss).

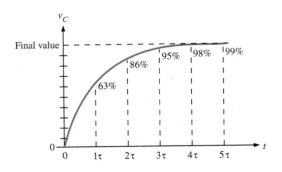

(a) Charging curve

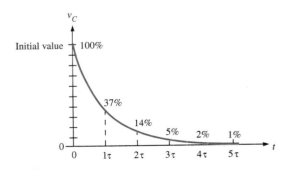

(b) Discharging curve

FIGURE 13–29
Charge and discharge curves for a capacitor.

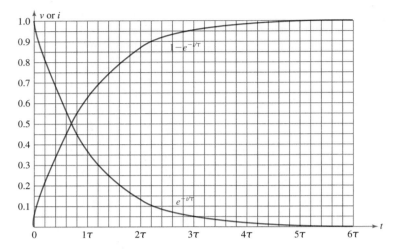

FIGURE 13–30
Universal exponential curves.

● *SKILL-BUILDER 13–6*

In Figure 13–31, when the switch is in position 1, the voltage source charges the capacitor through the resistor. When the switch is in position 2, the capacitor discharges through the resistor.

1. What is the time constant for the circuit if $R = 1.5$ kΩ and $C = 33$ μF?
2. How long will it take V_C to equal V_S when the switch moves to position 1?
3. How long will it take V_C to equal zero when the switch moves to position 2?

GIVEN:
  $R = 1.5$ kΩ
  $C = 33$ μF

FIND:
  τ
  $5\,\tau$

SOLUTION:
  1.

$$\tau = RC \qquad \text{Formula 13–6}$$
$$= 1.5 \text{ k}\Omega \times 33 \text{ }\mu\text{F}$$
$$= 49.5 \text{ ms} \qquad \text{49.5 milliseconds of time}$$

  2.
$$5\,\tau = 5 \times 49.5 \text{ ms} \qquad \text{It takes five time constants to reach}$$
$$\text{full charge.}$$
$$= 247.5 \text{ ms}$$

  3.
$$5\,\tau = 5 \times 49.5 \text{ ms} \qquad \text{It takes another five time constants to reach}$$
$$\text{full discharge.}$$
$$= 247.5 \text{ ms}$$

Repeat the Skill-Builder using the following values. You should obtain the results that follow.

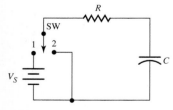

FIGURE 13–31
Circuit for Skill-Builder 13–6.

Given:

R	2.2 kΩ	560 Ω	360 kΩ	20 Ω	5.1 MΩ
C	0.047 µF	0.22 µF	220 pF	10 pF	33,000 µF

Find:

τ	103.4 µs	123.2 µs	79.2 µs	200 ps	1.95 days
5τ	517 µs	616 µs	396 µs	1 ns	1.39 week

- -

The last two items in Skill-Builder 13–6 deserve comment. In series with a 20-Ω resistor, a 10-pF capacitor reaches full charge in 1 nanosecond (one billionth of a second), about the time it takes for light to travel from the bottom of this page to the top. By contrast, a 33,000-microfarad capacitor charging through a 5.1-megaohm resistor takes almost a week and a half to reach full charge. As this extreme example illustrates, the charge/discharge time of capacitors can be controlled easily, accurately, and over a wide range of time values. This is why capacitors are used as timing devices in many circuits.

Transient Time

A capacitor requires five time constants to change completely from one voltage level to another.

In Figure 13–32, the switch places the resistor and capacitor in series with either of two voltage sources, V_1 or V_2. If V_C is less than the voltage source when the switch changes positions, the capacitor will charge to the higher voltage. Likewise, if V_C is more than the voltage source, the capacitor will discharge through the internal resistance of the voltage source, down to the lower voltage. In either case, the amount of time it takes the capacitor to reach the new voltage is exactly five time constants, regardless of the difference between the two voltages. It doesn't matter whether the difference between V_1 and V_2 is 1 volt, 100 volts, or any other value. It still takes the capacitor 5 τ to change from one voltage to the other when the switch changes position.

ac Current in a Capacitor

We know that a capacitor contains an insulator, the dielectric, which prevents charge from passing from one plate to the other. How then can we say that ac current passes through a capacitor, when we know that dc current cannot?

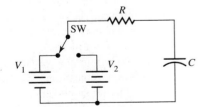

FIGURE 13–32
A capacitor requires five time constants to charge or discharge completely between any two voltage levels.

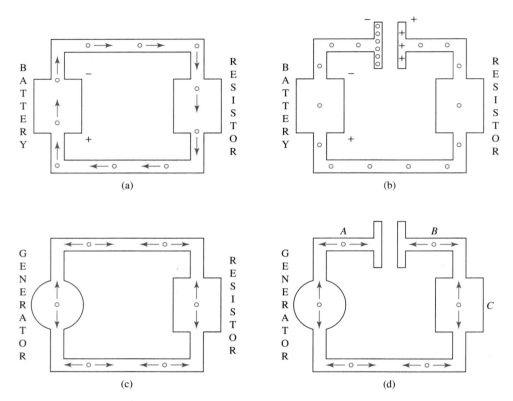

FIGURE 13–33
Comparison of dc and ac current through a resistor and a capacitor.

The answer is in Figure 13–33. In part (a) we see an exaggerated view of the movement of electrons through a resistor in a dc circuit. As the arrows indicate, the dc voltage requires the electrons to move continuously in the same direction.

Figure 13–33(b) shows why dc current cannot flow through a capacitor. The dielectric stops the forward movement of the electrons. They build up on one plate, creating holes on the other plate, until the capacitor is charged to the same voltage as the battery but with opposite polarity (series opposing). The equal but opposite voltages hold the free electrons in balance, and they do not move (no arrows in the picture).

Now examine Figure 13–33(c), which illustrates normal ac current through a resistor. Because the terminals of the ac generator reverse polarity every half-cycle, the movement of the electrons is a back-and-forth motion. Each electron remains more or less in the same spot, simply shifting position back and forth as the voltage reverses polarity.

In Figure 13–33(d), we can see that the capacitor allows this same kind of electron motion. When the top of the generator is negative, electron A is pushed onto the left plate of the capacitor. The electric field through the dielectric pushes electron B off the right plate, which pushes electron C downward in the resistor. Then, when the generator reverses polarity, the top of the generator is positive. This pulls electron A off the left plate and back to its original position. The reversal of the field in the dielectric pulls electron B back onto the right plate, which pulls electron C upward in the resistor. As far as electron C is concerned, it has the same back-and-forth motion through the resistor, whether the capacitor is in the circuit or not.

This is how a capacitor seems to allow ac current to pass through. In fact, no electron ever passes through the dielectric; but this does not matter. For every electron that moves on and off the left plate, a matching electron does the same thing on the right plate. The other electrons in the circuit can't tell the difference, and so they move back and forth as they would in a normal ac current.

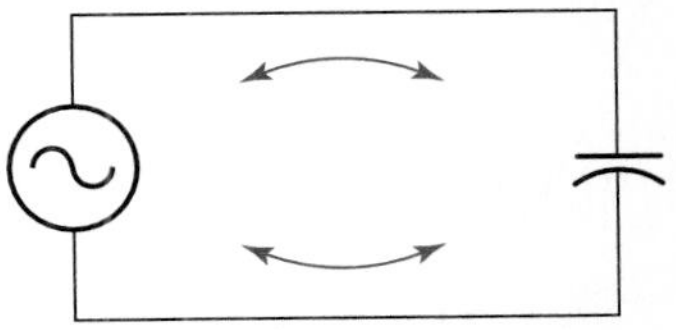

Capacitive Reactance

Capacitive reactance, symbolized X_C and measured in ohms, is the opposition of a capacitor to ac current due to the variations in the capacitor's plate voltage as it charges and discharges. Formula 13–7 allows us to calculate capacitive reactance, the equivalent ac ohmic value of the capacitor.

Formula 13–7

$$X_C = \frac{1}{2\pi f C}$$

FIGURE 13–34
Circuit for Skill-Builder 13–7.

● **SKILL-BUILDER 13–7**

An alternating voltage source is connected in series with a capacitor, as shown in Figure 13–34. What is the current if $V_S = 24\ V_{rms}$, $f = 250$ Hz, and $C = 0.47\ \mu F$?

GIVEN:
  $V_S = 24\ V_{rms}$
  $f = 250$ Hz
  $C = 0.47\ \mu F$

FIND:
  I

STRATEGY:
  1. To find the current, use Ohm's Law: $V_S \div X_C$. V_S is given. Find X_C.
  2. To find X_C, use Formula 13–7; f and C are given.

SOLUTION:
1.

$$X_C = \frac{1}{2\pi f C} \qquad\qquad \text{Formula 13–7}$$

$$= \frac{1}{2 \times \pi \times 250\ \text{Hz} \times 0.47\ \mu F}$$

$$= \frac{1}{7.3827 \times 10^{-4}}$$

$$= 1{,}354\ \Omega \qquad\qquad \text{At 250 Hz, a 0.47-}\mu\text{F capacitor acts like } 1{,}354\ \Omega.$$

2.

$$I = \frac{V_S}{X_C} \qquad\qquad \text{Ohm's Law using } X_C$$

$$= \frac{24\ V_{rms}}{1{,}354\ \Omega}$$

$$= 17.73\ \text{mA} \qquad\qquad \text{rms current}$$

Repeat the Skill-Builder using the following values. You should obtain the results that follow.

Given:

V_S	12 V	5 V	18 V	3 V	24 V
f	3 kHz	150 kHz	7.5 kHz	2 MHz	60 Hz
C	0.33 µF	220 pF	0.01 µF	47 pF	330 µF

Find:

X_C	161 Ω	4.82 kΩ	2.12 kΩ	1.69 kΩ	8.04 Ω
I	75 mA	1.04 mA	8.48 mA	1.77 mA	2.99 A

V_C Lags *I* by 90°

When ac current is applied to a capacitor, the capacitor's plate voltage V_C lags the current *I* by 90°. The phase shift happens because the current in a capacitor equals the rate of charge or discharge. Let's follow the capacitor through half a cycle, using the graph of Figure 13–35.

We begin at time t_1, when the plate voltage V_C is at its positive peak. At this moment, the capacitor is as charged up as it can be. Therefore, at t_1 the capacitor current is zero because the rate of charge is zero.

At time t_2 the capacitor has begun to discharge. Plate voltage V_C is falling but at a slow rate. Since the rate of voltage change is low, the rate of discharge is also low, which means the current is a small value. Notice in the second quarter-cycle that the polarity of the voltage is positive, but the polarity of the current is negative.

At time t_3, plate voltage V_C is falling rapidly. This allows a high rate of discharge, which means a high current.

At time t_4, plate voltage is zero, but passing through zero is when the rate of change of voltage is maximum. This is when the rate of discharge is greatest and, therefore, when the current is greatest.

It may sound strange at first that the current is maximum when the voltage is zero, but we must remember the difference between the amount of a quantity and the rate of change of that quantity. In ac theory, the rate of change is the determining factor.

At time t_5, the voltage has reversed polarity and is rising rapidly toward its negative peak. The capacitor now is charging at a high rate but not quite as high as it was at t_4. We

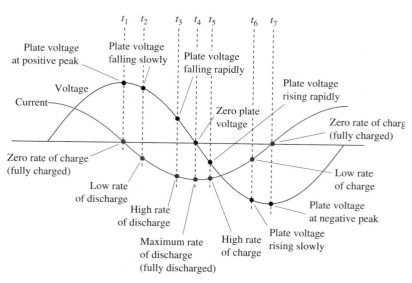

FIGURE 13–35
Phase relationship of capacitor voltage and current.

are now in the third quarter-cycle, and the polarity of the current remains negative. Notice that the direction of capacitor current is the same in the second and third quarter-cycles.

At time t_6, V_C is nearly at negative peak, and so its rate of change is less. This means the rate of charge is less and, therefore, the current is less.

Finally, at time t_7, the half-cycle is complete. Plate voltage V_C is at negative peak. Again the capacitor is as charged up as it can be. Therefore, the rate of charge is zero, which means the current is zero.

The next half-cycle is identical except for the polarity of voltage and current. This is shown on the graph by the voltage and current lines before t_1 and after t_7.

If we look again at time t_1, we can see that the plate voltage is at 90° on its cycle (positive peak). At the same time, the current is at 180° on its cycle (passing through zero going from positive polarity to negative). Therefore, the plate voltage cycle is 90° behind the current cycle. This is the information we were looking for: V_C lags I by 90°.

This relationship is shown as a phasor diagram in Figure 13–36. It is customary to use current as the reference phasor, and so phasor I is drawn out to the right, at 0°. Since V_C lags I by 90°, phasor V_C is drawn straight down, at 270°. However, it is customary to refer to this position as −90°, which emphasizes that this phasor is lagging the reference phasor.

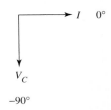

FIGURE 13–36
Phasor diagram of capacitor voltage and current.

→ *Self-Test*

10. What determines the amount of time required for a capacitor to charge and discharge?
11. How many time constants are required for a capacitor either to charge or discharge completely from any voltage to any other voltage?
12. What is capacitive reactance?
13. What is the phase angle between plate voltage and current when ac current is applied to a capacitor?

SECTION 13–5: Capacitor Applications

Power Supply Filters

Capacitors are widely used as filters in power supplies to smooth out the dc voltage required by electronic circuits. As shown in Figure 13–37, rectifier circuits in the power

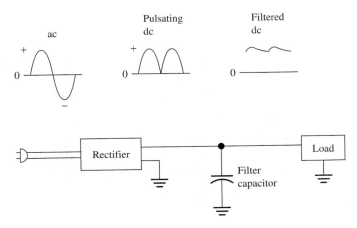

FIGURE 13–37
A filter capacitor changes pulsating dc into filtered dc.

supply change ac to pulsating dc. The pulsating dc quickly charges the filter capacitor to the dc peak-voltage value. The capacitor then discharges slowly into the load, providing the voltage and current to operate the circuits. The capacitor discharges only a little before the next peak of pulsating dc charges the capacitor up again. Thus, the dc voltage to the load is filtered by the capacitor.

Selective Frequency Response

Capacitors have selective frequency response, which means that they have more opposition to low-frequency currents and less opposition to high-frequency currents. Therefore, capacitors are often used in frequency filters, such as tone controls and equalizers in audio equipment.

Coupling and Decoupling

Coupling means allowing a signal to pass from one part of a circuit to another, and decoupling means preventing a signal from passing. Capacitors are often used to couple one part of a circuit with another, allowing ac signals to pass through while blocking dc voltages.

Transistors require dc voltage to operate, and the dc voltage needs to be slightly different at each point around the transistor. This is called bias voltage. You will study bias voltage in detail in your next course.

Capacitors are often used to connect one transistor with another, allowing the ac signals to pass through while separating the dc bias voltages.

A Capacitor Is a dc Open and an ac Short

Capacitors can be selected to have high reactance to low-frequency signals and low reactance to high-frequency signals. Technicians refer to this in a general way when they say that a capacitor is a dc open and an ac short. What this means is that dc current cannot pass through a capacitor (dc open), but ac current can; and if the ac reactance (X_C) is low, the capacitor is virtually the same as an ac short.

Timing

Capacitors are often used to provide timing in an electronic circuit. A capacitor can act as a timing device because it charges and discharges through a fixed resistance in an exact amount of time. Figure 13–38 shows how a capacitor and resistor can be used to make a strobe light flash automatically.

A strobe light uses a lamp filled with a gas such as neon or xenon. The gas is an insulator and requires a high voltage to produce current. A high dc voltage is applied, and the capacitor begins to charge up. The time required to reach full voltage is $5\ \tau$. We can control the amount of time by choosing appropriate values for R_1 and C_1.

When the capacitor reaches the necessary voltage, the gas in the lamp will break down. The lamp will conduct and quickly discharge the capacitor. The current through the lamp will be many times larger than the current provided by the battery because the capacitor has accumulated a large amount of charge and is now releasing it all at once. This is why a strobe light is bright and remains on for a short time.

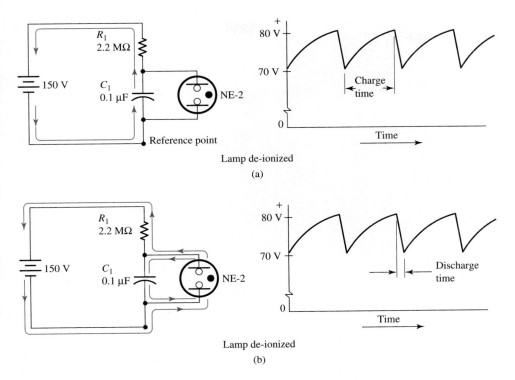

FIGURE 13–38
Capacitor as a timing element in an oscillator circuit.

Phase Shifting

As we have learned, capacitors produce a phase shift between source voltage and source current. The phase-shifting feature of a capacitor can be used to start certain types of electric motors.

There is a type of motor known as a single-phase induction motor, often nicknamed a *squirrel-cage* motor. We will study this motor in a later chapter. The problem with this motor is that it will not start turning by itself. The solution to the problem is to install an extra set of windings that assists the magnetic field inside the motor to start the rotor turning. For these windings to work properly, the current in them must be out of phase with the current in the main windings. A large capacitor is used to provide the necessary phase shift in the start windings. This capacitor is usually mounted on top of the motor, as shown in Figure 13–39, and is generally called a start capacitor.

Energy Storage

Capacitors may be used to store, absorb, or suppress large amounts of electrical energy. We have learned that capacitors store joules of energy in the dielectric due to dielectric stress, and that the amount of energy depends on the capacitance and the plate voltage. Thus, capacitors can be used to accumulate a large amount of energy, which then can be released all at once.

Capacitors are used in defibrillators, the machines doctors use to revive a patient whose heart has stopped beating. The defibrillator provides a short electrical shock of a precise voltage and frequency to stimulate the heart into beating again. A pair of hand-held paddles are touched to the patient's bare chest, and the electric shock causes the body to jump. Capacitors are used to accumulate the energy, and the controls on the

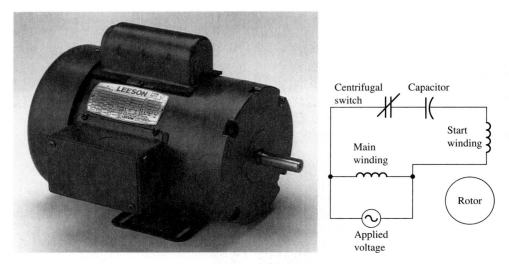

FIGURE 13–39
Start capacitor on a single-phase ac induction motor.

defibrillator are calibrated in joules, so the doctor can adjust the exact amount of energy to be given to the heart. Because the capacitors need time to charge, the doctor must wait a short time before using the defibrillator again.

Arc Suppression

An arc is a spark across open contacts. A capacitor may be used to suppress an arc by absorbing the voltage. A common example of arc suppression is the small capacitor inside the distributor of an automobile ignition system. This capacitor, usually called a condenser (the British name for a capacitor), suppresses the arc across the points.

SECTION 13–6: Testing and Troubleshooting Capacitors

In this section you will learn the following:

→ When applied to a normal capacitor, an ohmmeter will first indicate zero ohms, then indicate a high ohmic value.
→ When applied to a shorted capacitor, an ohmmeter will indicate zero ohms constantly.
→ When applied to an open capacitor, an ohmmeter will indicate infinity constantly.
→ When applied to a leaky capacitor, an ohmmeter will first indicate zero ohms, then indicate a low ohmic value.

Measuring Capacitance

Most meters which measure capacitance use an electronic circuit to generate a pulse waveform. This waveform is applied to the capacitor, and the circuit detects the rate of charge and discharge repeatedly. The circuit averages these measurements and interprets the results as the capacitance. This measuring technique, where a reading is taken over and over and the results automatically averaged, is generally called sampling. Sampling produces excellent results and lends itself well to digital test instruments. Its major drawback is an erratic display until enough samples have been taken to produce a consistent average value. To the technician using the test meter, the numbers in the window change quickly from one value to another with no apparent pattern. When measuring capacitors with sampling instruments, it may take some time for the display to become stable.

We have learned that temperature affects capacitance. You may verify this by mounting a small ceramic disk capacitor in a meter and waiting for the display to stabilize. Then hold the capacitor gently between your thumb and forefinger. You will see the value change as the capacitor responds to the heat from your fingers. If the value increases, the capacitor has a positive temperature coefficient.

Most meters measure capacitance in the nanofarad range, with perhaps a few upper ranges calibrated in microfarads. Since capacitors are labeled in microfarads or picofarads, not in nanofarads, you will have to move the decimal point around a bit to match the meter reading to the capacitor's nominal value.

Most capacitors have a tolerance range of either ±5%, ±10%, or ±20%.

Shorted Capacitor

If the plates come into contact with each other, the capacitor is shorted and cannot store charge. Moreover, a shorted capacitor may allow a large dc current to flow where no current was intended. This may produce a large amount of heat and seriously damage other components. Capacitors may become shorted from the dielectric breaking down after years of stress, from a high voltage cracking the dielectric, or from a high ac current breaking down the dielectric with heat.

Leaky Capacitor

A leaky capacitor has a low dielectric leakage resistance, allowing an excessive leakage current through the dielectric. A leaky cap is often on its way to becoming a shorted cap. The term *leaky* does not necessarily mean a physical leak, such as the electrolyte oozing out of the can.

Open Capacitor

An electrolyte may become open if its electrolyte dries out. Remember that the electrolyte is conductive and is part of one plate. When the electrolyte dries out, it becomes nonconductive, disconnecting its plate from electrical contact with the dielectric oxide layer on the other plate. Thus, a dry electrolyte creates an open in the capacitor.

Electrolytic capacitors may dry out from years of operation and the heat thus produced, or simply from sitting on the shelf waiting to be installed in a circuit. Thus, like batteries, replacement electrolytic capacitors need to be fresh: They have a limited shelf life.

Check the Capacitor with a Voltmeter First

Before testing or handling any capacitor, put a voltmeter across the capacitor's leads or terminals, as shown in Figure 13–40(a). If the capacitor is charged, the voltmeter will indicate the voltage. Leave the voltmeter in contact with the capacitor, and the capacitor will discharge safely through the large resistors inside the voltmeter. When the voltmeter reads zero, you may handle and test the capacitor.

Checking a Capacitor with an Ohmmeter

Remember that an ohmmeter has its own internal battery, which produces current in the component being tested. The ohmmeter measures the amount of current it can produce

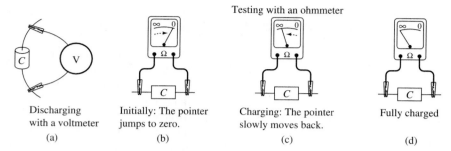

FIGURE 13–40
Testing a capacitor.

and interprets this amount as the resistance of the component. Therefore, if the ohmmeter produces a large amount of current, it figures the component has a resistance near zero; and if the meter produces only a small amount of current, it figures the component has a resistance near infinity.

When an ohmmeter is connected to a discharged capacitor, the meter's battery begins to charge the capacitor. At first, the current is large, and the ohmmeter responds by indicating a value near zero ohms, as shown in Figure 13–40(b).

As the charge cycle continues, the current decreases. The ohmmeter interprets this as an increasing resistance, as shown in Figure 13–40(c).

When the capacitor is fully charged, the only current flowing is the small leakage current through the dielectric. For electrolytic capacitors, this leakage current should produce a steady ohmmeter reading of about 500 kΩ (0.5 MΩ). For other capacitors, the reading should be well above 100 MΩ. Many meters are not sensitive enough to read such a small current, and will show a steady reading of infinity, as shown in Figure 13–40(d).

Ohmmeter Response of a Shorted Capacitor

If the ohmmeter reading goes to nearly zero ohms and stays there, the capacitor is shorted.

Ohmmeter Response of a Leaky Capacitor

If the ohmmeter reading becomes steady at a low ohmic value, the capacitor is leaky.

Ohmmeter Response of an Open Capacitor

If the ohmmeter remains at infinity and never moves toward zero or a lower value, the capacitor is open.

Testing Small Capacitors

Small capacitors may charge so quickly that you do not have time to observe the pattern of ohmmeter readings. In this case, put a large resistor in series with the capacitor to cause it to charge more slowly. If the capacitor is leaky or shorted, you will have to allow for the resistor value in interpreting the ohmmeter readings.

Limitations of Ohmmeter Readings

An ohmmeter produces a low-voltage dc current. The capacitor may have defects in the dielectric that do not show up when tested with an ohmmeter but occur when the capacitor is exposed to high voltages or frequencies in the actual circuit. For all components, including capacitors, the only conclusive test is replacing the component with one known to be good. The ohmmeter readings may indicate clearly that the old capacitor is bad, or they may not. In this case, you must decide whether to spend time and money replacing the capacitor in hopes of correcting the problem.

Precautions in Working with Capacitors

Capacitors present a real hazard to the technician, and we conclude this chapter with a summary of the safety precautions you should always follow in working with capacitors.

- Always check a capacitor with a voltmeter before handling it. If the voltmeter indicates the capacitor is charged, leave the voltmeter in contact until the capacitor has discharged.
- Never short a capacitor to discharge it. The high current and resulting heat can damage the capacitor and injure you.
- Capacitors have both a capacitance value and a voltage rating. Always replace a capacitor with one whose values are exactly the same as the original.
- Most electrolytic capacitors are polarized: One lead is negative, the other positive. If a polarized capacitor is mounted incorrectly, it may explode when power is applied. Check polarity carefully when replacing electrolytic capacitors.
- When working with electrolytic capacitors under power, always wear eye protection and keep your head to the side.
- If you see any sign of electrolyte leaking or oozing out of the can or capsule, replace the capacitor immediately and clean the electrolyte off the circuit.
- Electrolytic capacitors have a limited shelf life. Replace electrolytic capacitors with fresh capacitors, and check them with an ohmmeter before installing them.
- After replacing a capacitor, stand back when you turn on the power. Let the circuit operate for five or ten minutes to make sure there are no problems with the new capacitor.

→ *Self-Test*

14. What is the reading of an ohmmeter applied to a normal capacitor?
15. What is the reading of an ohmmeter applied to a shorted capacitor?
16. What is the reading of an ohmmeter applied to an open capacitor?
17. What is the reading of an ohmmeter applied to a leaky capacitor?

Formula List

Formula Diagram 13–1

$$\frac{Q}{C \mid V}$$

Formula 13–2

$$C = \frac{A \; \varepsilon_r \varepsilon_0}{d}$$

Formula 13–3

Total capacitance in series:

$$C_T = \frac{1}{\dfrac{1}{C_1} + \dfrac{1}{C_2} + \ldots + \dfrac{1}{C_n}}$$

Formula 13–4

$$V_x = \frac{C_T}{C_x \times V_T}$$

Formula 13–5

Total capacitance in parallel:

$$C_T = C_1 + C_2 + \ldots + C_n$$

Formula 13–6

$$\tau = RC$$

Formula 13–7

$$X_C = \frac{1}{2\pi f C}$$

● TROUBLESHOOTING: Applying Your Knowledge

The most likely problem is an open power supply filter capacitor. The output waveform appears to be pulsating dc coming straight from the rectifier diodes. The purpose of the filter capacitor is to eliminate most of the ripple, and the waveform indicates this is not happening. Therefore, the capacitor is not functioning normally.

If the capacitor were shorted, the power supply would blow fuses due to a high current to ground. If the capacitor were leaky, some filtering would occur. By process of elimination, an open capacitor is the most likely cause of the trouble.

CHAPTER SUMMARY: ANSWERS TO SELF-TESTS

1. Capacitance *(C)* is the ability to store charge.
2. The unit of capacitance is the farad (F). A capacitance of 1 farad is the ability to store 1 coulomb of charge at a potential of 1 volt.
3. An increase in plate area causes an increase in capacitance.
4. An increase in dielectric thickness causes a decrease in capacitance.
5. An increase in value of the dielectric constant causes an increase in capacitance.
6. When capacitors are charged in series, each plate stores an identical amount of charge.
7. When capacitors are connected in series, the total capacitance is the reciprocal of the sum of the reciprocals of each capacitor.
8. When capacitors are charged in series, plate voltage is inversely proportional to capacitance for each capacitor.
9. When capacitors are connected in parallel, the total capacitance equals the sum of the capacitance of each branch.

10. The amount of time required for a capacitor to charge and discharge depends on the value of capacitance and the value of series resistance.
11. A capacitor requires five time constants either to charge or discharge completely from any voltage to any other voltage. Therefore, 5τ (five time constants) is the transient time of a capacitor.
12. Capacitive reactance, symbolized X_C and measured in ohms, is the opposition of a capacitor to ac current due to the variations in the capacitor's plate voltage as it charges and discharges.
13. When ac current is applied to a capacitor, the capacitor's plate voltage V_C lags the current I by 90°.
14. When applied to a normal capacitor, an ohmmeter will first indicate zero ohms, then indicate a high ohmic value.
15. When applied to a shorted capacitor, an ohmmeter will indicate zero ohms constantly.
16. When applied to an open capacitor, an ohmmeter will indicate infinity constantly.
17. When applied to a leaky capacitor, an ohmmeter will first indicate zero ohms, then indicate a low ohmic value.

CHAPTER REVIEW QUESTIONS

Determine whether the following questions are true or false.

13–1. Capacitance is the ability to produce charge.

13–2. A capacitor consists of two conductive surfaces separated by an insulator.

13–3. The plates of a capacitor are called the dielectric.

13–4. A wide variety of materials are used to make the dielectric.

13–5. The principle of conservation of charge says that the total amount of charge in a group of objects remains constant as charge is transferred from one object to another.

13–6. Practical capacitors allow a certain amount of leakage current through the dielectric.

13–7. Axial and radial capacitors should not be used together in the same circuit.

13–8. Variable capacitors typically have an air dielectric.

13–9. Trimmer capacitors are usually large electrolytic types.

13–10. Power supply filter capacitors are usually small ceramic-disk types.

13–11. Paper and plastic film capacitors are often constructed in tubular form.

13–12. One advantage of metalized film capacitors is their ability to survive voltage overloads.

13–13. An electrolyte is any liquid or paste that is a good insulator.

13–14. The dielectric of an electrolytic capacitor is a thin oxide layer.

13–15. Titanium electrolytic capacitors have less leakage current and a higher breakdown voltage than aluminum electrolytic capacitors.

13–16. Electrolytic capacitors have an unlimited shelf life.

13–17. A polarized electrolytic capacitor installed incorrectly can be hazardous.

13–18. A ganged capacitor is one that has become defective.

13–19. To increase total capacitance, you may connect capacitors in series.

13–20. When capacitors are charged in series, the capacitor with the largest capacitance has the highest plate voltage.

13–21. The amount of time required for a capacitor to charge and discharge depends on the value of capacitance and the value of parallel resistance.

13–22. A capacitor can fully charge and then fully discharge in a total of five time constants.

13–23. Capacitive reactance equals 2 pi times capacitance times frequency.

13–24. Capacitor plate voltage lags current by 90°.

13–25. A capacitor physically disconnected from a circuit cannot remain charged.

13–26. The charge and discharge of a capacitor follows a pattern similar to the expansion and collapse of a magnetic field in an inductor.

13–27. Capacitors oppose changes in voltage by taking time to change from one voltage to another.

13–28. Only ac current can pass through the dielectric.

13–29. A capacitor will pass ac current and block dc current.

13–30. A capacitor has less reactance at a higher frequency than at a lower frequency.

13–31. A capacitor passes less current at a higher frequency than at a lower frequency.

13–32. A capacitor is frequency selective because the value of X_C depends on frequency.

Select the response that best answers the question or completes the statement.

13–33. The dielectric of a capacitor
   a. allows charge to pass between the plates.
   b. prevents charge from passing between the plates.
   c. creates charge on the plates.
   d. prevents charge from building up on the plates.

13–34. A solid dielectric is used primarily
   a. to make the capacitor more rugged.
   b. to reduce the danger of the capacitor exploding.
   c. to allow a strong electric field between the plates.
   d. to reduce the cost of the capacitor.

13–35. Electrostatic induction means that
   a. an uncharged object will become charged when a charged object is brought near.
   b. a charged object will become uncharged when an uncharged object is brought near.
   c. an uncharged object will become charged when a charged object is removed.
   d. a charged object will become uncharged when an uncharged object is removed.

13–36. One farad is the ability to store
   a. 1 amp at a potential of 1 volt.
   b. 1 volt at a potential of 1 amp.
   c. 1 coulomb at a potential of 1 volt.
   d. 1 volt at a potential of 1 coulomb.

13–37. Capacitance is directly proportional to
   a. plate voltage.
   b. plate area.
   c. dielectric thickness.
   d. dielectric strength.

13–38. Capacitance is inversely proportional to
   a. plate voltage
   b. plate area.
   c. dielectric thickness.
   d. dielectric strength.

13–39. Each plate stores an identical amount of charge when capacitors are charged
   a. with dc voltage.
   b. with ac voltage.
   c. in series.
   d. in parallel.

13–40. If two 16-μF capacitors are connected in series, the total capacitance is
   a. 16 μF.
   b. 8 μF.
   c. 32 μF.
   d. 4 μF.

13–41. If two 16-µF capacitors are connected in parallel, the total capacitance is
  a. 16 µF.
  b. 8 µF.
  c. 32 µF.
  d. 4 µF.

13–42. Which of the following is *not* a factor of capacitance?
  a. plate area
  b. plate material
  c. dielectric thickness
  d. dielectric material

13–34. If two plates overlap partially, the capacitance depends on
  a. total plate area.
  b. single plate area.
  c. effective plate area.
  d. stored plate area.

13–44. Which of the following will *not* intensify the electric field?
  a. increasing plate voltage
  b. decreasing plate thickness
  c. increasing dielectric permittivity
  d. decreasing dielectric thickness

13–45. Dielectric strength is
  a. the ability of the dielectric to withstand physical forces on the leads.
  b. the ability of the atoms in the dielectric to withstand the dislocating forces of an electric field.
  c. the same as dielectric constant.
  d. the same as relative permittivity.

13–46. Which of the following is *not* a common dielectric material?
  a. oil
  b. silicon
  c. mica
  d. plastic

13–47. A certain ac current flows through a capacitor. The current can be reduced 50% by
  a. doubling the frequency.
  b. doubling the capacitance.
  c. connecting another capacitor of equal value in series.
  d. connecting another capacitor of equal value in parallel.

13–48. The transient time of a capacitor is
  a. $5\,\tau$.
  b. $1\,\tau$.
  c. $10\,\tau$.
  d. $2\,\tau$.

13–49. A capacitor acts as
  a. a dc open and an ac open.
  b. a dc open and an ac short.
  c. a dc short and an ac open.
  d. a dc short and an ac short.

13–50. A capacitor may be used as a timing device in
  a. a power supply.
  b. an electric motor.
  c. an oscillator.
  d. a defibrillator.

13–51. Which feature of a capacitor may be used to start a single-phase induction motor?
  a. reactance
  b. frequency selectivity
  c. energy storage
  d. phase shifting

13–52. The technique of using large capacitors to offset the inductive reactance of large electric motors is called
  a. phase shifting.
  b. power factor correction.
  c. shielding.
  d. surge suppression.

13–53. A capacitor is used in a conventional automobile ignition system for
  a. timing.
  b. tuning.
  c. power correction.
  d. arc suppression.

13–54. An ohmmeter is applied to an uncharged capacitor. The ohmmeter first indicates zero ohms, then indicates 2.2 kΩ. Most likely the capacitor is
  a. normal.
  b. open.
  c. shorted.
  d. leaky.

Solve the following problems.

13–55. A 48-µF capacitor is charged to 12 V. How many coulombs of charge are stored in it?

13–56. A 3,300-µF capacitor contains 0.66 coulombs of charge. What is the voltage across its plates?

13–57. A capacitor is charged to 20 V and contains 0.015 coulombs. What is its capacitance?

13–58. A small capacitor has the digits "221" printed on it. What is its capacitance?

13–59. Three capacitors are connected in series: $C_1 = 15$ µF, $C_2 = 25$ µF, and $C_3 = 36$ µF. What is the total capacitance?

13–60. A 12-V source is connected to three capacitors in series. The values of the capacitors are $C_1 = 360$ µF, $C_2 = 220$ µF, and $C_3 = 100$ µF. What is the voltage potential across the plates of each capacitor?

13–61. What is the total capacitance of three capacitors in parallel when $C_1 = 19$ µF, $C_2 = 27$ µF, and $C_3 = 12$ µF?

13–62. A 6-V battery is applied to an uncharged 0.047-µF capacitor in series with a 360-Ω resistor. What is the plate voltage after 16.92 µs?

13–63. An ac voltage source is connected in series with a capacitor. What is the current if $V_S = 18\ V_{rms}$, $f = 320$ Hz, and $C = 0.68$ µF?

GLOSSARY
- - - - - - - -

axial leads Leads located on opposite ends of a component.
breakdown voltage The maximum working voltage that can be applied to a capacitor.
capacitance The ability to store charge.

capacitive reactance   The opposition in ohms of a capacitor to ac current.

capacitor   A device designed to store charge.

dielectric   The insulator material between the plates of a capacitor.

dielectric constant   A value that represents the ability of a dielectric material to support an electric field.

electrolytic capacitor   A capacitor whose dielectric consists of a thin layer of corrosion formed by chemical action between an electrolyte paste and aluminum foil plates.

farad   The unit of capacitance.

plates   The conductive surfaces of a capacitor.

radial leads   Leads located on the same end of a component.

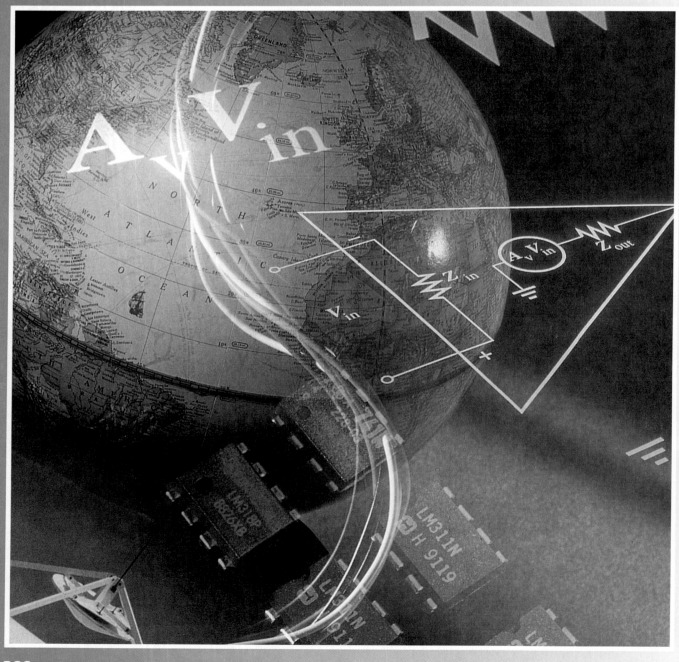

CHAPTER

14

AC INDUCTIVE AND CAPACITIVE CIRCUITS

● SECTION OUTLINE

14–1   Trigonometry

14–2   Series *RL* Circuits

14–3   Series *RC* Circuits

14–4   Series *RLC* Circuits

14–5   Parallel *RL* Circuits

14–6   Parallel *RC* Circuits

14–7   Parallel *RLC* Circuits

● TROUBLESHOOTING:
Applying Your Knowledge

Your fellow student is conducting a lab experiment with a circuit consisting entirely of resistors, coils, and capacitors. She comes to you in confusion and disbelief because her data indicate voltage drops within the circuit that are greater than the source voltage. This makes no sense to her, and she wants you to help her find out what she is doing wrong. How would you evaluate her problem?

● OVERVIEW AND LEARNING OBJECTIVES

Many electronic circuits contain combinations of resistors, inductors, and capacitors. Each has its own effect on voltage and current. We need a method of analyzing the combined effects of these components. The method involves phasor diagrams and the mathematics of trigonometry.

After completing this chapter, you will be able to:

1. Solve phasor diagrams for impedance and voltage in a basic series *RL* circuit.
2. Solve phasor diagrams for impedance and voltage in a basic series *RC* circuit.
3. Solve phasor diagrams for impedance and voltage in a basic series *RLC* circuit.
4. Solve phasor diagrams for current in a basic parallel *RL* circuit.
5. Solve phasor diagrams for current in a basic parallel *RC* circuit.
6. Solve phasor diagrams for current in a basic parallel *RLC* circuit.

SECTION 14–1: Trigonometry

In this section you will learn the following:

→ Trigonometry is the mathematics used to calculate the values of triangles.

→ A right triangle includes a 90° angle.

→ The sides of a right triangle are the adjacent, opposite, and hypotenuse.

→ The Pythagorean Theorem states that the square of the hypotenuse equals the sum of the squares of the other two sides.

→ The three principal trigonometric functions are called sine, cosine, and tangent. Each trigonometric function is a ratio of two sides of a right triangle.

→ The sine function is the ratio of the opposite to the hypotenuse.

→ The cosine function is the ratio of the adjacent to the hypotenuse.

→ The tangent function is the ratio of the opposite to the adjacent.

→ The three trigonometric ratios are functions of the angle theta. This means that the values of the ratios depend on the value of the angle and not on the values of the sides.

→ If theta changes, the values of all three functions also change.

→ An arc function is the angle associated with a value of one of the three functions. The value of the angle depends on the value of the function.

Introduction to ac Analysis

When ac voltage and current are out of phase, the best way to show this is to use phasor diagrams. We are particularly interested in phase angles of 90° because both inductance and capacitance produce a 90° phase shift between voltage and current.

Several of the basic rules of series and parallel circuits require addition. These rules are:

• In series circuits, the sum of the voltage drops equals the source voltage.
• In series circuits, the sum of the ohmic values of the components equals the total ohmic value.
• In parallel circuits, the sum of the branch currents equals the total current.
• In both series and parallel circuits, the sum of the power found in each component equals the total power.

Because ac quantities may be out of phase, we cannot simply add the numerical values to find totals. Instead, we must represent each value as a phasor, an arrow whose length represents the magnitude (amount) of the quantity, and whose direction represents the phase (timing) of the quantity. We add the phasors by placing the tail of one phasor on the arrowhead of another, then drawing a phasor from the origin (the tail of the first phasor) to the final coordinates (the arrowhead of the last phasor). This last phasor is the *phasor sum*.

If we have only two phasors to add, and if their phase angle is 90°, the two phasors and the phasor sum form a triangle, as shown in Figure 14–1. In this illustration, part (a) shows two phasors at a 90° phase angle. Part (b) shows the triangle formed when the phasors are redrawn to find the phasor sum.

We will add ac voltages, currents, ohmic values, and power values by finding phasor sums. In other words, we will draw triangle diagrams similar to Figure 14–1(b) and think of the sides of the triangle as representing various ac voltages, currents, and so on. This way we can account for the ac timing differences (phase angles) and find the correct total values.

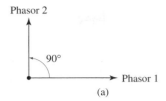

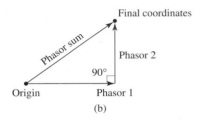

(b)

FIGURE 14–1
When two phasors are at a 90°
phase angle, the phasor sum is
the hypotenuse of a right
triangle.

Trigonometry is the mathematics used to calculate the values of triangles. Since our phasor diagrams will look like triangles, we must first learn basic trigonometry before we begin to calculate ac voltages and currents in inductive and capacitive circuits.

The Standard Right Triangle

If a triangle has one 90° corner, it is called a *right triangle*. Trigonometry is based on the relationships of the sides and angles of a right triangle.

The sides and angles have standard names. As shown in Figure 14–2, the horizontal side is called the *adjacent* side, the vertical side is called the *opposite* side, and the diagonal side is called the *hypotenuse*. The corner angle is the right angle (90°). The angle between the hypotenuse and the adjacent side is called *theta* (the Greek letter θ).

The Pythagorean Theorem

The ancient Greek mathematician Pythagoras discovered that for any right triangle, the square of the hypotenuse equals the sum of the squares of the other two sides. This basic rule is called the *Pythagorean Theorem* and is given in Formula 14–1.

Formula 14–1: The Pythagorean Theorem

$$\text{Hyp}^2 = \text{Adj}^2 + \text{Opp}^2$$

To demonstrate the Pythagorean Theorem, draw a right triangle with an adjacent side of 3 inches and an opposite side of 4 inches. Then draw the hypotenuse and measure it. It will be exactly 5 inches. If we then substitute these values in Formula 14–1, we get:

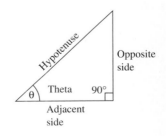

FIGURE 14–2
The standard right
triangle.

$$Hyp^2 = Adj^2 + Opp^2$$
$$5^2 = 3^2 + 4^2$$
$$25 = 9 + 16$$
$$25 = 25$$

The Pythagorean Theorem can be turned into three formulas, one for each side:

Formula 14–2: Hypotenuse

$$Hyp = \sqrt{Adj^2 + Opp^2}$$

Formula 14–3: Adjacent

$$Adj = \sqrt{Hyp^2 - Opp^2}$$

Formula 14–4: Opposite

$$Opp = \sqrt{Hyp^2 - Adj^2}$$

We will use most often the formula for the hypotenuse, Formula 14–2, because the hypotenuse will be the phasor sum of our ac electrical diagrams.

● SKILL-BUILDER 14–1

A right triangle has an adjacent side of 10 units and an opposite side of 14 units, as shown in Figure 14–3. What is the length of the hypotenuse?

SOLUTION:

$$Hyp^2 = \sqrt{Adj^2 + Opp^2} \qquad \text{Formula 14–2}$$
$$= \sqrt{10^2 + 14^2}$$
$$= \sqrt{100 + 196} \qquad \text{1. Square the two sides.}$$
$$= \sqrt{296} \qquad \text{2. Add the squares.}$$
$$= 17.2 \text{ units} \qquad \text{3. Take the square root.}$$

The calculator keystroke sequence is:

$$10 \ [x^2] \ [+] \ 14 \ [x^2] \ [=] \ [\sqrt{\ }]$$

Repeat the Skill-Builder using the following values. You should obtain the results that follow.

Given:

Adj	67	132	2.07	0.056	2,437
Opp	23	83	3.31	0.047	3,659

Find:

Hyp	70.84	155.93	3.90	0.073	4,396

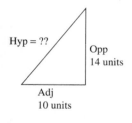

Hyp = ??

Opp
14 units

Adj
10 units

$$Hyp = \sqrt{Adj^2 + Opp^2}$$

FIGURE 14–3
Diagram and formula for Skill-Builder 14–1.

The Trigonometric Functions

The three principal trigonometric functions are called the *sine, cosine,* and *tangent.* Each function is a ratio of two sides of a right triangle.

The sine function is the ratio of the opposite side to the hypotenuse. The cosine function is the ratio of the adjacent to the hypotenuse. The tangent function is the ratio of the opposite to the adjacent. All three ratios are functions of the angle theta. Figure 14–4 illustrates a 3-4-5 right triangle and shows the value of the three functions for this triangle.

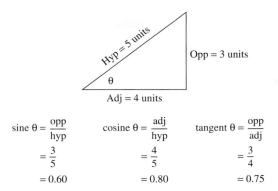

FIGURE 14–4
The trigonometric functions.

The sine function is the ratio of the opposite side to the hypotenuse. In Figure 14–4, the opposite side is 3 units and the hypotenuse is 5 units. The sine ratio is 3 divided by 5, which equals 0.60. In other words, in this triangle, the opposite side is 60% of the hypotenuse.

The cosine function is the ratio of the adjacent side to the hypotenuse. In Figure 14–4, the adjacent side is 4 units and the hypotenuse is 5 units. The cosine ratio is 4 divided by 5, which equals 0.80. In other words, in this triangle, the opposite side is 80% of the hypotenuse.

The tangent function is the ratio of the opposite side to the adjacent side. In Figure 14–4, the opposite side is 3 units and the adjacent is 4 units. The tangent ratio is 3 divided by 4, which equals 0.75. In other words, in this triangle, the opposite side is 75% of the adjacent.

Figure 14–5 shows the 3-4-5 triangle with its sides extended to form a 6-8-10 triangle. Even though the sides are longer, the three functions have the same value because the ratios (percentages) have not changed. For example, 3 divided by 5 gives a sine value of 0.60, and 6 divided by 10 gives the same value. Thus, the value of the sine function is the same both for the smaller triangle (3-4-5) and the larger triangle (6-8-10). The same is true for the cosine and tangent functions.

The values of the three functions depend on the angle theta (θ). In Figure 14–5, θ equals 36.87°. Any triangle with a theta of 36.87° will have a sine value of 0.60, a cosine value of 0.80, and a tangent value of 0.75.

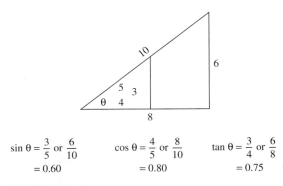

FIGURE 14–5
The values of the functions depend on the angle.

We have just explained that the three ratios are functions of the angle theta. This statement is the very heart of trigonometry. The whole system is built around this one fact: If the angle theta changes, the values of all three functions also change.

Figure 14–6 gives the values of the three functions for all angles from 0° to 90°.

Calculation Diagrams for the Trigonometric Functions

The three functions can be arranged into calculation diagrams:

$$\frac{\text{Opp}}{\text{Hyp} \mid \sin\theta} \qquad \frac{\text{Adj}}{\text{Hyp} \mid \cos\theta} \qquad \frac{\text{Opp}}{\text{Adj} \mid \tan\theta}$$

Angle	sin	cos	tan	Angle	sin	cos	tan	Angle	sin	cos	tan
0°	.0000	1.0000	.0000	30°	.5000	.86603	.57735	60°	.86603	.5000	1.73205
1	.01745	.99985	.01745	31	.51504	.85717	.60086	61	.87462	.48481	1.80404
2	.03490	.99939	.03492	32	.52992	.84805	.62487	62	.88295	.46947	1.88072
3	.05234	.99863	.05241	33	.54464	.83867	.64941	63	.89101	.45399	1.96261
4	.06976	.99756	.06993	34	.55919	.82904	.67451	64	.89879	.43837	2.05030
5	.08715	.99619	.08749	35	.57358	.81915	.70021	65	.90631	.42262	2.14450
6	.10453	.99452	.10510	36	.58778	.80902	.72654	66	.91354	.40674	2.24603
7	.12187	.99255	.12278	37	.60181	.79863	.75355	67	.92050	.39073	2.35585
8	.13917	.99027	.14054	38	.61566	.78801	.78128	68	.92718	.37461	2.47508
9	.15643	.98769	.15838	39	.62932	.77715	.80978	69	.93358	.35837	2.60508
10	.17365	.98481	.17633	40	.64279	.76604	.83910	70	.93969	.34202	2.74747
11	.19081	.98163	.19438	41	.65606	.75471	.86929	71	.94552	.32557	2.90421
12	.20791	.97815	.21256	42	.66913	.74314	.90040	72	.95106	.30902	3.07768
13	.22495	.97437	.23087	43	.68200	.73135	.93251	73	.95630	.29237	3.27085
14	.24192	.97029	.24933	44	.69466	.71934	.96569	74	.96126	.27564	3.48741
15	.25882	.96592	.26795	45	.70711	.70711	1.00000	75	.96592	.25882	3.73205
16	.27564	.96126	.28674	46	.71934	.69466	1.03553	76	.97029	.24192	4.01078
17	.29237	.95630	.30573	47	.73135	.68200	1.07236	77	.97437	.22495	4.33147
18	.30902	.95106	.32492	48	.74314	.66913	1.11061	78	.97815	.20791	4.70463
19	.32557	.94552	.34433	49	.75471	.65606	1.15036	79	.98163	.19081	5.14455
20	.34202	.93969	.36397	50	.76604	.64279	1.19175	80	.98481	.17365	5.67128
21	.35837	.93358	.38386	51	.77715	.62932	1.23489	81	.98769	.15643	6.31375
22	.37461	.92718	.40403	52	.78801	.61566	1.27994	82	.99027	.13917	7.11537
23	.39073	.92050	.42447	53	.79863	.60181	1.32704	83	.99255	.12187	8.14435
24	.40674	.91354	.44523	54	.80902	.58778	1.37638	84	.99452	.10453	9.51436
25	.42262	.90631	.46631	55	.81915	.57358	1.42814	85	.99619	.08715	11.4300
26	.43837	.89879	.48773	56	.82904	.55919	1.48256	86	.99756	.06976	14.3006
27	.45399	.89101	.50952	57	.83867	.54464	1.53986	87	.99863	.05234	19.0811
28	.46947	.88295	.53171	58	.84805	.52992	1.60033	88	.99939	.03490	28.6362
29	.48481	.87462	.55431	59	.85717	.51504	1.66427	89	.99985	.01745	57.29
								90	1.0000	.0000	∞

FIGURE 14–6

Table of trigonometric functions.

By using the table of trigonometric functions and these diagrams, we can calculate the sides and angles of a right triangle. The diagrams will show us how to calculate, and the table will give us the correct values to use in the calculation. Before we begin, however, there are two rules that we must follow.

1. Choose the diagram that has the two sides you are working with.
2. In using the diagrams, the sides are multiplied or divided by the value of the function, *not* the value of the angle.

Rule 1 tells us how to choose the correct diagram. Each diagram includes two of the three sides. Choose the diagram that contains the two sides you are working with. You may know the value of both sides, or you may know the value of one side and want to know the value of the other. Either way, you are working with two sides. If you are working with the opposite and the hypotenuse, use the sine diagram. If you are working with the adjacent and the hypotenuse, use the cosine diagram. Finally, if you are working with the opposite and the adjacent, use the tangent diagram.

Rule 2 warns us that the numerical value of the angle (the degrees of the angle) is *never used* in multiplication or division with the sides. The value of the angle is used to look up the correct value of the function in Figure 14–6, and this value (sine, cosine, or tangent) is used in the diagram.

PROBLEM 1: How do you find the opposite side if you know the angle and the hypotenuse? SOLUTION: Since you are working with the opposite and the hypotenuse, use the sine diagram. First, look up the sine of the angle in the function table. Then follow the diagram: The opposite equals the hypotenuse multiplied by the sine value. EXAMPLE: Find the opposite when θ equals 20° and the hypotenuse equals 15 units. From the function table, the sine of 20° is 0.34202. The hypotenuse multiplied by the sine value is 15 units × 0.34202, which equals 5.13 units. This is the length of the opposite side. See Figure 14–7.

PROBLEM 2: How do you find the adjacent side if you know the angle and the hypotenuse? SOLUTION: Since you are working with the adjacent and the hypotenuse, use the cosine diagram. First, look up the cosine of the angle in the function table. Then follow the diagram: The adjacent equals the hypotenuse multiplied by the cosine value. EXAMPLE: Find the adjacent when θ equals 32° and the hypotenuse equals 46 units. From the function table, the cosine of 32° is 0.84805. The hypotenuse multiplied by the cosine value is 46 units × 0.84805, which equals 39.01 units. This is the length of the adjacent side. See Figure 14–8.

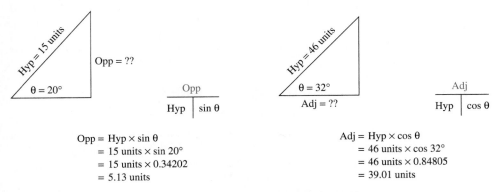

$$Opp = Hyp \times \sin \theta$$
$$= 15 \text{ units} \times \sin 20°$$
$$= 15 \text{ units} \times 0.34202$$
$$= 5.13 \text{ units}$$

$$Adj = Hyp \times \cos \theta$$
$$= 46 \text{ units} \times \cos 32°$$
$$= 46 \text{ units} \times 0.84805$$
$$= 39.01 \text{ units}$$

FIGURE 14–7
Using the sine function to find the opposite.

FIGURE 14–8
Using the cosine function to find the adjacent.

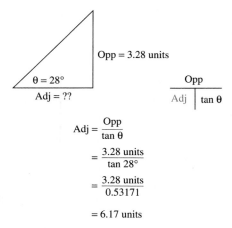

$$Adj = \frac{Opp}{\tan \theta}$$

$$= \frac{3.28 \text{ units}}{\tan 28°}$$

$$= \frac{3.28 \text{ units}}{0.53171}$$

$$= 6.17 \text{ units}$$

FIGURE 14–9
Using the tangent function to find
the adjacent.

PROBLEM 3: How do you find the adjacent side if you know the angle and the opposite side? SOLUTION: Since you are working with the adjacent and the opposite, use the tangent diagram. First, look up the tangent of the angle in the function table. Then follow the diagram: The adjacent equals the opposite divided by the tangent value. EXAMPLE: Find the opposite when θ equals 28° and the adjacent equals 3.28 units. From the function table, the tangent of 28° is 0.53171. The opposite divided by the tangent value is 3.28 units ÷ 0.53171, which equals 6.17 units. This is the length of the adjacent side. See Figure 14–9.

● *SKILL-BUILDER 14–2*

Develop your trigonometry skills by completing the following exercise. Each row represents a right triangle. Find the values in brackets by using the other values in the row with the function diagrams and table.

θ	Hyp	Opp	Adj
68°	17.70 units	[16.41 units]	[6.63 units]
17°	[5.56 units]	1.63 units	[5.32 units]
35°	[28.04 units]	[16.08 units]	[22.97 units]
54°	[417.65 units]	337.89 units	[245.49 units]
22°	[7.13 units]	[2.67 units]	6.61 units
83°	0.17 units	[0.169 units]	[0.02 units]

The Arc Functions: Using the Sides to Find the Angle

Figure 14–6 is a kind of directory: You look up the angle to find its three functions. The table can also be used in the reverse manner: You can find the angle by looking up one of its functions. For example, if you know that the sine is 0.25882, you could find this value in the table's sine column, look to the left, and see that the angle is 15°. Again, if you know that the cosine is 0.76604, you could find this value in the cosine column, look to the left, and see that the angle is 50°. Finally, if you know that the tangent is 2.14450, you could find this value in the tangent column, look to the left, and see that the angle is 65°.

When we use the functions to find the angle, we speak of the angle as the *arc function* of the functions. For example, using the values just given, sine 15° = 0.25882   15° = arc sin 0.25882

cos 50° = 0.76604   50° = arc cos 0.76604

tan 65° = 2.14450   65° = arc tan 2.14450

The arc function is often written with a superscript of $^{-1}$, as follows.

sin 15° = 0.25882   15° = $\sin^{-1}$ 0.25882

cos 50° = 0.76604   50° = $\cos^{-1}$ 0.76604

tan 65° = 2.14450   65° = $\tan^{-1}$ 2.14450

Whether we write "arc sin" or "$\sin^{-1}$," we always say "arc sine." This is also the case for arc cosine and arc tangent.

It is important to understand clearly the difference between a function and an arc function. A function is a ratio of two sides. The value of the ratio depends on the angle. An arc function is the angle associated with a ratio of two sides. The value of the angle depends on the ratio.

Each of the calculation diagrams can be restated to express the arc function, as shown in the following and in Figure 14–10.

An example of an arc sine is given in Figure 14–10(a).

$$\frac{Opp}{Hyp \mid \sin \theta} \qquad \sin \theta = \frac{Opp}{Hyp} \qquad \theta = \sin^{-1}\left[\frac{Opp}{Hyp}\right]$$

$$\theta = \sin^{-1} (Opp/Hyp)$$

$$= \sin^{-1} (5/16)$$

$$= \sin^{-1} (0.3125)$$

In the sine column of the function table (Figure 14–6), 0.3125 falls between 0.30902 and 0.32557. Therefore, θ is between 18° and 19°.

An example of an arc cosine is given in Figure 14–10(b).

$$\frac{Adj}{Hyp \mid \cos \theta} \qquad \cos \theta = \frac{Adj}{Hyp} \qquad \theta = \cos^{-1}\left[\frac{Adj}{Hyp}\right]$$

$$\theta = \cos^{-1} Adj/Hyp$$

$$= \cos^{-1} (17/23)$$

$$= \cos^{-1} (0.73913)$$

(a)	(b)	(c)

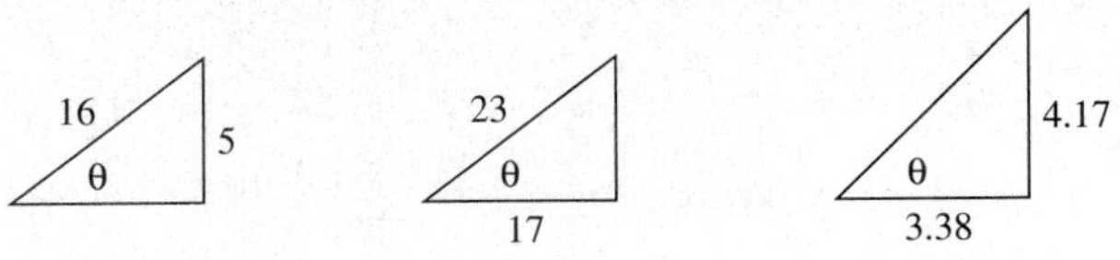

$$\frac{Opp}{Hyp \mid \sin \theta}$$

$$\sin \theta = \frac{Opp}{Hyp}$$

$$\theta = \sin^{-1}\left(\frac{Opp}{Hyp}\right)$$

$$= \sin^{-1}\left(\frac{5}{16}\right)$$

$$= \sin^{-1} (0.3125)$$

$$= 18.21°$$

$$\frac{Adj}{Hyp \mid \cos \theta}$$

$$\cos \theta = \frac{Adj}{Hyp}$$

$$\theta = \cos^{-1}\left(\frac{Adj}{Hyp}\right)$$

$$= \cos^{-1}\left(\frac{17}{23}\right)$$

$$= \cos^{-1} (0.73913)$$

$$= 42.34°$$

$$\frac{Opp}{Adj \mid \tan \theta}$$

$$\tan \theta = \frac{Opp}{Adj}$$

$$\theta = \tan^{-1}\left(\frac{Opp}{Hyp}\right)$$

$$= \tan^{-1}\left(\frac{4.17}{3.38}\right)$$

$$= \tan^{-1} (1.23373)$$

$$= 50.97°$$

FIGURE 14–10
The relationships of the functions and arc functions.

In the cosine column of the function table, 0.73913 falls between 0.74314 and 0.73135. Therefore, θ is between 42° and 43°.

Finally, an example of an arc tangent is given in Figure 14–10(b).

$$\frac{\text{Opp}}{\text{Adj}}\;\bigg|\;\tan \theta \qquad \tan \theta = \frac{\text{Opp}}{\text{Adj}} \qquad \theta = \tan^{-1}\left[\frac{\text{Opp}}{\text{Adj}}\right]$$

$$\theta = \tan^{-1}(\text{Opp/Adj})$$
$$= \tan^{-1}(4.17/3.38)$$
$$= \tan^{-1}(1.23373)$$

In the tangent column of the function table (Figure 14–6), 1.23373 is slightly less than 1.23489. Therefore, θ is slightly less than 51°.

Using a Scientific Calculator for Trigonometry

Scientific calculators contain the trigonometry function table. By using the calculator's trigonometry keys, you do not have to look up numbers in Figure 14–6, or enter the numbers manually into the calculator.

In this book, we measure angles in degrees. There are two other systems for measuring angles. Engineers often measure angle in radians. In this system, a circle contains 6.2832 radians, which is 2 π. The gradian system is used by the military for artillery calculations. In this system, a circle contains 400 grads.

Most scientific calculators will operate in any of the three angular systems. Your calculator must be set to the degree mode for your answers to agree with those in this book. Usually, the display window of the calculator will show the abbreviation "DEG" or "D" when the calculator is in the degree mode. Some calculators change mode by pressing a [MODE] key, followed by pressing a number, often the number 4. Other calculators have a [DRG] key which you press repeatedly until you get the mode you want.

We will now repeat the examples given earlier to illustrate the use of the calculation diagrams. This time we will indicate the calculator keystrokes using the trigonometry keys to speed the calculation.

PROBLEM 1:
Keystrokes:	15 [×] 20 [SIN] [=]
Display:	5.13

PROBLEM 2:
Keystrokes:	46 [×] 32 [COS] [=]
Display:	39.01

PROBLEM 3:
Keystrokes:	3.28 [÷] 28 [TAN] [=]
Display:	6.17

ARC SINE EXAMPLE:
Keystrokes:	5 [÷] 16 [=] [SIN^{-1}]
Display:	18.21
Meaning:	Theta is 18.21°.

ARC COSINE EXAMPLE
Keystrokes:	17 [÷] 23 [=] [COS^{-1}]
Display:	42.32
Meaning:	Theta is 42.32°.

ARC TANGENT EXAMPLE:

Keystrokes:	4.17 [÷] 3.38 [=] [TAN^{-1}]
Display:	50.97
Meaning:	Theta is 50.97°.

→ *Self-Test*

1. What is trigonometry?
2. What is a right triangle?
3. What are the sides of a right triangle?
4. What is the Pythagorean Theorem?
5. What are the three principal trigonometric functions?
6. What is the sine function?
7. What is the cosine function?
8. What is the tangent function?
9. What is the relationship between the three ratios and the angle theta?
10. What happens to the values of the three functions if the value of theta changes?
11. What is an arc function?

SECTION 14–2: Series *RL* Circuits

In this section you will learn the following:

→ Impedance *(Z)* is the total ohmic value of any circuit. Impedance is total voltage divided by total current.

→ In a series *RL* circuit, impedance is the phasor sum of resistance and inductive reactance.

→ In a series *RL* circuit, source voltage is the phasor sum of the resistive voltage drop and the reactive voltage drop.

→ Theta is the phase angle between source voltage and source current.

→ In a series *RL* circuit, an increase in frequency causes the circuit to become more reactive.

→ In a series *RL* circuit, an increase in inductance causes the circuit to become more reactive.

→ In a series *RL* circuit, an increase in resistance causes the circuit to become less reactive.

→ *"ELI* the *ICE* man" is a memory aid representing the relationships of voltage and current in reactive circuits. *ELI* represents voltage leading current in an inductor. *ICE* represents voltage lagging current in a capacitor.

ELI the *ICE* Man

We have learned that in an ac circuit both inductors and capacitors produce a voltage. The inductor produces a back voltage (CEMF) as its windings respond to changes in its magnetic field. The capacitor produces a voltage across its plates as it charges and discharges. The voltage produced by an inductor or capacitor is 90° out of phase with the current. However, in an inductor, voltage leads current; and in a capacitor, voltage lags current. There is an easy way to remember these facts. This memory aid is the phrase *"ELI* the *ICE* man." To understand "ELI the ICE man" you must know that *I* stands for current and *E* stands for voltage (electromotive force).

The symbol for inductance is *L*. In the case of an inductor, E_L (the inductor's voltage) leads I_L by 90°. In the word *ELI,* the middle *L* indicates inductance, and *E* comes before *I*. Thus, *ELI* reminds us that voltage leads current in an inductor.

The symbol for capacitance is *C*. In the case of a capacitor, E_C (the inductor's voltage) lags I_C by 90°. In the word *ICE,* the middle *C* indicates capacitance, and *E* comes after *I*. Thus, *ICE* reminds us that voltage lags current in a capacitor.

Impedance in a Series *RL* Circuit

Impedance (Z) is the total ohmic value of any circuit. Impedance is total voltage divided by total current, which accounts for all types of opposition to current.

In a series *RL* circuit, impedance is the phasor sum of resistance and inductive reactance.

We begin by finding the inductive reactance of the circuit using Formula 14–5.

Formula 14–5

$$X_L = 2\,\pi f L$$

If we have a series circuit that includes both resistance and inductance, and if the circuit is connected to an ac voltage source, then we have two kinds of ohms in the circuit: ohms of resistance and ohms of inductive reactance. Although both of these present an opposition to current, they are entirely different in nature.

Resistance is due to the atomic structure of the material in the resistor. Resistance dissipates power, produces heat, does not cause a phase shift with current, and has the same ohmic value in both dc and ac circuits.

Inductive reactance is the ohmic equivalent of the back voltage in the coil. Because X_L is based on V_L, inductive reactance has the same characteristics as back voltage. Inductive reactance stores and releases power, does not produce heat, causes a 90° phase shift with current, and appears only in ac circuits.

We have learned to find the total ohmic value of a series circuit by adding each ohmic value within the circuit. However, because *R* and X_L are so different in nature, we

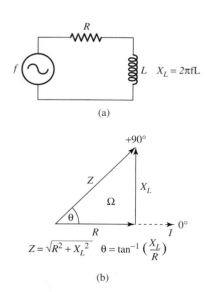

(a)

(b)

FIGURE 14–11
Impedance triangle for a series
RL circuit.

cannot just add the two numbers together. Instead, because of the phase shift caused by X_L, we must represent R and X_L as phasors, and add them by finding the phasor sum, as shown in Figure 14–11.

In a series circuit, there is only one current. Thus, it has the same value at all points in the circuit. Therefore, it makes sense to choose current as the reference quantity in a series circuit. When we draw phasor diagrams, the reference quantity is always drawn straight out to the right, which is the 0° position. In Figure 14–11, current is represented by a dashed arrow.

Resistance *(R)* is represented by a phasor also drawn to the right, in the 0° position. We place the resistance phasor at 0° because the voltage drop of the resistor is in phase with the current. Inductive reactance *(X$_L$)* is represented by a phasor drawn straight up because the inductive voltage *(V$_L$)* leads current by 90°.

As shown in Figure 14–11, in a series *RL* circuit with an ac source, impedance is the phasor sum of resistance and inductive reactance.

Because the natural phase angle between V_L and I is 90° of cycle time, the phasor diagram shows an angle of 90° between the phasors (arrows) for X_L and R. Therefore, impedance *(Z)*, which is the phasor sum of R and X_L, is represented as the hypotenuse of a right triangle, with R as the adjacent side and X_L as the opposite side. This is convenient, since it allows us to use the mathematics of trigonometry to calculate the exact value of Z.

Thus, Formula 14–6, the formula for Z in an ac series *RL* circuit, is the Pythagorean formula for the hypotenuse with the sides called Z, R, and X_L instead of hypotenuse, adjacent, and opposite.

Formula 14–6

$$Z = \sqrt{R^2 + X_L^2}$$

● SKILL-BUILDER 14–3

Suppose that in the circuit of Figure 14–11, the ac source has a voltage of 12 V$_{rms}$ and a frequency of 8 kHz. The resistance is 470 Ω and the inductance is 25 mH. Find the inductive reactance, impedance, and current.

GIVEN:
$V_S = 12 \text{ V}_{rms}$
$f = 8 \text{ kHz}$
$R = 470 \text{ } \Omega$
$L = 25 \text{ mH}$

FIND:
X_L
Z
I

$X_L = 2\pi fL$
$Z = \sqrt{R^2 + X_L^2}$
$I = V_S \div Z$ (Ohm's Law)

SOLUTION:
$X_L = 2\pi fL$
  $= 2 \times 3.1416 \times 8 \text{ kHz} \times 25 \text{ mH}$
  $= 1{,}257 \text{ } \Omega$ Reactance is measured in ohms.

$Z = \sqrt{R^2 + X_L^2}$ Impedance is the phasor sum of resistance and reactance.

  $= \sqrt{(470 \text{ } \Omega)^2 + (1{,}257 \text{ } \Omega)^2}$ keystrokes: 470 [X^2]
[+] 1257 [X^2] [=]
[√]

  $= 1{,}342 \text{ } \Omega$ total ohmic value of the circuit

$$I = V_S \div Z$$

Ohm's Law: total current equals total voltage divided by impedance (total ohms; total opposition to current)

$$= 12 \text{ V} \div 1,342 \text{ }\Omega$$
$$= 8.94 \text{ mA}$$

Repeat the Skill-Builder using the following values. You should obtain the results that follow.

Given:

V_S	6 V	9 V	15 V	3 V	5 V
f	20 kHz	25 kHz	5 kHz	38 kHz	1.7 kHz
R	510 Ω	3.6 kΩ	680 Ω	1.2 kΩ	200 Ω
L	3.2 mH	27 mH	75 mH	10 mH	24 mH

Find:

X_L	402 Ω	4,241 Ω	2,356 Ω	2,388 Ω	256 Ω
Z	649 Ω	5,563 Ω	2,452 Ω	2,672 Ω	325 Ω
I	9.24 mA	1.62 mA	6.12 mA	1.12 mA	15.38 mA

Voltage Drops in Series *RL* Circuits

The voltage rule for series circuits says that the sum of the voltage drops equals the source voltage (Kirchhoff's Voltage Law). We must follow this rule in the series *RL* circuit. However, because V_L and V_R are out of phase, we cannot just add the two values. Instead, we must use phasor addition, as shown in Figure 14–12.

Since current is the quantity that has a single value in a series circuit, we choose current as the reference phasor and place it at 0°. The voltage drop of the resistor V_R is in phase with current because resistance does not cause a phase shift. Therefore, the phasor for V_R is also drawn at the 0° position.

V_L is the back voltage (CEMF) of the inductor. Since V_L leads I by 90°, the phasor for V_L is drawn straight up at the +90° position. The source voltage V_S is the phasor sum of V_R and V_L. The three phasors form a right triangle with V_S as the hypotenuse. Therefore, the formula for V_S is the Pythagorean formula for the hypotenuse.

Formula 14–7

$$V_S = \sqrt{V_R^2 + V_L^2}$$

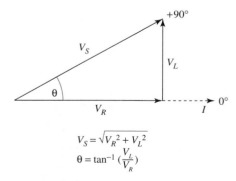

FIGURE 14–12
Voltage triangle for a series *RL* circuit.

Theta: The Phase Angle

Theta is the phase angle between source voltage and source current. Because of V_L, the source voltage V_S is out of phase with its current. The angle theta (θ) in Figure 14–12 is the amount of this phase shift. We can find theta by using the trigonometric arc functions. V_L is the opposite side in the triangle, and V_R is the adjacent side. Therefore, V_L divided by V_R is the tangent of theta; and theta itself is the arc tangent of V_L over V_R.

Because X_L is based on V_L, the value of theta is the same in both the impedance triangle (Figure 14–11) and the voltage triangle (Figure 14–12). Therefore, theta is also equal to the arc tangent of X_L over R.

Formulas 14–8(a) and (b) give both solutions for theta.

Formulas 14–8

(a)
$$\theta = \tan^{-1}\left[\frac{X_L}{R}\right]$$

(b)
$$\theta = \tan^{-1}\left[\frac{V_L}{V_R}\right]$$

● SKILL-BUILDER 14–4

In the series RL circuit of Figure 14–13, V_R is 3.4 V and V_L is 2.2 V. Find V_S and θ.

GIVEN:
$V_R = 3.4$ V
$V_L = 2.2$ V

FIND:
V_S
θ

SOLUTION:
$V_S = \sqrt{V_R^2 + V_L^2}$
$\quad = \sqrt{(3.4 \text{ V})^2 + (2.2 \text{ V})^2}$

keystrokes: 3.4 [x²]
[+] 2.2 [x²] [=]
[√]

$\quad = 4.05$ V

$\theta = \tan^{-1}(V_L \div V_R)$
$\quad = \tan^{-1}(2.2 \text{ V} \div 3.4 \text{ V})$

keystrokes: 2.2 [÷]
3.4 [=] [tan⁻¹]

$\quad = 32.9°$

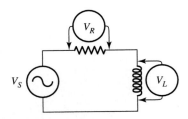

FIGURE 14–13
Circuit for Skill-Builder 14–7.

Repeat the Skill-Builder using the following values. You should obtain the results that follow.

Given:

V_R	1.58 V	6.43 V	2.84 V	1.15 V	3.28 V
V_L	0.68 V	3.75 V	1.82 V	0.95 V	4.51 V

Find:

V_S	1.72 V	7.44 V	3.37 V	1.49 V	5.58 V
θ	23.3°	30.3°	32.7°	39.6°	54.0°

We may summarize by saying that in a series RL circuit with an ac source, source voltage is the phasor sum of the resistive voltage drop and the reactive voltage drop.

Effect of Frequency on Series *RL* Circuits

In a series RL circuit, an increase in frequency causes an increase in X_L. This causes an increase in Z and a decrease in I.

Because X_L increases and R remains constant, more of the voltage is dropped across X_L and less across R.

The increase in X_L causes the circuit to become more reactive. As a result, the phase angle θ increases.

Effect of Inductance on Series *RL* Circuits

An increase in inductance L has the same effect on a series RL circuit as an increase in frequency.

Effect of Resistance on Series *RL* Circuits

An increase in resistance in a series RL circuit causes an increase in Z and a decrease in I.

Because R increases and X_L remains constant, more voltage is dropped across R and less across X_L.

The increase in R causes the circuit to become less reactive. As a result, the phase angle θ decreases.

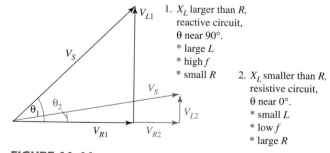

FIGURE 14–14
Voltage triangles for a series *RL* circuit indicate whether the circuit is characteristically resistive or reactive.

Summary of Series *RL* Circuits

In Figure 14–14, the two voltage triangles summarize what we have learned about series *RL* circuits. Notice that the value of source voltage, V_S, is the same in both triangles.

→ *Self-Test*

12. What is impedance?
13. What is the relationship between impedance, resistance, and inductive reactance in a series *RL* circuit?
14. What is the relationship between source voltage, resistive voltage drop, and reactive voltage drop in a series *RL* circuit?
15. What is the significance of theta?
16. What is the result of an increase in frequency in a series *RL* circuit?
17. What is the result of an increase in inductance in a series *RL* circuit?
18. What is the result of an increase in resistance in a series *RL* circuit?
19. What is the meaning of the phrase "*ELI* the *ICE* man"?

SECTION 14–3: Series *RC* Circuits

In this section you will learn the following:

→ In a series *RC* circuit, impedance is the phasor sum of resistance and capacitive reactance.

→ In a series *RC* circuit, source voltage is the phasor sum of the resistive voltage drop and the reactive voltage drop.

→ In a series *RC* circuit, an increase in frequency causes the circuit to become less reactive.

→ In a series *RC* circuit, an increase in capacitance causes the circuit to become less reactive.

→ In a series *RC* circuit, an increase in resistance causes the circuit to become less reactive.

The formulas, diagrams, and methods for solving series *RC* circuits are similar to those for series *RL* circuits. There are two chief differences, however.

First, Formula 14–9, the formula for capacitive reactance (X_C), involves a reciprocal. This means that a higher frequency or a larger capacitance produces less reactance. This is the opposite of inductive reactance, where higher frequency or larger inductance produced more reactance.

Formula 14–9

$$X_C = \frac{1}{2\pi f C}$$

Second, capacitive reactance produces a negative phase shift, which means the capacitor's voltage lags its current by 90°. Therefore, capacitive reactance causes source voltage to lag source current by some value between zero and 90 degrees. In the phasor diagrams for series *RC* circuits, X_C and V_C point down toward the −90° position. The phasor sums Z and V_S are below the 0° position, and theta is a negative angle.

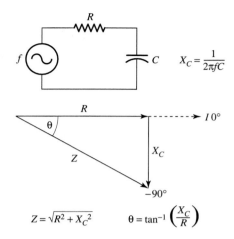

$$X_C = \frac{1}{2\pi f C}$$

$$Z = \sqrt{R^2 + X_C^2} \qquad \theta = \tan^{-1}\left(\frac{X_C}{R}\right)$$

FIGURE 14–15
Impedance triangle for a series *RC*
circuit.

MATH TIP

In Figure 14–15, X_C is represented as a negative phasor. There are two ways to apply this fact to the trigonometric formulas. If you put a negative value of X_C into the formulas, the results will be positive for Z and negative for theta. If you put a positive value of X_C into the formulas, the results will still be positive for Z. Theta will also be positive, and you will have to remember to write it as a negative value.

Impedance in a Series *RC* Circuit

The out-of-phase charging and discharging of the capacitor in a series *RC* circuit creates a second voltage in the circuit, V_C. This presents an opposition to current. We treat this opposition in the same way we treated the back voltage of a coil. We call it capacitive reactance, another kind of ac ohms. Capacitive reactance stores and releases power, does not produce heat, causes a 90° phase shift with current, and appears only in ac circuits. We then combine resistance and capacitive reactance into impedance, the overall ohmic value of the circuit.

Figure 14–15 illustrates impedance in a series *RC* circuit. Because of the phase shift caused by X_C, the impedance Z is the phasor sum of R and X_C.

Although X_C points down in the impedance triangle, we can see that X_C is the opposite side. Therefore, Formulas 14–10(a) and (b) for impedance and theta are similar to the trigonometric formulas we learned for series *RL* circuits.

Formulas 14–10

(a) $$Z = \sqrt{R^2 + X_C^2}$$

(b) $$\theta = \tan^{-1}\left[\frac{X_C}{R}\right]$$

● *SKILL-BUILDER 14–5*
- -
Suppose that in the circuit of Figure 14–15, the ac source has a voltage of 3 V_{rms} and a frequency of 28 kHz. The resistance is 470 Ω and the capacitance is 0.027 μF. Find the capacitive reactance, impedance, and current.

GIVEN:
$$V_S = 3\ V_{rms}$$
$$f = 28\ kHz$$
$$R = 470\ \Omega$$
$$L = 0.027\ \mu F$$

FIND:
  X_C

  Z
  I

$X_C = \dfrac{1}{2\pi f C}$

$Z = \sqrt{R^2 + X_C^2}$
$I = V_S \div Z$ (Ohm's Law)

SOLUTION:

$X_L = \dfrac{1}{2\pi f C}$

$= 1/(2 \times 3.1416$
$\quad \times 28 \text{ kHz} \times 0.027 \text{ μf})$

keystrokes:
2 [×] [π] [×]
28[EXP] 3 [×]
.027[EXP] 6 [±]
[=] [1/×]

$= 211 \; \Omega$

Reactance is measured in ohms.

$Z = \sqrt{R^2 + X_C^2}$

Impedance is the phasor sum of resistance and reactance.

$= \sqrt{(470 \; \Omega)^2 + (211 \; \Omega)^2}$

keystrokes: 470 [x²]
[+] 211 [x²] [=]
[√]

$= 515 \; \Omega$

total ohmic value of the circuit

$I = V_S \div \sqrt{Z}$

Ohm's Law: total current equals total voltage divided by impedance (total ohms; total opposition to current)

$= 3 \text{ V} \div 515 \; \Omega$
$= 5.83 \text{ mA}$

Repeat the Skill-Builder using the following values. You should obtain the results that follow.

Given:

V_S	6 V	12 V	8 V	15 V	5 V
f	4 kHz	37 kHz	13 kHz	68 kHz	125 kHz
R	560 Ω	5.1 kΩ	3.6 kΩ	6.8 kΩ	1 kΩ
C	0.033 μF	0.001 μF	0.047 μF	470 pF	680 pF

Find:

X_C	1,206 Ω	4,310 Ω	260 Ω	4,980 Ω	1,872 Ω
Z	1,329 Ω	6,672 Ω	3,609 Ω	8,428 Ω	2,123 Ω
I	4.51 mA	1.80 mA	2.22 mA	1.78 mA	2.36 mA

Voltage Drops in Series *RC* Circuits

Figure 14–16 shows the voltage triangle used to calculate voltage drops in a series *RC* circuit, along with the formulas obtained from the triangle (Formulas 14–11(a) and (b)). Again, the diagram and formulas resemble those for a series *RL* circuit, except that reactance is a lagging phasor and, therefore, theta is a negative value.

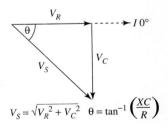

$V_S = \sqrt{V_R^2 + V_C^2} \qquad \theta = \tan^{-1}\left(\dfrac{XC}{R}\right)$

FIGURE 14–16
Voltage triangle for a series *RC* circuit.

Formula 14–11

(a) $$V_S = \sqrt{V_R^2 + V_C^2}$$

(b) $$\theta = \tan^{-1}\left[\frac{V_C}{V_R}\right]$$

● **SKILL-BUILDER 14–6**

In the series *RC* circuit of Figure 14–17, V_R is 3.57 V and V_C is 2.05 V. Find V_S and θ.

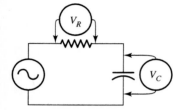

FIGURE 14–17
Circuit for Skill-Builder
14–12.

GIVEN:
$V_R = 3.57$ V
$V_C = 2.05$ V

FIND:
V_S
θ

SOLUTION:
$$V_S = \sqrt{V_R^2 + V_C^2}$$
$$= \sqrt{(3.57 \text{ V})^2 + (2.05 \text{ V})^2}$$

keystrokes: 3.57 [x²]
[+] 2.05 [x²] [=]
[√]

$$= 4.12 \text{ V}$$

$$\theta = \tan^{-1}(V_C \div V_R)$$
$$= \tan^{-1}(2.05 \text{ V} \div 3.57 \text{ V})$$

keystrokes: 2.05 [÷]
3.57 [=] [tan⁻¹]

$$= -29.9°$$

Remember to show theta as a negative
angle.

　　Repeat the Skill-Builder using the following values. You should obtain the results that follow.

Given:

V_R	6.25 V	1.52 V	8.24 V	2.27 V	5.23 V
V_C	1.45 V	7.42 V	6.28 V	5.75 V	0.85 V

Find:

V_S	6.42 V	7.57 V	10.36 V	6.18 V	5.30 V
θ	−13.1°	−78.4°	−37.3°	−68.5°	−9.2°

Effect of Frequency on Series *RC* Circuits

In a series *RC* circuit, an increase in frequency causes a decrease in X_C. This causes a decrease in Z and an increase in I.

　　Because X_C decreases and R remains constant, less of the voltage is dropped across X_C and more across R.

　　The decrease in X_C causes the circuit to become less reactive. As a result, the phase angle θ decreases.

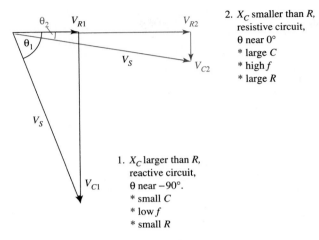

2. X_C smaller than R,
   resistive circuit,
   θ near 0°
   * large C
   * high f
   * large R

1. X_C larger than R,
   reactive circuit,
   θ near −90°.
   * small C
   * low f
   * small R

FIGURE 14–18
Voltage triangles for a series *RC* circuit indicate whether the circuit is characteristically resistive or reactive.

Effect of Capacitance on Series *RC* Circuits

An increase in capacitance C has the same effect on a series *RC* circuit as an increase in frequency.

Effect of Resistance on Series *RC* Circuits

An increase in resistance in a series *RC* circuit causes an increase in Z and a decrease in I.

Because R increases and X_C remains constant, more voltage is dropped across R and less across X_C.

The increase in R causes the circuit to become less reactive. As a result, the phase angle θ decreases.

Summary of Series *RC* Circuits

In Figure 14–18, the two voltage triangles summarize what we have learned about series *RC* circuits. Notice that the value of source voltage, V_S, is the same in both triangles.

→ *Self-Test*

20. What is the relationship between impedance, resistance, and reactance in a series *RC* circuit?
21. What is the relationship between source voltage, resistive voltage drop, and reactive voltage drop in a series *RC* circuit?

22. What is the result of an increase in frequency in a series *RC* circuit?
23. What is the result of an increase in capacitance in a series *RC* circuit?
24. What is the result of an increase in resistance in a series *RC* circuit?

- -

SECTION 14–4: Series *RLC* Circuits

In this section you will learn the following:

→ The total reactance X_T of a series *RLC* circuit is the absolute difference between inductive reactance X_L and capacitive reactance X_C.

→ In a series *RLC* circuit either V_L or V_C may be larger than V_S, due to increased current.

→ For every series *RLC* circuit, there is a frequency at which X_L equals X_C. This frequency is called the resonant frequency of the circuit (f_r).

→ Impedance is minimum in a series *RLC* circuit at the resonant frequency, f_r.

Total Reactance: X_T

The circuit shown in Figure 14–19(a) consists of a resistor, an inductor, and a capacitor in series with an ac source. This is a series *RLC* circuit.

We have learned that inductive reactance X_L leads resistance R by 90°, and capacitive reactance X_C lags R by 90°, as shown in Figure 14–19(b). This means that the phase angle between X_L and X_C is 180°, or half a cycle.

Remember that X_L represents V_L, the back voltage of the coil; and X_C represents V_C, the plate voltage of the capacitor. In an ac circuit, V_L and V_C are two ac voltages that are out of phase by half a cycle. This means that when one voltage is at its positive peak, the other voltage is at its negative peak, as shown in Figure 14–19(c). Also, V_L and V_C are both at zero volts at the same time, as shown.

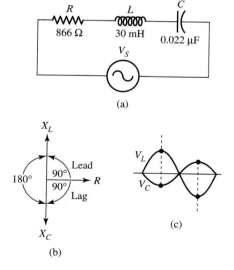

FIGURE 14–19
Series *RLC* circuit.

If V_L and V_C are in series, it is like having two voltage sources connected in series opposing: Their polarities are always working against each other. Therefore, the total reactive voltage in the circuit is the difference between V_L and V_C.

Applying this principle to reactance, in a series RLC circuit, the total reactance in the circuit is called X_T and is equal to the difference between X_L and X_C.

The total reactance X_T of a series RLC circuit is the absolute difference between inductive reactance X_L and capacitive reactance X_C. When we say *absolute difference,* we mean that we don't care whether X_L or X_C is larger in value. We only want the difference between them. We indicate this symbolically with vertical bars, as shown in Formula 14–12.

Formula 14–12

$$X_T = \mid X_L - X_C \mid$$

Impedance of a Series *RLC* Circuit

We can use X_T in the formulas for impedance and phase angle in series resistive-reactive circuits. Formulas 14–13 are identical to the formulas we have already learned for series RL and series RC circuits, with X_T substituted for X_L or X_C.

Formulas 14–13

(a)
$$Z = \sqrt{R^2 + X_T^2}$$

(b)
$$\theta = \tan^{-1} \left[\frac{X_T}{R} \right]$$

Is Theta Positive or Negative?

In using Formula 14–13(b) to find theta, you must pay attention to whether theta is a positive or negative angle. If X_L is larger than X_C, theta is a positive angle; but if X_C is larger than X_L, theta is a negative angle. The following examples will make this clear.

Case 1: X_C Larger Than X_L

Suppose that in the circuit of Figure 14–19(a) the frequency is 5,009 Hz. (The reason for the unusual value will become clear.) The phasor diagram for resistance and reactance and the resulting impedance triangle are shown in Figure 14–20(a).

At this frequency,

$$\begin{aligned}
X_L &= 2\,\pi f L \\
&= 2 \times 3.1416 \times 5{,}009 \text{ Hz} \times 30 \text{ mH} \\
&= 944 \ \Omega
\end{aligned}$$

$$\begin{aligned}
X_C &= \frac{1}{2\,\pi f C} \\
&= \frac{1}{2 \times 3.1416 \times 5{,}009 \text{ Hz} \times 0.022 \ \mu\text{F}} \\
&= 1{,}444 \ \Omega
\end{aligned}$$

$$\begin{aligned}
X_T &= \mid X_L - X_C \mid \\
&= \mid 944 \ \Omega - 1{,}444 \ \Omega \mid \\
&= 500 \ \Omega
\end{aligned}$$

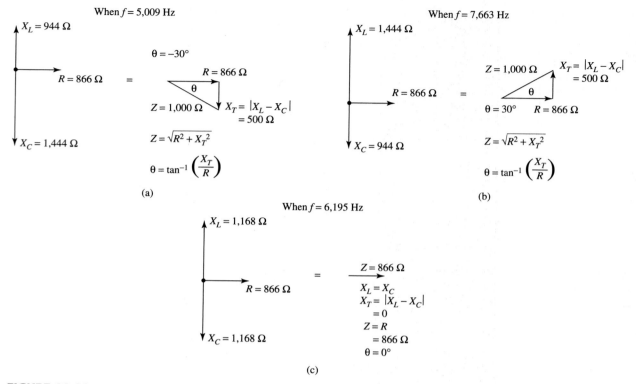

FIGURE 14–20

(a) At lower frequencies, X_C is larger than X_L, and the circuit is characteristically capacitive. (b) At higher frequencies, X_L is larger than X_C, and the circuit is characteristically inductive. (c) At the resonant frequency, X_L equals X_C, and the circuit is characteristically resistive.

Therefore,

$$Z = \sqrt{R^2 + X_T^2}$$
$$= \sqrt{(866\ \Omega)^2 + (500\ \Omega)^2}$$
$$= 1{,}000\ \Omega$$

$$\theta = \tan^{-1}(X_T \div R)$$
$$= \tan^{-1}(500\ \Omega \div 866\ \Omega)$$
$$= -30°$$

Theta is negative because X_C is larger than X_L. In other words, the 1,444 ohms of X_C are reduced by the 944 ohms of X_L; but there are still 500 Ω of X_C remaining. The circuit behaves overall like a series RC circuit. Therefore, theta is negative, indicating that source voltage is lagging source current by 30°.

Case 2: X_L Larger Than X_C

Now suppose that the frequency is 7,663 Hz. The phasor diagram and impedance triangle are shown in Figure 14–20(b).

By following the same sequence of calculations as in Case 1, we will find that now X_L is larger than X_C. This time X_L equals 1,444 Ω and X_C equals 944 Ω. You should verify this by calculating X_L and X_C for yourself.

However, the absolute difference between X_L and X_C is still 500 Ω, which is the total reactance, X_T. For this reason, the impedance Z is 1,000 Ω, the same value as in Case 1. Theta is 30° positive because X_L is larger than X_C, and the overall circuit acts like a series RL circuit, with source voltage leading source current.

Case 3: X_L Equals X_C

As shown in Figure 14–20(c), when the frequency is 6,195 Hz, X_L and X_C will both equal 1,168 Ω. You should verify this by calculating X_L and X_C for yourself.

Since X_L equals X_C, the difference between them is zero. Thus, the total reactance is zero. This does not mean that X_L and X_C are gone from the circuit. On the contrary, X_L and X_C are in the circuit and producing some interesting results, as we will soon see. When we say that X_T equals zero, we simply mean that X_L and X_C are exactly in balance, so the overall reactance appears to be zero ohms.

If the total reactance is zero ohms, then resistance is the only opposition to source current, and impedance (total ohms) equals resistance. Therefore, the phase angle is zero degrees, which means that source voltage is in phase with source current.

Voltage Drops Larger Than Source Voltage

Once again we must remember that V_L and V_C are really additional voltage sources within the circuit. However, to be consistent with the concept of reactance, we have decided to think of them as voltage drops and to calculate them with Ohm's Law: $V_L = I \times X_L$ and $V_C = I \times X_C$.

In a series *RLC* circuit, because of the balancing action between X_L and X_C, the impedance Z is smaller than it would be if either X_L or X_C were removed from the circuit. For example, in Figure 14–20(a), with X_L balancing X_C, Z equals 1,000 Ω. If X_L were removed, we would have a regular series *RC* circuit, and Z would equal $\sqrt{R^2 + X_C^2}$, which in this case works out to 1,684 Ω, considerably higher than the 1,000 Ω shown.

If the balancing action of X_L and X_C produces a smaller value of Z, then the source current I will be higher. This produces an unexpected result: V_L or V_C may turn out to be larger than the source voltage V_S.

For example, if the source voltage in the circuit of Figure 14–19(a) is 12 V and the frequency is 5,009 Hz, then from Figure 14–20(a) the impedance is 1,000 Ω. Therefore,

$$I = \frac{V_S}{Z}$$
$$= \frac{12 \text{ V}}{1 \text{ k}\Omega}$$
$$= 12 \text{ mA}$$

$$V_C = I \times X_C$$
$$= 12 \text{ mA} \times 1{,}444 \text{ }\Omega$$
$$= 17.33 \text{ V}$$

$$V_L = I \times X_L$$
$$= 12 \text{ mA} \times 944 \text{ }\Omega$$
$$= 11.33 \text{ V}$$

$$V_R = I \times R$$
$$= 12 \text{ mA} \times 866 \text{ }\Omega$$
$$= 10.39 \text{ V}$$

Notice that V_C equals 17.33 V, considerably higher than the source voltage of 12 V. This happens because V_L and V_C are actually independent voltage sources, and the balancing effect they have on each other produces large currents, which result in high voltages within the circuit.

At first you might think that Kirchhoff's Voltage Law is being violated. In fact, this can never happen. Our discussion so far has not taken into account the instantaneous polarities of the four voltages in the circuit. At all times, the algebraic sum of the voltages in the circuit equals zero, which means that when polarity is properly taken into account, Kirchhoff's Voltage Law is always fulfilled.

Resonant Frequency of a Series *RLC* Circuit

For every series *RLC* circuit, there is a frequency at which X_L equals X_C. This frequency is called the *resonant frequency* of the circuit (f_r).

The resonant frequency is part of a larger concept, which we will study in detail in the next chapter. For now, we only want to learn its name and formula. In a series *RLC* circuit, the resonant frequency may be calculated with Formula 14–14. This formula is derived by combining the formulas for X_L and X_C.

Formula 14–14

$$f_r = \frac{1}{2\,\pi\,\sqrt{L\,C}}$$

● *SKILL-BUILDER 14–7*

Use Formula 14–14 to find resonant frequency when $L = 20$ mH and $C = 0.033$ μF. The recommended calculator keystroke sequence is:

1. Multiply *L* and *C*, complete the multiplication by pressing equal, and take the square root.
   20 [EXP] 3 [+/−] [×] .033 [EXP] 6 [+/−] [=] [√]

2. Continue: Multiply by 2 and π, complete multiplication by pressing equal, and reciprocate.
   [×] 2 [×] [π] [=] [1/x]

The correct result is 6,195 Hz.

Repeat the Skill-Builder using the following values. You should obtain the results that follow.

Given:

L	16 mH	38 mH	7 mH	25 mH	10 mH
C	0.010 μF	0.033 μF	0.047 μF	0.056 μF	0.068 μF

Find:

f_r	12.58 kHz	4.49 kHz	8.77 kHz	4.25 kHz	6.10 kHz

At frequencies below f_r, X_C is larger than X_L. We describe this by saying that at low frequencies, a series *RLC* circuit is characteristically capacitive. This simply means there is some amount of X_C left over, since X_L is smaller and cannot balance X_C completely.

By similar reasoning, at frequencies higher than f_r, X_L is larger than X_C. Thus, we say that at high frequencies, a series *RLC* circuit is characteristically inductive.

To strengthen our understanding of series *RLC* concepts, let's work out a Skill-Builder that includes all we have learned about series resistive-reactive circuits.

● *SKILL-BUILDER 14–8*

For the series *RLC* circuit shown in Figure 14–21, find the phase angle and all values of reactance, current, and voltage at three frequencies: f_r, f_1, and f_2, where f_1 equals f_r minus 3 kHz, and f_2 equals f_r plus 3 kHz.

GIVEN:

$V_S = 15\ \text{V}_{\text{rms}}$

$R = 360\ \Omega$

$L = 15\ \text{mH}$

$C = 0.047\ \mu\text{F}$

PRELIMINARY STEP:

Find f_r, f_1, and f_2.

$$f_r = \frac{1}{2\pi \sqrt{LC}}$$

$$= \frac{1}{2 \times \pi \times \sqrt{15\ \text{mH} \times 0.047\ \mu\text{F}}}$$

$$= 5{,}994\ \text{Hz}$$

$f_1 = f_r - 3\ \text{kHz}$
  $= 5{,}994\ \text{Hz} - 3{,}000\ \text{Hz}$
  $= 2{,}994\ \text{Hz}$

$f_2 = f_r + 3\ \text{kHz}$
  $= 5{,}994\ \text{Hz} + 3{,}000\ \text{Hz}$
  $= 8{,}994\ \text{Hz}$

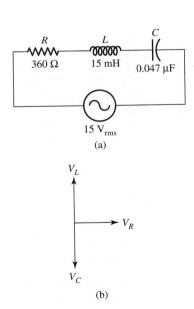

FIGURE 14–21
Circuit for Skill-Builder 14–8.

SOLUTION FOR f_r:

$X_L = 2 \pi f L$
$\quad = 2 \times \pi \times 5{,}994 \text{ Hz} \times 15 \text{ mH}$
$\quad = 565 \text{ } \Omega$

$X_C = \dfrac{1}{2 \pi f C}$

$\quad = \dfrac{1}{2 \times \pi \times 5{,}994 \text{ Hz} \times 0.047 \text{ } \mu\text{F}}$

$\quad = 565 \text{ } \Omega$

$X_T = |X_L - X_C|$
$\quad = |565 \text{ } \Omega - 565 \text{ } \Omega|$
$\quad = 0 \text{ } \Omega$

$Z = \sqrt{R^2 + X_T^2}$
$\quad = \sqrt{(360 \text{ } \Omega)^2 + (0 \text{ } \Omega)^2}$
$\quad = 360 \text{ } \Omega$

$\theta = \tan^{-1}(X_T \div R)$
$\quad = \tan^{-1}(0 \text{ } \Omega \div 360 \text{ } \Omega)$
$\quad = 0°$

$I = V_S \div Z$
$\quad = 15 \text{ V} \div 360 \text{ } \Omega$
$\quad = 41.67 \text{ mA}$

$V_R = I \times R$
$\quad = 41.67 \text{ mA} \times 360 \text{ } \Omega$
$\quad = 15.00 \text{ V}$

$V_L = I \times X_L$
$\quad = 41.67 \text{ mA} \times 565 \text{ } \Omega$
$\quad = 23.54 \text{ V}$

$V_C = I \times X_C$
$\quad = 41.67 \text{ mA} \times 565 \text{ } \Omega$
$\quad = 23.54 \text{ V}$

SOLUTION FOR f_1 AND f_2:

Follow the calculation sequence just given using a frequency of $f_1 = 2{,}994$ Hz. Then repeat using a frequency of $f_2 = 8{,}994$ Hz. You should obtain the following results.

	f_1	f_2
X_L	282 Ω	848 Ω
X_C	1,131 Ω	376 Ω
X_T	849 Ω	471 Ω
Z	922 Ω	593 Ω
θ	−67°	+53°
I	16.27 mA	25.30 mA
V_R	5.86 V	9.11 V
V_L	4.59 V	21.44 V
V_C	18.40 V	9.52 V

Impedance at the Resonant Frequency

Impedance is minimum in a series *RLC* circuit at the resonant frequency, f_r. This is true because X_T is zero at f_r. Current is at its maximum value, resulting in a minimum value of impedance.

25. What is the total reactance of a series *RLC* circuit?
26. What can cause either V_L or V_C to be larger than V_S?
27. What is resonant frequency?
28. When is impedance at a minimum in a series *RLC* circuit?

SECTION 14–5: Parallel *RL* Circuits

In this section you will learn the following:

> → The total current of a parallel *RL* circuit is the phasor sum of the branch currents.
> → The impedance of a parallel *RL* circuit may be found with Ohm's Law or through the conductance triangle.
> → In a parallel *RL* circuit, an increase in frequency causes the circuit to become less reactive.
> → In a parallel *RL* circuit, an increase in inductance causes the circuit to become less reactive.
> → In a parallel *RL* circuit, an increase in resistance causes the circuit to become more reactive.

Current in a Parallel *RL* Circuit

If a coil and resistor are in parallel with an ac source as shown in Figure 14–22, each branch will have its own current, I_R and I_L. The resistive current I_R will be in phase with the source voltage V_S. The reactive current I_L will lag the source current by 90°.

Remember the memory aid *ELI*, which symbolizes the phase relationship of voltage *E* and current *I* in an inductor *L*. We can describe the phase relationship in two ways: by saying that voltage leads current in an inductor, or by saying that current lags voltage. The words are different, but the fact is the same. When we studied the series *RL* circuit, we used current as our reference quantity and said that voltage leads current. Now we are studying a parallel circuit, using voltage as the reference quantity. Therefore, we now say that current lags voltage. The relationship is the same.

The total current I_T is the sum of I_R and I_L. However, since I_R and I_L are out of phase with each other, I_T must be the phasor sum of the two, as shown in Figure 14–23.

Notice that in Figure 14–23 I_L is drawn down, indicating its lagging phase relationship with I_R. You should recognize Formulas 14–15(a) and (b) as trigonometric formulas applied to this phasor diagram.

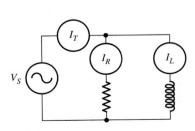

FIGURE 14–22
Currents in a parallel *RL* circuit.

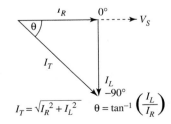

$$I_T = \sqrt{I_R^2 + I_L^2} \qquad \theta = \tan^{-1}\left(\frac{I_L}{I_R}\right)$$

FIGURE 14–23
Current triangle for a parallel *RL* circuit.

Formulas 14–15

(a) $$I_T = \sqrt{I_R^2 + I_L^2}$$

(b) $$\theta = \tan^{-1}\left[\frac{I_L}{I_R}\right]$$

● **SKILL-BUILDER 14–9**

Suppose that in Figure 14–22 I_R is 19.28 mA and I_L is 19.96 mA. Find I_T and theta.

GIVEN:
  $I_R = 19.28$ mA
  $I_L = 19.96$ mA

FIND:
  I_T
  θ

SOLUTION:

$$I_T = \sqrt{I_R^2 + I_L^2}$$
$$ = \sqrt{(19.28 \text{ mA})^2 + (19.96 \text{ mA})^2}$$
$$ = 27.75 \text{ mA}$$

$$\theta = \tan^{-1}\frac{I_L}{I_R}$$
$$ = \tan^{-1}\frac{19.96 \text{ mA}}{19.28 \text{ mA}}$$
$$ = -46.0° \qquad \text{Theta is negative because } I_L \text{ is lagging } I_R.$$

Repeat the Skill-Builder using the following values. You should obtain the results that follow.

Given:

I_R	21.59 mA	7.41 mA	21.68 mA	2.26 mA	25.87 mA
I_L	26.95 mA	3.00 mA	11.08 mA	16.04 mA	22.29 mA

Find:

I_T	34.53 mA	7.99 mA	24.34 mA	16.20 mA	34.15 mA
θ	51.3°	22.0°	27.1°	82.0°	40.7°

Impedance in a Parallel *RL* Circuit

When we studied parallel dc resistive circuits, we learned to find total resistance by a three-step process:

  1. Find branch currents with Ohm's Law.
  2. Find total current by adding branch currents.
  3. Find total resistance with Ohm's Law: total voltage divided by total current.

  We can follow a similar method in parallel ac resistive-reactive circuits.

● **SKILL-BUILDER 14–10**

Find the impedance of the circuit in Figure 14–24.

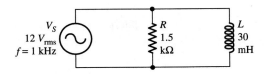

FIGURE 14–24
Circuit for Skill-Builder 14–19.

GIVEN:
$V_S = 12$ V$_{rms}$
$f = 1$ kHz
$R = 1.5$ kΩ
$L = 30$ mH

FIND:
Z
θ

SOLUTION:
1.

Find branch currents by Ohm's Law.

$I_R = V_S \div R$
$\quad = 12$ V $\div 1.5$ kΩ
$\quad = 8$ mA

$X_L = 2\pi f L$
$\quad = 2 \times 3.1416 \times 1$ kHz $\times 30$ mH
$\quad = 188$ Ω

$I_L = V_S \div X_L$
$\quad = 12$ V $\div 188$ Ω
$\quad = 63.83$ mA

2.

Find total current: the phasor sum of branch currents.

$I_T = \sqrt{I_R^2 + I_L^2}$
$\quad = \sqrt{(8 \text{ mA})^2 + (63.83 \text{ mA})^2}$
$\quad = 64.33$ mA

3.

Find impedance: total voltage divided by total current.

$Z = V_S \div I_T$
$\quad = 12$ V $\div 64.33$ mA
$\quad = 186$ Ω

4.

$\theta = \tan^{-1}(I_L \div I_R)$
$\quad = \tan^{-1}(63.83 \text{ mA} \div 8 \text{ mA})$
$\quad = 82.9°$

Repeat the Skill-Builder using the following values. You should obtain the results that follow.

Given:

V_S	6 V$_{rms}$	12 V$_{rms}$	8 V$_{rms}$	3 V$_{rms}$	9 V$_{rms}$
f	10 kHz	3 kHz	6 kHz	15 kHz	50 kHz
R	470 Ω	2,200 Ω	3,000 Ω	510 Ω	3,600 Ω
L	25 mH	60 mH	9 mH	12 mH	15 mH

Find:

X_L	1,571 Ω	1,131 Ω	339 Ω	1,131 Ω	4,712 Ω
I_R	12.77 mA	5.45 mA	2.67 mA	5.88 mA	2.50 mA
I_L	3.82 mA	10.61 mA	23.58 mA	2.65 mA	1.91 mA
I_T (mA)	13.33 mA	11.93 mA	23.73 mA	6.45 mA	3.15 mA
Z	450 Ω	1,006 Ω	337 Ω	465 Ω	2,861 Ω
θ	16.7°	62.8°	83.5°	24.3°	37.4°

Impedance and Phase-Angle Formulas for a Parallel *RL* Circuit

The reciprocal of resistance is called conductance *(G)*. As shown in Figure 14–25(a), the total conductance of a dc parallel resistive circuit is the sum of the conductance of each branch. If the total conductance G_T is then reciprocated, the result is the total resistance. This is the basis of the reciprocal method we learned to calculate total resistance in such circuits.

A similar approach can be used with parallel ac resistive-reactive circuits, as shown in Figure 14–25(b). The reciprocal of reactance is called susceptance *(B)*, and the reciprocal of impedance is called admittance *(Y)*. Admittance is the phasor sum of conductance and susceptance. Thus, the concept of admittance in ac circuits is similar to the concept of total conductance in dc circuits.

By applying the laws of algebra and trigonometry to the triangle in Figure 14–25(b), we obtain Formulas 14–16(a) and (b).

MATH TIP

When a formula such as Formula 14–16(a) involves division and contains a denominator of several terms, it is often easiest to calculate the denominator first, reciprocate the result, then multiply by the numerator. This works because:

$$\frac{A}{B} = \frac{1}{B} \times A$$

Formulas 14–16

(a)
$$Z = \frac{R \, X_L}{\sqrt{R^2 + X_L^2}}$$

(b)
$$\theta = \tan^{-1}\left[\frac{R}{X_L}\right]$$

● *SKILL-BUILDER 14–11*

In the circuit shown in Figure 14–25(b), let V_S equal 8 V, R equal 2 kΩ, and X_L equal 1,600 Ω. Solve for Z and θ first with Ohm's Law and the current triangle, then with Formulas 14–16.

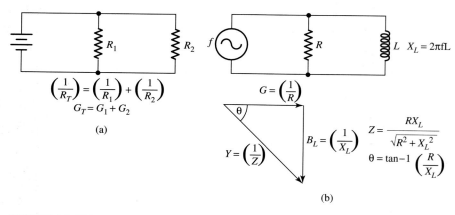

(a)

(b)

FIGURE 14–25
(a) Total conductance is the simple sum of branch conductances in a parallel resistive circuit. (b) Total conductance is the phasor sum of branch conductances in a parallel resistive-reactive circuit.

GIVEN:
  $V_S = 8$ V
  $R = 2$ kΩ
  $X_L = 1.6$ kΩ

FIND:
  Z
  θ

SOLUTION:

1.   by Ohm's Law

  $I_R = V_S \div R$
    $= 8$ V $\div 2$ kΩ
    $= 4$ mA

  $I_L = V_S \div X_L$
    $= 8$ V $\div 1.6$ kΩ
    $= 5$ mA

  $I_T = \sqrt{(4 \text{ mA})^2 + (5 \text{ mA})^2}$
    $= 6.4$ mA

  $Z = V_S \div I_T$
    $= 8$ V $\div 6.4$ mA
    $= 1{,}250$ Ω

  $\theta = \tan^{-1}(I_L \div I_R)$
    $= \tan^{-1}(5 \text{ mA} \div 4 \text{ mA})$
    $= 51.34°$

2.   by Formulas 14–16

  $Z = \dfrac{R\,X_L}{\sqrt{R^2 + X_L^2}}$

    $= \dfrac{2 \text{ kΩ} \times 1.6 \text{ kΩ}}{\sqrt{(2 \text{ kΩ})^2 + (1.6 \text{ kΩ})^2}}$

keystrokes using Math Tip:

2 [EXP] 3 [x²] [+]
1.6 [EXP] 3 [x²] [=]
[√] [1/x] [×] 2 [EXP] 3 [×]
1.6 [EXP] 3 [=]

    $= 1{,}249$ Ω

  $\theta = \tan^{-1}(R \div X_L)$
    $= \tan^{-1}(2 \text{ kΩ} \div 1.6 \text{ kΩ})$
    $= 51.34°$

- -

As you can see, it is quicker to use Formulas 14–16 for Z and theta.

● *SKILL-BUILDER 14–12*
- -

Calculate the impedance and phase angle of a 3.6-kΩ resistance and a 4.7-kΩ inductive reactance, first in series, then in parallel.

GIVEN:
  $R = 3.6$ kΩ
  $X_L = 4.7$ kΩ

FIND:
  Z and θ in a series circuit
  Z and θ in a parallel circuit

SOLUTION:
1. Series circuit:

$$Z = \sqrt{R^2 + X_L^2}$$
$$= \sqrt{(3.6 \text{ k}\Omega)^2 + (4.7 \text{ k}\Omega)^2}$$
$$= 5.92 \text{ k}\Omega$$

$$\theta = \tan^{-1}(X_L \div R)$$
$$= \tan^{-1}(4.7 \text{ k}\Omega \div 3.6 \text{ k}\Omega)$$
$$= 52.55°$$

2. Parallel circuit:

$$Z = \frac{R X_L}{\sqrt{R^2 + X_L^2}}$$
$$= \frac{3.6 \text{ k}\Omega \times 4.7 \text{ k}\Omega}{\sqrt{(3.6 \text{ k}\Omega)^2 + (4.7 \text{ k}\Omega)^2}}$$
$$= 2.86 \text{ k}\Omega$$

$$\theta = \tan^{-1}(R \div X_L)$$
$$= \tan^{-1}(3.6 \text{ k}\Omega \div 4.7 \text{ k}\Omega)$$
$$= 37.45°$$

Repeat the Skill-Builder using the following values. You should obtain the results that follow.

Given:

R	3,600 Ω	5,100 Ω	3,300 Ω	2,000 Ω	1,500 Ω
X_L	2,150 Ω	3,480 Ω	1,650 Ω	4,320 Ω	2,670 Ω

Find:

series:					
Z	4,193 Ω	6,174 Ω	3,690 Ω	4,760	3,062 Ω
θ ( )	30.8°	34.3°	26.6°	65.2°	60.7°
Parallel:					
Z	1,846 Ω	2,875 Ω	1,476 Ω	1,815 Ω	1,308 Ω
θ	59.2°	55.7°	63.4°	24.8°	29.3°

Effect of Frequency on Parallel *RL* Circuits

In a parallel *RL* circuit, an increase in frequency causes an increase in X_L. This causes an increase in Z and a decrease in I.

Because X_L increases and R remains constant, I_L decreases while I_R remains constant. The decrease in I_L causes the circuit to become less reactive. As a result, the phase angle θ decreases.

Effect of Inductance on Parallel *RL* Circuits

An increase in inductance L has the same effect on a parallel *RL* circuit as an increase in frequency.

Effect of Resistance on Parallel *RL* Circuits

An increase in resistance in a series *RL* circuit causes an increase in Z and a decrease in I_T.

Because R increases and X_L remains constant, I_R decreases while I_L remains constant. The decrease in I_R causes the circuit to become more reactive. As a result, the phase angle θ increases.

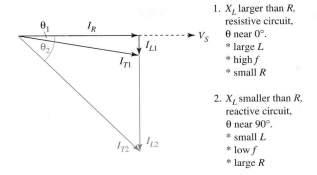

1. X_L larger than R, resistive circuit, θ near 0°.
   * large L
   * high f
   * small R

2. X_L smaller than R, reactive circuit, θ near 90°.
   * small L
   * low f
   * large R

FIGURE 14–26
Current triangles for a parallel *RL* circuit indicate whether the circuit is characteristically resistive or reactive.

Summary of a Parallel *RL* Circuit

The current triangles shown in Figure 14–26 summarize the characteristics of a parallel *RL* circuit. When I_R is larger than I_L, the circuit is resistive, as indicated by I_{L1} and I_{T1}. Theta is nearer to zero degrees, and source voltage is only slightly ahead of source (total) current. This occurs when the inductance is large, the frequency is high, or the resistance is small.

→ *Self-Test*

29. What is the total current in a parallel *RL* circuit?
30. What methods may be used to find the impedance of a parallel *RL* circuit?
31. What is the result of an increase in frequency in a parallel *RL* circuit?
32. What is the result of an increase in inductance in a parallel *RL* circuit?
33. What is the result of an increase in resistance in a parallel *RL* circuit?

SECTION 14–6: Parallel *RC* Circuits

In this section you will learn the following:

→ The total current of a parallel *RC* circuit is the phasor sum of the branch currents.
→ The impedance of a parallel *RC* circuit may be found with Ohm's Law or through the conductance triangle.
→ In a parallel *RC* circuit, an increase in frequency causes the circuit to become more reactive.
→ In a parallel *RC* circuit, an increase in capacitance causes the circuit to become more reactive.
→ In a parallel *RC* circuit, an increase in resistance causes the circuit to become more reactive.

Current in a Parallel *RC* Circuit

Figure 14–27 illustrates current in a parallel *RC* circuit, and Figure 14–28 illustrates the current triangle and trigonometric formulas for this circuit. These formulas are given as Formulas 14–17(a) and (b).

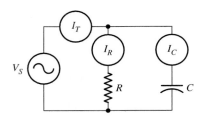

FIGURE 14–27
Currents in a parallel *RC* circuit.

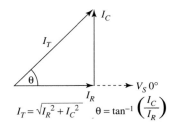

FIGURE 14–28
Current triangle for a
parallel *RC* circuit.

Formulas 14–17

(a) $$I_T = \sqrt{I_R^2 + I_C^2}$$

(b) $$\theta = \tan^{-1}\left[\frac{I_C}{I_R}\right]$$

● *SKILL-BUILDER 14–13*

Suppose that in the circuit of Figure 14–27 the circuit has the following values. Find branch currents, total current, impedance, and phase angle.

GIVEN:
$V_S = 6\ V_{rms}$
$f = 10\ kHz$
$R = 470\ \Omega$
$C = 0.033\ \mu F$

FIND:
X_C
I_C
I_R
Z
θ

SOLUTION:

$$X_C = \frac{1}{2\pi f C}$$ capacitive reactance formula

$$= \frac{1}{2 \times 3.1416 \times 10\ kHz \times 0.033\ \mu F}$$

$$= 482\ \Omega$$

$I_C = V_S \div X_C$ Ohm's Law
  $= 6\ V \div 482\ \Omega$
  $= 12.44\ mA$

$I_R = V_S \div R$ Ohm's Law
  $= 6\ V \div 470\ \Omega$
  $= 12.77\ mA$

$I_T = \sqrt{I_R^2 + I_C^2}$ phasor sum
  $= \sqrt{(12.77\ mA)^2 + (12.44\ mA)^2}$
  $= 17.83\ mA$

$Z = V_S \div I_T$ Ohm's Law

  $= 6\text{ V} \div 17.83\text{ mA}$

  $= 337\ \Omega$

$\theta = \tan^{-1}(I_C \div I_R)$ from current triangle

  $= \tan^{-1}(12.44\text{ mA} \div 12.77\text{ mA})$

  $= 44.3°$

Repeat the Skill-Builder using the following values. You should obtain the results that follow.

Given:

V_S	12 V	8 V	3 V	9 V	15 V
f	3 kHz	6 kHz	15 kHz	50 kHz	125 kHz
R	2,200 Ω	3,000 Ω	150 Ω	3,600 Ω	510 Ω
C	0.1 μF	0.022 μF	0.47 μF	0.00047 μF	0.00068 μF

Find:

X_C	531 Ω	1,206 Ω	23 Ω	6,773 Ω	1,872 Ω
I_R	5.45 mA	2.67 mA	20.00 mA	2.50 mA	29.41 mA
I_C	22.62 mA	6.64 mA	132.89 mA	1.33 mA	8.01 mA
I_T	23.27 mA	7.15 mA	134.39 mA	2.83 mA	30.48 mA
Z	516 Ω	1,119 Ω	22 Ω	3,179 Ω	492 Ω
θ	76.4°	68.1°	81.4°	28.0°	15.2°

Impedance and Phase-Angle Formulas for a Parallel *RC* Circuit

The impedance and phase angle for a parallel *RC* circuit can be found with Formulas 14–18. These formulas are similar to those for a parallel *RL* circuit.

Formulas 14–18

(a)
$$Z = \frac{R\,X_C}{\sqrt{R^2 + X_C^2}}$$

(b)
$$\theta = \tan^{-1}\left[\frac{R}{X_C}\right]$$

● SKILL-BUILDER 14–14

Calculate the impedance and phase angle of a 5.1-kΩ resistance and a 6.8-kΩ capacitive reactance, first in series, then in parallel.

GIVEN:

  $R = 5.1\text{ k}\Omega$

  $X_C = 6.8\text{ k}\Omega$

FIND:

  Z and θ in a series circuit

  Z and θ in a parallel circuit

SOLUTION:

1. Series circuit:

  $Z = \sqrt{R^2 + X_C^2}$

    $= \sqrt{(5.1\text{ k}\Omega)^2 + (6.8\text{ k}\Omega)^2}$

    $= 8.50\text{ k}\Omega$

$$\theta = \tan^{-1}(X_C \div R)$$
$$= \tan^{-1}(6.8 \text{ k}\Omega \div 5.1 \text{ k}\Omega)$$
$$= 53.13°$$

2. Parallel circuit:

$$Z = \frac{R X_C}{\sqrt{R^2 + X_C^2}}$$

$$= \frac{5.1 \text{ k}\Omega \times 6.8 \text{ k}\Omega}{\sqrt{(5.1 \text{ k}\Omega)^2 + (6.8 \text{ k}\Omega)^2}}$$

$$= 4.08 \text{ k}\Omega$$

$$\theta = \tan^{-1}(R \div X_C)$$
$$= \tan^{-1}(5.1 \text{ k}\Omega \div 6.8 \text{ k}\Omega)$$
$$= 36.87°$$

Repeat the Skill-Builder using the following values. You should obtain the results that follow.

Given:

R	2,200 Ω	5,100 Ω	4,700 Ω	1,000 Ω	3,300 Ω
X_L	2,150 Ω	3,740 Ω	2,830 Ω	3,450 Ω	6,980 Ω

Find:

Series:

Z	3,076 Ω	6,324 Ω	5,486 Ω	3,592 Ω	7,721 Ω
θ	44.3°	36.3°	31.1°	73.8°	64.7°

Parallel:

Z	1,538 Ω	3,016 Ω	2,424 Ω	960 Ω	2,983 Ω
θ	45.7°	53.7°	58.9°	16.2°	25.3°

We will use the circuit in Figure 14–29 to examine the effects of changes in frequency, capacitance, and resistance in a parallel *RC* circuit.

Effect of Frequency on Parallel *RC* Circuits

In a parallel *RC* circuit, an increase in frequency causes a decrease in X_C. This causes a decrease in Z and an increase in I.

Because X_C decreases and R remains constant, I_C increases while I_R remains constant. The increase in I_C causes the circuit to become more reactive. As a result, the phase angle θ increases.

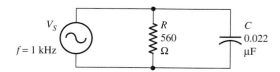

FIGURE 14–29
Circuit for evaluating effects of changes in frequency, capacitance, and resistance in a parallel *RC* circuit.

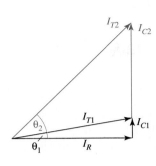

2. X_C smaller than R,
   resistive circuit,
   θ near 90°.
   * large C
   * high f
   * large R

1. X_C larger than R,
   resistive circuit,
   θ near 0°.
   * small C
   * low f
   * small R

FIGURE 14–30
Current triangles for a parallel *RC* circuit
indicate whether the circuit is
characteristically resistive or reactive.

Effect of Capacitance on Parallel *RC* Circuits

An increase in capacitance C has the same effect on a parallel *RC* circuit as an increase in frequency.

Effect of Resistance on Parallel *RC* Circuits

An increase in resistance in a parallel *RC* circuit causes an increase in Z and a decrease in I_T.

Because R increases and X_C remains constant, I_R decreases while I_C remains constant. The decrease in I_R causes the circuit to become more reactive. As a result, the phase angle θ increases.

Summary of a Parallel *RC* Circuit

The current triangles shown in Figure 14–30 summarize the characteristics of a parallel *RC* circuit. When I_R is larger than I_C, the circuit is resistive, as indicated by I_{C1} and I_{T1}. Theta is nearer to zero degrees, and source voltage is only slightly behind source (total) current. This occurs when the capacitance is small, the frequency is low, or the resistance is small.

→ *Self-Test*

34. What is the total current of a parallel *RC* circuit?
35. What methods may be used to find the impedance of a parallel *RC* circuit?
36. What is the result of an increase in frequency in a parallel *RC* circuit?
37. What is the result of an increase in capacitance in a parallel *RC* circuit?
38. What is the result of an increase in resistance in a parallel *RC* circuit?

SECTION 14–7: Parallel *RLC* Circuits

In this section you will learn the following:

→ The total reactive current I_T of a parallel *RLC* circuit is the absolute difference between inductive current I_L and capacitive current I_C.

→ The total current of a parallel *RLC* circuit is the phasor sum of resistive current and total reactive current.

→ Impedance is maximum in a parallel *RLC* circuit at the resonant frequency, f_r.

Total Reactive Current: I_X

The circuit shown in Figure 14–31(a) consists of a resistor, an inductor, and a capacitor in parallel with an ac source. This is a parallel *RLC* circuit.

In this circuit there is only one voltage, V_S. There are four currents: the three branch currents and the total current.

Since there is only one voltage, V_S is chosen as the reference phasor and placed at zero degrees. I_R is also at zero degrees, since it is in phase with V_S.

The expression "*ELI* the *ICE* man" reminds us that I_L lags voltage and I_C leads voltage. Therefore, I_L is placed at −90° and I_C at +90°.

In the previous section on series *RLC* currents, we learned that V_S, V_L, and V_C are typically of different values and always out of phase with each other. You may, therefore, wonder how the voltage across all three branches can be a single value. The answer is shown in Figure 14–32.

This illustration shows a Thevenin equivalent circuit within each branch except the resistive branch. You should recall that every practical voltage source can be expressed as a Thevenin voltage in series with an internal (Thevenin) resistance. The internal resistance of the ac source, the coil, and the capacitor provide the necessary voltage drops to allow the branch voltage to be a single value for all branches, fulfilling the voltage rule for parallel circuits.

Since the two reactive currents I_L and I_C are 180° (half a cycle) out of phase, they balance or offset each other. Thus, we will have a total reactive current I_X, which is the difference between the two, as stated in Formula 14–19.

Formula 14–19

$$I_X = |\,I_L - I_C\,|$$

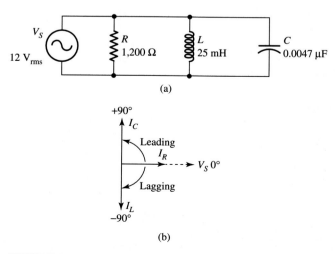

(a)

(b)

FIGURE 14–31
Parallel *RLC* circuit.

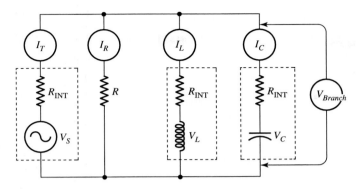

FIGURE 14–32
Thevenin circuits within parallel branches. Dashed
outlines enclose Thevenin equivalents to V_S, V_L, and V_C.

Total Current

The total current is the phasor sum of resistive and reactive current. Suppose a parallel
circuit has the values given in Figure 14–31(a). The resulting current triangle, shown in
Figure 14–33(a), produces Formulas 14–20.

Formulas 14–20

(a)
$$I_T = \sqrt{I_R^2 + I_X^2}$$

(b)
$$\theta = \tan^{-1}\left[\frac{I_X}{I_R}\right]$$

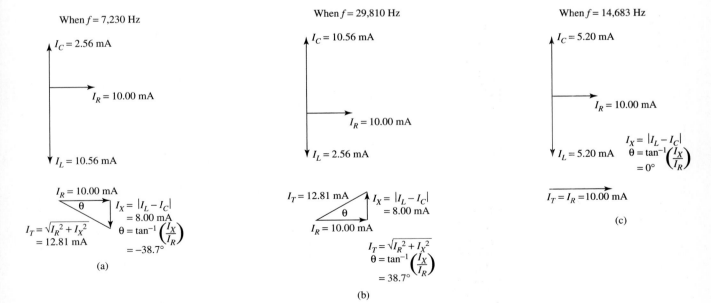

FIGURE 14–33
(a) At lower frequencies, I_L is larger than I_C, and the circuit is characteristically inductive. (b) At higher frequencies, I_C is
larger than I_L, and the circuit is characteristically capacitive. (c) At the resonant frequency, I_L equals I_C, and the circuit is
characteristically resistive.

Case 1: I_C Larger Than I_L

Suppose that in the circuit of Figure 14–31(a) the frequency is 7,230 Hz. The resulting current triangle is shown in Figure 14–33(a).

At this frequency,

$$\begin{aligned}
X_L &= 2\,\pi f L \\
&= 2 \times 3.1416 \times 7,230\ \text{Hz} \times 25\ \text{mH} \\
&= 1,136\ \Omega
\end{aligned}$$

$$\begin{aligned}
X_C &= \frac{1}{2\,\pi f C} \\
\\
&= \frac{1}{2 \times 3.1416 \times 7,230\ \text{Hz} \times 0.0047\ \mu\text{F}} \\
\\
&= 4,684\ \Omega
\end{aligned}$$

Therefore,

$$\begin{aligned}
I_L &= \frac{V_S}{X_L} \\
\\
&= \frac{12\ \text{V}}{1,136\ \Omega} \\
\\
&= 10.56\ \text{mA}
\end{aligned}$$

$$\begin{aligned}
I_C &= \frac{V_S}{X_C} \\
\\
&= \frac{12\ \text{V}}{4,684\ \Omega} \\
\\
&= 2.56\ \text{mA}
\end{aligned}$$

$$\begin{aligned}
I_X &= |\,I_L - I_C\,| \\
&= |\,10.56\ \text{mA} - 2.56\ \text{mA}\,| \\
&= 8.00\ \text{mA} \\
I_T &= \sqrt{I_R^2 + I_X^2} \\
&= \sqrt{(10.00\ \text{mA})^2 + (8.00\ \text{mA})^2} \\
&= 12.81\ \text{mA}
\end{aligned}$$

$$\begin{aligned}
Z &= \frac{V_S}{I_T} \\
\\
&= \frac{12\ \text{V}}{12.81\ \text{mA}} \\
\\
&= 937\ \Omega \\
\theta &= \tan^{-1}(I_X \div I_R) \\
&= \tan^{-1}(8.00\ \text{mA} \div 10.00\ \text{mA}) \\
&= -38.7°
\end{aligned}$$

Case 2: I_L Smaller Than I_C

Now suppose that the frequency is 29,810 Hz. The phasor diagram and current triangle is shown in Figure 14–33(b).

By following the same sequence of calculations as in Case 1, we will find that now I_C is larger than I_L. This time I_C equals 10.56 mA and I_L equals 2.56 mA. You should verify this by calculating I_C and I_L for yourself.

However, the absolute difference between I_C and I_L is still 8.00 mA, which is the total reactive current. For this reason, the total current is 12.81 mA, the same value as in Case 1. Theta also is again 38.7° and by Ohm's Law impedance is again 937 Ω.

Case 3: I_L Equals I_C

As shown in Figure 14–33(c), when the frequency is 14,683 Hz, I_L and I_C will both equal 5.20 mA. You should verify this by calculating I_L and I_C for yourself.

Since I_L equals I_C, the difference between them is zero. Thus, the total reactive current is zero. This does not mean that I_L and I_C are gone from the circuit. When we say that I_T equals zero, we simply mean that I_L and I_C are exactly in balance, so the overall reactive current appears to be zero.

If the total reactive current is zero, then the resistive current is the only current supplied by the source, and the impedance (total ohms) equals the value of the resistive branch, 1,200 Ω. Notice that this is the largest value of impedance in the four cases. The phase angle is zero degrees, which means that source voltage is in phase with source current.

We may summarize by saying that the total current of a parallel *RLC* circuit is the phasor sum of resistive current and total reactive current.

Impedance of a Parallel *RLC* Circuit

It is possible to derive a formula for Z in a parallel *RLC* circuit from the current triangle. However, the formula is complex. It is probably easier and faster to solve for Z with Ohm's Law as demonstrated earlier.

Effect of Frequency on a Parallel *RLC* Circuit

In Case 3, the circuit is at f_r, the resonant frequency: X_L equals X_C, and Z equals R. However, in a parallel *RLC* circuit, impedance is maximum at f_r. This is the opposite of a series *RLC* circuit, where impedance was minimum at f_r.

Remember that impedance is seen from the source voltage: Z equals V_S divided by I_T. At f_r, I_L equals I_C. This means that in effect the coil and capacitor are supplying each other with current and do not need any additional current from the source. Therefore, only the resistor draws current from the source at the resonant frequency. Thus, total current is minimum at f_r, and this is why impedance calculates out to its maximum value at the resonant frequency.

Formula for Resonant Frequency in a Parallel *RLC* Circuit

Formula 14–21 is the exact formula for the resonant frequency in a parallel *RLC* circuit. The formula includes the winding resistance R_W of the inductor.

Formula 14–21

$$f_r = \frac{\sqrt{1 - (R_W^2 C/L)}}{2\pi\sqrt{LC}}$$

However, in practical inductors with a low value of R_W, we can use Formula 14–14 for both series and parallel circuits. The difference in the value of f_r between the two formulas is typically less than 0.01%, not enough to matter for our purposes.

→ Self-Test

39. What is the total reactive current of a parallel *RLC* circuit?
40. What is the total current of a parallel *RLC* circuit?
41. When is impedance maximum in a parallel *RLC* circuit?

Formula List

Formula 14–1

$$\text{Hyp}^2 = \text{Adj}^2 + \text{Opp}^2$$

Formula 14–2

$$\text{Hyp} = \sqrt{\text{Adj}^2 + \text{Opp}^2}$$

Formula 14–3

$$\text{Adj} = \sqrt{\text{Hyp}^2 - \text{Opp}^2}$$

Formula 14–4

$$\text{Opp} = \sqrt{\text{Hyp}^2 - \text{Adj}^2}$$

Formula 14–5

$$X_L = 2\,\pi\,f\,L$$

Formula 14–6

$$Z = \sqrt{R^2 + X_L^2}$$

Formula 14–7

$$V_S = \sqrt{V_R^2 + V_L^2}$$

Formula 14–8

(a) $$\theta = \tan^{-1}\left[\frac{X_L}{R}\right]$$

(b) $$\theta = \tan^{-1}\left[\frac{V_L}{V_R}\right]$$

Formula 14–9

$$X_C = \frac{1}{2\,\pi\,f\,C}$$

Formula 14–10

(a) $$Z = \sqrt{R^2 + X_C^2}$$

(b) $$\theta = \tan^{-1}\left[\frac{X_C}{R}\right]$$

Formula 14–11

(a)
$$V_S = \sqrt{V_R^2 + V_C^2}$$

(b)
$$\theta = \tan^{-1}\left[\frac{V_C}{V_R}\right]$$

Formula 14–12

$$X_T = |X_L - X_C|$$

Formula 14–13

(a)
$$Z = \sqrt{R^2 + X_T^2}$$

(b)
$$\theta = \tan^{-1}\left[\frac{X_T}{R}\right]$$

Formula 14–14

$$f_r = \frac{1}{2\pi\sqrt{LC}}$$

Formula 14–15

(a)
$$I_T = \sqrt{I_R^2 + I_L^2}$$

(b)
$$\theta = \tan^{-1}\left[\frac{I_L}{I_R}\right]$$

Formula 14–16

(a)
$$Z = \frac{RX_L}{\sqrt{R^2 + X_L^2}}$$

(b)
$$\theta = \tan^{-1}\left[\frac{R}{X_L}\right]$$

Formulas 14–17

(a)
$$I_T = \sqrt{I_R^2 + I_C^2}$$

(b)
$$\theta = \tan^{-1}\left[\frac{I_C}{I_R}\right]$$

Formula 14–18

(a)
$$Z = \frac{RX_C}{\sqrt{R^2 + X_C^2}}$$

(b)
$$\theta = \tan^{-1}\left[\frac{R}{X_C}\right]$$

Formula 14–19

$$I_X = |I_L - I_C|$$

Formula 14–20

(a)

$$I_T = \sqrt{I_R^2 + I_X^2}$$

(b)

$$\theta = \tan^{-1}\left[\frac{I_X}{I_R}\right]$$

Formula 14–21

$$f_r = \frac{\sqrt{1 - (R_W^2 C / L)}}{2\pi\sqrt{LC}}$$

● TROUBLESHOOTING: Applying Your Knowledge

If the experimental circuit is a series *RLC* circuit, it is entirely possible that V_L or V_C could indeed exceed V_S due to large reactive currents. You ask your friend what kind of circuit she is testing, and what the parameters are of the components and voltage source. She gives you this information. After performing the proper calculations, you are able to show her that the reactive voltage drops recorded in her data are indeed about what they should be. You conclude that she is not doing anything wrong.

CHAPTER SUMMARY: ANSWERS TO SELF-TESTS

1. Trigonometry is the mathematics used to calculate the values of triangles.
2. A right triangle includes a 90° angle.
3. The sides of a right triangle are the adjacent, opposite, and hypotenuse.
4. The Pythagorean Theorem states that the square of the hypotenuse equals the sum of the squares of the other two sides.
5. The three principal trigonometric functions are called sine, cosine, and tangent. Each trigonometric function is a ratio of two sides of a right triangle.
6. The sine function is the ratio of the opposite to the hypotenuse.
7. The cosine function is the ratio of the adjacent to the hypotenuse.
8. The tangent function is the ratio of the opposite to the adjacent.
9. The three trigonometric ratios are functions of the angle theta. This means that the values of the ratios depend on the value of the angle and not on the values of the sides.
10. If theta changes, the values of all three functions also change.
11. An arc function is the angle associated with a value of one of the three functions. The value of the angle depends on the value of the function.
12. Impedance *(Z)* is the total ohmic value of any circuit. Impedance is total voltage divided by total current.

13. In a series *RL* circuit, impedance is the phasor sum of resistance and inductive reactance.
14. In a series *RL* circuit, source voltage is the phasor sum of the resistive voltage drop and the reactive voltage drop.
15. Theta is the phase angle between source voltage and source current.
16. In a series *RL* circuit, an increase in frequency causes the circuit to become more reactive.
17. In a series *RL* circuit, an increase in inductance causes the circuit to become more reactive.
18. In a series *RL* circuit, an increase in resistance causes the circuit to become less reactive.
19. "*ELI* the *ICE* man" is a memory aid representing the relationships of voltage and current in reactive circuits. *ELI* represents voltage leading current in an inductor. *ICE* represents voltage lagging current in a capacitor.
20. In a series *RC* circuit, impedance is the phasor sum of resistance and capacitive reactance.
21. In a series *RC* circuit, source voltage is the phasor sum of the resistive voltage drop and the reactive voltage drop.
22. In a series *RC* circuit, an increase in frequency causes the circuit to become less reactive.
23. In a series *RC* circuit, an increase in capacitance causes the circuit to become less reactive.
24. In a series *RC* circuit, an increase in resistance causes the circuit to become less reactive.
25. The total reactance X_T of a series *RLC* circuit is the absolute difference between inductive reactance X_L and capacitive reactance X_C.

26. In a series RLC circuit either V_L or V_C may be larger than V_S due to increased current.

27. For every series RLC circuit, there is a frequency at which X_L equals X_C. This frequency is called the resonant frequency of the circuit (f_r).

28. Impedance is minimum in a series RLC circuit at the resonant frequency, f_r.

29. The total current of a parallel RL circuit is the phasor sum of the branch currents.

30. The impedance of a parallel RL circuit may be found with Ohm's Law or through the conductance triangle.

31. In a parallel RL circuit, an increase in frequency causes the circuit to become less reactive.

32. In a parallel RL circuit, an increase in inductance causes the circuit to become less reactive.

33. In a parallel RL circuit, an increase in resistance causes the circuit to become more reactive.

34. The total current of a parallel RC circuit is the phasor sum of the branch currents.

35. The impedance of a parallel RC circuit may be found with Ohm's Law or through the conductance triangle.

36. In a parallel RC circuit, an increase in frequency causes the circuit to become more reactive.

37. In a parallel RC circuit, an increase in capacitance causes the circuit to become more reactive.

38. In a parallel RC circuit, an increase in resistance causes the circuit to become more reactive.

39. The total reactive current I_T of a parallel RLC circuit is the absolute difference between inductive current I_L and capacitive current I_C.

40. The total current of a parallel RLC circuit is the phasor sum of resistive current and total reactive current.

41. Impedance is maximum in a parallel RLC circuit at the resonant frequency, f_r.

CHAPTER REVIEW QUESTIONS

Select the response that best answers the question or completes the statement.

14–1. We are particularly interested in phase angles of 90° because
   a. the Pythagorean Theorem can be used.
   b. they are easier to work with
   c. X_L equals X_C at a 90° phase angle.
   d. both inductance and capacitance produce a 90° phase shift between voltage and current.

14–2. The length of a phasor represents
   a. magnitude.
   b. phase.
   c. voltage.
   d. current.

14–3. The direction of a phasor represents

   a. magnitude.
   b. phase.
   c. voltage.
   d. current.

14–4. If we have only two phasors to add, and if their phase angle is 90°,
   a. we may simply add their values.
   b. the phasors cancel each other and the sum is zero.
   c. the two phasors and the phasor sum form a triangle.
   d. the Pythagorean Theorem cannot be used.

14–5. The horizontal side of a right triangle is called
   a. the hypotenuse.
   b. theta.
   c. the opposite.
   d. the adjacent.

14–6. The vertical side of a right triangle is called
   a. the hypotenuse.
   b. theta.
   c. the opposite.
   d. the adjacent.

14–7. The diagonal side of a right triangle is called
   a. the hypotenuse.
   b. theta.
   c. the opposite.
   d. the adjacent.

14–8. The angle between the diagonal side and the horizontal side of a right triangle is called
   a. the hypotenuse.
   b. theta.
   c. the opposite.
   d. the adjacent.

14–9. The square root of the sum of the square of the opposite and the square of the adjacent is
   a. the sine function.
   b. the cosine function.
   c. the tangent function.
   d. a solution of the Pythagorean Theorem.

14–10. The ratio of the opposite side to the adjacent side is
   a. the sine function.
   b. the cosine function.
   c. the tangent function.
   d. a solution of the Pythagorean Theorem.

14–11. The ratio of the opposite side to the hypotenuse side is
   a. the sine function.
   b. the cosine function.
   c. the tangent function.
   d. a solution of the Pythagorean Theorem.

14–12. The ratio of the adjacent side to the hypotenuse side is
   a. the sine function.
   b. the cosine function.
   c. the tangent function.
   d. a solution of the Pythagorean Theorem.

14–13. Impedance is the phasor sum of resistance and inductive reactance in a
   a. series RL circuit.
   b. series RC circuit.
   c. parallel RL circuit.
   d. parallel RC circuit.

14–14. Impedance is the phasor sum of resistance and capacitive reactance in a
   a. series *RL* circuit.
   b. series *RC* circuit.
   c. parallel *RL* circuit.
   d. parallel *RC* circuit.

14–15. Impedance is the phasor sum of resistance and total reactance in a
   a. series *RLC* circuit.
   b. parallel *RLC* circuit.

14–16. Total voltage is the phasor sum of the resistive voltage drop and the total reactive voltage drop in
   a. a series *RL* circuit.
   b. a series *RC* circuit.
   c. a series *RLC* circuit.
   d. all of the above.

14–17. Total current is the phasor sum of resistive current and inductive current in a
   a. series *RL* circuit.
   b. series *RC* circuit.
   c. parallel *RL* circuit.
   d. parallel *RC* circuit.

14–18. Total current is the phasor sum of resistive current and capacitive current in a
   a. series *RL* circuit.
   b. series *RC* circuit.
   c. parallel *RL* circuit.
   d. parallel *RC* circuit.

14–19. Total current is the phasor sum of resistive current and total reactive current in a
   a. series *RLC* circuit.
   b. parallel *RLC* circuit.

14–20. An increase in inductance causes an increase in reactance in a
   a. series *RL* circuit.
   b. parallel *RL* circuit.
   c. series *RC* circuit.
   d. parallel *RC* circuit.

14–21. An increase in inductance causes a decrease in reactance in a
   a. series *RL* circuit.
   b. parallel *RL* circuit.
   c. series *RC* circuit.
   d. parallel *RC* circuit.

14–22. An increase in capacitance causes an increase in reactance in a
   a. series *RL* circuit.
   b. parallel *RL* circuit.
   c. series *RC* circuit.
   d. parallel *RC* circuit.

14–23. An increase in capacitance causes a decrease in reactance in a
   a. series *RL* circuit.
   b. parallel *RL* circuit.
   c. series *RC* circuit.
   d. parallel *RC* circuit.

14–24. An increase in resistance causes an increase in reactance in a
   a. series *RL* or series *RC* circuit.
   b. parallel *RL* or parallel *RC* circuit.
   c. series *RL* or parallel *RC* circuit.
   d. parallel *RL* or series *RC* circuit.

14–25. An increase in resistance causes a decrease in reactance in a
   a. series *RL* or series *RC* circuit.
   b. parallel *RL* or parallel *RC* circuit.
   c. series *RL* or parallel *RC* circuit.
   d. parallel *RL* or series *RC* circuit.

14–26. An increase in frequency causes an increase in reactance in a
   a. series *RL* or series *RC* circuit.
   b. parallel *RL* or parallel *RC* circuit.
   c. series *RL* or parallel *RC* circuit.
   d. parallel *RL* or series *RC* circuit.

14–27. An increase in frequency causes a decrease in reactance in a
   a. series *RL* or series *RC* circuit.
   b. parallel *RL* or parallel *RC* circuit.
   c. series *RL* or parallel *RC* circuit.
   d. parallel *RL* or series *RC* circuit.

14–28. Theta is the phase angle between
   a. resistive voltage and resistive current.
   b. reactive voltage and reactive current.
   c. resistive voltage and reactive current.
   d. reactive voltage and resistive current.
   e. source voltage and source current.

Solve the following problems.

14–29. A right triangle has an adjacent side of 11 units and an opposite side of 19 units. What is the length of the hypotenuse?

14–30. A right triangle has an opposite side of 37.9 units and a hypotenuse of 56.3 units. What is the length of the adjacent side?

14–31. A right triangle has an adjacent side of 73.6 units and a hypotenuse of 133.8 units. What is the length of the opposite side?

14–32. A right triangle has a hypotenuse of 27.6 units and an opposite of 5.9 units. Find the sine function and the value of theta.

14–33. A right triangle has a hypotenuse of 43.7 units and an adjacent of 30.3 units. Find the cosine function and the value of theta.

14–34. A right triangle has an opposite of 15.9 units and an adjacent of 11.6 units. Find the tangent function and the value of theta.

14–35. A right triangle has a hypotenuse of 25.4 units and a theta of 36.9°. Find the opposite.

14–36. A right triangle has a hypotenuse of 16.7 units and a theta of 71.5°. Find the adjacent.

14–37. A right triangle has an opposite of 42.9 units and a theta of 16.7°. Find the adjacent.

14–38. A right triangle has an opposite of 36.4 units and a theta of 43.7°. Find the hypotenuse.

14–39. A right triangle has an adjacent of 23.7 units and a theta of 22.5°. Find the hypotenuse.

14–40. A right triangle has an adjacent of 93.2 units and a theta of 51.3°. Find the opposite.

14–41. An ac source of 6.2 V_{rms} at 15 kHz is applied to a series RL circuit of 360 Ω and 14 mH. Find the inductive reactance, impedance, current, voltage drops, and phase angle.

14–42. An ac source of 11.4 V_{rms} at 28 kHz is applied to a series RC circuit of 470 Ω and 0.033 μF. Find the capacitive reactance, impedance, current, voltage drops, and phase angle.

14–43. An ac source of 4.1 V_{rms} at 37 kHz is applied to a parallel RL circuit of 2 kΩ and 25 mH. Find the inductive reactance, reactive current, resistive current, total current, impedance, and phase angle.

14–44. An ac source of 7.4 V_{rms} at 28 kHz is applied to a parallel RC circuit of 470 Ω and 0.022 μF. Find the capacitive reactance, reactive current, resistive current, total current, impedance, and phase angle.

14–45. An ac source of 4.2 V_{rms} at 37 kHz is applied to a series RLC circuit of 200 Ω, 1.5 mH, and 0.033 μF. Find the inductive reactance, capacitive reactance, total reactance, impedance, current, voltage drops, and phase angle. What is the resonant frequency for this circuit?

14–46. An ac source of 5.3 V_{rms} at 43 kHz is applied to a parallel RLC circuit of 680 Ω, 2.8 mH, and 0.0068 μF. Find the inductive reactance, inductive current, capacitive reactance, capacitive current, total reactive current, resistive current, total current, impedance, and phase angle. What is the resonant frequency for this circuit if $R_W = 1.2$ Ω?

GLOSSARY

adjacent The horizontal side of a right triangle.

arc function A value of theta associated with one of the three functions of theta (sine, cosine, or tangent).

cosine The ratio between the adjacent and hypotenuse.

hypotenuse The diagonal side of a right triangle.

impedance The total ohmic value of a circuit, symbolized Z.

opposite The vertical side of a right triangle.

phasor sum In phasor addition, the final phasor drawn from the beginning of the first phasor to the end of the last phasor.

Pythagorean Theorem The square of the hypotenuse of a right triangle equals the sum of the squares of the other two sides.

resonant frequency The frequency at which X_L equals X_C in an RLC circuit.

right triangle A triangle that includes a 90° angle.

sine The ratio between the opposite and hypotenuse.

tangent The ratio between the opposite and adjacent.

theta The angle between the adjacent and hypotenuse.

trigonometry The mathematics of triangles.

15

AC FREQUENCY RESPONSE

● SECTION OUTLINE

15–1 Waveforms
15–2 Frequency Bands
15–3 Filters
15–4 ac Frequency Response: *RL* and *RC* Pass Filters
15–5 ac Frequency Response: Resonant Band Filters

● OVERVIEW AND LEARNING OBJECTIVES

The major use of electronics is to process and communicate information, generally in the form of audio, video, and data. When converted into electronic form, such information is usually a wave of voltage and current. Typically, many such waves become mixed together. Therefore, we need electronic circuits that can select or sort out the waves we want from the waves we don't want.

For example, there are many radio and television stations sending signals to your antenna or cable. Since you want to receive only one station at a time, you need a way to separate or filter out all the others.

Because the impedance of inductors and capacitors depends in part on the frequency of the voltage and current, these components can be used in frequency-selective circuits to filter and tune electromagnetic waves.

After completing this chapter, you will be able to:

1. Describe the Fourier Series and explain its importance.
2. Perform simple calculations involving band, band width, center frequency, and cutoff frequency.
3. Perform simple calculations involving wavelength.
4. Classify filters based on output levels measured in decibels.
5. Classify filters based on inspection of schematics.
6. State the conditions that define resonance in an *LC* circuit and the impedance at resonance in both a series and parallel *LC* configuration.
7. Classify filters based on inspection of frequency-response curves.

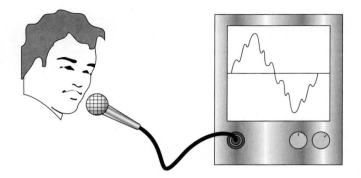

FIGURE 15–1
Audio signals produce complex electrical waveforms
containing many harmonics.

SECTION 15–1: Waveforms

In this section you will learn the following:

→ A harmonic is a wave whose frequency is an integer multiple of the fundamental
frequency of another wave.

→ Any waveform, however simple or complex, can be produced by a combination of
sine waves and cosine waves. This combination of waves is called the Fourier
Series of the waveform.

The shape of a wave pattern is called the *waveform*. The simplest and most common waveform is the sine wave. Most signals have a more complex waveform. For example, if you connect an oscilloscope to a microphone, the screen of the oscilloscope will display the electrical waveforms created by the microphone from the sound waves, as shown in Figure 15–1.

Harmonics

A *harmonic frequency* is a wave whose frequency is an integer multiple of the *fundamental frequency* of another wave. Figure 15–2 illustrates the concept of a harmonic. The basic frequency of a wave is called the fundamental frequency, f_1. For example, commer-

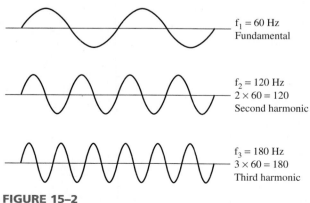

$f_1 = 60$ Hz
Fundamental

$f_2 = 120$ Hz
$2 \times 60 = 120$
Second harmonic

$f_3 = 180$ Hz
$3 \times 60 = 180$
Third harmonic

FIGURE 15–2
Harmonics of a wave.

cial ac voltage in North America has a fundamental frequency of 60 Hz. If we multiply the fundamental frequency f_1 by integers (whole numbers), we will calculate the frequencies of the various harmonics of f_1. For example, 2×60 Hz equals 120 Hz; therefore, the second harmonic of f_1 is another wave with a frequency of 120 Hz. By a similar process, the third harmonic is a wave with a frequency of 180 Hz (3×60 Hz), the fourth harmonic is 240 Hz, and so on.

The Fourier Series

Any waveform, however simple or complex, can be produced by a combination of sine waves and cosine waves at the waveform's fundamental frequency and at the various harmonic frequencies. A cosine wave is simply a sine wave 90° out of phase with another sine wave. This combination of waves is called the *Fourier Series* of the complex waveform.

The Fourier Series is a mathematical recipe for creating any waveform. The fact that any waveform can be produced by a combination of sine and cosine waves is extremely important in physics and engineering.

If we make a sound recording of an orchestra and then play the recording, the electrical signal going to the loudspeaker has an extremely complex waveform because it is the combination of the waveforms of many instruments, each of which produces a complex waveform. Amazingly, we can still hear each separate instrument, although its waveform is completely mixed in with the waveforms of the other instruments. In other words, the loudspeaker responds as if the electrical waveforms were separated from each other, when in fact they are completely blended together.

The Fourier Series and the example of an orchestra will help you understand a great deal of advanced electronic technology. For example, you may have cable television in your home. Have you ever wondered how one wire can carry so many different channels?

The answer is that the Fourier Series works both ways. Not only can you create any wave by combining other waves; you can also pick out one of the waves from the rest of them. This is the job of any type of electronic tuner, filter, tone control, and so on.

In this chapter you will learn how the series and parallel *RLC* circuits you studied in the last chapter can be applied as filters and tuners to pick out waves of certain frequencies from a combination of many other waves.

→ *Self-Test*

1. What is a harmonic?
2. What is a Fourier Series?

SECTION 15–2: Frequency Bands

In this section you will learn the following:

→ A frequency band is a group or range of frequencies.

→ Bandwidth is the size of a band, calculated by subtracting the lowest frequency in the band from the highest frequency in the band.

→ The center frequency is the frequency in the middle of a band.

→ The center frequency of a band is the frequency with maximum power.

→ The half-power points of a band are the frequencies f_1 and f_2 whose power is 50% of the center frequency.

Band and Bandwidth

A *band* is a range of frequencies. For example, suppose a circuit in a radio tunes in all frequencies between 540 kHz and 1,600 kHz. This range or group of frequencies is assigned to commercial AM radio stations. Therefore, we could say the circuit tunes in the AM band. Another circuit in the same radio might tune in frequencies between 88 MHz and 108 MHz. This group of frequencies is called the FM band.

The *bandwidth* is the highest frequency in the band minus the lowest frequency. The bandwidth of the AM band is 1,600 kHz minus 540 kHz, which equals 1,060 kHz. Likewise, the bandwidth of the FM band is 108 MHz minus 88 MHz, which equals 20 MHz.

The *center frequency* is the frequency in the middle of the band. For example, the AM bandwidth is 1,060 kHz. Half of this is 530 kHz. Knowing this, we can find the center frequency in two ways: add half the bandwidth to the lowest frequency, or subtract half the bandwidth from the highest frequency. In the case of the AM band, the center frequency is 1,070 kHz: 560 kHz + 530 kHz = 1,070 kHz, and 1,600 kHz − 530 kHz = 1,070 kHz. By similar reasoning, half the FM bandwidth is 10 MHz. Therefore, the center frequency of the FM band is 98 MHz.

The lowest frequency in the band is called f_1. The highest frequency is called f_2. The center frequency is called f_r. The bandwidth is symbolized as BW. The definitions of bandwidth and center frequency are expressed in Formulas 15–1 and 15–2.

Formula 15–1

$$BW = f_2 - f_1$$

Formula 15–2

$$f_r = f_1 + \frac{BW}{2}$$

$$f_r = f_2 - \frac{BW}{2}$$

● *SKILL-BUILDER 15–1*

What is the bandwidth and the center frequency of a band extending from 250 kHz to 350 kHz?

GIVEN:
  $f_1 = 250$ kHz
  $f_2 = 350$ kHz

FIND:
  BW
  f_r

SOLUTION:
  BW $= f_2 - f_1$
      $= 350$ kHz $- 250$ kHz
      $= 100$ kHz

$$f_r = f_1 + \frac{1}{2} BW \qquad \text{or, } f_c = f_2 - \frac{1}{2} BW$$

$$= 250 \text{ kHz} + \left(\frac{1}{2} \times 100 \text{ kHz} \right)$$

$$= 250 \text{ kHz} + 50 \text{ kHz}$$

$$= 300 \text{ kHz}$$

Repeat the Skill-Builder using the following values. You should obtain the results that follow.

Given:

f_1	5.3 MHz	10 kHz	450 MHz	300 Hz	1.2 GHz
f_2	5.9 MHz	14 kHz	500 MHz	600 Hz	3.0 GHz

Find:

BW	0.6 MHz	4 kHz	50 MHz	300 Hz	1.8 GHz
f_r	5.6 MHz	12 kHz	475 MHz	450 Hz	2.1 GHz

Half-Power Points

The center frequency of a band is the frequency with maximum power. The *half-power points* of a band are the frequencies f_1 and f_2 whose power is 50% of f_r.

Suppose a radio circuit tunes in a certain band of frequencies. What exactly defines the center frequency f_r and the edge frequencies f_1 and f_2?

The center frequency is the one whose signal comes through with the greatest amount of power, measured in watts. The other frequencies come through with less power. The band is defined as the range of frequencies whose signal strength in watts is 50% or more of the signal strength of the center frequency. Thus, f_1 is the frequency below f_r whose signal strength is 50% of f_r's signal strength. Similarly, f_2 is the frequency above f_r with 50% signal strength. Therefore, f_1 and f_2 are called the half-power points of the band centered on f_r.

Frequency-Response Curves

Figure 15–3 is a *frequency-response curve,* an important type of graph showing the relationship between frequency and power. Frequency-response curves are widely used to describe the performance of devices and circuits. Figure 15–3 indicates a frequency band with a center frequency of 300 kHz and half-power points at 220 kHz and 420 kHz.

A frequency-response curve may have only one half-power point, as shown in Figure 15–4. In this case, there is no center frequency, and the band includes all frequencies with signal strengths above 50% of maximum. The half-power point in this case is often called the *cutoff frequency.*

For example, in Figure 15–4(a), low frequencies near 0 Hz have maximum signal strength. As the frequency increases, the signal strength decreases until at 87 kHz it

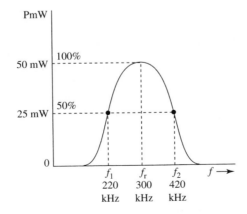

FIGURE 15–3
A frequency-response curve of output
power level at various frequencies.

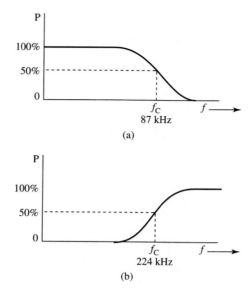

FIGURE 15–4
Frequency-response curves with cutoff frequencies (half-power points). (a) Low-pass response curve. (b) High-pass response curve.

reaches 50% of maximum. Thus, for this particular frequency-response curve, 87 kHz is the cutoff frequency or half-power point. The band is between 0 Hz and 87 kHz, since this group of frequencies has a signal strength of 50% or greater.

The other possibility is shown in Figure 15–4(b). Frequencies near 0 Hz have a low signal strength. As frequency increases, signal strength rises to a maximum value. Along the way to maximum, the signal reaches 50% strength at 224 kHz. Thus, for this particular frequency-response curve, 224 kHz is the cutoff frequency or half-power point. The band begins at 224 kHz and apparently has no upper limit. In reality, all circuits and devices have an upper limit to frequency response because of capacitance. However, this upper limit may be sufficiently high that it is of no importance to the job the circuit is designed to do. Therefore, the upper limit is often ignored and the frequency-response curve for the circuit resembles Figure 15–4(b).

Lopsided Frequency-Response Curves

In this section we have defined the center frequency f_r as exactly halfway between the cutoff frequencies, f_1 and f_2. This is approximately true in circuits that are designed by experienced engineers. However, in simple circuits intended for student experiments, the frequency-response curve may be somewhat lopsided, as shown in Figure 15–5, where one cutoff frequency is closer to f_r than the other.

→ *Self-Test*

3. What is a frequency band?
4. What is bandwidth?
5. What is the center frequency?

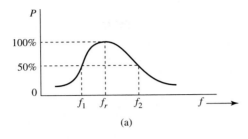

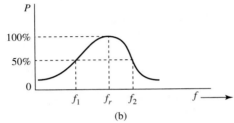

FIGURE 15–5
Lopsided frequency-response curves.

6. Which frequency in a band has maximum power?
7. What are the half-power points?

- -

SECTION 15–3: Filters

In this section you will learn the following:

→ A low-pass filter is a circuit that passes signals below a certain frequency and rejects signals above that frequency.

→ A high-pass filter is a circuit that passes signals above a certain frequency and rejects signals below that frequency.

→ A band-pass filter is a circuit that passes signals within a certain band and rejects signals outside that band.

→ A band-reject filter is a circuit that rejects signals within a certain band and passes signals outside that band.

→ A decibel is a logarithmic unit that compares one power level to another.

Signals and Noise

If we examine any circuit in an actual machine such as a computer, stereo, radio, and so on, we will usually find a complex waveform that is a mixture of currents at several different frequencies. Some of these are signals, which represent information. The others are noise, created by defects in the circuit or by interference from the environment.

Passing and Rejecting Signals

To deal with unwanted signals and with noise, we need electronic circuits that respond differently to the mixture of frequencies in a complex waveform. Remember that although the frequencies are all mixed together, the circuit will respond as if the signals were still separate. What we want the circuit to do is to pass signals at certain frequencies and reject signals (including noise) at all other frequencies. A circuit that can do this is called a filter. Certain filters can act as tuners. We will now learn the different classifications of filters, then see how the *RLC* circuits we studied in the last chapter can act as filters and tuners.

The way we decide whether a signal is passed or rejected is by evaluating the power of the signal going in and coming out of the circuit. If the signal comes out at 50% or greater of the power it had going in, we say the filter has passed the signal. On the other hand if the signal comes out below 50% of its power going in, we say the filter has rejected the signal.

Low-Pass Filter

A *low-pass filter* is a circuit that passes signals below a certain frequency and rejects signals above that frequency. Figure 15–4(a) is the frequency-response curve of a low-pass filter.

High-Pass Filter

A *high-pass filter* is a circuit that passes signals above a certain frequency and rejects signals below that frequency. Figure 15–4(b) is the frequency-response curve of a high-pass filter.

Band-Pass Filter

A *band-pass filter* is a circuit that passes signals within a certain band and rejects signals outside that band. Figure 15–6(a) is the frequency-response curve of a band-pass filter. A tuning circuit is generally a band-pass filter.

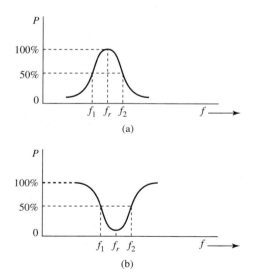

FIGURE 15–6
(a) Band-pass frequency-response curve.
(b) Band-reject frequency-response curve.

Band-Reject Filter

A *band-reject filter* is a circuit that rejects signals within a certain band and passes signals outside that band. Figure 15–6(b) is the frequency-response curve of a band-reject filter. A band-reject filter is also called a *band-stop filter*. Notice that the frequency-response curve of a band-reject filter is simply a band-pass curve turned upside down (inverted).

Logarithms and Logarithmic Scales

A *logarithm* is an exponent of one number used to represent another number. In Figure 15–7, in the equation $10^2 = 100$, 10 is called the base, 2 is an exponent, and 100 is the number represented by the equation. When the equation is rewritten, the terminology changes: 2 is the logarithm to the base 10 of the number 100. In other words, if the base is 10, then the logarithm 2 represents the number 100 because when the logarithm is used as an exponent of the base, the number is produced.

As shown in Figure 15–7(b), increasing the value of the logarithm by 1 amounts to multiplying by the value of the base. For example, if the logarithm changes by one (from 1 to 2), the number represented by the logarithm multiplies by the base (10×10 equals 100).

If 10 is represented by a logarithm of 1, and 100 is represented by a logarithm of 2, then numbers between 10 and 100 must be represented by logarithms between 1 and 2. Figure 15–8 lists the common logarithms for values of ten between 10 and 100.

Logarithmic Scales

Whenever we draw a graph, we must first select a scale. Until now, all our graph scales have been linear, as shown in Figure 15–9(a). Linear scales are easy to recognize. They begin with 0 and the increments are evenly spaced.

It is often better to draw a graph using a logarithmic scale (Figure 15–9(b)). Logarithmic scales begin with 1 and the increments are unevenly spaced. The distance between each increment corresponds to the difference in the logarithms of the two numbers.

An extended logarithmic scale is shown in Figure 15–9(c). Notice that the major divisions of the scale represent an increase of ten times the value. On a logarithmic frequency-response curve, each major division is a *decade*, a band whose upper frequency is ten times the lower frequency. Thus, the first frequency decade is 1 kHz to 10 kHz; the second decade is 10 kHz to 100 kHz; and so on.

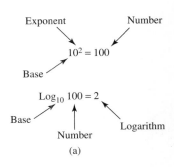

FIGURE 15–7
Relationships among bases, exponents, and logarithms.

$$\text{Log}_{10} \quad 10 = 1.000$$
$$\text{Log}_{10} \quad 20 = 1.301$$
$$\text{Log}_{10} \quad 30 = 1.477$$
$$\text{Log}_{10} \quad 40 = 1.602$$
$$\text{Log}_{10} \quad 50 = 1.699$$
$$\text{Log}_{10} \quad 60 = 1.778$$
$$\text{Log}_{10} \quad 70 = 1.845$$
$$\text{Log}_{10} \quad 80 = 1.903$$
$$\text{Log}_{10} \quad 90 = 1.954$$
$$\text{Log}_{10} \quad 100 = 2.000$$

FIGURE 15–8
Logarithms of numbers between 10 and 100.

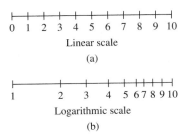

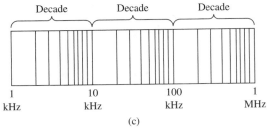

FIGURE 15–9
(a) Linear scale, (b) logarithmic scale, (c) 3-decade logarithmic scale.

The Decibel

A *decibel* is a logarithmic unit that compares one power level to another. The decibel is abbreviated dB.

The decibel is 1/10th of a bel, a logarithmic unit developed by Alexander Graham Bell to measure the performance of telephone circuits. The decibel has proven to be a more convenient unit, and today the bel is rarely used.

The decibel is a relative unit, and a reference level must be defined. When the decibel is used to measure the performance of an electronic filter or amplifier, the input power level is chosen as the reference. The output of the circuit in decibels is calculated by Formula 15–3.

Formula 15–3

$$dB = 10 \log \frac{\text{output power}}{\text{input power}}$$

● SKILL-BUILDER 15–2

A circuit has an input power level of 25 milliwatts and an output of 15 milliwatts. What is the output power level in dB?

GIVEN:
  $P_{in} = 25$ mW
  $P_{out} = 15$ mW

FIND:
  dB

SOLUTION:
  $dB = 10 \log \dfrac{P_{out}}{P_{in}}$

  $= 10 \log \dfrac{15 \text{ mW}}{25 \text{ mW}}$

  $= 10 \log (0.6)$
  $= 10 \times -0.2218$
  $= -2.22$ dB

keystrokes:
15 [EXP] 3 [+/−]
[÷] 25 [EXP] 3 [+/−]
[=] [log] [×] 10 [=]

Repeat the Skill-Builder using the following values. You should obtain the results that follow.

Given:

P_{in}	60 mW	7.5 W	18 mW	22 W	4.6 W
P_{out}	2 mW	15 W	24 mW	11 W	0.6 W
Find:					
dB	−14.8	+3.0	+1.2	−3.0	−8.8

If the output power level is the same as the input level, the ratio of input power to output is the number 1. The common logarithm of 1 is zero. This means that if a circuit has an output of 0 dB, the input and output power levels are the same.

If the output power level is less than the input level, the logarithm will be negative and the decibel value will also be negative. An output power level less than the input level is called *loss* or *attenuation*. Thus, a negative decibel value indicates a loss in power.

If the output power level is greater than the input level, the logarithm will be positive and the decibel value will also be positive. An output greater than an input is called *gain* or *amplification*. Thus, a positive decibel value indicates a gain in power.

Gain or amplification requires an active device in the circuit: a transistor or integrated circuit. You will study active filters in your next course. The components we are studying in this course are passive components, which cannot produce amplification or gain.

The logarithm for the number 2 is approximately 0.3, which becomes a decibel value of 3 when multiplied by 10 as required by Formula 15–3. Thus, an output of +3 dB means the output power is twice the input; and an output of −3 dB means the output power is half the input.

To determine the number of times the power has doubled or been cut in half, divide the decibel value by 3. Thus, if an audio amplifier has a signal-to-noise ratio of −72 dB, we divide 72 by 3 and get 24. This means that the noise level is 2^{24} (16.7 million) times less than the signal level. If we are listening to the music at normal power levels, the noise level is so small that we cannot hear it. A compact disc player typically has a signal-to-noise ratio of −90 dB, which makes the noise level a billion times smaller than the signal. This is why compact discs sound so clear.

Since an output of −3 dB represents a 50% reduction in power, the cutoff frequencies (half-power points) that define the edge of a frequency band are also called the −3 dB frequencies.

Decibel Levels Based On Input and Output Voltages

In unloaded circuits it is often better to measure gain or attenuation with voltage readings using Formula 15–4.

Formula 15–4

$$dB = 20 \log \frac{V_{out}}{V_{in}}$$

● *SKILL-BUILDER 15–3*

- -

At a certain frequency, a filter has an input voltage of 3.59 V and an output voltage of 2.54 V. What is the attenuation in decibels?

GIVEN:
  $V_{in} = 3.59$ V
  $V_{out} = 2.54$ V

FIND:
  dB

SOLUTION:
  $dB = 20 \log \dfrac{V_{out}}{V_{in}}$

  $= 20 \log \dfrac{2.54 \text{ V}}{3.59 \text{ V}}$

  $= 20 \log (0.707)$

  $= 20 \times -0.15$

  $= -3$ dB

keystrokes:
2.54 [÷] 3.59
[=][log]
[×] 20 [=]

Repeat the Skill-Builder using the following values. You should obtain the results that follow.

Given:

V_{in}	2.25 V	6.04 V	9.17 V	0.13 V	4.16 V
V_{out}	1.67 V	0.43 V	11.28 V	2.07 V	1.15 V

Find:

dB	−2.6	−22.9	+1.8	+24.0	−11.2

At the cutoff frequency f_c, V_{out} will equal 0.707 of the maximum value of V_{out}, as shown in Figure 15–10. Since V_{out} would equal V_{in} when the output level is 0 dB (neither gain nor loss), it follows that at the cutoff frequency, V_{out} equals 0.707 V_{in}.

Rolloff

The *rolloff* of a filter is the rate at which the output power level changes with frequency. Rolloff is usually specified in decibels per decade. The normal rolloff rate of a passive filter is −20 dB per decade.

Figure 15–11 illustrates a rolloff of −20 dB per decade. Suppose a frequency-response curve has a maximum output power at 10 kHz. This would become the reference level and be assigned a value of 0 dB. One decade below 10 kHz is 1 kHz. At this frequency the output power is down −20 dB. Two decades below 10 kHz is 100 Hz. Here the output level is −40 dB. The rest of the graph may be interpreted in a similar way.

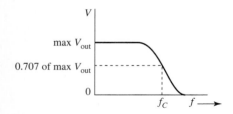
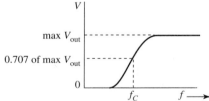
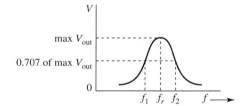

FIGURE 15–10
At the half-power point, output voltage is 0.707 of input voltage.

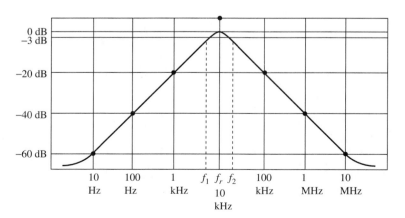

FIGURE 15–11
−20 dB/decade natural rolloff of passive components.

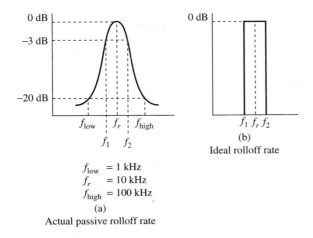

f_{low} = 1 kHz
f_r = 10 kHz
f_{high} = 100 kHz

(a)
Actual passive rolloff rate

(b)
Ideal rolloff rate

FIGURE 15–12
(a) Actual rolloff. (b) Ideal rolloff.

This is far from the ideal rate, as shown in Figure 15–12. An ideal filter would allow full power output within the frequency band, then instantly drop the output to zero at each cutoff frequency. Active filters and multisection filters are used to increase the rolloff rate toward this ideal performance.

The Bode Plot

A *Bode plot* is a frequency-response curve drawn on a logarithmic scale, as shown in Figure 15–13. The logarithmic scale allows us to show a much wider range of frequencies. It also allows us to show the rolloff as a straight line rather than a curve. Most frequency-response curves in technical data sheets are drawn this way.

Each decade in Figure 13–15 is marked by a vertical line labeled with a frequency. Thus, we can identify the decade between 100 Hz and 1 kHz, because these values are marked on the graph. Between each decade we see five unlabeled vertical lines. Each of these indicates one-fifth of the decade. Thus, between 100 Hz and 1 kHz the first vertical line is 200 Hz, the second line is 400 Hz, and so on.

Figure 15–13 describes a circuit with a lower cutoff frequency of 30 Hz and an upper cutoff frequency of 15 kHz. Each end of the graph shows a drop of −60 dB in less than a decade. This is an extremely sharp rolloff rate.

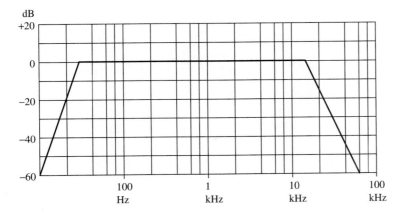

FIGURE 15–13
A Bode plot of rolloff.

→ *Self-Test*

8. What is a low-pass filter?
9. What is a high-pass filter?
10. What is a band-pass filter?
11. What is a band-reject filter?
12. What is a decibel?

SECTION 15–4: ac Frequency Response: *RL* and *RC* Pass Filters

In this section you will learn the following:

→ A series *RL* circuit acts as a low-pass filter when the resistor is grounded and used to develop the output voltage.

→ A series *RL* circuit acts as a high-pass filter when the coil is grounded and used to develop the output voltage.

→ A series *RC* circuit acts as a low-pass filter when the capacitor is grounded and used to develop the output voltage.

→ A series *RC* circuit acts as a high-pass filter when the resistor is grounded and used to develop the output voltage.

In the last chapter we learned how the impedance of *RL* and *RC* circuits changes with frequency. Now we will apply these characteristics to see how these circuits can act as pass filters. First, let's review some of these basic characteristics and some basic electrical principles.

If two devices are in series and there is a difference in their impedances, the device with more impedance will have a larger voltage drop. This principle comes from the basic rules for series circuits.

A coil has low impedance at low frequencies and high impedance at high frequency. This principle comes from the inductive reactance of the coil due to its back voltage.

A capacitor has high impedance at low frequencies and low impedance at high frequencies. This principle comes from the capacitive reactance of the capacitor due to its action of charging and discharging.

Now we must add two new principles, as shown in Figure 15–14. Passive filters generally have some components located in the signal path directly between input and output, and other components located to create a path from the signal line to ground. The

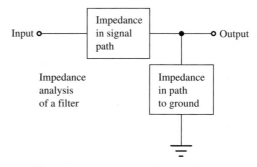

FIGURE 15–14
The concept of impedance in the signal path and in the path to ground.

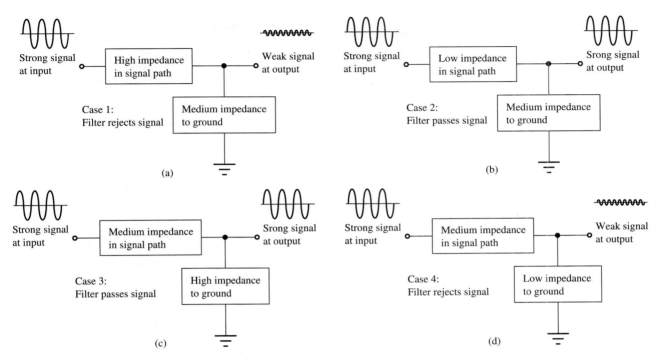

FIGURE 15–15
(a) Case 1: filter rejects signal. (b) Case 2: filter passes signal. (c) Case 3: filter passes signal. (d) Case 4: filter rejects signal.

purpose of the component to ground is to develop an output voltage. In other words, the output voltage is the voltage drop across the component connected to ground. We analyze the action of the filter by examining the impedances in these two locations. There are four possible cases, as shown in Figure 15–15.

CASE 1: In Figure 15–15(a), there is high impedance in the signal path and a medium impedance to ground. If a strong signal voltage appears at the input, most of the voltage will be dropped across the high impedance in the signal path. Only a small amount of voltage will be dropped across the medium impedance to ground. Thus, the signal voltage which was strong at the input is weak at the output. The filter has rejected (reduced or attenuated) the signal.

CASE 2: In Figure 15–15(b), there is low impedance in the signal path and a medium impedance to ground. If a strong signal voltage appears at the input, only a small amount of voltage will be dropped across the low impedance in the signal path. Most of the voltage will be dropped across the medium impedance to ground. Thus, the signal voltage which was strong at the input is still strong at the output. The filter has passed the signal.

CASE 3: In Figure 15–15(c), there is medium impedance in the signal path and a high impedance to ground. If a strong signal voltage appears at the input, only a small amount of voltage will be dropped across the medium impedance in the signal path. Most of the voltage will be dropped across the high impedance to ground. Thus, the signal voltage which was strong at the input is still strong at the output. The filter has passed the signal.

CASE 4: In Figure 15–15(d), there is medium impedance in the signal path and a low impedance to ground. If a strong signal voltage appears at the input, most of the voltage will be dropped across the medium impedance in the signal path. Only a small amount of voltage will be dropped across the low impedance to ground. Thus, the signal voltage which was strong at the input is weak at the output. The filter has rejected (reduced or attenuated) the signal.

Now all we have to do is draw a series *RL* circuit and then a series *RC* circuit in the arrangement shown in Figure 15–14 and analyze the circuit first at a low frequency and then at a high frequency to see which of the four cases of Figure 15–15 applies at each frequency. Then we will know exactly what sort of filtering action each circuit provides.

Cutoff Frequency f_c

When a series *RL* or *RC* circuit is used as a pass filter, the following facts are true at the cutoff frequency f_c:

1. $V_{out} = 0.707$ of V_{in}
2. $V_L = V_R$ for an *RL* circuit
   $V_C = V_R$ for an *RC* circuit
3. $\theta = \pm 45°$

The cutoff frequency for an *RL* circuit may be calculated with Formula 15–5(a), and for an *RC* circuit with Formula 15–5(b).

Formula 15–5

(a)
$$f_c = \frac{R}{2\pi L}$$

(b)
$$f_c = \frac{1}{2\pi R C}$$

● SKILL-BUILDER 15–4

- -

What is the cutoff frequency for an *RL* pass filter with a 3.6-kΩ resistor and a 7-mH inductor?

GIVEN:
 $R = 3.6$ kΩ
 $L = 7$ mH

FIND:
 f_c

SOLUTION:

$f_c = \dfrac{R}{2\pi L}$

$= \dfrac{3.6 \text{ k}\Omega}{2 \times \pi \times 7 \text{ mH}}$

$= 8,185$ Hz

keystrokes:
2 [×] [π] [×]
7 [EXP] 3 [+/−]
[=] [1/x] [×]
3.6 [EXP] 3 [=]

Repeat the Skill-Builder using the following values. You should obtain the results that follow.

Given:					
R	2.2 kΩ	680 Ω	4.7 kΩ	330 Ω	15 kΩ
L	2.6 mH	12 mH	37 mH	0.6 mH	22 mH
Find:					
f_c	134.7 kHz	9.0 kHz	20.2 kHz	87.5 kHz	108.5 kHz

- -

Series *RL* Circuit as a Low-Pass Filter

Let us analyze the circuit in Figure 15–16 as indicated earlier.

At $f = 2$ kHz:

$$X_L = 2 \pi f L$$
$$= 2 \times \pi \times 2 \text{ kHz} \times 5 \text{ mH}$$
$$= 63 \ \Omega$$

$$Z = \sqrt{R^2 + X_L^2}$$
$$= \sqrt{(510 \ \Omega)^2 + (63 \ \Omega)^2}$$
$$= 514 \ \Omega$$

$$I = V_S \div Z$$
$$= 5 \text{ V} \div 514 \ \Omega$$
$$= 9.73 \text{ mA}$$

$$V_{\text{out}} = V_R$$
$$= I \times R$$
$$= 9.73 \text{ mA} \times 510 \ \Omega$$
$$= 4.96 \text{ V}$$

$$\text{dB} = 20 \log (V_{\text{out}} \div V_{\text{in}})$$
$$= 20 \log (4.96 \text{ V} \div 5.00 \text{ V})$$
$$= -0.07 \text{ dB}$$

At $f = 120$ kHz:

$$X_L = 2 \pi f L$$
$$= 2 \times \pi \times 120 \text{ kHz} \times 5 \text{ mH}$$
$$= 3.77 \text{ k}\Omega$$

$$Z = \sqrt{R^2 + X_L^2}$$
$$= \sqrt{(510 \ \Omega)^2 + (3.77 \text{ k}\Omega)^2}$$
$$= 3.8 \text{ k}\Omega$$

$$I = V_S \div Z$$
$$= 5 \text{ V} \div 3.8 \text{ k}\Omega$$
$$= 1.31 \text{ mA}$$

$$V_{\text{out}} = V_R$$
$$= I \times R$$
$$= 1.31 \text{ mA} \times 510 \ \Omega$$
$$= 0.67 \text{ V}$$

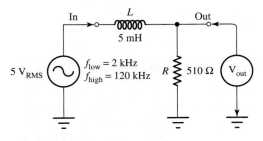

FIGURE 15–16
Series *RL* circuit as low-pass filter.

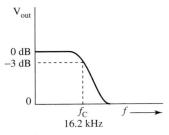

FIGURE 15–17
Frequency-response curve of Figure 15–16.

$$dB = 20 \log (V_{out} \div V_{in})$$
$$= 20 \log (0.67 \text{ V} \div 5.00 \text{ V})$$
$$= -17.45 \text{ dB}$$

ANALYSIS: At the lower frequency of 2 kHz, the circuit responded as in Case 2: low impedance in the signal path and medium impedance to ground. The attenuation was only −0.07 dB. The filter passed the 2-kHz signal.

At the higher frequency of 120 kHz, the circuit responded as in Case 1: high impedance in the signal path and medium impedance to ground. The attenuation was −17.45 dB. The filter rejected the 120-kHz signal.

CONCLUSION: If a complex waveform consisting of a 2-kHz signal combined with a 120-kHz signal appeared at the input of the circuit in Figure 15–16, the circuit would pass the low-frequency signal and reject the high-frequency signal. Therefore, the circuit acts as a low-pass filter.

By Formula 15–5(a), the cutoff frequency for this circuit is:

$$f_c = \frac{R}{2 \pi L}$$
$$= \frac{510 \ \Omega}{2 \times \pi \times 5 \text{ mH}}$$
$$= 16.2 \text{ kHz}$$

The frequency-response curve for this circuit is given in Figure 15–17.

● **SKILL-BUILDER 15–5**

Repeat the analysis of the circuit using the following values. You should obtain the results that follow.

V_S	5 V
R	1,200 Ω
L	11 mH
f_{low}	800 Hz
X_L	55 Ω
Z	1,201 Ω
I	4.16 mA
V_R	4.99 V
dB	−0.01
f_{high}	150 kHz
X_L	10,367 Ω
Z	10,436 Ω
I	0.48 mA
V_R	0.57 V
dB	−18.79

Series *RL* Circuit as a High-Pass Filter

We can change the circuit in Figure 15–16 from a low-pass to a high-pass filter simply by switching the positions of the components, as shown in Figure 15–18. Because the component values, source voltage, and frequencies are the same, the reactance, impedance, and current are also the same. The only difference is that now the output voltage is developed across the coil to ground instead of across the resistor. Thus:

$$V_{out} = I \times X_L$$

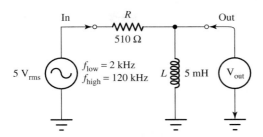

FIGURE 15–18
Series *RL* circuit as high-pass filter.

At $f = 2$ kHz:

$$V_{out} = 9.73 \text{ mA} \times 63 \text{ } \Omega$$
$$= 0.61 \text{ V}$$
$$dB = 20 \log (V_{out} \div V_{in})$$
$$= 20 \log (0.61 \text{ V} \div 5.00 \text{ V})$$
$$= -18.25$$

At $f = 120$ kHz:

$$V_{out} = 1.31 \text{ mA} \times 3.77 \text{ k}\Omega$$
$$= 4.95 \text{ V}$$
$$dB = 20 \log (V_{out} \div V_{in})$$
$$= 20 \log (4.95 \text{ V} \div 5.00 \text{ V})$$
$$= -0.08$$

ANALYSIS: At the lower frequency of 2 kHz, the circuit responded as in Case 4: medium impedance in the signal path and low impedance to ground. The attenuation was −18.25 dB. The filter rejected the 2-kHz signal.

At the higher frequency of 120 kHz, the circuit responded as in Case 3: medium impedance in the signal path and high impedance to ground. The attenuation was only −0.08 dB. The filter passed the 120-kHz signal.

CONCLUSION: If a complex waveform consisting of a 2-kHz signal combined with a 120-kHz signal in equal strengths appeared at the input of the circuit in Figure 15–16, the circuit would reject the low-frequency signal and pass the high-frequency signal. Therefore, the circuit acts as a high-pass filter.

Because the component values have not changed, the cutoff frequency is still 16.2 kHz. However, the frequency-response curve now has the appearance shown in Figure 15–19.

● *SKILL-BUILDER 15–6*
- -

Repeat the analysis of the circuit using the following values. You should obtain the results that follow.

V_S	5 V
R	3,600 Ω
L	0.72 mH
f_{low}	140 kHz
X_L	633 Ω
Z	3,655 Ω
I	1.37 mA
V_L	0.87 V
dB	−15.23

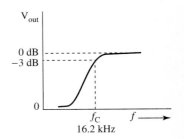

FIGURE 15–19
Frequency-response curve of Figure 15–18.

f_{high}	4.5 MHz
X_L	20,358 Ω
Z	20,673 Ω
I	0.24 mA
V_L	4.92 V
dB	-0.13

- -

Series *RC* Circuit as a Low-Pass Filter

Let us analyze the circuit in Figure 15–20 as indicated earlier.

At $f = 1.5$ kHz:

$$X_C = \frac{1}{2\pi f C}$$

$$= \frac{1}{2 \times \pi \times 1.5 \text{ kHz} \times 0.022 \text{ μF}}$$

$$= 4,823 \text{ Ω}$$

$$Z = \sqrt{R^2 + X_C^2}$$

$$= \sqrt{(200 \text{ Ω})^2 + (4.823 \text{ Ω})^2}$$

$$= 4,827 \text{ Ω}$$

$$I = V_S \div Z$$
$$= 9 \text{ V} \div 4,827 \text{ Ω}$$
$$= 1.86 \text{ mA}$$

$$V_{out} = V_C$$
$$= I \times X_C$$
$$= 1.86 \text{ mA} \times 4,823 \text{ Ω}$$
$$= 8.99 \text{ V}$$

$$dB = 20 \log (V_{out} \div V_{in})$$
$$= 20 \log (8.99 \text{ V} \div 9.00 \text{ V})$$
$$= -0.01 \text{ dB}$$

At $f = 400$ kHz:

$$X_C = \frac{1}{2\pi f C}$$

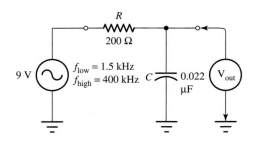

FIGURE 15–20
Series *RC* circuit as low-pass filter.

$$= \frac{1}{2 \times \pi \times 400 \text{ kHz} \times 0.022 \text{ }\mu\text{F}}$$

$$= 18 \text{ }\Omega$$

$$Z = \sqrt{R^2 + X_L^2}$$

$$= \sqrt{(200 \text{ }\Omega)^2 + (18 \text{ }\Omega)^2}$$

$$= 201 \text{ }\Omega$$

$$I = V_S \div Z$$

$$= 9 \text{ V} \div 201 \text{ }\Omega$$

$$= 44.82 \text{ mA}$$

$$V_{\text{out}} = V_C$$

$$= I \times X_C$$

$$= 44.82 \text{ mA} \times 18 \text{ }\Omega$$

$$= 0.81 \text{ V}$$

$$\text{dB} = 20 \log (V_{\text{out}} \div V_{\text{in}})$$

$$= 20 \log (0.81 \text{ V} \div 9.00 \text{ V})$$

$$= -20.91 \text{ dB}$$

ANALYSIS: At the lower frequency of 1.5 kHz, the circuit responded as in Case 3: medium impedance in the signal path and high impedance to ground. The attenuation was only −0.01 dB. The filter passed the 1.5-kHz signal.

At the higher frequency of 400 kHz, the circuit responded as in Case 4: medium impedance in the signal path and low impedance to ground. The attenution was −20.91 dB. The filter rejected the 400-kHz signal.

CONCLUSION: If a complex waveform consisting of a 1.5-kHz signal combined with a 400-kHz signal in equal strengths appeared at the input of the circuit in Figure 15–20, the circuit would pass the low-frequency signal and reject the high-frequency signal. Therefore, the circuit acts as a low-pass filter.

By Formula 15–5(b), the cutoff frequency for this circuit is:

$$f_c = \frac{1}{2 \pi R C}$$

$$= \frac{1}{2 \times \pi \times 200 \text{ }\Omega \times 0.022 \text{ }\mu\text{F}}$$

$$= 36.2 \text{ kHz}$$

The frequency-response curve for this circuit is given in Figure 15–21.

● *SKILL-BUILDER 15–7*

Repeat the analysis of the circuit using the following values. You should obtain the results that follow.

V_S	9 V
R	360 Ω
C	0.01 μF
f_{low}	2.8 kHz
X_C	5,684 Ω
Z	5,695 Ω
I	1.58 mA
V_C	8.98 V
dB	−0.02

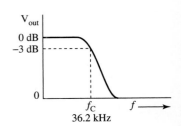

FIGURE 15–21
Frequency-response curve of Figure 15–20.

f_{high}	600 kHz
X_C	27 Ω
Z	361 Ω
I	24.93 mA
V_C	0.66 V
dB	-22.68

Series *RC* Circuit as a High-Pass Filter

We can change the circuit in Figure 15–20 from a low-pass to a high-pass filter simply by switching the positions of the components, as shown in Figure 15–22. Because the component values, source voltage, and frequencies are the same, the reactance, impedance, and current are also the same. The only difference is that now the output voltage is developed across the resistor to ground instead of across the capacitor. Thus:

$$V_{out} = I \times R$$

At $f = 1.5$ kHz:

$$V_{out} = 1.86 \text{ mA} \times 200 \text{ Ω}$$
$$= 0.37 \text{ V}$$

$$dB = 20 \log (V_{out} \div V_{in})$$
$$= 20 \log (0.37 \text{ V} \div 9.00 \text{ V})$$
$$= -27.65$$

At $f = 400$ kHz:

$$V_{out} = 44.82 \text{ mA} \times 200 \text{ Ω}$$
$$= 8.96 \text{ V}$$

$$dB = 20 \log (V_{out} \div V_{in})$$
$$= 20 \log (8.96 \text{ V} \div 9.00 \text{ V})$$
$$= -0.04$$

ANALYSIS: At the lower frequency of 1.5 kHz, the circuit responded as in Case 1: high impedance in the signal path and medium impedance to ground. The attenuation was −27.65 dB. The filter rejected the 1.5-kHz signal.

At the higher frequency of 400 kHz, the circuit responded as in Case 2: low impedance in the signal path and medium impedance to ground. The attenuation was only −0.04 dB. The filter passed the 400-kHz signal.

CONCLUSION: If a complex waveform consisting of a 1.5-kHz signal combined with a 400-kHz signal in equal strengths appears at the input of the circuit in Figure

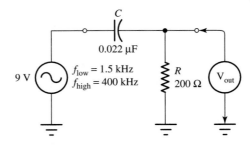

FIGURE 15–22
Series *RC* circuit as high-pass filter.

15–22, the circuit would reject the low-frequency signal and pass the high-frequency signal. Therefore, the circuit acts as a high-pass filter.

Because the component values have not changed, the cutoff frequency is still 36.2 kHz. However, the frequency-response curve now has the appearance shown in Figure 15–23.

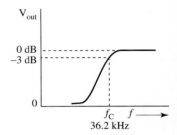

FIGURE 15–23
Frequency-response curve of Figure 15–22.

● *SKILL-BUILDER 15–8*

Repeat the analysis of the circuit using the following values. You should obtain the results that follow.

V_S	9 V
R	360 Ω
C	0.0068 μF
f_{low}	3.2 kHz
X_C	7,314 Ω
Z	7,323 Ω
I	1.23 mA
V_R	0.44 V
dB	−26.17
f_{high}	1.2 MHz
X_C	20 Ω
Z	361 Ω
I	24.96 mA
V_R	8.99 V
dB	−0.01

Summary of *RL* and *RC* Pass Filters

A series *RL* and a series *RC* circuit can each be used as either a low-pass or high-pass filter, as indicated next:

Low-Pass Filter:
   Series *RL* circuit with output voltage developed across resistor, or series *RC* circuit with output voltage developed across capacitor.
High-Pass Filter:
   Series *RL* circuit with output voltage developed across coil, or series *RC* circuit with output voltage developed across resistor.

Applications of Pass Filters

A tone control on a stereo amplifier is a good example of a pass filter. Low sounds are called bass (pronounced like base) frequencies, and high sounds are called treble frequencies. Increasing the proportion (volume) of a frequency band is called *boosting* the signal, and decreasing the volume is called *cutting* the signal. Most tone controls have a central position where they have no effect on the band. Setting the control off-center then produces a boost or cut in the frequency band. Inexpensive audio equipment may have only two tone controls, labeled bass and treble. More expensive equipment will have a set of filters whose bands overlap. This allows the user to adjust the sound more precisely, boosting or cutting the low and high frequencies. Such a set of filters may be called either an equalizer or a parametric filter. The purpose of an equalizer is to compensate for the acoustics of a particular room, but users often adjust the controls to suit their individual musical taste.

→ *Self-Test*

13. How can a series *RL* circuit act as a low-pass filter?
14. How can a series *RL* circuit act as a high-pass filter?
15. How can a series *RC* circuit act as a low-pass filter?
16. How can a series *RC* circuit act as a high-pass filter?

SECTION 15–5: ac Frequency Response: Resonant Band Filters

In this section you will learn the following:

- → Resonance is the tendency of a circuit to oscillate at a specific frequency, called the resonant frequency.
- → An *LC* circuit at resonance stores a significant amount of energy.
- → In a series *LC* circuit, the reactive voltage drop at resonance will be approximately *Q* times the source voltage.
- → Selectivity is the ability of a circuit to pass or reject a narrow frequency band.
- → In a well-designed circuit with high *Q*, the bandwidth is approximately equal to the resonant frequency divided by the value of *Q*.
- → A series *LC* circuit has low impedance at its resonant frequency.
- → A parallel *LC* circuit has high impedance at its resonant frequency.
- → A series resonant circuit acts as a band-pass filter when the *LC* section is in the signal path.
- → A parallel resonant circuit acts as a band-pass filter when the *LC* tank is grounded and used to develop the output voltage.
- → A series resonant circuit acts as a band-reject filter when the *LC* section is grounded and used to develop the output voltage.
- → A parallel resonant circuit acts as a band-reject filter when the *LC* tank is in the signal path.

Oscillation

The word *oscillate* means to swing back and forth. In general, an *oscillation* is a transfer of energy back and forth between two objects or locations.

A pendulum or swing is a basic example of oscillation. At the top of its swing, a pendulum has minimum motion: All its energy is potential (stored). At the bottom of its swing, a pendulum has maximum motion: All its energy is kinetic (moving).

A pendulum must swing an equal distance on both sides in order for its energy to transfer properly from potential to kinetic and back. Its physical oscillation is merely a result of its energy transfer between potential and kinetic. If a pendulum is not allowed to swing an equal distance in both directions, it quickly loses its energy and stops moving.

Each time the pendulum swings, it loses a small amount of energy due to friction, which dissipates energy by producing heat. Thus, in time the pendulum comes to a stop.

An oscillation that dies down like this is called a *damped* oscillation, one where the energy is gradually reduced to zero by heat losses.

To compensate for heat losses and keep the pendulum oscillating, we must feed energy back into the pendulum. Every child who has ever been on a swing knows this: Either you have to kick a little each time you go back, or else get someone to push you. Otherwise you will soon stop swinging.

When we feed energy back into an oscillating system, we must provide just the right amount of energy, and we must do so in phase with the oscillations. Using the playground swing as an example, if you push too hard, the swing goes out of control; and if you push at the wrong time, the swing jerks and you lose even more energy.

Resonance

All mechanical vibrations are a form of oscillation, from the ringing of a bell to the rattle of loose screws in an old car. Just as a pendulum or swing moves at a natural rhythm, all mechanical systems have a natural rate of vibration. This natural rate of vibration, where the energy losses due to friction are at a minimum, is called the *resonant frequency* of the system. When a system is vibrating at its resonant frequency, we say it is at *resonance*.

An electronic circuit has a natural tendency to oscillate at its resonant frequency. Sometimes this is useful, and sometimes it is unwanted and undesirable. In electronic circuits, there are two principal methods to create resonant oscillations. One way is by using quartz crystals in CB radios, digital computers, wrist watches, and many other devices. Another way of creating resonance is with a combination of an inductor and a capacitor.

In the last chapter we learned that the resonant frequency f_r of an RLC circuit is the frequency where X_L equals X_C. A series RLC circuit has minimum impedance at f_r, and a parallel RLC circuit has maximum impedance at f_r. Now we will combine the ideas of damped oscillation and impedance and see how they go together.

Transfer of Reactive Energy

Remember that resistance dissipates energy as heat. Thus, resistance in a circuit is like friction in a pendulum: It dissipates energy as heat and damps (reduces) the oscillations.

Reactance does not dissipate energy (produce heat). The kinetic energy of current flowing in an inductor builds up a magnetic field around the inductor, and the kinetic energy becomes stored as potential energy in the magnetic field. When the field collapses, the induced voltage creates current in the inductor. Thus, the potential energy becomes kinetic energy once again.

In a similar manner, when the kinetic energy of current charges a capacitor, the energy does not turn into heat. Instead, it becomes stored as potential energy in the electric field within the dielectric. When the capacitor discharges, the potential energy becomes kinetic energy once again as it produces the discharge current.

In order for a pendulum to swing, it needs room on both sides of center. A pendulum cannot swing on only one side. In a similar manner, neither an inductor alone nor a capacitor alone can produce strong oscillations. However, the combination of an inductor and capacitor can produce very strong oscillations. Like the two sides of a see-saw or the space on both sides of a pendulum, the combination of a capacitor and inductor allows a natural back-and-forth transfer of energy, from kinetic to potential, which is what produces all oscillations.

We have learned about the Fourier Series: A complex waveform acts like a mixture of many simple waveforms. The most complex waveform of all is that of a sudden event. Thus, the waveform of a sudden event is like a wave that contains virtually all frequencies. If a sudden burst of energy is applied to any physical system, the system will oscillate because it will find within the waveform of the sudden event a wave of energy at its resonant frequency.

For example, if you strike different bells with the same hammer, each bell will ring at its own frequency. The sudden event of the hammer blow produces a complex waveform of energy within the bell. However, the bell quickly and naturally damps out all frequencies within the burst except its own resonant frequency. It accepts the input of energy at that frequency and begins to ring.

In electronic circuits, turning on the power is a sudden event. Simply closing the switch and applying source voltage to the circuit creates a sudden-event waveform, a complex burst of energy across a wide frequency band. If the circuit is a *tuned circuit* (a circuit designed to have a resonant frequency), the circuit will act like the bell and damp out energy at all frequencies except at its resonant frequency. The circuit will begin to oscillate. We would say the circuit has been shocked into oscillation.

In a similar manner, if a complex waveform is applied continuously to a tuned circuit, the circuit will reject energy at nonresonant frequencies and accept energy only at resonance. However, there are some important details to the process we must consider in this section.

Energy Storage in a Resonant Circuit

The two circuits in Figure 15–24 both resonate at 250 kHz but have significantly different electrical parameters (values or measurements). According to the manufacturer's catalog, the 4,700-μH inductor is designed to operate at 250 kHz with a minimum Q of 50. In Chapter 12, we learned that the Q (quality factor) of a coil equals its reactance divided by its resistance. Therefore, the resistance of the coil R_W can be found from knowing X_L and Q.

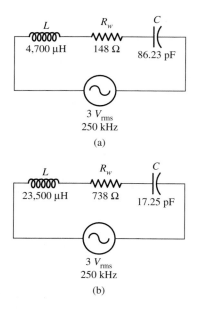

FIGURE 15–24
Series *RLC* circuit with different electrical parameters at the same resonant frequency.

$$X_L = 2\,\pi f L$$

At $f_r = 250$ kHz,

$$X_L = 2 \times \pi \times 250 \text{ kHz} \times 4,700 \text{ μH}$$
$$= 7,383 \text{ Ω}$$

$$\text{If } Q = \frac{X_L}{R_W}$$

$$\text{then } R_W = \frac{X_L}{Q}$$

$$= \frac{7,383 \text{ Ω}}{50}$$

$$= 148 \text{ Ω}$$

In a series *RLC* circuit at resonance, $Z = R$. In this case, $Z = R_W$. Therefore,

$$I = \frac{V_S}{R_W}$$

$$= \frac{3 \text{ V}}{148 \text{ Ω}}$$

$$= 20.32 \text{ mA}$$

By Ohm's Law:

$$V_L = I \times X_L$$
$$= 20.32 \text{ mA} \times 7,383 \text{ Ω}$$
$$= 150 \text{ V}$$

Why is the coil's voltage so high? Because the coil and capacitor, acting together, form a resonant system capable of storing a substantial amount of energy and passing this energy back and forth. The circuit is oscillating.

Without the capacitor, the circuit impedance would be approximately equal to the reactance, and the voltage across the coil would be approximately 3 V. However, with the capacitor to act as a reservoir, energy builds up and passes back and forth between the coil and capacitor, producing extraordinarily high voltages across both of them. In this circuit at resonance, the reactive energy is given by:

$$W = \tfrac{1}{2}L\,I^2$$
$$= 0.5 \times 4,700 \text{ μH} \times (20.32 \text{ mA})^2$$
$$= 0.97 \text{ μJ}$$

An energy of 97 microjoules may not sound like much, but it puts 150 volts across both the coil and capacitor.

In Figure 15–24(b), the coil is five times larger and the capacitor five times smaller. The resonant frequency is exactly the same, 250 kHz. You can produce resonance at any frequency with any coil simply by matching it up with a capacitor of the proper value.

In this circuit, if the new coil also has a *Q* of 50,

$$X_L = 2 \times \pi \times 250 \text{ kHz} \times 23,500 \text{ μH}$$
$$= 36,914 \text{ Ω}$$

$$R_W = \frac{X_L}{Q}$$

$$= \frac{36{,}914\ \Omega}{50}$$

$$= 738\ \Omega$$

$$= \frac{3\ \text{V}}{738\ \Omega}$$

$$= 4.06\ \text{mA}$$

Therefore,

$$V_L = I \times X_L$$
$$= 4.06\ \text{mA} \times 36{,}914\ \Omega$$
$$= 150\ \text{V}$$

$$W = \tfrac{1}{2}\, L\, I^2$$
$$= \tfrac{1}{2} \times 23{,}500\ \mu\text{H} \times (4.06\ \text{mA})^2$$
$$= 0.19\ \mu\text{J}$$

Notice that the second circuit had five times more reactance, five times less current, and five times less energy stored as reactive energy; yet the voltage drop was still 150 V. This is because of Q, the relationship between X_L and R_W.

In a series LC circuit, the reactive voltage drop at resonance will be approximately Q times the source voltage. For this reason, Q is also called the *magnification factor* of a series LC circuit.

● *SKILL-BUILDER 15–9*

A 350-μH inductor with a winding resistance of 2.3 Ω is placed in series with a 0.0047-μF capacitor. The source voltage is 1.3 V_{rms}. Find the resonant frequency, reactance, quality, current, and reactive voltage drop.

GIVEN:
$L = 350\ \mu\text{H}$
$R_W = 2.3\ \Omega$
$C = 0.0047\ \mu\text{F}$
$V_S = 1.3\ \text{V}$

FIND:
f_r
X_L
Q
I
V_X

SOLUTION:

1. $f_r = \dfrac{1}{2\,\pi\,\sqrt{LC}}$ series resonance

$$= \frac{1}{2\,\pi\,\sqrt{350\ \mu\text{H} \times 0.0047\ \mu\text{F}}}$$

keystrokes: 350 [EXP] 6
[+ −] [×] .0047 [EXP] 6
[+ −] [=] [√] [×] 2 [×]
[π] [=] [1/x]

$$= 124.1\ \text{kHz}$$

2. $X_L = 2\pi f L$
$\quad = 2 \times \pi \times 124.1 \text{ kHz}$ calculated at f_r
$\quad\quad \times 350 \ \mu\text{H}$
$\quad = 272.89 \ \Omega$

3. $Q = \dfrac{X_L}{R_W}$

$\quad = \dfrac{272.89 \ \Omega}{2.3 \ \Omega}$

$\quad = 118.65$

4. $I = \dfrac{V_S}{R_W}$ at f_r, $Z = R_W$

$\quad = \dfrac{1.3 \text{ V}}{2.3 \ \Omega}$

$\quad = 565 \text{ mA}$

5. $V_L = I \times X_L$
$\quad = 565 \text{ mA} \times 272.89 \ \Omega$
$\quad = 154.2 \text{ V}$ V_L is Q times larger than V_S

Repeat the Skill-Builder using the following values. You should obtain the results that follow.

Given:

L	470 μH	900 μH	1,320 μH	500 μH	125 μH
R_W	3.3 Ω	6.3 Ω	7.1 Ω	3.6 Ω	1.3 Ω
C	0.0022 μF	0.001 μF	0.0022 μF	0.0047 μF	0.0075 μF
V_S	0.7 V	2.1 V	0.4 V	1.8 V	1.1 V

Find:

f_r	156.5 kHz	167.7 kHz	93.4 kHz	103.8 kHz	164.4 kHz
X_L	462 Ω	949 Ω	775 Ω	326 Ω	129 Ω
Q	140	151	109	91	99
I	212 mA	333 mA	56 mA	500 mA	846 mA
V_L	98.0 V	316.2 V	43.6 V	163.1 V	109.2 V

Why Series Resonance Has Low Impedance and Parallel Resonance Has High Impedance

There are several ways to explain this, depending on the viewpoint that seems most familiar both to instructor and student. We will choose an approach that is intuitive rather than formal.

Earlier we compared an inductor and capacitor at resonance to the two sides of a swinging pendulum. Imagine an ideal pendulum with zero friction in the system. If this were true, then once the pendulum starts to swing, it would swing forever. The only reason you need to keep giving the pendulum a little extra push is to make up for the energy losses due to friction.

We have said that in an *RLC* circuit, resistance is similar to friction: It dissipates energy as heat. Suppose an inductor and capacitor are placed in parallel and the capacitor is charged with a battery through a switch, as shown in Figure 15–25(a). When the switch is opened (b), the capacitor will discharge through the coil, creating a magnetic

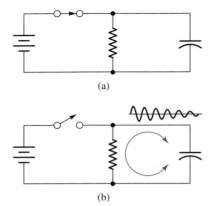

FIGURE 15–25
Opening or closing a switch
shocks the *LC* tank into ringing
(damped oscillation) at the
resonant frequency.

field. Then the magnetic field will collapse, inducing a voltage in the coil which re-charges the capacitor. According to the size of the coil and capacitor, there is a certain frequency at which the two devices are in step or phase with each other. This is what the resonant frequency really is. The coil and capacitor will find their natural resonant frequency, and they will oscillate as shown, producing a sine wave. This is the way in which the coil and capacitor resemble the swing of a pendulum or the ringing of a bell: finding a natural frequency and then transferring energy back and forth at this frequency.

Because of resistance in the devices, each cycle of the oscillation loses a small amount of energy as heat. This dampens (gradually reduces) the oscillations as shown. In fact, *ringing* is a name for this particular waveform, a damped sine wave. You will see many examples of ringing as you progress in your study of electronics. With no feedback of energy into the system to make up for the resistance losses, eventually all the energy is dissipated as heat, and the oscillations die away.

In Figure 15–26(a), the coil and capacitor must charge and discharge through the voltage source and its internal resistance. If the source is at the resonant frequency, the coil and capacitor can sustain their oscillations, passing energy back and forth to each other without any losses, because they have the help of the source to provide the extra current demanded by the resistance. This is why the source current is equal to source voltage divided by resistance.

The fact that V_L equals Q multiplied by V_S is simply a rearrangement of Ohm's Law using the definition of Q:

$$V_L = X_L \times I$$

$$= X_L \times \frac{V_S}{R_W}$$

$$= \frac{X_L}{R_W} \times V_S$$

$$= Q \times V_S$$

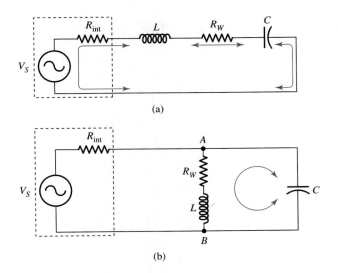

FIGURE 15–26
Comparison of current path in series and parallel resonance.

Why Parallel Resonance Has High Impedance

Kirchhoff's Current Law is an easy way to see why parallel resonance has high impedance: The sum of all currents entering a junction equals the sum of all currents leaving a junction.

Figure 15–26(b) shows a coil and capacitor in parallel with an ac source. I_L and I_C are always 180° out of phase, which means they always flow in opposite directions. At resonance, I_L equals I_C. Suppose that I_L flows from the coil into junction A. Then I_C must flow out of junction A toward the capacitor. If they are equal, then I_L and I_C fulfill Kirchhoff's Current Law at junction A, and no current can flow between junction A and the source, either in or out. A similar line of reasoning applies to junction B.

This behavior of I_L and I_C at parallel resonance reminds us of a water tank that is full: No more water can enter it. Accordingly, a parallel resonant circuit is generally called a *tank circuit*.

Parallel Equivalent Resistance of R_W

Actually a small amount of current from the source does flow through a tank circuit to make up for the energy losses caused by R_W. As shown in Figure 15–27, although the coil's winding resistance R_W is in series with the coil's inductance in one branch, it is possible to mathematically convert the practical circuit into an ideal equivalent *RLC* parallel circuit, as shown. In other words, we pretend that we have removed the resistance from the coil and placed it in a separate branch with a different ohmic value. Formula 15–6 is used to calculate this equivalent ohmic value.

Formula 15–6

$$R_{eq} = R_W (Q^2 + 1)$$

This equivalent ohmic value R_{eq} is the impedance of the tank circuit. In other words, a perfect tank of pure inductance and capacitance at resonance would act like an open to source current at the resonant frequency. At resonance, the tank's impedance would be infinity, as shown in Figure 15–27(b). However, because of winding resistance,

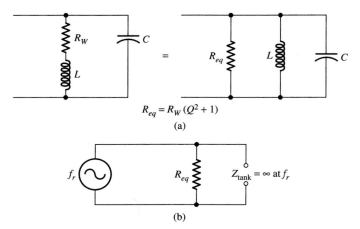

FIGURE 15–27
Equivalent circuits of a practical *LC* tank.

a little source current would still flow through the tank. This is represented by R_{eq} as a separate parallel branch.

● *SKILL-BUILDER 15–10*

A tank circuit consists of a 0.022-μF capacitor in parallel with a coil that has an inductance of 3 mH and a winding resistance of 5.2 Ω. What is the impedance of the tank?

GIVEN:
  $C = 0.022\ \mu F$
  $L = 3\ mH$
  $R_W = 5.2\ \Omega$

FIND:
  R_{eq}

STRATEGY:

This is a multistep problem. The best approach is to select an equation for the value we want to find (R_{eq}), and then plan a sequence of equations that will lead up to the answer. At each step, we look at what value we need to find, then select an equation for that value.

Equation for R_{eq}:

  $R_{eq} = R_W (Q^2 + 1)$ R_W is given.
                             We must find Q.

Equation for Q:

  $Q = \dfrac{X_L}{R_W}$ R_W is given.
                             We must find X_L.

Equation for X_L:

  $X_L = 2\pi f L$ L is given.
                         We must find f.
                         In this case, f should be f_r.

Equation for f_r:

$$f_r = \frac{1}{2\pi\sqrt{LC}}$$

L and *C* are given. Now we can begin our calculations.

SOLUTION:

Solve the equations from the first to the last.

$$f_r = \frac{1}{2\pi\sqrt{LC}}$$

$$= \frac{1}{2\times\pi\times\sqrt{(3\text{ mH}\times 0.022\text{ }\mu F)}}$$

keystrokes:
3 [EXP] 3 [+/−] [×]
.022 [EXP] 6 [+/−] [=]
[√] [×] 2 [×] [π]
[=] [1/x]

$$= 19.59\text{ kHz}$$

$$X_L = 2\pi f L$$
$$= 2\times\pi\times 19.59\text{ kHz}\times 3\text{ mH}$$
$$= 369.3\text{ }\Omega$$

$$Q = \frac{X_L}{R_W}$$

$$= \frac{369.3\text{ }\Omega}{5.2\text{ }\Omega}$$

$$= 71.01$$

$$R_{eq} = R_W(Q^2+1)$$
$$= 5.2\text{ }\Omega\times(71.01^2+1)$$

keystrokes:
71.01 [x²] [+] 1
[=] [×] 5.2 [=]

$$= 26.23\text{ k}\Omega$$

Repeat the Skill-Builder using the following values. You should obtain the results that follow.

Given:

L	5 mH	7 mH	11 mH	15 mH	20 mH
R_W	6.4 Ω	7.5 Ω	8.1 Ω	9.6 Ω	10.4 Ω
C	0.015 μF	0.010 μF	0.0068 μF	0.0047 μF	0.0033 μF

Find:

f_r	18.38 kHz	19.02 kHz	18.40 kHz	18.96 kHz	19.59 kHz
X_L	577.4 Ω	836.7 Ω	1,271.9 Ω	1,786.5 Ω	2,461.8 Ω
Q	90.21	111.55	157.02	186.09	236.71
R_{eq}	52.09 kΩ	93.34 kΩ	199.72 kΩ	332.46 kΩ	582.76 kΩ

Selectivity

Selectivity is the ability of a circuit to pass or reject a narrow frequency band. A tuner is a selective circuit. For example, a radio receives a complex waveform containing signals

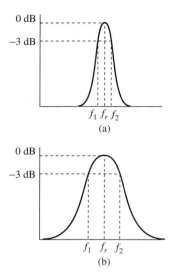

FIGURE 15–28
(a) High selectivity: high Q, narrow bandwidth. (b) Low selectivity: low Q, broad bandwidth.

from many stations at many different frequencies. It selects a narrow frequency band containing the signals of a particular station, passes this band on for further processing, and rejects all other signals at frequencies outside the selected band.

Figure 15–28(a) is the frequency-response curve of a circuit with high selectivity. The cutoff frequencies f_1 and f_2 are near each other, which means the circuit has a narrow bandwidth. The rolloff is sharp or steep, meaning frequencies just outside the band are at very low power levels compared to frequencies within the band.

By comparison, Figure 15–28(b) is the frequency-response curve of a circuit with low selectivity. The cutoff frequencies are farther apart, which means the circuit has a wide or broad bandwidth. The rolloff is more gradual, which means signals at frequencies outside the band would still pass through the circuit at power levels high enough to be detected.

High Q Produces High Selectivity

The selectivity of a resonant LC circuit depends mainly on the Q factor of the coil. A coil with a high value of Q at f_r will cause an LC circuit to have high selectivity.

Q is defined as the ratio of reactance to resistance. Thus, we must realize that the Q of the coil itself may be higher than the overall Q of the entire circuit. Any additional resistance in the circuit tends to lower the overall Q and cause the circuit to be less selective.

Relationship between Q, Resonant Frequency, and Bandwidth

In a well-designed circuit with high Q, the bandwidth is approximately equal to the resonant frequency divided by the value of Q, as given by Formula Diagram 15–7.

Formula Diagram 15–7

$$\frac{f_r}{BW \mid Q}$$

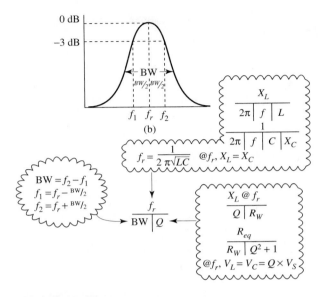

FIGURE 15–29
The relationship of f_r, BW, and Q ties together
many other formulas.

If we assume that the resonant frequency is in the center of the band, then Formula Diagram 15–7 can help us find the cutoff frequencies f_1 and f_2.

Figure 15–29 shows how Formula Diagram 15–7 ties together three groups of other formulas. In solving multistep problems involving Q, BW, and f_r, the formula diagram acts like a bridge to help set up the correct sequence of equations to solve the problem. The following Skill-Builder is an example of this process.

● **SKILL-BUILDER 15–11**
- -

A tank circuit produces a frequency-response curve with cutoff points at 252 kHz and 258 kHz. The impedance of the tank is 400 kΩ. What are the values of inductance and capacitance in the tank?

GIVEN:
   $f_1 = 252$ kHz
   $f_2 = 258$ kHz
   $R_{eq} = 400$ kΩ

FIND:
   L
   C

STRATEGY:

$$C = \frac{1}{2 \times \pi \times f_r \times X_C}$$ From Formula Diagram.
We need f_r and X_C.

At f_r, $X_L = X_C$ At resonance, X_C has the
same value as X_L. We need f_r and X_L.

$$L = \frac{X_L}{2 \times \pi \times f_r}$$ From Formula Diagram.
We need f_r and X_L

At this point, we should look for other formulas containing f_r and X_L to see if somehow we can use these formulas to find f_r and X_L.

$$f_r = f_1 + \frac{\text{BW}}{2}$$

The frequency-response curve shows f_r in the center of the band, halfway between f_1 and f_2. We know f_1. We need to find BW.

$$\text{BW} = f_2 - f_1$$

Here is our starting point. We know f_1 and f_2.

$$= 258\ \text{kHz} - 252\ \text{kHz}$$
$$= 6\ \text{kHz}$$

$$f_r = f_1 + \frac{\text{BW}}{2}$$

$$= 252\ \text{kHz} + \frac{6\ \text{kHz}}{2}$$

$$= 255\ \text{kHz}$$

We have found f_r. Now all we need is X_L.

$$X_L = Q \times R_W$$

From the Formula Diagram. We could find X_L if we knew Q and R_W.

$$R_W = \frac{R_{\text{eq}}}{Q^2 + 1}$$

From the Formula Diagram. We know R_{eq}. The key to the whole problem is finding Q.

$$Q = \frac{f_r}{\text{BW}}$$

Here is how we use this Formula Diagram as the bridge that links the groups of formulas.

$$= \frac{255\ \text{kHz}}{6\ \text{kHz}}$$

Since we know f_r and BW, we can find Q. The rest of the problem will then fall easily into place.

$$= 42.5$$

$$R_W = \frac{R_{\text{eq}}}{Q^2 + 1}$$

$$= \frac{400\ \text{k}\Omega}{42.5^2 + 1}$$

keystrokes:
42.5 [x^2] [+] 1
[=] [1/x] [×]
400 [EXP] 3 [+/−] [=]

$$= 221.33\ \Omega$$

$$X_L = Q \times R_W$$

Now that we have all the necessary values, we can solve the set of equations we have already selected.

$$= 42.5 \times 221.33\ \Omega$$

$$= 9{,}407\ \Omega$$

$$L = \frac{X_L}{2\,\pi f_r}$$

$$= \frac{9{,}407\ \Omega}{2 \times \pi \times 255\ \text{kHz}}$$

$$= 5.87\ \text{mH}$$

$$C = \frac{1}{2\ \pi\ f_r\ X_C}$$

$$= \frac{1}{2 \times \pi \times 255\ \text{kHz} \times 9{,}407\ \Omega} \qquad \text{At } f_r,\ X_C = X_L.$$

$$= 66.35\ \text{pF}$$

Repeat the Skill-Builder using the following values. You should obtain the results that follow.

Given:

f_1	356 kHz	1,512 kHz	163 kHz	2,650 kHz	870 kHz
f_2	361 kHz	1,540 kHz	175 kHz	2,780 kHz	920 kHz
R_{eq}	600 kΩ	750 kΩ	320 kΩ	430 kΩ	620 kΩ

Find:

BW	5 kHz	28 kHz	12 kHz	130 kHz	50 kHz
f_r	358.5 kHz	1,526 kHz	169 kHz	2,715 kHz	895 kHz
Q	71.7	54.5	14.1	20.9	17.9
R_W	116.7 Ω	252.4 Ω	1,605.3 Ω	983.6 Ω	1,929 Ω
X_L	8,367 Ω	13,757 Ω	22,608 Ω	20,542 Ω	34,529 Ω
L	3.71 mH	1.43 mH	21.29 mH	1.20 mH	6.14 mH
C	53.1 pF	7.6 pF	41.7 pF	2.9 pF	5.2 pF

Series Resonant Band-Pass Filter

A series resonant circuit acts as a band-pass filter when the *LC* section is in the signal path, as shown in Figure 15–30. The coil and capacitor in series present low impedance to signals at frequencies within the resonant band. Thus, signals within the resonant band pass through the coil and capacitor and develop a relatively large voltage across the resistor at the output.

The capacitor presents high impedance to signals at frequencies below the resonant band. Similarly, the coil presents high impedance to signals above the band. Thus, signals outside the band drop most of their voltage across either the coil or the capacitor and develop a relatively small voltage across the resistor at the output.

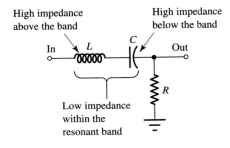

FIGURE 15–30
Series *LC* circuit as a band-pass filter.

MATH TIP

When you realize that a math problem involves several steps (a sequence of equations), a good strategy is to write out on a single sheet of paper every formula and diagram you can think of that has anything to do with the quantities you are given or asked to find. Circle the quantities you are asked to find in every formula or diagram where they appear, and write down the quantities you are given wherever they appear.

Now you have a kind of "bird's-eye view" of the problem. Like fitting together the pieces of a jigsaw puzzle, you hope to discover a path from one formula or diagram to another, a path that connects the values you want to find with the values you are given. Remember to build the path in reverse—from what you want to what you have. If you try to work forward (from what you have to what you want), you may go down any number of blind trails and not discover a successful sequence of equations.

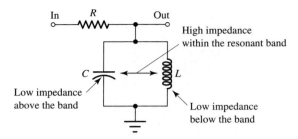

FIGURE 15–31
Parallel *LC* circuit as a band-pass filter.

Since signals within the resonant band develop a relatively large voltage at the output and signals outside the band do not, the circuit acts as a band-pass filter.

Parallel Resonant Band-Pass Filter

A parallel resonant circuit acts as a band-pass filter when the *LC* tank is grounded and used to develop the output voltage, as shown in Figure 15–31. The coil and capacitor in parallel present a high impedance to ground for signals at frequencies within the resonant band. Thus, signals within the resonant band develop a relatively large voltage at the output.

The capacitor presents a low-impedance path to ground to signals at frequencies above the resonant band. Similarly, the coil presents a low-impedance path to ground to signals below the band. Thus, signals at frequencies outside the band develop a relatively low voltage at the output because of low impedance to ground through either the coil or the capacitor.

Since signals within the resonant band develop a relatively large voltage at the output and signals outside the band do not, the circuit acts as a band-pass filter.

Series Resonant Band-Reject Filter

A series resonant circuit acts as a band-reject filter when the *LC* section is grounded and used to develop the output voltage, as shown in Figure 15–32. The coil and capacitor in series present a low-impedance path to ground to frequencies within the resonant band. Thus, signals within the resonant band pass to ground through the coil and capacitor and develop a relatively small voltage at the output.

The capacitor presents high impedance to signals at frequencies below the resonant band. Similarly, the coil presents high impedance to signals above the band. Thus, signals

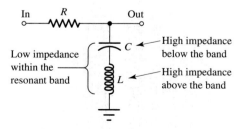

FIGURE 15–32
Series *LC* circuit as a band-reject filter.

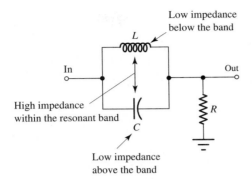

FIGURE 15–33
Parallel *LC* circuit as a band-reject filter.

outside the band drop most of their voltage across either the coil or the capacitor and develop a relatively large voltage at the output.

Since signals outside the resonant band develop a relatively large voltage at the output and signals within the band do not, the circuit acts as a band-reject filter.

Parallel Resonant Band-Reject Filter

A parallel resonant circuit acts as a band-reject filter when the *LC* tank is in the signal path, as shown in Figure 15–33. The coil and capacitor in parallel form a tank circuit that presents high impedance to frequencies within the resonant band. Thus, signals within the resonant band are rejected by the tank circuit and develop a relatively small voltage across the resistor at the output.

The capacitor presents a low impedance to signals at frequencies above the resonant band. Similarly, the coil presents a low impedance to signals below the band. Thus, signals at frequencies outside the band pass through either the coil or the capacitor and develop a relatively large voltage across the resistor at the output.

Since signals outside the resonant band develop a relatively large voltage at the output and signals within the band do not, the circuit acts as a band-reject filter.

Applications of Resonant Filters

The most common application of resonant filters is tuning circuits in radios and televisions. In the case of a television signal each channel is actually a group of bands. Each of these bands contains particular kinds of signals: audio, video, synchronization (timing), and so on. Thus, after the channel is tuned (separated from other channels), it must be broken down into its various bands. Other resonant band filters in the television circuits will do this and send each of the bands to the proper section of the television for further processing.

An *oscillator* is a circuit that produces electrical waves at a specific frequency. Resonant band filters are also used to establish and control the frequencies of oscillators. The filter (usually a tank circuit) is shocked into oscillation when power is applied. The output sine waves are amplified, and a small portion of the amplified signal is fed back into the tank to make up for energy losses due to resistance.

→ *Self-Test*

17. What is resonance?
18. What can we say about the energy in a resonant *LC* circuit?
19. What is the relationship between reactive voltage and source voltage in a resonant series *LC* circuit?
20. What is selectivity?
21. What is the relationship between bandwidth, resonant frequency, and *Q?*
22. What is the impedance of a series *LC* circuit at its resonant frequency?
23. What is the impedance of a parallel *LC* circuit at its resonant frequency?
24. How can a series resonant *LC* circuit act as a band-pass filter?
25. How can a parallel resonant *LC* circuit act as a band-pass filter?
26. How can a series resonant *LC* circuit act as a band-reject filter?
27. How can a parallel resonant *LC* circuit act as a band-reject filter?

Formula List

Formula 15-1

$$BW = f_2 - f_1$$

Formula 15–2

$$f_c = f_1 + \frac{BW}{2}$$

$$f_c = f_2 - \frac{BW}{2}$$

Formula 15–3

$$dB = 10 \log \frac{\text{output power}}{\text{input power}}$$

Formula 15–4

$$dB = 20 \log \frac{V_{\text{out}}}{V_{\text{in}}}$$

Formula 15–5

(a)
$$f_c = \frac{R}{2 \pi L}$$

(b)
$$f_c = \frac{1}{2 \pi R C}$$

Formula 15–6

$$R_{\text{eq}} = R_W (Q^2 + 1)$$

Formula Diagram 15–7

$$\frac{f_r}{BW \mid Q}$$

CHAPTER SUMMARY: ANSWERS TO SELF-TESTS

1. A harmonic is a wave whose frequency is an integer multiple of the fundamental frequency of another wave.
2. Any waveform, however simple or complex, can be produced by a combination of sine waves and cosine waves at the waveform's fundamental frequency and at the various harmonic frequencies. This combination is called the Fourier Series of the waveform.
3. A frequency band is a group or range of frequencies.
4. Bandwidth is the size of a band, calculated by subtracting the lowest frequency in the band from the highest frequency in the band.
5. The center frequency is the frequency in the middle of a band.
6. The center frequency of a band is the frequency with maximum power.
7. The half-power points of a band are the frequencies f_1 and f_2 whose power is 50% of the center frequency.
8. A low-pass filter is a circuit that passes signals below a certain frequency and rejects signals above that frequency.
9. A high-pass filter is a circuit that passes signals above a certain frequency and rejects signals below that frequency.
10. A band-pass filter is a circuit that passes signals within a certain band and rejects signals outside that band.
11. A band-reject filter is a circuit that rejects signals within a certain band and passes signals outside that band.
12. A decibel is a logarithmic unit that compares one power level to another.
13. A series RL circuit acts as a low-pass filter when the resistor is grounded and used to develop the output voltage.
14. A series RL circuit acts as a high-pass filter when the coil is grounded and used to develop the output voltage.
15. A series RC circuit acts as a low-pass filter when the capacitor is grounded and used to develop the output voltage.
16. A series RC circuit acts as a high-pass filter when the resistor is grounded and used to develop the output voltage.
17. Resonance is the tendency of a circuit to oscillate at a specific frequency, called the resonant frequency.
18. An LC circuit at resonance stores a significant amount of energy.
19. In a series LC circuit, the reactive voltage drop at resonance will be approximately Q times the source voltage.
20. Selectivity is the ability of a circuit to pass or reject a narrow frequency band.
21. In a well-designed circuit with high Q, the bandwidth is approximately equal to the resonant frequency divided by the value of Q.
22. A series LC circuit has low impedance at its resonant frequency.
23. A parallel LC circuit has high impedance at its resonant frequency.
24. A series resonant circuit acts as a band-pass filter when the LC section is in the signal path.
25. A parallel resonant circuit acts as a band-pass filter when the LC tank is grounded and used to develop the output voltage.
26. A series resonant circuit acts as a band-reject filter when the LC section is grounded and used to develop the output voltage.
27. A parallel resonant circuit acts as a band-reject filter when the LC tank is in the signal path.

CHAPTER REVIEW QUESTIONS

Determine whether the following questions are true or false.

15–1. A harmonic is a wave whose frequency is a fraction of the fundamental frequency of another wave.

15–2. Any waveform, however simple or complex, can be produced by a Fourier Series of sine waves and cosine waves at the waveform's fundamental frequency and at the various harmonic frequencies.

15–3. If several waves combine into a complex wave, an electronic circuit will respond as if the waves were still separate.

15–4. A band is a group of half-power points.

15–5. Bandwidth is the size of a frequency band.

15–6. The half-power points of a band are the frequencies f_1 and f_2 whose power is 50% of f_r.

15–7. A frequency-response curve shows the relationship between frequency and an electrical quantity.

15–8. A half-power point is also called a cutoff frequency.

15–9. Logarithmic scales begin with 1 and the increments are unevenly spaced.

15–10. Each section of a logarithmic scale is called a decade.

15–11. The rolloff of a filter is the rate at which the output power level changes with frequency.

15–12. An electronic circuit will not oscillate at its resonant frequency.

15–13. Selectivity is the ability of a circuit to pass or reject a narrow frequency band.

Select the response that best answers the question or completes the statement.

15–14. The fourth harmonic of a 60-Hz wave is
 a. 64 Hz.
 b. 60^4 Hz.
 c. 240 Hz.
 d. 400 Hz.

15–15. 3 kHz is the third harmonic of a fundamental frequency of
 a. 3 Hz.
 b. 1 kHz.
 c. 9 kHz.
 d. 6 kHz.

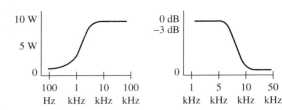

FIGURE 15–34

15–16. In Figure 15–34(a), the cutoff frequency is
   a. 100 Hz.
   b. 1 kHz.
   c. 10 kHz.
   d. 100 kHz.

15–17. In Figure 15–34(b), the cutoff frequency is
   a. 1 kHz.
   b. 5 kHz.
   c. 10 kHz.
   d. 50 kHz.

15–18. A decibel is
   a. a logarithmic unit that compares one power level to another.
   b. a linear unit that compares one power level to another.
   c. a power unit that compares one logarithm to another.
   d. an output unit that compares one input to another.

15–19. The output of a circuit is −1 dB. The signal is
   a. amplified.
   b. boosted.
   c. unchanged.
   d. attenuated.

15–20. An output of +3 dB means
   a. the output level is twice the input level.
   b. the input level is twice the output level.
   c. the output level is three times the input level.
   d. the input level is three times the output level.

15–21. In a series *LC* circuit, the reactive voltage drop at resonance will be approximately
   a. *Q* times the source current.
   b. *Q* divided by the source current.
   c. *Q* times the source voltage.
   d. *Q* divided by the source voltage.

15–22. In a well-designed circuit with high *Q*, the bandwidth is approximately equal to
   a. the resonant frequency divided by the value of *Q*.
   b. the value of *Q* divided by the resonant frequency.
   c. the cutoff frequency divided by the value of *Q*.
   d. the value of *Q* divided by the cutoff frequency.

15–23. A series *LC* circuit has
   a. low impedance at its cutoff frequency.
   b. high impedance at its cutoff frequency.
   c. low impedance at its resonant frequency.
   d. high impedance at its resonant frequency.

15–24. A parallel *LC* circuit has
   a. low impedance at its cutoff frequency.

   b. high impedance at its cutoff frequency.
   c. low impedance at its resonant frequency.
   d. high impedance at its resonant frequency.

15–25. When the *LC* section is in the signal path, a series resonant circuit acts as a
   a. low-pass filter.
   b. band-pass filter.
   c. high-pass filter.
   d. band-reject filter.

15–26. When the *LC* tank is grounded and used to develop the output voltage, a parallel resonant circuit acts as a
   a. low-pass filter.
   b. band-pass filter.
   c. high-pass filter.
   d. band-reject filter.

15–27. When the *LC* section is grounded and used to develop the output voltage, a series resonant circuit acts as a
   a. low-pass filter.
   b. band-pass filter.
   c. high-pass filter.
   d. band-reject filter.

15–28. When the *LC* tank is in the signal path, a parallel resonant circuit acts as a
   a. low-pass filter.
   b. band-pass filter.
   c. high-pass filter.
   d. band-reject filter.

15–29. A resonant circuit with high *Q* has
   a. high impedance.
   b. low impedance.
   c. high selectivity.
   d. low selectivity.

Solve the following problems.

15–30. What is the bandwidth and the center frequency of a band extending from 400 kHz to 600 kHz?

15–31. A circuit has an input power level of 12 milliwatts and an output of 9 milliwatts. What is the attenuation in dB?

15–32. At a certain frequency, a filter has an input voltage of 3.59 V and an output voltage of 2.54 V. What is the attenuation in decibels?

15–33. What is the cutoff frequency for an *RL* pass filter with a 4.7-kΩ resistor and a 5-mH inductor?

The following problems are more challenging.

15–34. Suppose the circuit in Figure 15–16 has a 3.6-V_{rms} source, a 2.5-mH inductor, and a 470-Ω resistor. Find the cutoff frequency, and then find the output voltage level one decade below f_c and one decade above f_c.

15–35. Suppose the circuit in Figure 15–18 has a 4.3-V_{rms} source, a 3.1-mH inductor, and a 680-Ω resistor. Find the cutoff frequency, and then find the output voltage level one decade below f_c and one decade above f_c.

15–36. Suppose the circuit in Figure 15–20 has a 2.7-V_{rms} source, a 0.022-μF capacitor, and a 470-Ω resistor. Find the cutoff frequency, and then find the output voltage level one decade below f_c and one decade above f_c.

15–37. Suppose the circuit in Figure 15–22 has a 1.9-V_{rms} source, a 0.033-μF capacitor, and a 510-Ω resistor. Find the cut-off frequency, and then find the output voltage level one decade below f_c and one decade above f_c.

15–38. A 575-μH inductor with a winding resistance of 4.3 Ω is placed in series with a 0.0068-μF capacitor. The voltage drop across the coil at resonance is 53.7 V_{rms}. What is the source voltage?

15–39. A tank circuit consists of a 0.015-μF capacitor in parallel with a coil that has an inductance of 2.5 mH and a winding resistance of 4.6 Ω. What is the impedance of the tank?

GLOSSARY
- - - - - - -

amplification   An increase in power; same as gain.

attenuation   A decrease in power; same as loss.

band   A range of frequencies.

band-pass filter   A circuit that passes signals within a certain band and rejects signals outside that band.

band-reject filter   A circuit that rejects signals within a certain band and passes signals outside that band.

band-stop filter   Same as band-reject filter.

bandwidth   The size of a band calculated by subtracting the lowest frequency from the highest.

Bode plot   A frequency-response curve drawn on a logarithmic scale.

boost   Same as amplification or gain.

center frequency   The frequency in the middle of a band.

cut   Same as attenuation or loss.

cutoff frequency   Same as half-power point.

damp   To absorb energy at a specific frequency.

decade   A band whose upper frequency is ten times the lower frequency.

decibel   A logarithmic unit that compares one power level to another.

Fourier Series   A combination of sine and cosine waves. Any complex waveform can be produced by an appropriate Fourier Series.

frequency-response curve   A graph of the relationship between frequency and power.

fundamental frequency   The lowest frequency in a series of harmonics.

gain   An increase in power; same as amplification.

half-power points   The lowest and highest frequencies of a band, where the power is half that of the center frequency.

harmonic frequency   A frequency that is an integer multiple of a fundamental frequency.

high-pass filter   A circuit that passes signals above a certain frequency and rejects signals below that frequency.

logarithm   An exponent of one number used to represent another number.

loss   A decrease in power; same as attenuation.

low-pass filter   A circuit that passes signals below a certain frequency and rejects signals above that frequency.

magnification factor   Same as Q.

oscillation   A vibration.

oscillator   A circuit that produces electrical waves at a specific frequency.

Q The ratio of reactance to resistance.

resonance   The tendency of a circuit to oscillate at a specific frequency.

resonant frequency   The frequency at which a circuit tends to oscillate, when X_L equals X_C.

ringing   A damped sine wave.

rolloff   The rate at which the output power level of a filter changes with frequency.

selectivity   The ability of a circuit to pass or reject a narrow frequency band.

tank circuit   A parallel resonant circuit.

tuned circuit   A circuit designed to have a resonant frequency.

waveform   The shape of a wave; the graph of voltage versus time as seen on an oscilloscope.

16

DC FREQUENCY RESPONSE

● SECTION OUTLINE

16–1 Pulses

16–2 Integration

16–3 Differentiation

16–4 Using Pulses to Test Circuits

● TROUBLESHOOTING:
Applying Your Knowledge

Your father tells you his company is having a lot of trouble with a computer system at work. It seems they decided to connect two microcomputers located in buildings half a mile apart. The plant electrician ran a telephone line between the two buildings, but the computers could not recognize each other's signal. A lab technician tested the hookup with an oscilloscope, and saw a clean pulse going into the line and a steady dc voltage coming out of the line. What do you think might be wrong?

● OVERVIEW AND LEARNING OBJECTIVES

Modern digital circuits such as computers use dc pulses as signals. Inductance and capacitance can alter the shapes of these pulses and, thus, can affect the accuracy and reliability of the circuits. In this chapter you will learn the details of how this can happen and how to recognize this type of problem.

After completing this chapter, you will be able to:

1. Predict response of a circuit to a periodic dc pulse based on inspection of schematics.

2. Recognize waveforms of dc integration and differentiation.

3. Describe how low-pass filters produce integration.

4. Describe how high-pass filters produce differentiation.

SECTION 16-1: Pulses

In this section you will learn the following:

→ A pulse is a waveform that changes from one dc voltage level to another.

→ The period of a pulse, symbolized T and measured in seconds, is the time required for one repetition of the pulse pattern.

→ The pulse repetition rate, symbolized PRR and measured in hertz (Hz), is the number of repetitions per second of the pulse pattern.

→ The active state of a pulse is the voltage level that represents a signal, causing another circuit to respond.

→ The resting state of a pulse is the voltage level that has no effect on another circuit or device.

→ The duty cycle of a pulse is the ratio of its active-state time to its period. Duty cycle is expressed as a percentage.

Pulses

The waveforms in Figure 16-1 are examples of a *pulse,* a waveform that changes quickly from one dc voltage level to another. A pulse may be single, repetitive, or irregular, as shown. In digital equipment such as computers, compact disc players, cellular telephones, and so on, the signals are usually pulses. Therefore, we must understand how pulses are affected by the electrical characteristics of resistance, inductance, and capacitance.

An ideal pulse remains at one dc voltage level for a certain length of time, changes instantly to another dc voltage level, remains at that level for a certain length of time, then changes instantly back to the original dc voltage level.

Resistance affects the ability of the pulse to reach the required voltage level. For example, a pulse intended to change from 0 V to +5 V may actually change from perhaps +0.2 V to +4.3 V because of voltage drops due to resistance.

Inductance and capacitance affect the ability of the pulse to change instantly and to remain at a steady voltage level. For example, a pulse may need a few microseconds to change levels, or it may gain or lose voltage after reaching the new level.

All these changes are classified as pulse-shaping effects. Sometimes we intentionally change the pulse shape, but much of the time the changes in pulse shape are unwanted distortions of the signal, which may result in faulty operation of digital equipment.

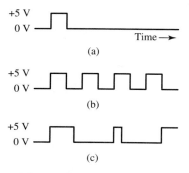

FIGURE 16-1
dc pulses.

In this section we will study the effects of *RL* and *RC* circuits on pulses. These are the same circuits which we studied as ac high-pass and low-pass filters in Chapter 15. Now we will examine their frequency response to dc repetitive pulses.

Pulse Repetition Rate

The *period* of a pulse, symbolized *T* and measured in seconds, is the time required for one repetition of the pulse pattern. The *pulse repetition rate,* symbolized PRR and measured in hertz (Hz), is the number of repetitions per second of the pulse pattern.

Formula Diagram 16–1 gives the relationship between period and pulse repetition rate. This is similar to the relationship between period and frequency in ac waves.

Formula Diagram 16–1

$$\frac{1}{T \mid \text{PRR}}$$

Duty Cycle

The *active state* of a pulse is the voltage level that represents a signal, causing another circuit to respond. The *resting state* of a pulse is the voltage level that causes the pulse to have no effect on another circuit or device. The *duty cycle* of a pulse is the ratio of its active state to its period. Duty cycle is expressed as a percentage.

Active state and resting state describe the effects of a pulse on a digital circuit. The pulse carries data or information to the circuits by changing states in some predetermined manner.

For example, many digital circuits need some sort of timing or synchronization signal to tell them when they may accept or transfer information. This timing signal is usually a repetitive pulse referred to as the clock pulse. When the pulse is at its active state (at one voltage level), the digital circuits are enabled (allowed to operate). Similarly, when the pulse is at its inactive state (at the other voltage level), the digital circuits are disabled (not allowed to operate).

Figure 16–2 illustrates these ideas. A pulse has a repetitive pattern, changing between voltage levels V_1 and V_2. It remains at V_1 for 9 μs, then remains at V_2 for 6 μs. Therefore, its period *T* equals 15 μs and its pulse repetition rate PRR equals 1/15 μs, which is 66.7 kHz. If we declare V_1 to be the active state, then the duty cycle equals 9 μs ÷ 15 μs, which equals 0.6 or 60%. On the other hand, if we declare V_2 the active state, then the duty cycle is 40% (6 μs ÷ 15 μs).

Formula 16–2 and Formula 16–3 express the relationships of active state, resting state, period, and duty cycle.

Formula 16–2

$$T = t_{\text{active}} + t_{\text{resting}}$$

Formula 16–3

$$\text{Duty cycle} = \frac{t_{\text{active}}}{T} \times 100$$

$$t_{\text{active}} = T \times \frac{\text{duty cycle}}{100}$$

$$T = \frac{t_{\text{active}} \times 100}{\text{duty cycle}}$$

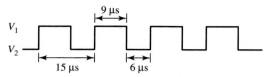

FIGURE 16–2
Relationships among period, active state,
resting state, and duty cycle.

● *SKILL-BUILDER 16–1*

A pulse remains at +5 V for 2.58 μs and at 0 V for 1.16 μs. If the active state is the +5-V
level, what is the period, pulse repetition rate, and duty cycle?

GIVEN:
$t_{active} = 2.58$ μs
$t_{resting} = 1.16$ μs

FIND:
T
PRR
duty cycle

SOLUTION:
$$T = t_{active} + t_{resting}$$
$$= 2.58 \text{ μs} + 1.16 \text{ μs}$$
$$= 3.74 \text{ μs}$$
$$PRR = 1 / T$$
$$= 1 / 3.74 \text{ μs}$$
$$= 267 \text{ kHz}$$

$$\text{Duty cycle} = \frac{t_{active}}{T} \times 100$$

$$= \frac{2.58 \text{ μs}}{3.74 \text{ μs}} \times 100$$

$$= 69\%$$

Repeat the Skill-Builder using the following values. You should obtain the results
that follow.

Given:

t_{active}	3.43 μs	0.20 μs	1.77 μs	0.75 μs	0.37 μs
$t_{resting}$	1.24 μs	1.79 μs	0.37 μs	0.59 μs	1.60 μs

Find:

T	4.67 μs	1.99 μs	2.14 μs	1.34 μs	1.97 μs
PRR	214 kHz	503 kHz	466 kHz	747 kHz	508 kHz
duty cycle	73%	10%	83%	56%	19%

Average Voltage of a Pulse

The average voltage of a waveform is the voltage that a dc voltmeter would read. The formula for average voltage depends on the waveform. In other words, each waveform
would have its own formula for average voltage.

The average voltage for a repetitive pulse is given by Formula 16–4.

Formula 16–4

$$V_{average} = \frac{\text{duty cycle} \times (V_{active} - V_{resting})}{100} + V_{resting}$$

● **SKILL-BUILDER 16–2**

As shown in Figure 16–3, an oscilloscope screen displays the waveform of a pulse that is active at –7 V and resting at –4 V, with a 65% duty cycle. What is the average voltage that a dc voltmeter would display? We assume that the frequency of the pulse is within the frequency-response range of the voltmeter, so that the meter's measurement will be accurate.

GIVEN:

$V_{active} = -7 \text{ V}$
$V_{resting} = -4 \text{ V}$
duty cycle = 65%

FIND:

$V_{average}$

SOLUTION:

$$V_{average} = \frac{\text{duty cycle} \times (V_{active} - V_{resting})}{100} + V_{resting}$$

$$= \frac{65 \times [(-7 \text{ V}) - (-4 \text{ V})]}{100} + (-4 \text{ V})$$

$$= \frac{65 \times [-7 \text{ V} + 4 \text{ V}]}{100} - 4 \text{ V}$$

$$= \frac{65 \times (-3 \text{ V})}{100} - 4 \text{ V}$$

$$= \frac{-195 \text{ V}}{100} - 4 \text{ V}$$

$$= -1.95 \text{ V} - 4 \text{ V}$$

$$= -5.95 \text{ V}$$

keystrokes:
7 [+/–]
[–] 4 [+/–]
[=]
[×] 65
[÷] 100 [=]
[+] 4 [+/–] [=]

Note: If a voltage value is positive, do not press the [+/–] key after entering the voltage value.

Repeat the Skill-Builder using the following values. You should obtain the results that follow.

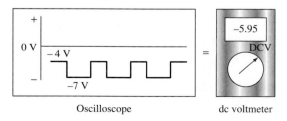

| Oscilloscope | dc voltmeter |

FIGURE 16–3
Indications of pulse in Skill-Builder 16–2.

Given:

V_{active}	8 V	−12 V	3.5 V	−6.8 V	9.3 V
$V_{resting}$	0 V	2.2 V	1.6 V	−10.4 V	14.7 V
duty cycle	23%	35%	87%	16%	42%

Find:

V_{avg}	1.84 V	−2.77 V	3.25 V	−9.82 V	12.43 V

- -

→ *Self-Test*
- - - - - - - -

1. What is a pulse?
2. What is the period of a pulse?
3. What is the pulse repetition rate?
4. What is the active state of a pulse?
5. What is the resting state of a pulse?
6. What is the duty cycle of a pulse?

- -

SECTION 16–2: Integration

In this section you will learn the following:

> → An integrator is a circuit that smooths the waveform of a pulse.
> → A low-pass filter acts as an integrator.

RL Integrator

An *integrator* is a circuit that smooths the waveform of a pulse. Figure 16–4 shows an *RL* low-pass filter used as an integrator.

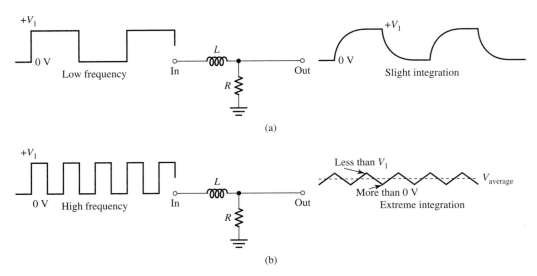

FIGURE 16–4
Slight and extreme integration produced by an *RL* circuit.

Electrical Analysis of the Waveform

When the input voltage rises sharply from zero to $+V_1$, the rapid change in current causes the coil's magnetic field to expand quickly. This induces a back voltage in the coil equal to $+V_1$ but opposite in polarity. Thus, at the moment the input rises from zero to $+V_1$, the coil's back voltage prevents current from flowing. There is no voltage drop across the resistor, and the output voltage remains at zero.

However, because the dc current in the coil is not attempting to change, the coil's induced back voltage dies down to zero. The time required for this is the transient time of the circuit, 5 τ. During the transient time, the coil's back voltage is falling from a value equal to $+V_1$ to a value of zero. Current begins to flow, and the voltage drop across the resistor begins to increase. After 5 τ, the coil has no voltage and almost zero dc impedance. For all practical purposes, the coil is simply a piece of wire connecting the input to the top of the resistor. Thus, the output voltage reaches $+V_1$.

When the input suddenly drops from $+V_1$ back to zero, the magnetic field around the coil begins to collapse, once again inducing a back voltage in the coil, but one of opposite polarity, which tends to keep the current flowing in the same direction. Thus, at the time the input drops suddenly to zero volts, the output remains momentarily at $+V_1$ and then falls to zero in a time of 5 τ. After 5 τ, the output remains at zero until the pulse changes state again.

Relationship Among Pulse Period, Circuit Transient Time, and Extent of Integration

To be clear in your thinking, consider the transient time (5 τ) as the time the reactive component *needs* to gain or lose all of its reactive voltage. Then think of the state time (time in active or resting state) as the time the reactive component *has* in order to gain or lose all of its voltage.

We emphasize the words *need* and *has* because the time the circuit needs is completely independent of the time the circuit has. Understanding this difference is the key to understanding how the pulse's waveform is changed by the circuit.

For example, if the transient time is shorter than the state time, then the reactive component (the coil or capacitor) needs less time than it has, and so it will gain or lose all of its voltage. Following this line of thinking, we reason as follows.

At lower frequencies, the active-state and resting-state times are longer than the transient time of the circuit. In this case, the magnetic field of the coil has enough time to fully expand and then fully collapse. As a result, the back voltage of the coil reaches zero before the pulse changes state. Thus, the output has enough time to reach the two input voltage levels. The integration is slight.

On the other hand, at higher frequencies, the active-state and resting-state times are shorter than the transient time of the circuit. In this case, the magnetic field of the coil does not have enough time to fully expand and then fully collapse. As a result, the back voltage of the coil cannot reach zero before the pulse changes states. Thus, the output cannot reach the two input voltage levels and tends toward the average voltage level. The integration is extreme.

If the pulse is a square wave (50% duty cycle), the active-state and resting-state times are equal to each other and to half the period. We will use the square wave for our analysis.

Visual Analysis of the Waveform

Visually, the leading and trailing edges of the pulse become rounded or smoothed, as shown in the output of Figure 16–4(a). This change is a distortion of the waveform which increases as the PRR increases. At some PRR the integration is so much that the output waveform can no longer reach either of the two input voltage levels. As the PRR continues to increase, the two output voltage levels become closer and closer to the average voltage of the input waveform. As shown in Figure 16–4(b), the output waveform begins to resemble a zigzag between two voltage levels. Eventually, the zigzag almost flattens out into a straight line at the average voltage.

The slight integration at a low frequency is a small distortion of the waveform, and the extreme integration at a high frequency is a large distortion. If the pulse carries some sort of information (data, timing, or control) to a digital circuit, then the large distortion of extreme integration could cause a problem: a loss of data or an error in circuit function.

From this point of view, we could say that the circuit passes low-frequency pulses (no loss of data or error in circuit function) and rejects high-frequency pulses (problems are likely to occur). Thus, the *RL* integrator responds to dc pulses in a manner similar to the way an *RL* low-pass filter responds to sinusoidal ac signals.

To determine whether the integration will be slight or extreme, we must compare the transient time of the circuit (5 τ) with half the period of the pulse. For a square wave (50% duty cycle), if half the period is longer than the transient time, the coil has more time than it needs for its back voltage to reach zero. Since the output is developed by the resistor, the output will reach the input voltage level early in each half-cycle. Therefore, the output waveform resembles the input, and the integration is slight.

On the other hand, if half the period is shorter than the transient time, the coil has less time than it needs for its back voltage to reach zero. Since some back voltage remains in the coil at all times, the resistor cannot reach the input voltage levels. Therefore, the output also cannot reach the input levels. The output waveform does not resemble the input, and the integration is extreme.

● *SKILL-BUILDER 16–3*

Suppose that in the circuit of Figure 16–4 the resistance is 480 Ω, the inductance is 4.5 mH, and the pulse is a square wave with a PRR of 339.4 kHz. Predict whether the integration will be slight or extreme.

GIVEN:

$R = 480 \; \Omega$

$L = 4.5 \; \text{mH}$

$\text{PRR} = 339.4 \; \text{kHz}$

FIND:

$5 \, \tau$ Transient time: the time the output needs to reach the input voltage levels.

$\dfrac{T}{2}$ Half the period: the time the output has to reach the input voltage levels.

SOLUTION:

$$5 \, \tau = 5 \times \frac{L}{R}$$ Formula 12–4

$$= 5 \times \frac{4.5 \; \text{mH}}{480 \; \Omega}$$

$$= 46.88 \; \mu\text{s}$$

$$T = \frac{1}{339.4 \text{ kHz}}$$

$$= 2.95 \text{ μs}$$

$$\frac{1}{2} T = 1.48 \text{ μs}$$

ANALYSIS:

The integrator output needs 46.88 μs to reach the voltage levels of the input. It has 1.48 μs to do this, which is not enough time. The conclusion is that the output cannot reach the levels of the input. The output waveform has a large amount of distortion, and the integration is extreme.

Repeat the Skill-Builder using the following values. You should obtain the results that follow.

Given:

R	530 Ω	330 Ω	650 Ω	580 Ω
L	4.5 mH	3.6 mH	0.7 mH	2.1 mH
PRR	1.8 kHz	340.1 kHz	12.4 kHz	873.8 kHz

Find:

5 τ	42.45 μs	54.55 μs	5.38 μs	18.1 μs
$\frac{1}{2} T$	277.78 μs	1.47 μs	40.33 μs	0.57 μs

Integration:

	slight	extreme	slight	extreme

- -

RC Integrator

Figure 16–5 shows an *RC* low-pass filter used as an integrator.

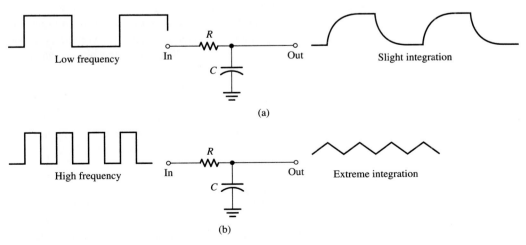

FIGURE 16–5
Slight and extreme integration produced by an *RC* circuit.

Electrical Analysis of the Waveform

Assume that the capacitor is initially discharged. This means that the plate voltage of the capacitor is zero. Thus, the output voltage is also zero. When the input voltage rises sharply from zero to $+V_1$, the capacitor needs time to charge. In fact, the capacitor needs exactly $5\,\tau$ (the transient time) to reach $+V_1$. Thus, the output voltage level begins to rise from zero volts as the capacitor charges. After $5\,\tau$, the capacitor is fully charged. The output reaches $+V_1$ and remains there until the pulse changes state.

When the input suddenly drops from $+V_1$ back to zero, the capacitor begins to discharge. At the time the input drops suddenly to zero volts, the output remains momentarily at $+V_1$ and then falls to zero in a time of $5\,\tau$. After $5\,\tau$, the output remains at zero until the pulse changes state again.

Relationships Among Pulse Period, Circuit Transient Time, and Extent of Integration

At lower frequencies the period of the pulse is longer than the transient time of the circuit. In this case, the circuit has more time than it needs to reach steady state. In other words, the capacitor has enough time to fully charge and then fully discharge. As a result, the plate voltage of the capacitor reaches either $+V_1$ or zero before the pulse changes state. The output has enough time to reach the two input voltage levels. The integration is slight.

On the other hand, at higher frequencies the period of the pulse is shorter than the transient time of the circuit. In this case, the circuit has less time than it needs to reach steady state. In other words, the capacitor does not have enough time to fully charge and then fully discharge. As a result, the plate voltage of the capacitor cannot reach either $+V_1$ or zero before the pulse changes states. Thus, the output cannot reach the two input voltage levels and tends toward the average voltage level. The integration is extreme.

Visual Analysis of the Waveform

The visual analysis of the output waveform of this *RC* integrator is identical to the waveform analysis just conducted for the *RL* integrator of Figure 16–4. Thus, the *RC* integrator responds to pulses in a manner similar to the way an *RC* low-pass filter responds to sinusoidal signals.

To determine whether the integration will be slight or extreme, we apply standards similar to those we used with the *RL* integrator. For a square wave (50% duty cycle), if half the period is longer than the transient time, the capacitor has more time than it needs to charge and discharge completely. Since the capacitor voltage is the output, the output reaches both of the input voltage levels. Therefore, the output waveform resembles the input, and the integration is slight.

On the other hand, if half the period is shorter than the transient time, the capacitor has less time than it needs to charge and discharge completely. The voltage on the plates of the capacitor varies slightly above and below the average voltage of the input pulse. Therefore, the output waveform does not resemble the input, and the integration is extreme.

● *SKILL-BUILDER 16–4* –

Suppose that in the circuit of Figure 16–5 the resistance is 2.7 kΩ, the capacitance is 0.0015 μF, and the pulse is a square wave with a PRR of 4 kHz. Predict whether the integration will be slight or extreme.

GIVEN:
  $R = 2.7 \text{ k}\Omega$
  $C = 0.0015 \text{ }\mu\text{F}$
  $\text{PRR} = 4 \text{ kHz}$

FIND: Transient time: the time the output needs
  $5\,\tau$ to reach the input voltage levels.

  $\dfrac{T}{2}$ Half the period: the time the output has to
                                     reach the input voltage levels.

SOLUTION:
  $5\,\tau = 5 \times R \times C$ Formula 13–6
  $\quad = 5 \times 2.7 \text{ k}\Omega \times 0.0015 \text{ }\mu\text{F}$
  $\quad = 20.25 \text{ }\mu\text{s}$

  $T = \dfrac{1}{4 \text{ kHz}}$

  $\quad = 250 \text{ }\mu\text{s}$

  $\dfrac{1}{2}\,T = 125 \text{ }\mu\text{s}$

ANALYSIS:

The integrator output needs 20.25 μs to reach the voltage levels of the input. It has 125 μs to do this, which is more than enough time. The conclusion is that the output reaches the levels of the input. The output waveform has little distortion and the integration is slight.

Repeat the Skill-Builder using the following values. You should obtain the results that follow.

Given:

R	2.4 kΩ	4.5 kΩ	1.6 kΩ	6.0 kΩ
C	0.0010 μF	0.0052 μF	0.0030 μF	0.0053 μF
PRR	5 kHz	144 kHz	2 kHz	107 kHz

Find:

$5\,\tau$	12 μs	117 μs	24 μs	159 μs
$\dfrac{1}{2}\,T$	100 μs	3.47 μs	250 μs	4.68 μs
Integration:	slight	extreme	slight	extreme

→ *Self-Test*

7. What is an integrator?
8. What circuit acts as an integrator?

SECTION 16–3: Differentiation

In this section you will learn the following:

→ A differentiator is a circuit that sharpens the waveform of a pulse.
→ A high-pass filter acts as a differentiator.

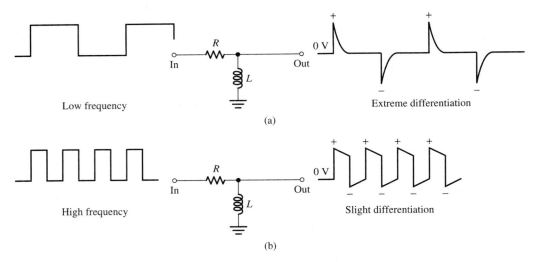

FIGURE 16–6

Extreme and slight differentiation produced by an *RL* circuit.

RL Differentiator

A *differentiator* is a circuit that sharpens the waveform of a pulse. Figure 16–6 shows an *RL* high-pass filter used as a differentiator.

Electrical Analysis of the Waveform

When the input voltage rises sharply from zero to $+V_1$, the rapid change in current causes the coil's magnetic field to expand quickly. This induces a back voltage in the coil equal to $+V_1$ but opposite in polarity. Since the output voltage is identical to the back voltage of the coil, and since the back voltage equals $+V_1$ at the moment the pulse changes state, the output voltage rises quickly to a value equal to $+V_1$. Thus, the edge of the output waveform appears the same as the edge of the input waveform.

However, because the dc current is not attempting to change, the coil's induced back voltage falls to zero. The time required for this is the transient time of the circuit: $5\ \tau$. During the transient time, the coil's back voltage (which is also the output voltage) is falling from a value equal to $+V_1$ to a value of zero. After $5\ \tau$, the coil has no voltage and almost zero dc impedance. For all practical purposes, the coil is simply a piece of wire connecting the output to ground. Thus, the output reaches zero and remains there until the pulse changes state.

When the input suddenly drops from $+V_1$ back to zero, the magnetic field around the coil begins to collapse, once again inducing a back voltage in the coil, but of opposite polarity, which tends to keep the current flowing in the same direction. Since the top of the coil had a positive polarity before and has an opposite polarity now, the top of the coil must now be negative with respect to ground. This is consistent with the principle that the coil is now trying to keep current flowing in the same direction. It is the reversal of the coil's polarity that produces the negative spike in the output waveform.

Thus, at the time the input drops suddenly to zero volts, the output rises quickly to a voltage equal to $-V_1$ and then falls to zero in a time of $5\ \tau$. After $5\ \tau$, the output remains at zero until the pulse changes state again.

Relationships Among Pulse Period, Circuit Transient Time, and Extent of Differentiation

At lower frequencies, the period of the pulse is longer than the transient time of the circuit. In this case, the magnetic field of the coil has enough time to fully expand and then fully collapse. As a result, the back voltage of the coil reaches zero before the pulse changes state. Thus, the output has enough time to reach zero volts whenever the pulse changes state. The differentiation is extreme.

On the other hand, at higher frequencies, the period of the pulse is shorter than the transient time of the circuit. In this case, the magnetic field of the coil does not have enough time to fully expand and then fully collapse. As a result, the back voltage of the coil cannot reach zero before the pulse changes states. Thus, the output voltage cannot reach zero on the trailing edge of the pulse, and the negative spike is small. The differentiation is slight.

As shown in Figure 16–7, the trailing edge of the output waveform of a differentiator falls from V_1 to a lower level V_2. The value of the negative spike (V_3) is simply the difference between V_1 and V_2. The magnetic field is only partly expanded, and V_3 represents the back voltage in the coil at the moment the pulse changes state.

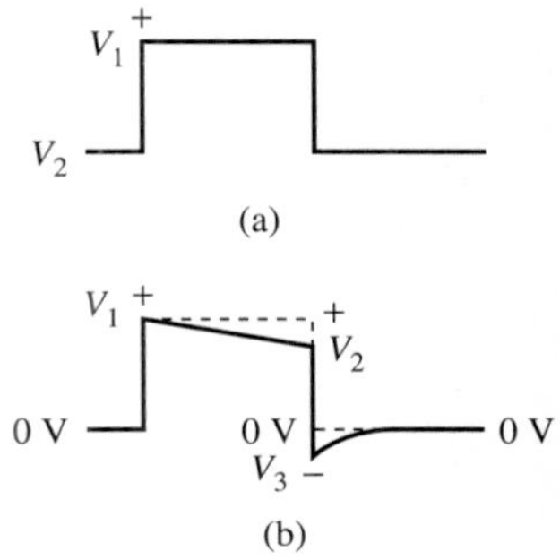

FIGURE 16–7
The negative spike equals the difference between initial voltage and final voltage.

Visual Analysis of the Waveform

Visually, the leading and trailing edges of the pulse are maintained, but the top of the pulse is distorted as the output waveform quickly progresses toward zero, as shown in the output of Figure 16–6(a). This distortion decreases as the PRR increases. At some PRR the output waveform can no longer reach zero. As the PRR continues to increase, the output waveform becomes closer and closer to the two voltage levels of the input waveform. As shown in Figure 16–6(b), the output waveform begins to resemble the input pulse. Eventually, the differentiation is so small that it no longer matters.

The slight differentiation at a high frequency is a small distortion of the waveform, and the extreme differentiation at a low frequency is a large distortion. Any extreme distortion can cause loss of data or errors in function. From this point of view, we could say that the circuit passes high-frequency pulses and rejects low-frequency pulses. Thus, the *RL* differentiator responds to pulses in a manner similar to the way an *RL* high-pass filter responds to sinusoidal signals.

To determine whether the differentiation will be slight or extreme, we must compare the transient time of the circuit (5 τ) with the period of the pulse. For a square wave (50% duty cycle), if half the period is longer than the transient time, the coil has more time than it needs for its back voltage to reach zero. Since the output is developed by the coil, the output will fall to zero early in each half-cycle. Therefore, the output waveform does not resemble the input, and the differentiation is extreme.

On the other hand, if half the period is shorter than the transient time, the coil has less time than it needs for its back voltage to reach zero. If the back voltage in the coil is still large when the input changes state, the magnetic field is still small and has not had time to expand very much. Therefore, the back voltage will be small as the small field collapses. Thus, with the back voltage first remaining large and then dropping to nearly zero, the output levels are approximately the same as the input levels. The output waveform resembles the input, and the differentiation is slight.

● *SKILL-BUILDER 16–5*

Suppose that in the circuit of Figure 16–6 the resistance is 550 Ω, the inductance is 3.1 mH, and the pulse is a square wave with a PRR of 561.4 kHz. Predict whether the differentiation will be slight or extreme.

GIVEN:

$$R = 550 \ \Omega$$
$$L = 3.1 \ \text{mH}$$
$$\text{PRR} = 561.4 \ \text{kHz}$$

FIND:

$$5 \ \tau$$
$$\frac{T}{2}$$

Transient time: the time the output needs to reach the input voltage levels.

Half the period: the time the output has to reach the input voltage levels.

SOLUTION:

$$5 \ \tau = 5 \times \frac{L}{R}$$

$$= 5 \times \frac{3.1 \ \text{mH}}{550 \ \Omega}$$

$$= 28.18 \ \mu s$$

$$T = \frac{1}{561.4 \ \text{kHz}}$$

$$= 1.78 \ \mu s$$

$$\frac{1}{2} \ T = 0.89 \ \mu s$$

ANALYSIS:

The differentiator output needs 28.18 μs to reach zero. It has 0.89 μs to do this, which is not enough time. The conclusion is that the output cannot reach zero during the active state. The output waveform has a small amount of distortion, and the differentiation is slight.

Repeat the Skill-Builder using the following values. You should obtain the results that follow.

Given:

R	360 Ω	330 Ω	610 Ω	190 Ω
L	2.7 mH	2.9 mH	3.5 mH	0.6 mH
PRR	421 kHz	391 kHz	2.1 kHz	3.7 kHz

Find:

$5 \ \tau$	37.5 μs	43.94 μs	28.69 μs	15.79 μs
$\frac{1}{2} \ T$	1.19 μs	1.28 μs	238.1 μs	135.14 μs
Differentiation:	slight	slight	extreme	extreme

RC Differentiator

Figure 16–8 shows an *RC* high-pass filter used as a differentiator.

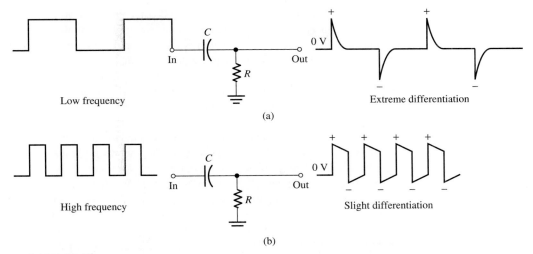

FIGURE 16–8
Extreme and slight differentiation produced by an *RC* circuit.

Electrical Analysis of the Waveform

Suppose that the capacitor is discharged when the input voltage rises sharply from zero to $+V_1$. Remember that, as shown in Figure 16–9, a pulse generator (a dc voltage source) is connected between the input and ground. Therefore, the capacitor will be charged by this input voltage source. If the input is positive with respect to ground, then current will flow from the voltage source into ground as shown in Figure 16–9(a) and then out of ground toward the capacitor through the resistor. If current is flowing into the resistor from ground, then the top of the resistor is positive with respect to ground. In other words, since the output voltage is equal to the voltage drop of the resistor, the polarity of the output voltage will be the same as the polarity of the top of the resistor with respect to ground.

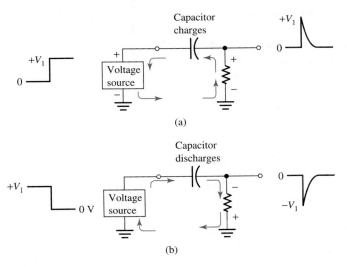

FIGURE 16–9
Output polarity of *RC* differentiator.

According to the Fourier Series, a sudden change in voltage is mathematically equal to a very high frequency. Given this fact, we may reason as follows: At a very high frequency, the reactance of the capacitor is almost zero ohms, and the capacitor acts like a short. Therefore, at the moment the pulse changes state, the resistor appears to be the only impedance in the circuit, and $+V_1$, the full voltage of the source, is dropped across the resistor. Thus, when the pulse changes state, the output voltage rises quickly to a value equal to $+V_1$, and the edge of the output waveform appears the same as the edge of the input waveform.

However, as the capacitor begins to charge, the increasing voltage on its plates is in series opposing polarity to the voltage source. The increasing voltage and opposing polarity of the capacitor reduce the current through the resistor. This reduces the resistor's voltage drop, which appears as the output voltage. Thus, the output voltage drops quickly, reaching zero when the capacitor is fully charged and no more current is flowing through the resistor.

When the input suddenly drops from $+V_1$ back to zero, the capacitor begins to discharge as shown in Figure 16–9(b). The internal resistance of a practical voltage source is nearly zero; thus, we can assume that now the voltage source acts like a short. Therefore, the full plate voltage of the capacitor is initially dropped across the resistor. The current is now flowing from the right plate of the capacitor through the top of the resistor into ground, and then out of ground and through the source to the left plate of the capacitor. It is the reversal of the current flow through the resistor during discharge that produces the negative spike in the output waveform.

Thus, at the time the input drops suddenly to zero volts, the output rises quickly to a voltage equal to $-V_1$ and then falls to zero in a time of $5\ \tau$. After $5\ \tau$, the output remains at zero until the pulse changes state again.

Relationships Among Pulse Period, Circuit Transient Time, and Extent of Differentiation

At lower frequencies, the period of the pulse is longer than the transient time of the circuit. In this case the capacitor has enough time to fully charge and then fully discharge. As a result, voltage drop across the resistor reaches zero before the pulse changes state. Thus, the output has enough time to reach zero volts whenever the pulse changes state. The differentiation is extreme.

On the other hand, at higher frequencies, the period of the pulse is shorter than the transient time of the circuit. In this case the capacitor does not have enough time to fully charge and then fully discharge. As a result, voltage drop across the resistor cannot reach zero before the pulse changes states. Thus, the output voltage cannot reach zero on the trailing edge of the pulse, and the negative spike is small. The differentiation is slight.

Visual Analysis of the Waveform

The visual analysis of the waveform is the same as for the *RL* differentiator. The leading and trailing edges of the pulse are maintained, but the top of the pulse is distorted as the output waveform quickly progresses toward zero, as shown in the output of Figure 16–8(a). This distortion decreases with the PRR. At some PRR the output waveform can no longer reach zero. As the PRR continues to increase, the output waveform becomes closer and closer to the two voltage levels of the input waveform. As shown in Figure 16–8(b), the output waveform begins to resemble the input pulse. Eventually, the differentiation is so small that it no longer matters.

The slight differentiation at a high frequency is a small distortion of the waveform, and the extreme differentiation at a low frequency is a large distortion. Any extreme distortion can cause loss of data or errors in function. From this point of view, we could say that the circuit passes high-frequency pulses and rejects low-frequency pulses. Thus, the *RC* differentiator responds to pulses in a manner similar to the way an *RC* high-pass filter responds to sinusoidal signals.

To determine whether the differentiation will be slight or extreme, we apply standards similar to those we used with the *RL* differentiator. For a square wave (50% duty cycle), if half the period is longer than the transient time, the capacitor has more time than it needs to charge and discharge completely. This means that the current through the resistor stops flowing early in the half-cycle, and the voltage drop across the resistor reaches zero. Since the resistor develops the output voltage, the output waveform does not resemble the input, and the differentiation is extreme.

On the other hand, if half the period is shorter than the transient time, the capacitor has less time than it needs to charge and discharge completely. Because a substantial current is flowing through the resistor during charge and only a slight current during discharge, the voltage drops across the resistor are nearly the same as the input voltage levels. The output waveform resembles the input, and the differentiation is slight.

● *SKILL-BUILDER 16–6*

Suppose that in the circuit of Figure 16–8 the resistance is 4.9 kΩ, the capacitance is 0.0059 μF, and the pulse is a square wave with a PRR of 95 kHz. Predict whether the differentiation will be slight or extreme.

GIVEN:
$R = 4.9 \text{ k}\Omega$
$C = 0.0059 \text{ μF}$
$PRR = 95 \text{ kHz}$

FIND:
5τ

$\dfrac{T}{2}$

Transient time: the time the output needs to reach the input voltage levels.

Half the period: the time the output has to reach the input voltage levels.

SOLUTION:
$5\tau = 5 \times R \times C$
$\quad = 5 \times 4.9 \text{ k}\Omega \times 0.0059 \text{ μF}$
$\quad = 144.55 \text{ μs}$

$T = \dfrac{1}{95 \text{ kHz}}$

$\quad = 10.53 \text{ μs}$

$\dfrac{1}{2}T = 5.27 \text{ μs}$

ANALYSIS:
The differentiator output needs 144.55 μs to reach zero. It has 5.27 μs to do this, which is not enough time. The conclusion is that the output does not reach zero during the active state. The output waveform has little distortion and the differentiation is slight.

Repeat the Skill-Builder using the following values. You should obtain the results that follow.

Given:

R	1.7 kΩ	3.5 kΩ	1.5 kΩ	1.3 kΩ
C	0.0056 µF	0.0030 µF	0.0038 µF	0.0028 µF
PRR	274 kHz	293 kHz	2 kHz	3 kHz

Find:

5τ	47.6 µs	52.5 µs	28.5 µs	18.2 µs
$\dfrac{1}{2}T$	1.83 µs	1.71 µs	250 µs	333 µs

Differentiation:

slight	slight	extreme	extreme

→ *Self-Test*

9. What is a differentiator?
10. What circuit acts as a differentiator?

SECTION 16–4: Using Pulses to Test Circuits

In this section you will learn the following:

→ The voltage levels of a dc pulse are equivalent to a band of low-frequency ac waves.

→ The voltage transitions of a dc pulse are equivalent to a band of high-frequency ac waves.

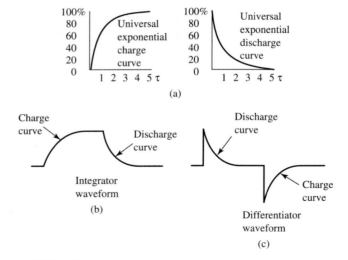

(a)

(b)

(c)

FIGURE 16–10
Integrator and differentiator waveforms have the same shape of universal exponential charge and discharge curves.

Relationship Between Integrator and Differentiator Waveforms and Universal Exponential Curves

When we studied the dc characteristics of inductors and capacitors, we learned that the graph of the back voltage of a coil during the expansion and collapse of the coil's magnetic field is identical to the graph of the plate voltage of a capacitor during charge and discharge. Both of these graphs follow a precise mathematical pattern called the universal exponential curve. These graphs are shown in Figure 16–10(a).

As shown in Figure 16–10(b) and (c), the curved sections of integrator and differentiator waveforms are identical to the universal exponential curve for the simple reason that the waveforms are produced by the charge and discharge of inductor and capacitor voltages.

The two curves are mirror images of each other. In the *RL* and *RC* circuits we have just studied, during the transient time of the circuit (the interval of 5 τ after the input changes state), the graph of the voltage across the coil or capacitor will follow one of these curves, and the graph of the voltage across the resistor will follow the other curve. In fact, because 5 τ is typically a short interval of time, a good way to observe these curves on an oscilloscope is by applying a square wave input pulse at an appropriate frequency and examining these sections of the output waveform.

Relationship Between Regions of the Waveform and Frequencies Within the Fourier Series

The Fourier Series tells us that any waveform, including a dc pulse, can be created by mixing together sine waves of various harmonic frequencies. Therefore, a dc square wave can be regarded as a mixture of ac sine waves of many frequencies.

When we analyze an integrator, we notice two things. First, a dc integrator is the same circuit as an ac low-pass filter. Second, an integrator distorts the leading and trailing edges of the pulse. The tops and bottoms of the waveform remain undistorted until the frequency becomes very high.

When we analyze a differentiator, we notice that it is the same as a high-pass filter. We also notice that it distorts the top and bottom of the pulse and does not distort the edges.

From this observation and our knowledge of the Fourier Series, we realize that the upper and lower voltage levels of the pulse are equivalent to a band of low-frequency ac waves. Similarly, the leading and trailing edges (voltage transitions) of the pulse are equivalent to a band of high-frequency ac waves. Technicians take advantage of these relationships when they use a square wave as a test signal for audio equipment. By observing the distortions the circuit may produce on the edges and tops of the waveform, the technician understands how the circuit would respond to low- and high-frequency sine waves. These relationships are illustrated in Figure 16–11.

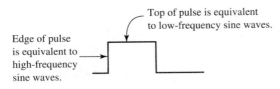

FIGURE 16–11
Voltage levels (top of pulse) consist of low-frequency harmonics. Voltage transitions (edges) consist of high-frequency harmonics.

Applications of Integration and Differentiation

Integration and differentiation are natural features of the dc pulse response of *RL* and *RC* circuits. Applying these features in a useful way depends on both the aim and the ingenuity of a circuit designer. For example, if an unusual duty cycle is required, it may be easier to integrate or differentiate a square wave than to redesign the oscillator. The differentiator can be particularly useful, since it produces both positive and negative spikes.

The distributed resistance, inductance, and capacitance of a long data transmission line may cause it to act as an integrator or differentiator to the digital signal pulses traveling in the line. This could cause a communication failure between two computers, for example. Understanding the process of integration and differentiation may help the technician understand and correct the trouble.

→ *Self-Test*

11. What is the equivalent of the voltage levels of a dc pulse?
12. What is the equivalent of the voltage transitions of a dc pulse?

Formula List

Formula Diagram 16–1

$$\frac{1}{T \mid \text{PRR}}$$

Formula 16–2

$$T = t_{\text{active}} + t_{\text{resting}}$$

Formula 16–3

$$\text{Duty cycle} = \frac{t_{\text{active}}}{T} \times 100$$

$$t_{\text{active}} = T \times \frac{\text{duty cycle}}{100}$$

$$T = \frac{t_{\text{active}} \times 100}{\text{duty cycle}}$$

Formula 16–4

$$V_{\text{average}} = \frac{\text{duty cycle} \times (V_{\text{active}} - V_{\text{resting}})}{100} + V_{\text{resting}}$$

● TROUBLESHOOTING: Applying Your Knowledge

The dc pulse going into the transmission line is turning into a nearly steady dc voltage because the inductance and/or capacitance in the line is integrating the signal. The telephone line is intended for low-frequency voice signals and is not the proper type for high-frequency digital data and should be replaced.

CHAPTER SUMMARY: ANSWERS TO SELF-TESTS

1. A pulse is a waveform that changes from one dc voltage level to another.
2. The period of a pulse, symbolized *T* and measured in seconds, is the time required for one repetition of the pulse pattern.
3. The pulse repetition rate, symbolized PRR and measured in hertz (Hz), is the number of repetitions per second of the pulse pattern.
4. The active state of a pulse is the voltage level that represents a signal, causing another circuit to respond.
5. The resting state of a pulse is the voltage level that has no effect on another circuit or device.
6. The duty cycle of a pulse is the ratio of its active state to its period. Duty cycle is expressed as a percentage.
7. An integrator is a circuit that smooths the waveform of a pulse.
8. A low-pass filter acts as an integrator.
9. A differentiator is a circuit that sharpens the waveform of a pulse.
10. A high-pass filter acts as a differentiator.
11. The voltage levels of a dc pulse are equivalent to a band of low-frequency ac waves.
12. The voltage transitions of a dc pulse are equivalent to a band of high-frequency ac waves.

CHAPTER REVIEW QUESTIONS

Determine whether the following questions are true or false.

16–1. A pulse is a waveform that changes from one ac voltage level to another.
16–2. The duty cycle of a pulse is the ratio of its active state to its period.

Select the response that best answers the question or completes the statement.

16–3. The pulse repetition rate
   a. is the dc equivalent of period.
   b. is the number of repetitions per second of the pulse pattern.
   c. equals the reciprocal of the duty cycle.
   d. equals the reciprocal of bandwidth.
16–4. An integrator
   a. smooths the waveform of a pulse.
   b. sharpens the waveform of a pulse.
   c. is a band-pass filter.
   d. is a band-reject filter.
16–5. A differentiator
   a. smooths the waveform of a pulse.
   b. sharpens the waveform of a pulse.
   c. is a band-pass filter.
   d. is a band-reject filter.

16–6. A low-pass filter also acts as
   a. a differentiator.
   b. a resonator.
   c. an integrator.
   d. an oscillator.
16–7. A high-pass filter also acts as
   a. a differentiator.
   b. a resonator.
   c. an integrator.
   d. an oscillator.
16–8. A technician uses a dc square wave as a test signal for an audio amplifier. She does this because
   a. an audio signal is a square wave.
   b. a square wave is equivalent to a band of low-frequency ac waves and a band of high-frequency ac waves.
   c. an audio amplifier acts as an integrator.
   d. an audio amplifier acts as a differentiator.

Solve the following problems.

16–9. A pulse remains at 0 V for 4.32 μs and at −5 V for 7.46 μs. If the active state is the −5-V level, what is the period, pulse repetition rate, and duty cycle?
16–10. An oscilloscope screen displays the waveform of a pulse that is active at +6 V and resting at −2 V, with a 75% duty cycle. What is the average voltage that a dc voltmeter would display?

The following problems are more challenging.

16–11. An *RL* circuit consists of a 2-kΩ resistor in series with a 5.5-mH coil. If the circuit is connected as an integrator and a 3-kHz square wave is applied to the input, will the output wave have slight or severe distortion?
16–12. An *RC* circuit consists of a 680-Ω resistor in series with a 0.022-μF capacitor. If the circuit is connected as a differentiator and a 60-kHz square wave is applied to the input, will the output wave have slight or severe distortion?

GLOSSARY

active state   The voltage level of a pulse that represents a signal, causing another circuit to respond.
differentiator   A circuit that sharpens the waveform of a pulse.
duty cycle   The ratio of active state to period.
integrator   A circuit that smooths the waveform of a pulse.
period   Time required for one repetition of the pulse pattern.
pulse   A waveform that changes from one dc voltage level to another.
pulse repetition rate   The number of repetitions per second of the pulse pattern.
resting state   The voltage level of a pulse that has no effect on another circuit or device.

17

TRANSFORMERS

● SECTION OUTLINE

17–1   Mutual Inductance

17–2   Basic Transformer Characteristics

17–3   Transformer Applications

17–4   Transformer Variations

17–5   Troubleshooting Transformers

● TROUBLESHOOTING:
Applying Your Knowledge

Your friend sings in a band. He complains that his new microphone does not produce as much volume as his old one, although it worked fine when he tried it out at the music store. He wonders if his sound system amplifier may be going bad. What do you think?

● OVERVIEW AND LEARNING OBJECTIVES

Transformers use induction to change voltage without wasting power. They are used throughout ac power systems and are found in the power supply of every electronic appliance that operates on ac power. They have several other important applications, including tuned and frequency-selective circuits.

After completing this chapter, you will be able to:

1. Describe mutual inductance, the basis of transformer operation.
2. Describe the basic construction of a transformer.
3. Explain the relationship between turns ratio and the values of voltage and current on both sides of a transformer.
4. Explain reflected impedance and the use of a transformer for impedance matching.
5. Describe the way in which an isolation transformer protects service personnel from shock hazard.
6. Describe the construction and applications of multiple-tap and multiple-winding transformers.

SECTION 17–1: Mutual Inductance

In this section you will learn the following:

→ Mutual inductance is the ability of one circuit to transfer magnetic energy into another circuit.

→ Mutual inductance produces transformer action, the process where current in one coil produces magnetism, which produces voltage in another coil.

→ The coefficient of coupling (k) is the part of the flux created by the primary winding that induces voltage in the secondary winding.

We have learned that when current flows through a coil, a magnetic field develops around the coil. Energy is stored in the magnetic field. When the current increases, the magnetic field draws more energy from the circuit, and when the current decreases, the magnetic field returns energy to the circuit. The back voltage of the coil is the means by which energy is conserved during the transfer between the circuit and the magnetic field. The ability of a coil to create a back voltage within itself is inductance (L), measured in henries. We have already studied inductance in considerable detail.

However, when another coil in a separate circuit is nearby, as shown in Figure 17–1, the situation changes. It is now possible for the magnetic energy to transfer into the other circuit. In other words, instead of the magnetic energy creating a back voltage in the first coil, it can create a voltage in the second coil. This voltage can then produce current in a load in the second circuit, thus transferring the magnetic energy. This form of inductance is called *mutual inductance,* symbolized L_M. Mutual inductance is the principle that produces transformer action, the process where current in one coil produces magnetism, which produces voltage in another coil.

A basic transformer consists of two windings and an iron core. Current in the first winding creates voltage in the second winding through mutual inductance. Each winding is a coil of wire, wrapped into many turns. The windings are wrapped around the same iron core. Current in the first winding creates magnetic flux, which travels through the core and induces voltage in the second winding.

As shown in Figure 17-2, the transformer winding connected to the ac voltage source is called the *primary winding.* The other winding, connected to the load, is called the *secondary winding.* Thus, we can also speak of primary voltage and current, and sec-

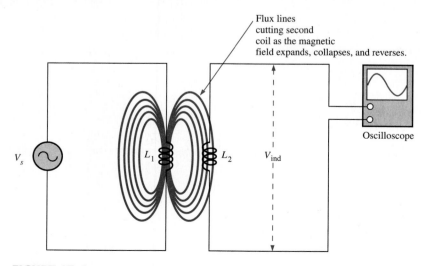

FIGURE 17–1
Mutual inductance produces transformer action. As current changes in the first winding, magnetic flux induces a voltage in the second winding.

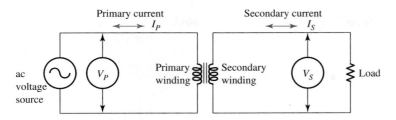

FIGURE 17–2
The primary winding is connected toward the voltage source.
The secondary winding is connected toward the load.

ondary voltage and current. Notice that the primary winding acts as a load to the voltage source, and the secondary winding acts as a voltage source to the load.

Coefficient of Coupling

As shown in Formula 17–1, the amount of mutual inductance depends on three factors: the inductance of the two windings $(L_1$ and $L_2)$, and a value called the *coefficient of coupling,* symbolized k.

The coefficient of coupling is the part of the flux created by the primary winding that induces voltage in the secondary winding. If the secondary winding is exposed to all the magnetic flux created by the first winding, then k has a value of 1, which means 100%. A fractional value of k indicates the percentage of flux that links (flows between and thus connects) the two windings.

Formula 17–1

$$L_M = k \sqrt{L_1 L_2}$$

Much of the effort in transformer design is devoted to improving the magnetic coupling between the windings to bring k closer to an ideal value of 1. Remember that when we say the two windings are linked or coupled by the magnetic flux, we are not speaking of physical connections but of an invisible magnetic connection. Both windings are surrounded by the same magnetic field, and this allows energy to pass from one winding to the other.

Core Type

The core material affects the amount of coupling between the two windings. The three commonly used core materials are iron, ferrite, and air. Ferrite is a ceramic containing powdered iron. Ferrite and air-core transformers are generally small, low-power, high-frequency transformers. Iron is the core material for most transformers. Figure 17–3 shows the schematic symbols for these three transformer types. Figure 17–4 shows several small iron-core transformers of the type commonly used in electronic equipment.

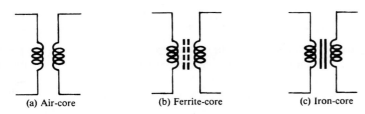

(a) Air-core   (b) Ferrite-core   (c) Iron-core

FIGURE 17–3
Transformer schematic symbols indicate the type of core.

FIGURE 17–4
Typical small iron-core transformers.

→ *Self-Test*

1. What is mutual inductance?
2. What is transformer action?
3. What is the coefficient of coupling?

SECTION 17–2: Basic Transformer Characteristics

In this section you will learn the following:

- → A transformer changes voltage without wasting power.
- → The turns ratio of a transformer is the number of primary turns divided by the number of secondary turns.
- → The voltage ratio of a transformer is equal to the turns ratio.
- → In an ideal transformer, primary power equals secondary power.
- → The current ratio of a transformer is the reciprocal of the turns ratio.
- → Reflected impedance is the apparent ohmic value of the primary winding as seen by the voltage source.
- → The maximum amount of power is transferred from a source to a load when the input impedance of the load equals the output impedance of the source.
- → Impedance matching is the process of making Z_{out} and Z_{in} equal at every power transfer point in an interconnected system, producing maximum power transfer.

A transformer is an ac device. In other words, a dc current will not produce transformer action for the same reason that a dc current will not produce self-inductance (back voltage) in a coil. Induced voltage occurs only when a magnetic field changes.

However, it is possible to produce a brief transformer action with dc current by using a switch to turn on and off the current flowing in the first winding. This is how an automobile ignition coil works. The points in the distributor are actually a switch that controls a dc current in the low-voltage winding of the ignition coil. When the points open, the collapsing magnetic field in the coil produces a brief high voltage (18 to 20 kV) in the other winding. This is the voltage that produces a spark in the spark plugs. Thus, an ignition coil is really a transformer.

Basic Function of a Transformer

A transformer changes voltage without wasting power. First, we will see how a transformer changes voltage. Then we will see how it does this without wasting power.

Turns Ratio

Each loop of wire in a winding is called a turn. We wrap windings into turns for two reasons: to expose more wire to the magnetic flux, and to produce a flux pattern similar to that of a bar magnet with a definite north and south pole.

A ratio in mathematics is a comparison of two numbers. We make the comparison by dividing one number by the other. In Figure 17–5(a), the primary winding has 200 turns and the secondary has 100 turns. Therefore, the turns ratio is 2 to 1. Instead of writing the word *to,* we use the colon punctuation mark and write the turns ratio as 2:1. This means that the primary winding has two turns of wire for every one turn on the secondary winding.

In Figure 17–5(b), notice that the turns ratio is still 2:1 even though the actual number of turns has changed. Figure 17–5(c) shows how a fraction can be written as a turns ratio.

Let N_P represent the turns of the primary winding, and N_S the turns of the secondary. We will symbolize the turns ratio with the letter n. Thus, $n = 1:5$ means a turns ratio of 1 to 5. Formula 17–2 defines the turns ratio.

Formula 17–2

$$n = \frac{N_P}{N_S}$$

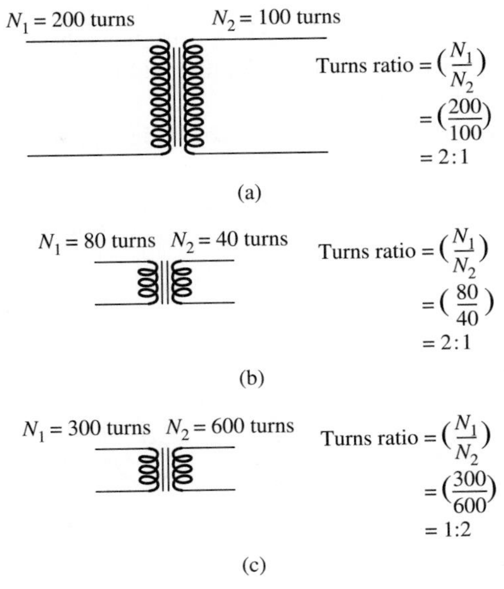

$N_1 = 200$ turns $N_2 = 100$ turns

$$\text{Turns ratio} = \left(\frac{N_1}{N_2}\right)$$
$$= \left(\frac{200}{100}\right)$$
$$= 2:1$$

(a)

$N_1 = 80$ turns $N_2 = 40$ turns

$$\text{Turns ratio} = \left(\frac{N_1}{N_2}\right)$$
$$= \left(\frac{80}{40}\right)$$
$$= 2:1$$

(b)

$N_1 = 300$ turns $N_2 = 600$ turns

$$\text{Turns ratio} = \left(\frac{N_1}{N_2}\right)$$
$$= \left(\frac{300}{600}\right)$$
$$= 1:2$$

(c)

FIGURE 17–5
Examples of turns ratios.

To calculate and write the turns ratio *(n)*, follow these steps:

1. Divide N_P by N_S.
2. If the result is larger than 1, write the result on the left side of the colon and the number 1 on the right side. Example: $153 \div 39 = 3.92$, turns ratio $= 3.92{:}1$.
3. If the result is smaller than 1, reciprocate the result. Write this value on the right of the colon, and write 1 on the left. Example: $623 \div 948 = 0.65717$, reciprocal of $0.65717 = 1.52$, turns ratio $= 1{:}1.52$.

● *SKILL-BUILDER 17–1*

Following the steps just given, determine the turns ratios for the following combinations of windings:

GIVEN:

N_P	567	348	1,654	876	439	2,157
N_S	287	1,673	5,852	227	1,984	396

FIND:

n	1.96:1	1:4.81	1:3.54	3.86:1	1:4.52	5.45:1

Voltage Ratio

The voltage ratio of a transformer is equal to the turns ratio. This important principle is illustrated in Figure 17–6. In part (a), 240 V is applied to the primary of a 2:1 turns ratio transformer. Since the voltage ratio is the same as the turns ratio, this transformer will produce 1 volt on its secondary winding for every 2 volts applied to its primary. Therefore, the secondary voltage is 120 V.

In Figure 17–6(b), the primary voltage is again 240 V. However, this transformer has a 1:2 turns ratio, which means the secondary will produce 2 volts for every 1 volt on the primary. Thus, the secondary voltage is 480 V.

Each turn of wire in a primary winding acts like a tiny electromagnet, producing a certain amount of flux and contributing a certain amount of energy to the magnetic field. Thus, the primary winding is like a number of electromagnets oriented in the same direction. The total flux and thus the total energy in the field is the sum of the flux contributed by each turn in the primary winding.

In a secondary winding exposed to the changing magnetic flux, each turn of wire acts like a tiny voltage source, picking up a certain amount of voltage from the energy in the flux. Thus, the secondary winding is like a number of voltage sources in series. The total secondary voltage is the sum of the voltage produced by each turn in the secondary winding.

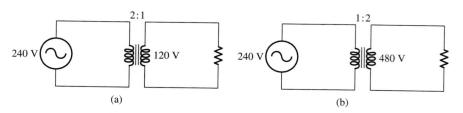

(a)   (b)

FIGURE 17–6
The voltage ratio is the same value as the turns ratio.

When we combine these two principles, we can see why the turns ratio and the voltage ratio are the same. Each primary turn produces flux from current, and each secondary turn produces voltage from flux. The exact relationship between turns ratio and voltage ratio can be proven by equations which are beyond the scope of this text.

Step-Up and Step-Down Transformers

A *step-up* transformer has a turns ratio that produces a secondary voltage higher than the primary voltage, as shown in Figure 17–6(a). Similarly, a *step-down* transformer has a turns ratio that produces a secondary voltage lower than the primary voltage, as shown in Figure 17–6(b). If you associate the two numbers in the turns ratio with the primary and secondary windings, you can easily identify whether the transformer is acting as a step-up or step-down transformer. Thus, 2:1 is a step-down ratio, and 1:2 is a step-up ratio.

● SKILL-BUILDER 17–2

A transformer with a turns ratio of 20:1 has 4.8 kV on its primary. Is this transformer acting as a step-up or step-down transformer? What is the secondary voltage?

GIVEN:
 $V_P = 4.8$ kV
 $n = 20:1$

FIND:
 V_S

SOLUTION:
 The ratio factor is 20, and it is on the primary side. This is a step-down transformer. The secondary voltage will be smaller than the primary. We should divide by the ratio factor.

$$V_S = \frac{V_P}{\text{ratio factor}}$$

$$= \frac{4.8 \text{ kV}}{20}$$

$$= 240 \text{ V}$$

 Repeat the Skill-Builder using the following values. For each exercise, identify the ratio factor, the type of action (step-up or step-down), and the correct arithmetic (multiply or divide by the ratio factor). Then calculate the answer. You should obtain the results that follow.

Given:					
Voltage	18 kV	120 V	2.4 kV	240 V	600 V
Winding	Pri	Pri	Pri	Sec	Sec
Turns ratio	1:7.5	9.524:1	10:1	56.25:1	1:5
Find:					
Action	Up	Down	Down	Down	Up
Ratio factor	7.5	9.524	10	56.25	5
Arithmetic	Mult	Div	Div	Mult	Div
Voltage of other winding	135 kV	12.6 V	240 V	13.5 kV	120 V

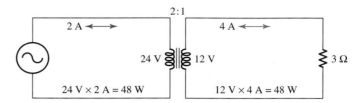

FIGURE 17–7
The power is the same value in both the primary and
secondary circuits of an ideal transformer.

Primary and Secondary Power

Earlier we said that the main idea of a transformer is to change voltage without wasting power. We have seen how it changes voltage. How does it do this without wasting power?

When we say that a transformer does not waste power, we mean that all the power produced by the voltage source in the primary circuit passes through the transformer without any loss into the secondary circuit.

In an ideal transformer, primary power equals secondary power. From Watt's Law, we know that electrical power is the combination of voltage and current. If a transformer changes voltage without changing power, the only way it can do this is to also change current. For example, in Figure 17–7 a 2:1 step-down transformer changes 24 V to 12 V. The secondary load is 3 Ω, and by Ohm's Law, 12 V applied to 3 Ω produces a current of 4 A. Therefore, by Watt's Law the secondary power is 12 V × 4 A = 48 W.

If all of this power has passed into the secondary circuit from the primary circuit, then the power in the primary circuit must also be 48 W. However, the voltage is 24 V. By Watt's Law, if 24 V produces 48 W, the current must be 48 W ÷ 24 V = 2 A. Thus, we see that the transformer apparently changes current along with voltage but in the opposite manner. In other words, if the voltage steps down, the current steps up by the same ratio to keep power the same on both sides; or if the voltage steps up, then the current steps down. Mathematically, this means that the current ratio of a transformer is the reciprocal of the turns ratio.

This happens because of secondary flux. Remember that whenever current flows in a coil or winding, flux is produced. When there is current in the secondary winding, this winding produces a magnetic flux of its own, which induces a voltage in the primary winding. This is not exactly the same as back voltage, but it has a similar effect: It opposes primary current. As a result of this extra voltage, the primary current will adjust itself to exactly the right amount to keep the power the same in both circuits.

● *SKILL-BUILDER 17–3*

Suppose a 240-V source is connected to the primary of a 12:1 step-down transformer, with a 5-Ω load connected to the secondary. Find the power and current on both sides of the transformer.

GIVEN:
  $V_{\text{pri}} = 240$ V
    $n = 12\!:\!1$
  $Z_{\text{sec}} = 5$ Ω Z_{sec} is secondary impedance

FIND:
  V_{sec}
  I_{sec}

P_{sec}
P_{pri}
I_{pri}

SOLUTION:

V_{sec}	$= V_{\text{pri}} \div 12$	V_{sec} is smaller than V_{pri}, so
	$= 240 \text{ V} \div 12$	divide by ratio factor.
	$= 20 \text{ V}$	
I_{sec}	$= V_{\text{sec}} \div Z_{\text{sec}}$	Ohm's Law
	$= 20 \text{ V} \div 5 \, \Omega$	
	$= 4 \text{ A}$	
P_{sec}	$= V_{\text{sec}} \times I_{\text{sec}}$	Watt's Law
	$= 20 \text{ V} \times 4 \text{ A}$	
	$= 80 \text{ W}$	
P_{pri}	$= P_{\text{sec}}$	principle of transformer
	$= 80 \text{ W}$	power transfer
I_{pri}	$= P_{\text{pri}} \div V_{\text{pri}}$	Watt's Law
	$= 80 \text{ W} \div 240 \text{ V}$	
	$= 0.33 \text{ A}$	

Notice that the voltage went down by a factor of 12 (240 V ÷ 12 = 20 V), and the current went up by a factor of 12 (0.33 A × 12 = 4 A). This is what keeps the power the same on both sides.

Repeat the Skill-Builder using the following values. You should obtain the results that follow.

Given:

V_{pri}	18 kV	240 V	6.8 V	288 V	15 V
n	1:7.5	12:1	1:25	16:1	1:50
Z_{sec}	3 kΩ	8 Ω	850 Ω	4.5 Ω	3.75 kΩ

Find:

V_{sec}	135 kV	20 V	170 V	18 V	750 V
I_{sec}	45 A	2.5 A	200 mA	4 A	200 mA
P_{sec}	6,075 kW	50 W	34 W	72 W	150 W
P_{pri}	6,075 kW	50 W	34 W	72 W	150 W
I_{pri}	337.5 A	208.3 mA	5 A	250 mA	10 A

Reflected Impedance

The primary voltage source sees the primary winding as its load. Since the actual load is in the secondary, we say that its impedance has been reflected into the primary winding. Therefore, *reflected impedance* is the apparent ohmic value of the primary winding as seen by the voltage source.

We can calculate the ohmic value of the reflected impedance by Ohm's Law: $Z_{\text{pri}} = V_{\text{pri}} \div I_{\text{pri}}$. For example, in the previous Skill-Builder, the 5-Ω load in the secondary resulted in a primary current of 0.33 A with a primary voltage of 240 V. Therefore, Z_{pri}, the reflected impedance, equals 240 V ÷ 0.33 A, or 720 Ω.

By rearranging Ohm's Law and the turns ratio formula, we obtain Formula 17–3 for reflected impedance.

Formula 17–3

$$Z_{pri} = n^2 \times Z_{sec}$$

To use the turns ratio in this formula, we simply divide where we see the colon. For example, if $n = 20:1$ then $n = 20 \div 1$, or simply 20. On the other hand, if $n = 1:20$, then $n = 1 \div 20$, or 0.05. This reduced expression of n can then be used in Formula 17–3.

● SKILL-BUILDER 17–4

What is the reflected impedance of a 120-Ω load connected to the secondary winding of a 1:40 step-down transformer?

GIVEN:
   $Z_{sec} = 120\ \Omega$
   $n = 1:40$

FIND:
   Z_{pri}

SOLUTION:

$n = 1:40$ the reciprocal of 40
$\quad = 1 \div 40$
$\quad = 0.025$

$Z_{pri} = n^2 \times Z_{sec}$ keystrokes:
$\quad = 0.025^2 \times 120\ \Omega$ 40 [1/x] [x²]
$\quad = 0.075\ \Omega$ [×] 120 [=]

If the transformer had been a step-up transformer ($n = 40:1$), we would not have reciprocated 40.

Repeat the Skill-Builder using the following values. You should obtain the results that follow.

Given:

Z_{sec}	510 Ω	1.6 Ω	2 kΩ	0.015 Ω	75 Ω
n	30:1	15:1	1:50	60:1	2:1

Find:

Z_{pri}	459 kΩ	360 Ω	0.8 Ω	54 Ω	300 Ω

Impedance Matching

The maximum amount of power is transferred from a source to a load when the input impedance of the load equals the output impedance of the source. This principle is known as the Maximum Power Transfer Theorem.

The Maximum Power Transfer Theorem applies to all circuits. Remember from earlier chapters that every voltage source is equivalent to an ideal voltage source in series with an internal impedance (Thevenin's Theorem), and that every load is equivalent to a specific amount of impedance, as viewed from the input terminals. These concepts are illustrated in Figure 17–8. When an antenna is connected to the input of a radio receiver, the internal impedance of the antenna (Z_{out}) is in series with the impedance of the receiver's input circuits (Z_{in}). The voltage developed in the antenna sends current through

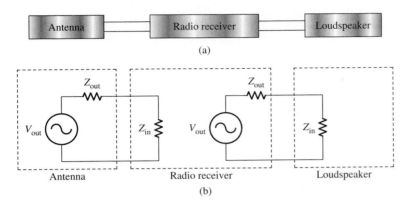

FIGURE 17–8

Examples of impedance matching between sources and loads.

both Z_{out} and Z_{in}, developing power in each of them. The receiver responds only to the power developed in Z_{in}. Any power developed in Z_{out} is wasted. We will get the maximum possible power in Z_{in} when Z_{in} and Z_{out} are equal in value.

The output circuit of the receiver and the input of the loudspeaker can be analyzed in a similar way. Maximum power will transfer from the receiver to the loudspeaker when Z_{out} of the receiver equals Z_{in} of the loudspeaker.

Impedance matching is the process of making Z_{out} and Z_{in} equal at every power transfer point in an interconnected system. Often the best way to do this is with a transformer.

We have learned that in the secondary circuit of a transformer, the true impedance of the load is reflected into the primary winding. The value of the reflected impedance depends on the value of the secondary impedance and the value of the turns ratio. Thus, by adjusting the turns ratio, we can match any load to any source, as shown in Figure 17–9.

In this illustration, an antenna with a characteristic impedance of 300 Ω is matched through a transformer with a receiver whose input impedance is 75 Ω. Formula 17–4 gives the turns ratio necessary to match an output impedance on the primary winding with an input impedance on the secondary.

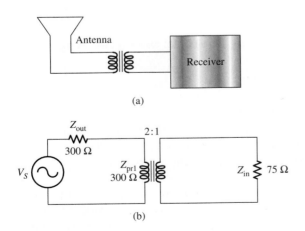

FIGURE 17–9

A transformer is often used to match the impedance of a source and a load.

Formula 17–4

$$n = \sqrt{\frac{Z_{pri}}{Z_{sec}}}$$

● *SKILL-BUILDER 17–5*
- -

What turns ratio is required to match an output impedance of 500 Ω to an input impedance of 16 Ω?

GIVEN:
$Z_{pri} = 500\ \Omega$
$Z_{sec} = 16\ \Omega$

FIND:
n

SOLUTION:

$$n = \sqrt{\frac{Z_{pri}}{Z_{sec}}}$$

$$= \sqrt{\frac{500\ \Omega}{16\ \Omega}}$$

keystrokes:
500 [÷] 16 [=]
[√]

$$= 5.59{:}1$$

Repeat the Skill-Builder using the following values. You should obtain the results that follow.

Given:

Z_{pri}	20 Ω	200 Ω	600 Ω	1 kΩ	8 Ω
Z_{sec}	1.2 kΩ	4 Ω	3 kΩ	600 Ω	1 kΩ
Find:					
n	1:7.75	7.07:1	1:2.24	1.29:1	1:125

- -

Effect of Transformer Loading on Phase Angle

When the secondary circuit is open, we say the transformer is unloaded, as shown in Figure 17–10. In this case, the primary winding draws a small amount of current from the source. This current is just enough to create the magnetic field in the core and is called the energizing current. Since the secondary is open, no energy can transfer from the magnetic field to the load. Thus, the energy in the magnetic field passes back into the primary winding. This is the back voltage that we have already studied. It keeps the energizing current at a low value and causes the current to lag primary voltage by 90°, as we see in Figure 17–10(a).

You have probably used an ac adapter (converter) that plugs into a wall receptacle to operate small battery-powered appliances such as small portable tape recorders. You may know from experience that you can leave the adapter plugged into the wall all the time, whether or not the load (the tape recorder) is connected or turned on. The adapter stays warm but does not cause any problem. Now you can understand why.

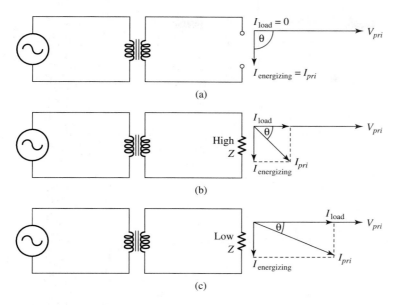

FIGURE 17–10
Effect of transformer loading on phase angle.

The adapter contains a transformer. When the load is disconnected or turned off, the secondary is open. Energizing current continues to flow through the primary, producing a slight amount of heat. The back voltage in the primary keeps the current down to this low value, which is why we can ignore it. Without the back voltage, the primary winding, which is just a long piece of wire, would act like a short to the voltage source, and the primary would draw a high current, requiring us to unplug the adapter.

In Figure 17–10(b), a high-impedance load has been connected to the secondary, drawing power from the secondary. As the magnetic field begins to transfer power into the secondary, more current flows in the primary. This current is in phase with primary voltage, as shown. Thus, the total primary current has two components: the energizing current which lags by 90° and the in-phase current due to the secondary load. The result is a decrease in the phase angle.

With a low secondary impedance (Figure 17–10(c)), the secondary draws enough power from the transformer to create a large in-phase current in the primary. As shown, this reduces the phase angle still further. The phase angle may become close to 0° but can never reach zero because of the energizing current.

→ *Self-Test*

4. What is the main purpose of a transformer?
5. What is the turns ratio of a transformer?
6. What is the relationship between the turns ratio and voltage ratio of a transformer?
7. What is the relationship between primary power and secondary power in an ideal transformer?
8. What is the relationship between the current ratio and turns ratio in a transformer?
9. What is reflected impedance?
10. When is maximum power transferred from a source to a load?
11. What is impedance matching, and what is its result?

SECTION 17–3: Transformer Applications

In this section you will learn the following:

→ A commercial power distribution system uses high-voltage lines to transmit large amounts of power at low current levels to minimize power losses in the lines.

→ An electronic power supply uses a transformer to step down ac voltage before rectifying it into dc voltage.

→ The isolation feature of a transformer is used to eliminate a current path to ground from the chassis of an electrical machine, protecting service personnel from shock hazard.

→ A tuned coupled transformer in a radio-frequency circuit provides frequency selectivity while coupling the ac signal and blocking dc bias voltages.

Commercial Power Distribution

A commercial power distribution system uses high-voltage lines to transmit large amounts of power at low current levels to minimize power losses in the lines. An interesting bit of history will help us appreciate this principle.

The first commercial electrical generating station was built by Thomas Edison in 1882 in Manhattan. The station produced a relatively low-voltage direct current for 400 electric lamps, all located within a mile of the station. Edison believed that commercial electrical power could not be distributed over a long distance because of power losses in the transmission lines (wires).

A young engineer named Nikolai Tesla convinced George Westinghouse that the problem of power losses in the lines could be solved by using a combination of alternating current and transformers. Tesla realized that the power loss in the lines was due to the high current necessary to provide large amounts of power at a low voltage. If the line voltage was high, the same amount of power could be transmitted at a low current level, and the line losses would be small. However, this would require raising the voltage produced by the generators, transmitting the power through the lines, then lowering the voltage again near the customer's location.

Tesla knew that transformers could step up and step down ac voltage without wasting power but could not do so with dc voltage. Therefore, he concluded that a commercial electric power system had to be an ac system. He convinced George Westinghouse to build large ac generators and transformers and thus began the modern electrical power industry.

Although it is extremely expensive to construct high-voltage power lines and install millions of transformers to step the voltage up and down, as well as hazardous to work on such a system, at the present time there is simply no other practical way to transmit large amounts of electrical power over long distances. At every point in a power distribution system, from the generating station to the lamp in a customer's home, the goal is to keep the current in the lines as low as possible. This means keeping the line voltage high right up to the customer's building. Thus, the power lines behind your house may carry power at 7.2 kV or higher, dropping it down to 240 V with the transformer on the pole before running the lines to your meter.

Power Supplies in Electronic Equipment

An electronic power supply uses a transformer to step down ac voltage before rectifying it into dc voltage.

All modern electronic equipment such as computers, televisions, audio systems, and so on rely on transistorized circuits to perform their many tasks. Transistors, in turn, operate on low-voltage direct current. Therefore, if we wish to operate electronic equipment on commercial electric power, we must change 120-V ac into dc voltages between 5 V and 24 V. The first step in this process is a transformer to step down the 120-V ac to perhaps 24-V dc before rectifying it into a dc voltage. Thus, a step-down transformer is inside every kind of electronic equipment that plugs into an ac wall receptacle.

Isolation

The isolation feature of a transformer is used to eliminate a current path to ground from the chassis of an electrical machine, protecting service personnel from shock hazard.

Isolation refers to the fact that the primary and secondary circuits are electrically separate. In other words, there is no electrical connection (no current path) between the two circuits. They are linked only by the magnetic flux in the transformer core.

The isolation feature of a transformer is often used as a safety feature to protect technicians. Remember that in residential ac wiring, one of the current-carrying conductors, the neutral wire, has been grounded to the earth. The other current-carrying conductor is the hot wire and is at a potential of 120 V to the neutral. In an ac receptacle, the hot wire is supposed to be the short slot, and the neutral is supposed to be the long slot. We say "supposed to be" because the receptacle could be wired incorrectly.

The ac plug is supposed to be polarized, which in this case means one of its blades is wider than the other, to match the difference in width of the receptacle slots. The idea is to allow the plug to go into the receptacle in only one way. However, if the plug is not polarized, if the blades have been modified, or if the technician is simply stubborn, it is possible to put the plug into the receptacle the wrong way, so that the blade intended to contact the neutral wire instead contacts the hot wire.

Inside the equipment, the metal chassis (frame) is usually used as circuit ground. In this case, one of the wires from the ac plug may be connected directly to the metal chassis. This wire would normally plug into the neutral wire, thus connecting the chassis to earth through the neutral wire.

However, if the plug has been reversed as described previously, the chassis would be connected to the hot wire. A technician who touches a hot chassis while standing on a grounded surface would receive a dangerous shock, as shown in Figure 17–11(a).

An isolation transformer protects the technician against this shock hazard, as shown in Figure 17–11(b). Because the secondary circuit has no connection to earth ground, no current will flow from the chassis through the technician to earth. We would say the chassis has a floating ground, which means the chassis ground is relative only to the chassis circuit and has no physical connection to earth ground. The isolation transformer has a 1:1 turns ratio, since we do not wish to change the voltage.

Tuned Coupling Transformers

A tuned coupled transformer in a radio-frequency circuit provides frequency selectivity while coupling the ac signal and blocking dc bias voltages.

A radio or television processes the incoming signal in several stages. Each stage consists of a circuit of one or more transistors or integrated circuits with accompanying resistors, capacitors, and so on. Each stage has its own particular requirements for dc voltage to operate the chips and transistors correctly. Within the stage, the ac signal becomes combined with these dc bias voltages.

The signal must be passed from one stage to the next without allowing the dc bias voltages of one stage to interfere with those of the next stage. At the same time, the signal may need filtering to separate the desired band of frequencies from other frequencies in the signal.

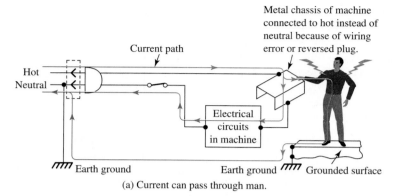

Metal chassis of machine
connected to hot instead of
neutral because of wiring
error or reversed plug.

Current path

Hot
Neutral

Electrical
circuits
in machine

Earth ground   Earth ground   Grounded surface

(a) Current can pass through man.

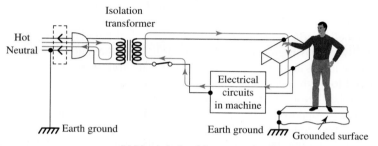

Isolation
transformer

Hot
Neutral

Electrical
circuits
in machine

Earth ground   Earth ground   Grounded surface

(b) Man is isolated from current path.

FIGURE 17–11
The isolation feature of a transformer is used to protect service
personnel from shock hazards while working on electrical
equipment.

A transformer can be used to couple stages and filter the signal at the same time, as
shown in Figure 17–12(a). Notice that capacitors are placed in parallel with both primary
and secondary windings. This turns the transformer into a double-tank circuit and provides band filtering. At the same time, only the ac component of the primary current will
be transformed to the secondary. Remember that dc currents do not produce transformer

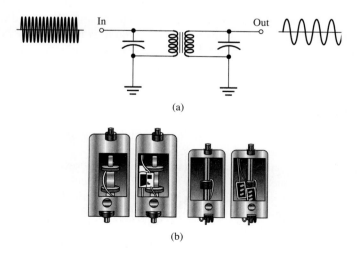

In   Out

(a)

(b)

FIGURE 17–12
Miniature transformers used for tuned coupling in radio-
frequency circuits.

action and that if a signal contains a mixture of frequencies (including a dc component counted as a frequency of 0 Hz), a circuit will respond to the mixed signal as it would to each component separately. Thus, the dc bias currents of the primary stage do not appear in the secondary stage, but the ac signal currents do. Figure 17–12(a) is an example of tuned coupling.

Figure 17–12(b) shows the type of transformer used for tuned coupling. These small transformers are called by several names: interstage coupling transformers, RF (radio-frequency) transformers, and so on. They are enclosed by an aluminum cover which acts as a shield to prevent stray magnetic RFI (radio-frequency interference) from being induced into the signal. Any surrounding RFI will induce current in the shield, not in the transformer windings. The shield is grounded to carry away these unwanted noise currents.

These RF transformers have a threaded ferrite core called a slug, shaped something like a small bolt without a head. The technician uses a special miniature plastic screwdriver to adjust the slug slightly in and out of the core. This changes the inductance of the windings and tunes the tank circuits to the desired resonant band. In radio and television work, this adjusting process is called aligning the stages (tuning them to the appropriate bands).

→ *Self-Test*

12. Describe a commercial power distribution system.
13. What is the purpose of a transformer in an electronic power supply?
14. What is the use of the isolation feature of a transformer?
15. What is the purpose of a tuned coupled transformer?

SECTION 17–4: Transformer Variations

In this section you will learn the following:

- → A multiple-winding transformer has more than one primary or secondary winding.
- → A multiple-tap transformer has additional connections to its windings.
- → Secondary voltage may be regulated by changing taps on the primary winding.
- → If a secondary winding is center-tapped and the center tap is grounded, the two ends of the winding produce two voltages 180° out of phase with each other, a technique called phase splitting.
- → Instantaneous polarity is the relationship between the polarity of one winding and the polarity of another winding.
- → An autotransformer is a transformer that uses one winding as both its primary and secondary.

Multiple-Winding Transformers

A multiple-winding transformer has more than one primary or secondary winding. This feature gives us several options:

1. We can have several secondary circuits, each isolated from the others.
2. We can connect the primary windings in series or in parallel to allow either 120 V or 240 V on the primary with the same voltage on the secondary.
3. We can regulate secondary voltage by adjusting the connection on the primary.
4. We can have many secondary voltages.

Sometimes it is helpful to isolate (float) the ground in different sections of the same circuit. As shown in Figure 17–13, we can do this by putting each circuit on its own secondary winding so there is no electrical connection between the two circuit sections.

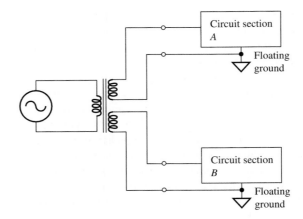

FIGURE 17–13
Multiple secondaries used to isolate circuit sections.

Electronic equipment made to sell worldwide must operate on either the 120-V voltage level used in North America or the 240-V level used in the rest of the world. Figure 17–14 shows how multiple primaries can be connected in either series or parallel to adapt to either voltage.

Suppose that the desired secondary voltage is 12 V and that each primary has a 10:1 ratio to the secondary. If the two primaries are wired in parallel as shown in Figure 17–14(b), the turns ratio remains at 10:1. In other words, wiring the primaries in parallel did not change the turns ratio. Thus, 120 V on the primaries will step down 10:1 to 12 V on the secondary. Each primary will deliver half the power. Therefore, each primary carries half the total primary current.

If the primaries are wired in series as shown in Figure 17–14(c), the turns ratio now becomes 20:1. If 240 V is applied to the primary, it will step down 20:1 to the required 12 V. Each primary carries the same current as before, but now this is the total primary current since the windings are in series.

Remember that transformer windings follow many of the same rules as other voltage sources. Like batteries, transformer windings are connected in parallel to get more current and in series to get more voltage.

Multiple-Tap Transformers

A multiple-tap transformer has additional connections to its windings. A tap is a wire that connects to a winding somewhere along its length.

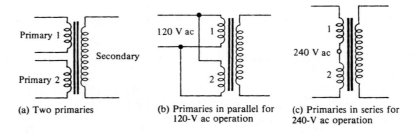

(a) Two primaries

(b) Primaries in parallel for 120-V ac operation

(c) Primaries in series for 240-V ac operation

FIGURE 17–14
A multiple primary allows transformer to be wired for either 120-V or 240-V operation.

Figure 17–15 shows a transformer with multiple taps on its primary. The device that looks like a rotary switch is called a tap changer. It selects which of the primary taps connects to the voltage source.

Secondary voltage may be regulated by changing taps on the primary winding. Selecting a different tap effectively changes the voltage ratio from the primary side. This feature can be used to regulate secondary voltage.

For example, suppose the entire primary consisted of 1,000 turns and the secondary consisted of 50 turns. Thus, the maximum turns ratio is 20:1. Now suppose the taps are connected to the primary ten turns apart. By selecting various taps, we can change the turns ratio to 19.8:1, 19.6:1, 19.4:1, and so on.

Secondary voltage regulation is the reason for doing this. From Thevenin's Theorem, we know that the output voltage of any voltage source drops as the load draws more current from the source. We can regulate secondary voltage by selecting a different primary tap, changing the turns ratio to keep secondary voltage constant. This is how utility companies regulate output voltage to the customer. Large utility transformers have automatic tap-changers. Electronic monitoring circuits detect changes in secondary voltage and signal a motor-driven arm to move to another tap.

Figure 17–16(a) illustrates a secondary with many taps. Since each tap represents a different turns ratio with the primary, each tap produces a different output voltage. Certain electronic appliances, particularly televisions, require a number of dc voltages ranging from 5 V to over 20 kV. Such equipment generally includes a transformer with either multiple secondaries or a single multiple-tapped secondary.

Figure 17–16(b) shows a center-tapped transformer, a common and important type. Each half of the winding produces half the total secondary voltage.

One common use of the center-tapped transformer is in residential power distribution and wiring. As shown in Figure 17–17, the final transformer from the utility company to your house has a center-tapped secondary. The center-tapped conductor is grounded to become the neutral (zero-volts) conductor. The two ends of the winding become the two hot wires. Thus, any large electrical appliance that operates on 240 V (such as a stove or clothes dryer) will be connected across the two hot wires. All other 120-V circuits will be connected across the neutral and one or the other of the hot wires. This wiring arrangement is generally called single-phase, three-wire service. Notice in the illustration that the transformer's primary has multiple taps at one end for voltage regulation with a tap changer.

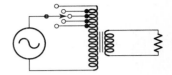

FIGURE 17–15
Multiple-tap primary with tap changer used to regulate secondary voltage.

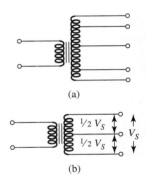

(a)

(b)

FIGURE 17–16
(a) Multiple secondary taps offer several voltages. (b) Center-tapped secondary is widely used.

Phase Splitting

If a secondary is center-tapped and the center tap is grounded, the two ends of the winding produce two voltages 180° out of phase with each other, a technique called phase splitting. Figure 17–18 illustrates this technique.

We begin by analyzing the full output of the secondary with an oscilloscope. With the lower terminal selected as ground and the probe placed on the upper terminal, the oscilloscope displays the familiar sine wave pattern. The polarity of the secondary is marked for emphasis. During the first half-cycle, the upper terminal is positive with respect to the lower, and negative during the second half-cycle.

We can split the phase by selecting the center terminal as ground. Notice that the polarity of the secondary is exactly the same in parts (a) and (c) of Figure 17–18. Because the center terminal is now ground, the waves displayed on the oscilloscopes are each half the voltage of the secondary and out of phase with each other by 180°. This is only because of the choice of ground. Nothing has really changed in the secondary. If ground is on the center tap, then during the first half-cycle, the upper terminal is more positive than the center, and the lower terminal more negative. The oscilloscopes display this

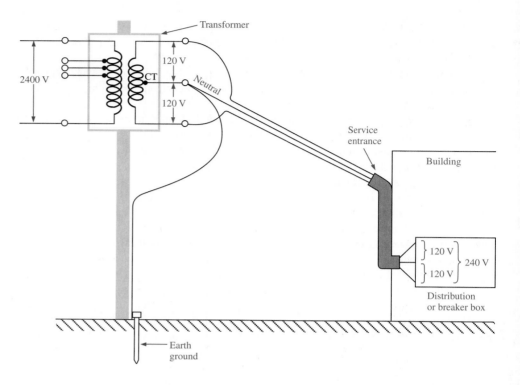

FIGURE 17–17
Center-tapped utility transformer in standard residential single-phase, three-wire service.

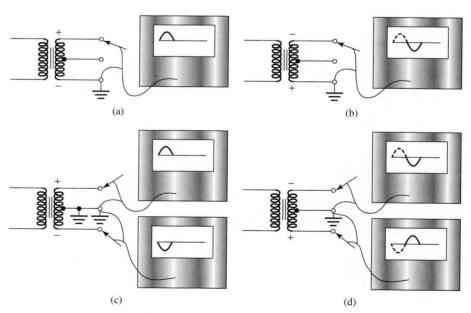

FIGURE 17–18
A grounded center tap produces phase splitting.

because of the way they are connected. A similar line of reasoning applies to the second half of the cycle. The secondary has the same polarity in (d) as in (b). The phase is split by placing ground at the center.

Figure 17–19 illustrates a common application of phase splitting. We have learned that the power supply is the part of the circuitry that turns the incoming 120-V ac into low-voltage dc to operate the transistors in the rest of the circuitry. The first step in the process is to step the voltage down. However, it is still ac. The center-tapped transformer is used to split the phase, and two ac waveforms enter the next section consisting of an arrangement of rectifier diodes. These devices, which you will study in your next course, act as one-way valves for current. This means they can be arranged to pass a positive polarity and reject a negative polarity, or vice versa.

In Figure 17–19, the diodes are arranged to pass only a positive polarity. The diodes receive positive half-cycle *A* and negative half-cycle *C* at the same time. They pass *A* and reject *C,* as shown at the output. By similar reasoning, during the second half-cycle, the diodes pass *D* and reject *B*. Thus, the output consists of pulsating dc. Although the waveform requires further processing to smooth it into a steady dc voltage, it is now classified as dc because the polarity no longer reverses. Thus, we say the waveform has been rectified by the diodes.

Instantaneous Polarity

Instantaneous polarity is the relationship between the polarity of one winding and the polarity of another winding. This subject is of great importance in learning how to connect transformer windings and is one of the first things a utility company teaches its technicians. It is of equal importance to electronic technicians.

We cannot label the terminals of a transformer winding as either positive or negative because ac is constantly reversing polarity. Yet somehow we must know which terminals match other terminals in order to connect terminals correctly.

Figure 17–20 shows the solution to this labeling problem. Dot notation is a system used to indicate the instantaneous polarity of windings. The phase dots near the winding terminals indicate which terminals always match each other's polarity. In Figure 17–20(a) the phase dots are by the upper terminals. This means that these terminals will always have the same polarity, either positive or negative, throughout the ac cycle. Likewise, in Figure 17–20, the upper left and lower right terminals will always have matching polarity. A dual-trace oscilloscope, which can show two waves at once, could be used to determine instantaneous polarity in an actual transformer.

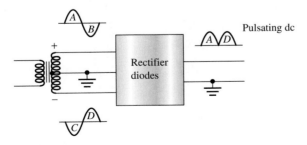

FIGURE 17–19
Phase splitting is widely used to produce full-wave rectification in electronic power supplies.

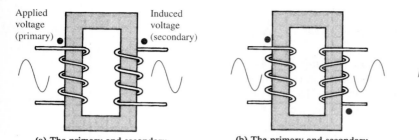

Applied voltage (primary)   Induced voltage (secondary)

(a) The primary and secondary voltages are in phase when the windings are in the same effective direction around the magnetic path.

(b) The primary and secondary voltages are 180° out of phase when the windings are in the opposite direction.

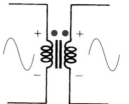

(c) Voltages are in phase.

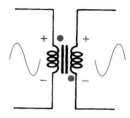

(d) Voltages are out of phase.

FIGURE 17–20
Instantaneous polarity and dot notation.

Additive and Subtractive Polarity

The polarity relationships of the primary and secondary may also be determined with a special voltmeter test, as shown in Figure 17–21. The secondary is open with a jumper wire connected between a primary terminal and the corresponding secondary terminal. The voltmeter is then placed as shown. If the voltmeter reads the difference in primary and secondary voltage, we say the transformer has subtractive polarity. The phase dots would be as in Figure 17–20(a). On the other hand, if the voltmeter reads the sum of the voltages, we say the transformer has additive polarity. The phase dots would be as in Figure 17–20(b).

Sometimes you may want to connect the secondaries of two transformers. You can determine the instantaneous polarity of the secondary terminals by using a voltmeter as shown in Figure 17–22. If the secondary voltages are equal, the meter will read zero volts when connected as shown in Figure 17–22(b).

The Autotransformer

An autotransformer is a transformer that uses one winding as both its primary and secondary, as shown in Figure 17–23. In part (b), part of the winding is used as the primary and all of it is used as the secondary. Thus, the voltage is stepped up. By contrast, in part

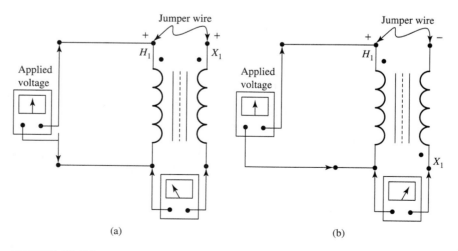

FIGURE 17–21
Determining additive or subtractive polarity.

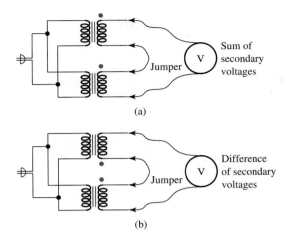

(a)

(b)

FIGURE 17–22
Testing instantaneous polarity on secondary
windings in series.

(c) all of the winding is used as the primary and part of it is used as the secondary. The voltage is stepped down.

Part (d) illustrates a sliding contact arm, which amounts to an adjustable tap. Thus, the output voltage can be varied. Part (a) is a photograph of a widely used adjustable autotransformer generally called a variac (variable ac).

The autotransformer can usually produce an output voltage between 0% and 120% of the input voltage. The major disadvantage of the autotransformer is that it has no isolation feature. This creates a possible hazard when the autotransformer is used as a step-down transformer, as shown in Figure 17–24(b). If the winding opens between the output slider and ground, the full input voltage is applied to the load, which may damage or destroy the load.

Practical Application

The autotransformer is extremely helpful in repairing electronic equipment, and many repair shops have both an isolation transformer and an autotransformer permanently installed at every repair workstation, as shown in Figure 17–25.

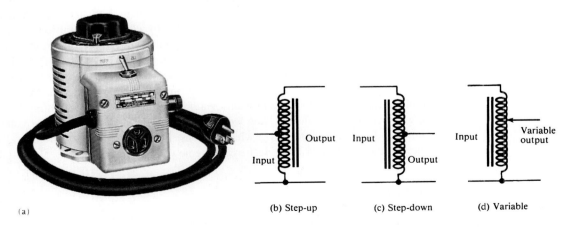

(a)

(b) Step-up

(c) Step-down

(d) Variable

FIGURE 17–23
Autotransformer and schematic symbols.

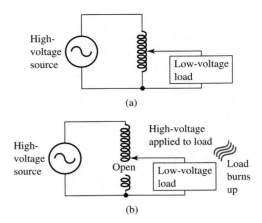

FIGURE 17–24
An autotransformer does not isolate
primary and secondary circuits. Therefore,
an open winding can apply high voltage
to the load.

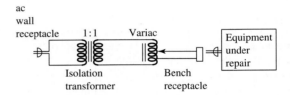

FIGURE 17–25
Combination of isolation and autotransformer
installed at a repair workstation.

Often the equipment under repair has a short circuit somewhere that drew enough current to burn up a number of diodes, transistors and chips before the fuse or circuit breaker could open. To repair the circuit, you must find the short. Otherwise, the new components you install will all burn up when you apply power, and you will be back where you started, having wasted time and money.

How can you find the short if the circuit won't draw current long enough for you to test it? The autotransformer is the solution to the problem. Let's think it through: Components burn up because of heat, heat is produced by current, and current is produced by voltage. If the input voltage is small, even with the short in place, there won't be enough current to burn anything up. Therefore, we do not apply full line voltage (120 V) after replacing the burned-up components. Instead, we set the autotransformer to zero volts, plug the equipment into the autotransformer, and bring the voltage up slowly. We would apply perhaps 5 or 10 volts and wait for a minute or so to see if anything is already getting hot at this low voltage. Then we would repeat the process: Raise the voltage a little more, and wait a minute to see what happens. In this manner we can generally find the short if it's still there without destroying the new components.

→ *Self-Test*

16. Describe a multiple-winding transformer.
17. Describe a multiple-tap transformer.

18. How may voltage be regulated with a multiple-tap transformer?
19. How may a transformer be used to produce phase splitting?
20. What is instantaneous polarity?
21. What is an autotransformer?

SECTION 17–5: Troubleshooting Transformers

In this section you will learn the following:

→ A partial short in a transformer winding causes the output voltage to vary from its normal value.

→ A full short in a transformer winding will cause a high overcurrent when power is applied.

→ An open in a transformer winding will produce a secondary voltage of zero.

→ A transformer may develop high impedance in either winding because of deterioration of internal connections.

Partial Short in a Winding

A special wire is used in windings for transformers, relays, contactors, electromagnets, generators, and motors. This wire is coated with varnish to provide insulation. With age and heat, the varnish begins to melt or crack and flake off. This creates partial shorts between adjacent turns in the windings, which increase current, produce more heat, melt and crack more varnish, and create more partial shorts. In other words, once the varnish begins to deteriorate, it goes quickly. When this is happening, the machine runs hotter. Thus, an increase in operating temperature may be a warning that the windings are beginning to go bad.

In the case of transformers, a partial short in a winding affects the turns ratio. Therefore, the secondary voltage will begin to vary from its normal value. The variation could be either above or below normal, depending on whether the partial short is in the primary or secondary winding. To the user, the variation would appear as poor voltage regulation.

Full Short in a Winding

A full short in a transformer winding will cause a high overcurrent when power is applied. This will cause the overcurrent protection device (fuse or circuit breaker) to open. The best way to detect the problem is to connect the equipment to a variac and bring the supply voltage up slowly from zero, with the secondary winding disconnected from the rest of the circuit. A full short in the transformer will result in abnormally high current at almost zero supply voltage. This gives you good reason to remove the transformer for further testing.

The reason for disconnecting the secondary in this test is to see whether the short is in the transformer or in the circuitry connected to the transformer.

Open in a Winding

An open in a transformer winding will produce a secondary voltage of zero. If the primary is open as shown in Figure 17–26(a) and the equipment is plugged in and turned on, the primary voltmeter will read line voltage but the primary ammeter will read zero cur-

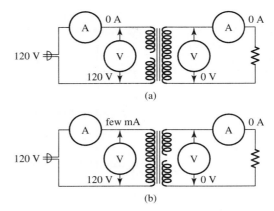

FIGURE 17–26
Troubleshooting open windings.

rent because of the open primary winding. Of course, the secondary voltmeter and ammeter will both read zero, since there is no current in the primary to produce a magnetic field in the transformer.

If the secondary winding is open as shown in Figure 17–26(b), the primary voltmeter will read line voltage and the primary ammeter will read energizing current, the minimum amount of current that maintains the magnetic field in the transformer's core. However, the secondary voltmeter and ammeter will read zero because of the open secondary.

High Impedance

A transformer may develop high impedance in either winding because of deterioration of internal connections. The result will be low current in the transformer, resulting in low-power operation of the circuit.

Applying Your Knowledge

High impedance in a transformer is a problem that may be difficult to diagnose. Ohmmeter readings may give uncertain results. When you have taken all possible measurements and are still confused as to what is causing a problem, your only practical option is to start replacing components. The rule of thumb is to replace components that are (1) most likely to cause the problem, (2) easy to replace, and (3) inexpensive if you destroy the component in replacing it. Transformers generally take more time to dismount than other components and are comparatively expensive. Therefore, unless you have strong reason to suspect the transformer, check other components thoroughly before going after the transformer.

→ *Self-Test*

22. What is the result of a partial short in a transformer winding?
23. What is the result of a full short in a transformer winding?
24. What is the result of an open in a transformer winding?
25. What is the result of a deterioration in the internal connections of a transformer?

Formula List

Formula 17–1

$$L_M = k\sqrt{L_1 L_2}$$

Formula 17–2

$$n = \frac{N_P}{N_S}$$

Formula 17–3

$$Z_{\text{pri}} = n^2 \times Z_{\text{sec}}$$

Formula 17–4

$$n = \sqrt{\frac{Z_{\text{pri}}}{Z_{\text{sec}}}}$$

● TROUBLESHOOTING: Applying Your Knowledge

The problem could be an impedance mismatch between the new microphone and the old amplifier. Before sending the amplifier to the shop, try the old microphone in the old amp. If the volume sounds okay, check the manufacturer's specification sheets for the amplifier and the new microphone to see if the impedances match. If not, get an in-line impedance matching transformer for the microphone.

CHAPTER SUMMARY: ANSWERS TO SELF-TESTS

1. Mutual inductance is the ability of one circuit to transfer magnetic energy into another circuit.
2. Mutual inductance produces transformer action, the process where current in one coil produces magnetism, which produces voltage in another coil.
3. The coefficient of coupling (k) is the part of the flux created by the primary winding that induces voltage in the secondary winding.
4. A transformer changes voltage without wasting power.
5. The turns ratio of a transformer is the number of primary turns divided by the number of secondary turns.
6. The voltage ratio of a transformer is equal to the turns ratio.
7. In an ideal transformer, primary power equals secondary power.
8. The current ratio of a transformer is the reciprocal of the turns ratio.
9. Reflected impedance is the apparent ohmic value of the primary winding as seen by the voltage source.
10. The maximum amount of power is transferred from a source to a load when the input impedance of the load equals the output impedance of the source.
11. A commercial power distribution system uses high-voltage lines to transmit large amounts of power at low current levels to minimize power losses in the lines.
12. An electronic power supply uses a transformer to step down ac voltage before rectifying it into dc voltage.
13. The isolation feature of a transformer is used to eliminate a current path to ground from the chassis of an electrical machine, protecting service personnel from shock hazard.
14. A tuned coupled transformer in a radio-frequency circuit provides frequency selectivity while coupling the ac signal and blocking dc bias voltages.
15. A multiple-winding transformer has more than one primary or secondary winding.
16. A multiple-tap transformer has additional connections to its windings.
17. Secondary voltage may be regulated by changing taps on the primary winding.
18. If a secondary winding is center-tapped and the center tap is grounded, the two ends of the winding produce two voltages 180° out of phase with each other, a technique called phase splitting.
19. Instantaneous polarity is the relationship between the polarity of one winding and the polarity of another winding.
20. An autotransformer is a transformer that uses one winding as both its primary and secondary.
21. A partial short in a transformer winding causes the output voltage to vary from its normal value.
22. A full short in a transformer winding will cause a high overcurrent when power is applied.
23. An open in a transformer winding will produce a secondary voltage of zero.
24. A transformer may develop high impedance in either winding because of deterioration of internal connections.

CHAPTER REVIEW QUESTIONS

Determine whether the following questions are true or false.

17–1. Transformer action is the process where voltage in one coil produces magnetism in another coil.

17–2. A basic transformer consists of two cores and a winding.

17–3. The voltage ratio of a transformer is the same as the turns ratio.

17–4. Reflected impedance is the apparent ohmic value of the primary winding as seen by the voltage source.

17–5. All electronic equipment includes a transformer.

17–6. A tap changer may be used to regulate secondary voltage.

17–7. A power supply transformer typically has a center-tapped secondary.

17–8. A final step-down transformer for residential power distribution typically has a center-tapped secondary.

17–9. A transformer with a center-tapped secondary may be used for phase splitting.

17–10. When connecting together the windings of several transformers, instantaneous polarity is not important.

17–11. Winding wire is insulated with thin plastic.

17–12. A partial short is not possible in a transformer winding.

Select the response that best answers the question or completes the statement.

17–13. The transformer winding connected toward the load is the
   a. primary winding.   **b.** secondary winding.

17–14. Reflected impedance occurs in the
   a. primary winding.   **b.** secondary winding.

17–15. The transformer winding connected toward the voltage source is the
   a. primary winding.   **b.** secondary winding.

17–16. Which of the following is *not* used for transformer cores?
   a. air   **c.** carbon
   b. iron   **d.** ferrite

17–17. The turns ratio is
   a. the number of primary turns minus the number of secondary turns.
   b. the number of primary turns multiplied by the number of secondary turns.
   c. the number of primary turns divided by the number of secondary turns.
   d. the number of secondary turns plus the number of primary turns.

17–18. In an ideal transformer,
   a. primary voltage equals secondary voltage.
   b. primary power equals secondary power.
   c. primary current equals secondary current.
   d. primary impedance equals secondary impedance.

17–19. The maximum amount of power is transferred from a source to a load when
   a. the input impedance of the load equals the output impedance of the source.
   b. the output impedance of the load equals the input impedance of the source.
   c. the input impedance of the load equals the input impedance of the source.
   d. the output impedance of the load equals the output impedance of the source.

17–20. Energizing current is
   a. zero when the load is open.
   b. the result of secondary flux.
   c. zero when the impedances are matched.
   d. the current necessary to produce the magnetic field in the core.

17–21. Which of the following is a good reason for an ac power distribution system?
   a. ac alternators use less fuel than dc generators.
   b. ac power plants can be located closer than dc power plants to large cities.
   c. The smooth rise and fall of an ac wave improves the efficiency of a motor.
   d. Alternating current allows the use of transformers to transmit large amounts of power at low current levels.

17–22. Which of the following statements best expresses the benefit of an isolation transformer?
   a. An isolation transformer prevents a current path from equipment chassis to ground.
   b. An isolation transformer prevents a service technician from touching ground.
   c. An isolation transformer places the equipment chassis at ground potential.
   d. An isolation transformer insulates a service technician from voltage.

17–23. An example of tuned coupling is using a transformer
   a. to match impedances.
   b. between circuit stages in a radio receiver.
   c. to isolate a power supply from a radio signal.
   d. to step down noise and step up signals.

17–24. A stereo system intended for international use is likely to have a power supply transformer with
   a. a tuned core.
   b. multiple secondary windings.
   c. matched impedances.
   d. multiple primary windings.

17–25. Dot notation is used to indicate
   a. instantaneous polarity.   **c.** secondary windings.
   b. primary windings.   **d.** turns ratio.

17–26. Additive and subtractive polarity is usually determined by testing a transformer with
   a. an ohmmeter.   **c.** an ammeter.
   b. a voltmeter.   **d.** an oscilloscope.

17–27. An autotransformer
   a. is used in an automobile ignition system.
   b. automatically adjusts its turns ratio.
   c. uses one winding as both its primary and secondary.
   d. operates with either ac or dc voltage sources.

Solve the following problems.

17–28. A transformer has a primary winding of 2,385 turns and a secondary winding of 436 turns. What is the turns ratio?

17–29. A transformer has a primary winding of 957 turns and a secondary winding of 36,493 turns. What is the turns ratio?

17–30. A transformer with a turns ratio of 17:1 has 600 V on its primary. What is the secondary voltage?

17–31. A transformer with a turns ratio of 1:15 has 18.5 kV on its primary. What is the secondary voltage?

17–32. A transformer with a turns ratio of 16:1 and a load of 6 Ω has a primary voltage of 140 V. Find secondary voltage, secondary current, primary current, power, and reflected impedance.

17–33. A transformer with a turns ratio of 1:5 and a load of 900 Ω has a secondary current of 20 mA. Find primary voltage, primary current, power, and reflected impedance.

17–34. What turns ratio is required to match an output impedance of 200 Ω to an input impedance of 8 Ω?

17–35. A transformer has an input of 67 A at 277 V. The output is 17.8 kVA. What should be the output kVA rating?

GLOSSARY

autotransformer A transformer that uses one winding as both its primary and secondary.

coefficient of coupling The part of the flux created by the primary winding that induces voltage in the secondary winding.

impedance matching The process of making the output impedance of a source equal to the input impedance of a load.

instantaneous polarity The relationship between the polarity of one winding and the polarity of another winding.

isolation The technique of using a transformer to prevent the possibility of a chassis being connected to a hot wire.

mutual inductance The ability of one circuit to transfer magnetic energy into another circuit.

primary winding The winding connected to the voltage source.

reflected impedance The apparent ohmic value of the primary winding as seen by the voltage source.

secondary winding The winding connected to the load.

step-down A decrease in voltage.

step-up An increase in voltage.

tap A connection to a winding.

transformer A device consisting of two windings and an iron core. Current in the first winding creates voltage in the second winding through mutual inductance.

turns ratio The number of primary turns divided by the number of secondary turns.

18 ELECTRIC MOTORS AND ac POWER

● SECTION OUTLINE

18–1   Principles of Electric Motors
18–2   Types of Electric Motors
18–3   ac Power
18–4   Miscellaneous Power Ratings

● TROUBLESHOOTING: Applying Your Knowledge

Your neighbor has a 2.5-ton 240-V single-phase central air-conditioning unit that draws 40 A. Your neighbor says his house is too hot and thinks something might be wrong with the unit. What do you think?

● OVERVIEW AND LEARNING OBJECTIVES

Many machines include electric motors and, thus, you need a general understanding of the basic principles and categories of motors. You will also benefit from a basic understanding of ac power for two reasons. First, we are all consumers of ac power, particularly for heating and air conditioning. Second, many industrial processes involving large motors, heaters, and chillers are controlled electronically with digital equipment. If you become employed in this branch of industry either as an electrician or electronics technician, you will need a basic understanding of the principles of ac power.

After completing this chapter, you will be able to:

1. Define the horsepower unit and convert between watts and horsepower.
2. Define torque and perform simple torque calculations.
3. Perform simple calculations relating horsepower, torque, and rpm in electric motors.
4. Name the basic parts of an electric motor and explain their purpose.
5. State the most commonly used methods of fulfilling the three requirements of all electric motors.
6. Name the most common types of electric motors.
7. Explain the power triangle.
8. Define the thermal ratings used in heating and air-conditioning equipment and convert these ratings to watts.

SECTION 18–1: Principles of Electric Motors

In this section you will learn the following:

→ One horsepower (hp) is 550 foot-pounds of work per second.
→ One horsepower equals 746 watts.
→ Torque is rotational energy or work.
→ Horsepower is a combination of torque and the speed of shaft rotation.
→ An electric motor consists of a stator and a rotor.
→ In a slip-ring arrangement, each end of the armature always contacts the same brush as the slip rings rotate with the shaft.
→ In a commutator arrangement, each end of the armature contacts first one brush, then the other brush, as the commutator rotates with the shaft.
→ The first requirement of an electric motor is to create at least two magnetic fields, one in the stator and the other in the rotor.
→ The second requirement of an electric motor is to create current in the rotor.
→ The third requirement of an electric motor is to continuously reposition one or the other of the magnetic fields to keep the rotor turning.

Horsepower

The *horsepower* is the traditional unit of mechanical power in the British (Imperial) system. As we have learned in earlier chapters, all units of power indicate a certain amount of work done in a certain amount of time, and the definition of work is a force acting through a distance. In the British system, force is measured in pounds, distance in feet, and time in seconds. The horsepower is defined in these units. One horse power equals 550 foot-pounds of work per second.

Relationship Between Horsepower and Watts

The horsepower and the watt are both units of power. One hp equals 746 W, as indicated in Formula Diagram 18–1.

Formula Diagram 18–1

$$\frac{\text{Watts}}{\text{Horsepower} \mid 746}$$

Torque

Torque is rotational work or energy. Thus, torque is measured in foot-pounds. A good example of torque is using a wrench to tighten a bolt. When you pull on the wrench, you apply a force (your strength) through a distance (the length of the wrench). If we measure the strength of your pull in pounds and the length of the wrench in feet, we multiply the two numbers to get the torque: the foot-pounds of rotational energy you are applying to the bolt. For example, if you pull with a force of 80 pounds on a 6-inch (0.5 foot) wrench, you apply $80 \times 0.5 = 40$ ft-lb of torque to the bolt, as shown in Figure 18–1(a).

To increase the torque, you can increase the force, the distance, or both. For example, if your 40 ft-lb of torque was not enough to tighten the bolt, you have two options. You could get a stronger hand to apply more force, as shown in Figure 18–1(b), where 200 pounds acting through 6 inches produces 100 ft-lb of torque. Your other choice is to get a longer wrench, as shown in Figure 18–1(c). Here your strength of 80 pounds of force acting through 15 inches (1.25 feet) produces 100 ft-lb of torque.

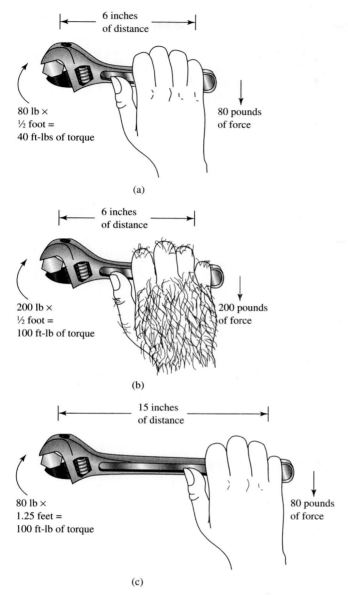

6 inches
of distance

80 lb ×
½ foot =
40 ft-lbs of torque

80 pounds
of force

(a)

6 inches
of distance

200 lb ×
½ foot =
100 ft-lb of torque

200 pounds
of force

(b)

15 inches
of distance

80 lb ×
1.25 feet =
100 ft-lb of torque

80 pounds
of force

(c)

FIGURE 18–1
Torque is rotational work measured in foot-pounds.

Any source of mechanical power is called a prime mover. In our example, the hand is the prime mover. Rotating prime movers include water wheels, steam engines, steam and gas turbines, reciprocating engines (gasoline and diesel), and electric motors. All rotating prime movers produce torque: rotational energy that can perform rotational work.

rpm

Our example said nothing about time: how quickly the work was done. However, in rotational prime movers such as electric motors, time is involved because these machines rotate at various speeds. The rotational speed is usually measured in revolutions per minute (*rpm*).

Relationship Among Horsepower, Torque, and rpm

When we combine torque and rpm, we are combining work and time. Therefore, the combination of torque and rpm is mechanical power: a certain amount of work (torque) being done in a certain amount of time (rpm).

In other words, the horsepower of a motor is a combination of torque and the speed of shaft rotation, measured in rpm. This is why many motors operate their load through a gearbox or transmission, or through belts turning pulleys of different sizes. The purpose of the gears or pulleys is to change the motor's combination of torque and rpm into a different combination more suitable for the load.

The relationships among hp, torque, and rpm are given in Formulas 18–1.

Formula 18–1

a.
$$hp = \frac{rpm \times torque}{5{,}252}$$

b.
$$torque = \frac{5{,}252 \times hp}{rpm}$$

c.
$$rpm = \frac{5{,}252 \times hp}{torque}$$

● *SKILL-BUILDER 18–1*

A table saw has a 120-V single-phase motor that produces 2.5 hp. What is the motor's current?

GIVEN:
$V_S = 120$ V
$P = 2.5$ hp

FIND:
I

STRATEGY:

1. $I = \dfrac{P}{V_S}$ Watt's Law. Source voltage is given. Find power in watts.

2. Watts = horsepower × 746   Formula Diagram 18–1. Horsepower is given.

SOLUTION:

Watts = horsepower × 746
     = 2.5 hp × 746
     = 1,865 W

$$I = \frac{P}{V_S}$$

$$= \frac{1{,}865 \text{ W}}{120 \text{ V}}$$

$$= 15.54 \text{ A}$$

Repeat the Skill-Builder using the following values. You should obtain the results that follow.

Given:

V_S	240 V	277 V	208 V	120 V	240 V
hp	3.8 hp	4.6 hp	2.9 hp	1.3 hp	1.7 hp

Find:

watts	2,835 W	3,432 W	2,163 W	970 W	1,268 W
I	11.81 A	12.39 A	10.40 A	8.08 A	5.28 A

● SKILL-BUILDER 18–2

A 240-V single-phase motor draws 12.8 A and runs at 1,500 rpm. What is the motor's torque?

GIVEN:
  $V_S = 240$ V
  $I = 12.8$ A
  rpm $= 1,500$

FIND:
  torque

STRATEGY:
  1.
  $$torque = \frac{5,252 \times hp}{rpm}$$

  Formula 18–1(b)
  We have rpm. We need to find hp.

  2.
  $$hp = \frac{watts}{746}$$

  Formula Diagram 18–1.
  We need to find watts.

  3.
  $$P = I \times V_S$$

  Watt's Law. We have I and V_S.

SOLUTION:
  $P = I \times V_S$
    $= 12.8$ A $\times 240$ V
    $= 3,072$ W

  $hp = \dfrac{watts}{746}$

    $= \dfrac{3,072\ W}{746}$

    $= 4.12$ hp

  $torque = \dfrac{5,252 \times hp}{rpm}$

    $= \dfrac{5,252 \times 4.12\ hp}{1,500\ rpm}$

    $= 14.42$ ft-lb

  Repeat the Skill-Builder using the following values. You should obtain the results that follow.

Given:

V_S	120 V	208 V	277 V	120 V	240 V
I	7.3 A	13.2 A	6.9 A	3.5 A	11.6 A
rpm	800	2,200	3,450	8,500	2,150

Find:

P	876 W	2,746 W	1,911 W	420 W	2,784 W
hp	1.17 hp	3.68 hp	2.56 hp	0.56 hp	3.73 hp
torque	7.71 ft-lb	8.79 ft-lb	3.90 ft-lb	0.35 ft-lb	9.12 ft-lb

---

Electric Motor Terminology

All electric motors consist of two basic parts: the *stator* and the *rotor*.

As shown in Figure 18–2, the stator is the outer stationary part of the motor, the part that is attached to the case of the motor and therefore cannot move. The stator has a large hole in the middle. The rotor fits inside this hole. The rotor includes the shaft and is the part of the motor that rotates (turns).

As you can see from the illustrations, both stator and rotor typically have wire windings. In general, the stator winding is called the *field coil* and the rotor winding is called the *armature*. However, these terms are sometimes used inconsistently, and so you must be careful to interpret the terms correctly in technical literature on motors.

Figure 18–3 illustrates two mechanical arrangements commonly found in both motors and generators. Figure 18–3(a) shows *slip rings*. The illustration shows a single loop of wire labeled as the armature. An actual armature would be many hundreds of loops of wire wrapped in a complex fashion. The illustration has simplified the armature to reveal its basic principles.

In the drawing, each end of the armature is soldered to a slip ring. The slip rings are made of brass and are mounted on the shaft, which is not shown. Small blocks called *brushes* press against the slip rings, held by springs which are not shown. The brushes are made of carbon, an element similar to coal or pencil leads. As the rotor shaft turns, the soft carbon rubs smoothly against the hard brass slip rings. Carbon is a conductor, and by attaching wires to the brushes we can make electrical contact with the moving slip rings and the armature winding.

Figure 18–3(b) shows an arrangement called a *commutator*. Here a single long ring has been split lengthwise into two segments. The gap between the segments is filled with insulation, and the shaft passes through the hole. Each end of the armature is soldered to one segment of the commutator.

FIGURE 18–2
An electric motor consists of a rotor and a stator.

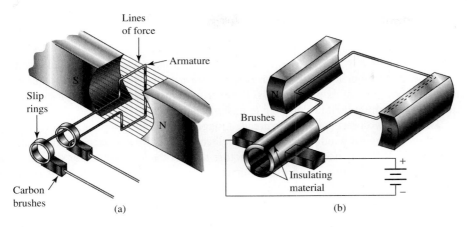

FIGURE 18–3
(a) Slip rings and brushes. (b) Commutator and brushes.

To understand how slip rings and commutators are used in generators and motors, you must recognize the difference in the way each arrangement connects the armature to the brushes. In a slip-ring arrangement, each end of the armature always contacts the same brush as the slip rings rotate with the shaft. In a commutator arrangement, each end of the armature contacts first one brush, then the other brush, as the commutator rotates with the shaft.

In other words, the slip-ring arrangement provides only electrical contact between the armature and the brushes, whereas the commutator provides both contact and switching action, connecting the brushes to different sections of the armature.

Slip rings are generally found in generators or simple electrical devices such as revolving electrical signs, where electrical contact is the only requirement. Commutators are found mostly in motors, where the switching action is important, as we will soon learn.

Finally, there is a particular rotor design called a squirrel cage, shown in Figure 18–4. The term is a nickname because of the appearance of the rotor frame, which is made of aluminum bars mounted between two aluminum rings. The open cage is then filled in with iron laminations (thin slices or disks), creating a solid rotor.

The squirrel-cage rotor does not have an armature made of a wire winding. Instead, the aluminum frame is the armature, and current circulates through the rods and rings. The iron laminations are insulated with an oxide glaze so that rotor current does not enter them. We will discuss the operation of the squirrel-cage rotor later in this chapter.

FIGURE 18–4
Squirrel-cage rotor of an ac
induction motor.

The Three Requirements of All Electric Motors
Requirement 1

The first requirement of an electric motor is to create at least two magnetic fields, one in the stator and the other in the rotor. An electric motor turns because of the pushing and pulling between the two magnetic fields. If either field disappears, the motor cannot turn because the remaining field has nothing to push or pull against.

There are two practical ways to create a magnetic field: Use a permanent magnet or pass current through a winding. Some small electric motors use permanent magnets for the stator field. These motors include toy motors such as those found in toy slot cars, and certain high-tech miniature motors in military or aerospace equipment. Most common motors use current in a field winding to create the stator field, and virtually all motors use current in a rotor winding or squirrel cage to create the rotor field.

Requirement 2

The second requirement of an electric motor is to create current in the rotor. This is necessary to create the rotor magnetic field. There are two general methods to create rotor current.

The first method is a commutator connected to a rotor winding. Current is fed directly onto the rotor through the brushes.

The second method is induction. Induction motors use the stator winding to create current in the rotor through transformer action. In other words, the stator winding is doing two jobs. First, it is creating the stator magnetic field. Second, it is acting like the primary of a transformer, inducing voltage into the rotor, which acts like the secondary of a transformer.

Induction motors usually have a squirrel-cage rotor and do not have brushes or commutators, although there are some exceptions to this rule. Also, induction motors are ac motors, since dc does not produce transformer action.

Requirement 3

The third requirement of an electric motor is to continuously reposition one or the other of the magnetic fields to keep the rotor turning. As shown in Figure 18–5(a), suppose two permanent magnets are placed near each other. One magnet is mounted so it cannot move, and acts as the stator field. The other magnet can spin on a shaft, and acts as the rotor field. Because of magnetic attraction and repulsion, the rotor magnet moves as shown. But when the rotor field lines up with the stator field as shown in Figure 18–5(b), the rotor stops turning. In order to keep the rotor turning, we have to reposition one or the other of the fields to keep the two fields at an angle to each other. This is what produces torque in an electric motor.

One solution to the problem is shown in Figure 18–5(c). The rotor winding is actually a continuous loop, divided into sections by connections to a commutator of many segments. Current is fed into the armature through the brushes. The pattern of current in the armature depends on which segments of the commutator are touching the brushes. The brushes are positioned so the current pattern will produce a magnetic field at a large angle to the stator field. This creates torque, and the rotor begins to turn.

As the rotor turns, one set of commutator segments slides off the brushes and another set of segments slides on. This is the switching action that maintains the pattern of current flow in the armature so that the rotor field remains at a large angle to the stator field, which maintains the torque necessary to keep the rotor turning.

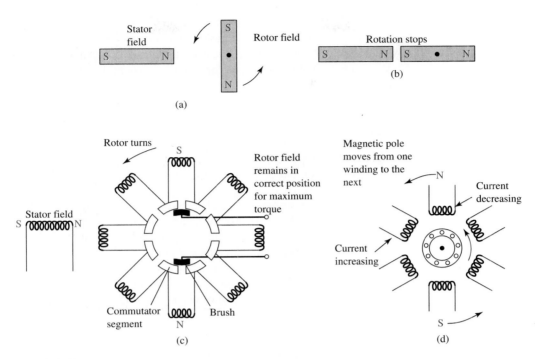

FIGURE 18–5
Either the rotor field or the stator field must be continuously repositioned for rotation to continue.

This action is similar to a socket wrench. The handle of the wrench moves back and forth through a small distance, but the torque is always in the same direction. Therefore, the socket continues to turn in the same direction. The handle is simply repositioned for the next little pull.

In summary, Figure 18–5(c) shows how the switching action of the commutator controls the pattern of current flow in the armature to reposition the rotor field and maintain torque.

In Figure 18–5(d) we see a method for repositioning the stator field. Remember that the stator field is not the same thing as the stator. The stator is a physical object: copper wire wound around an iron core. The physical stator does not ever move. On the other hand, the stator field is an invisible magnetic force which seems to be concentrated at two locations, the north and south poles of the field. If we can make the stator field's magnetic poles come from different parts of the stator, the rotor field will simply chase the stator field, and we will always have torque. This method is called a revolving stator field.

The easy way to do this is to have several sets of stator windings interconnected in some way so that the currents in the windings are out of phase. This means that at any moment one winding will have more current than the others, and the magnetic field will be strongest near that winding. Then, because the winding currents are out of phase, as the current decreases in that winding, the current will be increasing in another winding. The result is that the poles, the strong points of magnetism, will move to another part of the stator. This is the invisible action described as the revolving stator field. Current in the rotor will produce a rotor field that will chase the moving stator field, and the rotor will continue to turn. As shown in the illustration, a motor that uses a revolving stator field is typically an induction motor with a squirrel-cage rotor, whose current is induced by transformer action from the stator windings.

Although the movement of the stator field is invisible, it moves at a certain speed, which can be measured in rpm. The speed at which the stator field revolves is called the *synchronous speed* of the motor. Since the rotor field is chasing the revolving stator field, the rotor can never go faster than the synchronous speed. The synchronous speed is usually a multiple of the ac line frequency, 60 Hz in North America and 50 Hz overseas. Thus, a typical synchronous speed for American motors is 3,600 rpm, and 3,000 rpm for European motors.

→ *Self-Test*

1. What is 1 horsepower?
2. What is the relationship between horsepower and watts?
3. What is torque?
4. What is the relationship between horsepower, torque, and the speed of shaft rotation?
5. What are the two basic parts of an electric motor?
6. What is the function of a slip-ring arrangement?
7. What is the function of a commutator arrangement?
8. What is the first requirement of an electric motor?
9. What is the second requirement of an electric motor?
10. What is the third requirement of an electric motor?

SECTION 18–2: Types of Electric Motors

In this section you will learn the following:

- → A series dc motor has its field coil and armature in series.
- → A shunt dc motor has its field coil and armature in parallel.
- → A compound dc motor is a combination of a series and a shunt arrangement.
- → The most common type of single-phase ac motor is the series ac motor.
- → The repulsion-start induction motor is a single-phase ac motor that has a commutator and brushes similar to the series ac motor.
- → Squirrel-cage rotor, single-phase ac induction motors do not have a commutator.
- → The three-phase induction motor is the most common type of large industrial motor.
- → Synchronous speed is the speed at which the stator field revolves in a three-phase motor.
- → A synchronous motor is a three-phase motor that runs at synchronous speed.

dc Motors

Direct current motors generally have a commutator and wound rotor. Small dc motors may have a permanent magnet for the stator field. Larger motors have a field coil.

As shown in Figure 18–6, dc motors are classified according to the electrical connection between the field coil and the armature. A *series dc motor* has its field coil and armature in series, as shown in Figure 18–6(a). A *shunt dc motor* has its field coil and armature in parallel, as shown in Figure 18–6(b). The word *shunt* sometimes means a normal parallel connection, and sometimes means an intentional short.

A *compound dc motor,* shown in Figure 18–6(c) and (d), is a combination of a series and a shunt arrangement. Figure 18–6(c) is called a long shunt, and Figure 18–6(d) is called a short shunt.

Each type of dc motor has its advantages and disadvantages. A series motor has very high starting torque, but if it is operated without a mechanical load, it may run so

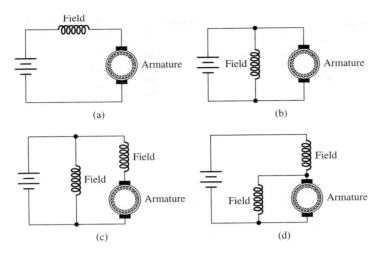

FIGURE 18–6
Schematics of dc motors.

fast that it flies apart. A shunt motor will not overspeed if it loses its mechanical load, but it does not produce as much torque as a series motor. A compound motor is a compromise design with some of the benefits of both motors.

Single-Phase ac Motors

The most common type of single-phase ac motor is the *series ac motor,* which is similar in design to the series dc motor: The field coil and armature are in series, and a commutator is used. This is the motor that operates electric drills, blenders, small fans, and so on. It is inexpensive and reliable.

A series ac motor produces very high starting torque even when its rotor cannot turn. We say the motor has a high locked-rotor torque. Thus, the motor is used in applications where high locked-rotor torque is necessary, from drills to railroad locomotives. In a railroad locomotive, a diesel engine drives an ac generator, which powers a series ac motor that turns the wheels. The high locked-rotor torque of a series ac motor allows the motor to get the train moving without having to use a clutch or automatic transmission.

If you operate a drill or blender in a dark room and look inside the ventilation slots, you can see sparks as the commutator slides beneath the brushes. In time, the brushes wear down and have to be replaced. When you open the motor to do this, you will see a lot of black carbon powder inside the motor. Because of the sparks, a brush-and-commutator motor cannot be used near flammable or explosive materials. Because of the brushes wearing out, these motors are not used in locations where maintenance would be difficult or dangerous, such as the top of a tower or the bottom of a deep shaft.

Single-Phase Induction Motors

The *repulsion-start induction motor* is a single-phase ac motor that has a commutator and brushes similar to the series ac motor. The difference is that the brushes are shorted to each other instead of bringing current to the armature from outside the motor. This motor uses the brushes and commutator only for the switching action to direct the pattern of current in the rotor so the rotor field maintains torque with the stator field.

Squirrel-cage rotor single-phase ac induction motors do not have a commutator. A motor without a commutator has several advantages: It is inexpensive to manufacture, it needs little maintenance, and it does not produce sparks. Its chief disadvantage is the need for extra components and windings to get the stator field to revolve. Ac induction motors use several methods to produce the revolving stator field.

To make the stator field revolve, we use more than one stator winding, and then get the currents out of phase in these windings. The *split-phase motor* simply uses windings of different inductance and resistance, which produce the necessary phase shift. The *capacitor-start motor* uses a large capacitor to produce the phase shift. This type of motor is quite common in medium-size single-phase appliances such as washing machines. Figure 18–7 shows a typical capacitor-start motor, with the capacitor mounted on top of the case.

Figure 18–8 shows some details of a *shaded-pole motor.* This type is widely used in phonograph turntables, tape decks, VCRs, and so on. It is unusual in its construction. The stator core has two iron arms that extend toward the rotor. These extensions are called *salient poles.* Each pole has a slot cut in its face. The slot is cut off-center, and a ring of heavy copper is mounted in the slot as shown. This single ring acts like an additional stator winding. Current is induced in the ring from the main stator winding (the field coil). Because the ring and slot are off-center, the current induced in the ring is out of phase with the current in the field coil. The small magnetic field created by the ring interacts with the main magnetic field to produce a revolving stator field.

Applying Your Knowledge

If its capacitor becomes defective, a capacitor-start induction motor will hum or buzz when power is turned on but will not start to turn. Sometimes you can spin the shaft by hand, and the motor will then start to run. However, a dirty or rusty motor will behave in a similar manner. You should clean and lubricate the motor before replacing the capacitor.

Three-Phase ac Motors

Three-phase power is widely used in industrial applications. We will discuss three-phase power in the next section.

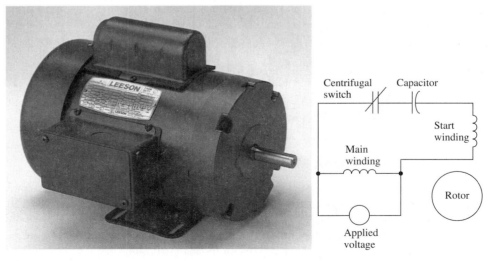

FIGURE 18–7
Capacitor-start single-phase induction motor.

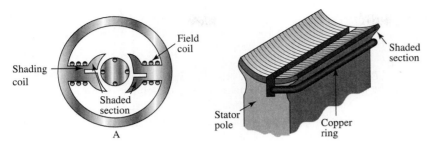

FIGURE 18–8
Shaded-pole motor.

The *three-phase induction motor* is the most common type of large industrial motor used in countless applications requiring medium to high mechanical power. This motor has many advantages. It is extremely rugged and simple in its construction. It has a solid squirrel-cage rotor and does not need any extra features like capacitors to get it started because it is already getting current out of phase on its three incoming power lines. Like all three-phase loads, it receives almost continuous power, as explained in the next section. Because its maintenance requirement is so low, you can put it in difficult locations and forget about it for several years. It produces no sparks and so is suitable for refineries, chemical plants, and so on. It has high locked-rotor torque, and in fact gains torque automatically when a heavy load slows down its rotor. Thus, it is ideal for applications where the mechanical load can vary, such as a passenger elevator in a building.

However, the three-phase induction motor cannot run at a steady speed under varying load conditions. It changes torque by changing speed.

Synchronous speed is the speed at which the stator field revolves in a three-phase motor. An ordinary induction motor always runs slower than synchronous speed. If the rotor ever caught up with the revolving stator field, there would be no relative motion between the stator field and the rotor. This would eliminate the transformer action between stator and rotor. The rotor would have no current and thus no magnetic field of its own. Therefore, the induction motor must run slower than synchronous speed to maintain its rotor current.

There are many industrial applications where a steady speed under varying loads is required, such as driving a conveyor belt on an assembly line. The motor used for this job is a *synchronous motor,* which runs exactly at synchronous speed. A synchronous motor has a rotor which includes both a squirrel cage and a separate winding. The motor begins to rotate as a normal induction motor and picks up speed until it is running just below the synchronous speed. At that point, an external dc current is applied to the rotor winding through brushes and slip rings. This establishes a separate rotor field that allows the rotor to catch up to the revolving stator field. In other words, the rotor can run at synchronous speed because it no longer depends on induction to provide rotor current. Note that the rotor has slip rings instead of the usual commutator, and the rotor current is dc, even though the stator windings are running on three-phase ac. The reason for the dc rotor current is to maintain a constant magnetic polarity on the rotor, allowing it to lock onto the poles of the stator field. If the load changes and the rotor needs more torque, we simply increase the dc rotor current to create a stronger rotor field, which will have enough strength to stay locked onto the stator field under the greater load.

If we need to control the speed of a three-phase induction motor, we use a wound rotor instead of a squirrel-cage rotor. This rotor has a three-section winding, which connects through slip rings to an external adjustable resistor. By controlling the resistance of the rotor winding, we control the current and, therefore, control the torque and rpm of the rotor.

→ *Self-Test*

11. What is a series dc motor?
12. What is a shunt dc motor?
13. What is a compound dc motor?
14. What is the most common type of single-phase ac motor?
15. What is a repulsion-start induction motor?
16. What feature characterizes squirrel-cage motors?
17. What is the most common type of large industrial motor?
18. What is synchronous speed?
19. What is a synchronous motor?

SECTION 18–3: ac Power

In this section you will learn the following:

→ Apparent power equals the voltage applied to a circuit multiplied by the total current in the circuit. Apparent power is measure in volt-amperes (VA).

→ True power equals the power dissipated by the circuit. True power is measured in watts (W).

→ Reactive power equals the power drawn from the source and then returned to the source by reactive components (coils and capacitors) in a circuit. Reactive power is measured in volt-amperes reactive (VAR).

→ When ac voltage and current have the same polarity, the circuit is drawing power from the voltage source.

→ When ac voltage and current have opposite polarity, the circuit is returning power to the voltage source.

→ Power factor is the ratio of true power to apparent power.

→ Three-phase power consists of three separate ac waves out of phase by 120°.

→ Three-phase power uses two special connections, delta and wye, to carry the three ac waves on three wires.

→ Three-phase apparent power is calculated as line-to-line voltage multiplied by line current multiplied by 1.73.

When we analyze motors, we treat the mechanical output power of the motor as if it were additional resistance in the circuit. We do this for two reasons. First, mechanical power is dissipated: The energy is used up and gone. Second, the current associated with mechanical work is in phase with voltage. In these two ways, mechanical work resembles resistance.

Figure 18–9(a) is the equivalent circuit of a single-phase ac motor. The resistance value of 4.157 Ω represents both the actual electrical resistance of the motor windings and the mechanical work performed by the motor. The windings also have an inductance of 6.37 mH. Therefore, the motor acts like a series *RL* circuit.

Figure 18–9(b) shows a *power triangle*. The power triangle is an important concept in industrial electricity where motors are involved. The purpose of this section is to explain the relationship of the circuit and the power triangle in Figure 18–9.

Apparent Power

The word *apparent* means "looks like" or "seems to be." The *apparent power* of a motor is the power it seems to have. According to Watt's Law, electrical power equals volts multiplied by amps. Therefore, if we measure the voltage applied to the motor and the current flowing into the motor, and then multiply the two values, we have the apparent power. Apparent power is measured in volt-amperes (VA). In Figure 18–9(a), with 120 V_{rms} applied and 25 A_{rms} flowing, the motor's power appears to be $120 \times 25 = 3,000$ VA.

True Power

True power equals the power dissipated by the circuit. True power is measured in watts (W) and represents both mechanical work and heat. True power is also called effective, average, or active power.

Reactive Power

Reactive power equals the power drawn from the source and then returned to the source by reactive components (coils and capacitors) in a circuit. Reactive power is measured in volt-amperes reactive (VAR).

Reactive power is not dissipated. This means it does not turn into heat or mechanical work. Instead, reactive power is drawn from the source to create the magnetic field around a coil or to charge the plates of a capacitor, then returned to the voltage source when the magnetic field collapses or the plates discharge.

Large motors typically have power figures in thousands of units, and so the three symbols for power often have the prefix *k* in front of them. Thus, you will often see ac power values given as kVA, kW, and kVAR.

The power triangle for Figure 18–9 indicates a true power of 2,598 W and a reactive power of 1,500 VAR. We will now see where these numbers come from.

Let's begin by calculating some values for the circuit in Figure 18–9, using the formulas and methods we have already learned.

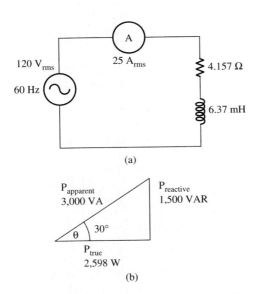

(a)

(b)

FIGURE 18–9
Equivalent motor circuit and power triangle.

$$X_L = 2\,\pi f L$$
$$= 2 \times \pi \times 60\ \text{Hz} \times 6.37\ \text{mH}$$
$$= 2.4\ \Omega$$

$$Z = \sqrt{R^2 + X_L^{\,2}}$$
$$= \sqrt{4.157^2 + 2.4^2}$$
$$= 4.8\ \Omega$$

$$I = \frac{V_S}{Z}$$
$$= \frac{120\ \text{V}_{\text{rms}}}{4.8\ \Omega}$$
$$= 25\ \text{A}_{\text{rms}}$$

$$V_R = I \times R$$
$$= 25\ \text{A}_{\text{rms}} \times 4.157\ \Omega$$
$$= 103.93\ \text{V}_{\text{rms}}$$

$$V_L = I \times X_L$$
$$= 25\ \text{A}_{\text{rms}} \times 2.4\ \Omega$$
$$= 60\ \text{V}_{\text{rms}}$$

$$\theta = \tan^{-1}(X_L \div R)$$
$$= \tan^{-1}(2.4\ \Omega \div 4.157\ \Omega)$$
$$= 30°$$

The three power values are calculated using ordinary Watt's Law formulas with the appropriate values:

$$P_{\text{apparent}} = I \times V_S$$
$$= 25\ \text{A} \times 120\ \text{V}$$
$$= 3{,}000\ \text{VA}$$

$$P_{\text{true}} = I \times V_R$$
$$= 25\ \text{A} \times 104\ \text{V}$$
$$= 2{,}598\ \text{W}$$

$$P_{\text{reactive}} = I \times V_L$$
$$= 25\ \text{A} \times 60\ \text{V}$$
$$= 1{,}500\ \text{VAR}$$

Now we will see exactly what is going on in the motor to produce these numbers.

The ac Power Cycle

When ac voltage and current have the same polarity, the circuit is drawing power from the voltage source. When ac voltage and current have opposite polarity, the circuit is returning power to the voltage source.

Figure 18–10 is a graph of current and resistive voltage. Notice that because the two quantities are in phase, they always have the same polarity: Both are positive from 0° to 180°, and both are negative from 180° to 360°. According to the rules we just learned, this means that the resistive component is always drawing power from the source. This is what we mean when we say the resistive component (heat and mechanical work) dissipates power. It always takes energy from the source and never gives any energy back.

In the preceding calculations, we found I equals 25 A and V_R equals 104 V. These are rms values. We can convert them to peak values by dividing by 0.707. Thus, V_R has a peak value of 147 V and I has a peak of 35.36 A, as shown in the graph.

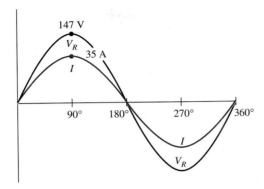

FIGURE 18–10
Resistive voltage and current.

Figure 18–11 is the power cycle for true power: the power dissipated as heat and mechanical work. The graph is obtained by taking the instantaneous values of I and V_R from Figure 18–10 and multiplying them. For example, if we multiply the peak voltage and the peak current, we get the peak power: 147 V × 35.36 A equals 5,198 W. The graph never goes below zero, which means the resistive component never returns power to the source.

Figure 18–12 is a graph of current and reactive voltage. As we have learned, V_L leads I by 90°. Therefore, V_L is already at its peak value of 84.87 V when I is at zero.

Notice the following points in the graph during the first half-cycle. The second half-cycle is identical to the first.

1. From 0° to 90°, current and voltage have the same polarity. During this time, the coil is drawing energy from the source to build up its magnetic field. The polarity of the back voltage opposes the attempt of the current to increase, and energy is needed to overcome the effects of the back voltage.

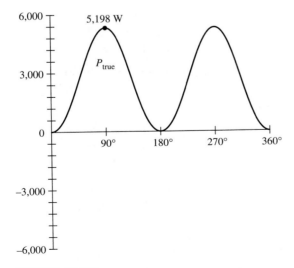

FIGURE 18–11
True power cycle. Power is drawn from source continuously.

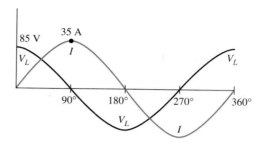

FIGURE 18–12
Reactive voltage and current.

2. From 90° to 180°, current and voltage have opposite polarity. During this time, the coil is returning energy to the source. Its collapsing magnetic field now induces a back voltage in the circuit with a polarity that tends to keep the current flowing in the same direction, even though the source voltage is dropping during this time.

Figure 18–13 is a reactive power cycle, which clearly shows the coil drawing power from the source (positive power) during the first and third quarter-cycles, and returning an equal amount to the source (negative power) during the second and fourth quarter-cycles. In this particular motor, the winding draws a peak reactive power of 1,500 VAR, and then returns it.

Figure 18–14 is a graph of all three power cycles: true, reactive, and apparent. This graph is the key to understanding ac power.

The power cycles for true and reactive power are plotted first. The apparent power cycle is simply the instantaneous sum of the other two. When we work with instantaneous values, time (phase angle) does not matter. The whole idea of instantaneous values is like stopping the clock and seeing what the circuit is doing at that exact moment.

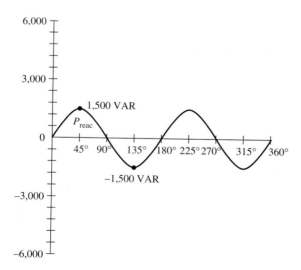

FIGURE 18–13
Reactive power cycle. Power is drawn from source for a quarter-cycle, then returned to the source during the next quarter-cycle.

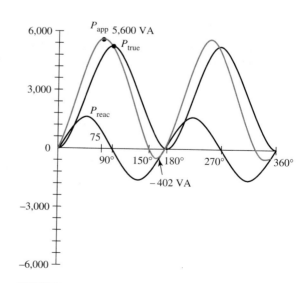

FIGURE 18–14
Apparent power is the combination of resistive and reactive cycles.

We analyze this graph as follows:

1. From 0° to 90°, apparent power is larger than true power. During the first quarter-cycle, the source must provide energy both for heat and mechanical work (true power) and to expand the magnetic field against the opposition of the back voltage. In other words, apparent power is larger than true power because of the additional demand of reactive power.

2. From 90° to 150°, apparent power is smaller than true power. During this time, the coil is returning power to the circuit as its magnetic field collapses. However, this power does not go back to the source. Instead, it goes into the resistive load, which reduces the load's need for apparent power. In other words, during this time the returning power from the coil is helping to run the load, and the voltage source provides only the difference between the power that the load demands and the power that the coil provides.

3. Apparent power reaches zero at 150° in this particular circuit because at that moment the coil is providing exactly the amount of power the load demands. Between 150° and 180°, the returning reactive power is greater than the power required by the load, and the difference, the extra reactive power, is actually returned to the voltage source. Therefore, apparent power is negative during this time.

As shown on the graph, the peak value of power drawn from the source is 5,600 VA, and the peak value returned to the source is 402 VA.

Figure 18–15 shows the relationship between source voltage and current that results from Figure 18–14.

If the coil did not demand reactive power and then return it, the apparent power from the source would always equal the true power required by the load. On the other hand, if the load did not dissipate power, then all the reactive power drawn from the source would be returned to the source, and the total energy consumed would be zero. This means you would not owe the light company any money because nothing happened: no work was done. The energy merely bounced back and forth between the generator and the magnetic field of the coil.

How Do the Power Values Fit Together?

Our calculations using rms values gave us an apparent power of 3,000 VA_{rms}, a true power of 2,598 W_{rms}, and a reactive power of 1,500 VAR_{rms}. Then our graphs gave us

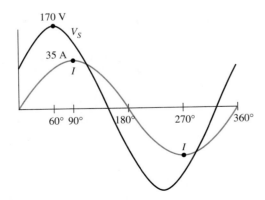

FIGURE 18–15
Source voltage and current.

peak values: Peak apparent power equals +5,600 VA and −402 VA, peak true power equals 5,198 W, and peak reactive power equals ± 1,500 VAR. The reactive power value seems to be the same for both rms and peak, the true power peak value is twice the rms, and the apparent power values make no sense at all. What is going on here?

The rms values represent the average value of the areas under the curves in the instantaneous value graphs, as shown in Figure 18–16. Because the graph is a curve and not a straight line, the problem is solved using an advanced mathematical method called integral calculus, which is beyond the coverage of this book. For this reason, the relationship between the rms values and the peak values is not obvious. You must trust the rms values to give you correct results in your calculations.

The Power Triangle

The power triangle is used to describe ac power for several reasons. First, it is less confusing to look at. Second, trigonometry can be used to calculate values based on the triangle. Third, by using rms values instead of peak values, we can compare an ac circuit with an equivalent dc circuit.

What Is the Problem with ac Power?

The problem is that the motor draws more energy than it uses, so the apparent power is higher than it really has to be. This means that the current in the lines is larger than necessary. This costs money in two ways.

First, the wires, conduit, breakers, and so on have to be larger to carry the higher current. For example, if the motor had no reactive power, a current of 21.65 A would produce the same amount of true power: 21.65 A × 120 V = 2,598 W. Because of reactive power, the lines carry 25 A when only 21.65 A is really being used. This may not seem like much of a difference, but on a large scale such as a factory or refinery, the wires would have to carry hundreds of amps more than necessary, requiring larger and more expensive wires, boxes, and so on.

Second, the transmission lines have resistance. A higher current in the lines results in a higher power loss in the lines, which increases the monthly electric bill.

Overall in a large industrial installation, the extra current due to reactive power can cost the user hundreds of thousands of dollars a year. To avoid this expense, companies are willing to install large banks (groups) of special capacitors to balance the reactive power requirement of the motor windings. Instead of being dissipated in the load, the

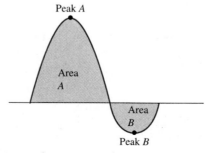

rms value = average value (area *A* − area *B*)

FIGURE 18–16
The rms value is the average area
under the curves.

reactive power can bounce back and forth between the motor windings and the capacitors. All the current drawn from the voltage source goes to the load, which is a more economical operation.

To see the problem more clearly, look at Figure 18–17. We have found that our motor draws 3 kVA to produce 2,598 W of true power, which equals 3.48 hp. However, 3 kVA equals 4.02 hp. If we run our electric motor off a generator powered by a gasoline engine, the engine must provide 4.02 hp to the generator to eventually get 3.48 hp out of the motor. Now you can see how reactive power in an electric motor creates a situation that is not efficient or economical.

Power Factor

In Figure 18–9(b), we put 3,000 VA into a system and got 2,598 W out of it. The *power factor* is the ratio of these two numbers:

$$PF = \frac{P_{out}}{P_{in}}$$

$$= \frac{2,598 \text{ W}}{3,000 \text{ VA}}$$

$$= 0.867$$

In any of the phasor-sum triangles appropriate to this system, the power factor would equal the cosine of theta because the adjacent side of any triangle always represents resistive values (true power), and the hypotenuse always represents total values (apparent power). In our example, theta equals 30°, and the cosine of 30° is 0.867, the power factor.

Three-Phase Power

Three-phase power consists of three separate ac waves out of phase by 120°. All commercial ac power is three phase. A three-phase generator similar to those used by utility companies is shown in Figure 18–18.

At first, you might think that six wires would be required to carry three-phase power, since the generator contains three windings and each winding has two ends. However, three-phase power uses two special connections to carry the three ac waves on three wires.

The *delta* connection is shown in Figure 18–19. Since each line is connected to two windings, the line current is higher than the current in a single winding. Because the windings are 120° out of phase, the line current is 1.73 times the winding current. The number 1.73 is the square root of 3 and is always used in three-phase calculations.

The *wye* connection is shown in Figure 18–20. One end of each winding is connected to a common point, and each of the other ends is connected to a line. Here the fac-

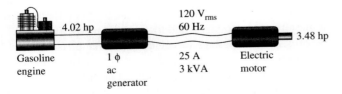

FIGURE 18–17
Power loss due to reactive power.

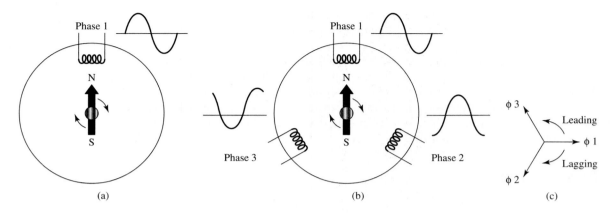

FIGURE 18–18
(a) Single-phase alternator. (b) Three-phase alternator. (c) Phasor diagram of three-phase system.

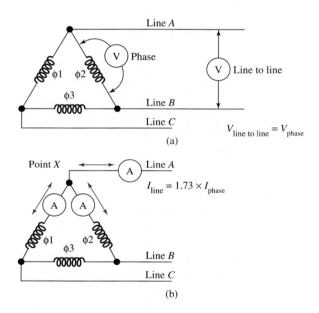

FIGURE 18–19
Delta voltage and current relationships.

tor 1.73 is applied to line-to-line voltage, since there are two windings between any two lines.

In a delta connection, there is no place to put a grounded neutral wire. However, in a wye connection, it is possible to put a neutral where the three windings connect together, as shown in Figure 18–21. This is the most common three-phase arrangement, since it allows a mixture of single-phase and three-phase loads, as shown.

The factor of 1.73 is important in calculating three-phase power. We know that Watt's Law says power is the product of voltage and current. However, in three-phase power we have three ac waves, out of phase by 120°. For this reason, three-phase apparent power is calculated as line-to-line voltage multiplied by line current multiplied by 1.73.

The factor of 1.73 makes a big difference. For example, in Figure 18–22, a 240-V single-phase motor draws 20 A, which figures out to 4.8 kVA, approximately 6.43 hp. By comparison, a 240-V three-phase motor drawing 11.56 A on each of its three hot lines

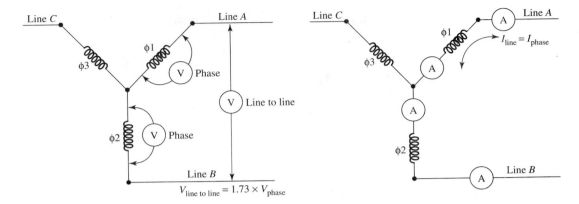

FIGURE 18–20
Wye voltage and current relationships.

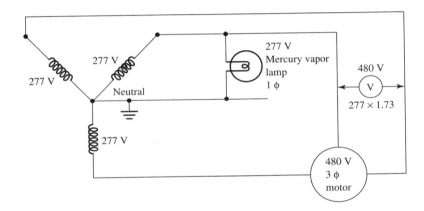

FIGURE 18–21
Wye-connected three-phase four-wire system widely used for
industrial applications.

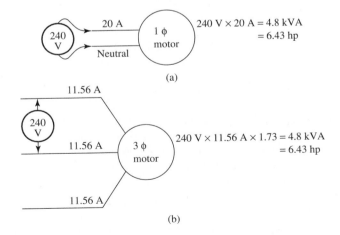

FIGURE 18–22
Comparison of single-phase power and three-phase
power.

produces exactly the same amount of power. Since power is voltage multiplied by current, the factor 1.73 appears in the power formula regardless of whether the machine is wired in delta or in wye. This is one more advantage of a three-phase motor: It can produce the same horsepower at lower line currents.

Formula Diagram 18–2 is Watt's Law with the three-phase factor included.

Formula Diagram 18–2

$$\frac{Power}{Voltage \mid Current \mid 1.73}$$

● *SKILL-BUILDER 18–3*

Compare the current necessary to produce 8 hp with a 240-V single-phase motor and a 240-V three-phase motor.

GIVEN:
$V_S = 240 \text{ V}$
$P = 8 \text{ hp}$

FIND:
$I_{1\phi}$
$I_{3\phi}$

SOLUTION:
$P = 8 \text{ hp} \times 746$
    $= 5{,}968 \text{ W}$

$I_{1\phi} = \dfrac{5{,}968 \text{ W}}{240 \text{ V}}$

    $= 24.87 \text{ A}$

$I_{3\phi} = \dfrac{5{,}968 \text{ W}}{240 \text{ V} \times 1.73}$

    $= 14.37 \text{ A}$

Repeat the Skill-Builder using the following values. You should obtain the results that follow.

Given:

V_S	208 V	120 V	240 V	208 V	120 V
P	3 hp	5.5 hp	4.2 hp	7.6 hp	2.6 hp

Find:

P	2,238 W	4,103 W	3,133 W	5,670 W	1,940 W
$I_{1\phi}$	10.76 A	34.19 A	13.06 A	27.26 A	16.16 A
$I_{3\phi}$	6.22 A	19.76 A	7.55 A	15.76 A	9.34 A

→ *Self-Test*

20. What is apparent power?
21. How is apparent power measured?
22. What is true power?
23. How is true power measured?

24. What is reactive power?
25. How is reactive power measured?
26. When is a circuit drawing power from the voltage source?
27. When is a circuit returning power to the voltage source?
28. What is power factor?
29. How is three-phase apparent power calculated?

SECTION 18–4: Miscellaneous Power Ratings

In this section you will learn the following:

→ One Btu is the energy necessary to change the temperature of 1 pound of water by 1° Fahrenheit.
→ The Btu rating of a heating or refrigerating machine is the number of Btus the machine can produce or remove in a time of one hour.
→ The ton rating of a refrigerating machine is the number of tons of ice the machine can produce in 24 hours by removing latent heat (144 Btus per pound) from water already cooled to 32° F.

We conclude our introductory study of electricity with some basic information on heating and refrigeration power ratings, since you may occasionally need to calculate current based on these ratings.

The British Thermal Unit (Btu)

The *Btu* is a unit of heat energy. One Btu is the energy necessary to increase the temperature of 1 pound of water by 1° Fahrenheit.

The Btu Rating

The Btu rating of a heating or refrigerating machine is the number of Btus the machine can produce or remove in a time of one hour.

Thus, a 10,000-Btu window air conditioner can remove 10,000 Btus of heat from the air in the room in a time of one hour.

Formula Diagram 18–3 gives the relationship between watts and Btus.

Formula Diagram 18–3

$$\frac{\text{Watts}}{\text{Btu rating} \mid 0.2928}$$

How many amps would a 240-V 12,000-Btu air conditioner draw? According to Formula Diagram 18–3.

$$\text{Watts} = \text{Btu rating} \times 0.2928$$
$$= 12{,}000 \text{ Btu} \times 0.2928$$
$$= 3{,}514 \text{ W}$$

$$I = \frac{3{,}514 \text{ W}}{240 \text{ V}}$$

$$= 14.64 \text{ A}$$

The Ton Rating

The *ton* rating of a refrigerating machine is the number of tons of ice the machine can produce in 24 hours by removing latent heat (144 Btus per pound) from water already cooled to 32°F.

As you can see, the ton rating is more complicated than the Btu rating. The ton rating is also a unit of power: a certain amount of energy (Btus) in a certain amount of time. To understand the ton rating, we must learn an interesting fact from chemistry: *latent heat.*

Water freezes at 32°. If we have a pound of water at 33° and cool it down one degree to 32°, we have removed one Btu. You might think that removing one more Btu would turn the water to ice at 31°, but this is not true. Like most substances, water contains a certain amount of latent (hidden) heat which it must give up in order to freeze. It is possible to have water at 32°, and it is also possible to have ice at 32°, the same temperature. The difference is whether or not the water has given up its latent heat in order to turn into ice.

The latent heat of water is 144 Btus per pound. Therefore, a ton of water (2,000 pounds) contains 288,000 Btus of latent heat. A one-ton refrigerating unit can remove this much heat in one day (24 hours).

Formula Diagram 18–4 gives the relationship between watts and ton rating.

Formula Diagram 18–4

$$\frac{\text{Watts}}{\text{Ton rating} \mid 3{,}517}$$

How much current would you estimate is necessary for a 240-V, 2½-ton central air-conditioning unit?

$$\text{Watts} = \text{ton rating} \times 3{,}517$$
$$= 2.5 \times 3{,}517$$
$$= 8{,}793 \text{ W}$$

$$I = \frac{8{,}793 \text{ W}}{240 \text{ V}}$$

$$= 36.6 \text{ A}$$

Formula Diagram 18–5

$$\frac{\text{Btu rating}}{\text{Ton rating} \mid 12{,}000}$$

Formula Diagram 18–5 allows us to convert quickly between a Btu rating and a ton rating. For example, a 3-ton rating equals a 36,000-Btu rating. Therefore, a 3-ton central air-conditioning unit cools the same as three 12,000-Btu window air conditioners.

→ *Self-Test*

30. What is 1 Btu?
31. What is the Btu rating?
32. What is the ton rating?

Formula List

Formula 18–1

(a)
$$hp = \frac{rpm \times torque}{5{,}252}$$

(b)
$$Torque = \frac{5{,}252 \times hp}{rpm}$$

(c)
$$rpm = \frac{5{,}252 \times hp}{torque}$$

Formula Diagram 18–1

Watts	
Horsepower	746

Formula Diagram 18–2

Power		
Voltage	Current	1.73

Formula Diagram 18–3

Watts	
Btu rating	0.2928

Formula Diagram 18–4

Watts	
Ton rating	3,517

Formula Diagram 18–5

Btu rating	
Ton rating	12,000

● TROUBLESHOOTING: Applying Your Knowledge

It is possible that something might be wrong with the central air-conditioning unit, but you cannot determine this based on the information you have. A 2.5-ton unit is equivalent to 8,793 watts (2.5 × 3,517); 240 V multiplied by 40 A is an apparent power of 9,600 VA, and 8,793 W divided by 9,600 VA is a power factor of 0.916, which is reasonable. Thus, there is no indication that the current is abnormal. Further investigation is needed to find the problem.

CHAPTER SUMMARY: ANSWERS TO SELF-TESTS

1. One horsepower is 550 foot-pounds of work per second.
2. One horsepower equals 746 watts.
3. Torque is rotational energy or work.
4. Horsepower is a combination of torque and the speed of shaft rotation.
5. An electric motor consists of a stator and a rotor.
6. In a slip-ring arrangement, each end of the armature always contacts the same brush as the slip rings rotate with the shaft.
7. In a commutator arrangement, each end of the armature contacts first one brush, then the other brush, as the commutator rotates with the shaft.
8. The first requirement of an electric motor is to create at least two magnetic fields, one in the stator and the other in the rotor.
9. The second requirement of an electric motor is to create current in the rotor.
10. The third requirement of an electric motor is to continuously reposition one or the other of the magnetic fields to keep the rotor turning.
11. A series dc motor has its field coil and armature in series.

12. A shunt dc motor has its field coil and armature in parallel.
13. A compound dc motor is a combination of a series and a shunt arrangement.
14. The most common type of single-phase ac motor is the series ac motor.
15. The repulsion-start induction motor is a single-phase ac motor that has a commutator and brushes similar to the series ac motor.
16. Squirrel-cage rotor, single-phase ac induction motors do not have a commutator.
17. The three-phase induction motor is the most common type of large industrial motor.
18. Synchronous speed is the speed at which the stator field revolves in a three-phase motor.
19. A synchronous motor is a three-phase motor that runs at synchronous speed.
20. Apparent power equals the voltage applied to a circuit multiplied by the total current in the circuit.
21. Apparent power is measure in volt-amperes (VA).
22. True power equals the power dissipated by the circuit.
23. True power is measured in watts (W).
24. Reactive power equals the power drawn from the source and then returned to the source by reactive components (coils and capacitors) in a circuit.
25. Reactive power is measured in volt-amperes reactive (VAR).
26. When ac voltage and current have the same polarity, the circuit is drawing power from the voltage source.
27. When ac voltage and current have opposite polarity, the circuit is returning power to the voltage source.
28. Power factor is the ratio of true power to apparent power.
29. Three-phase apparent power is calculated as line-to-line voltage multiplied by line current multiplied by 1.73.
30. One Btu is the energy necessary to increase the temperature of 1 pound of water by 1° Fahrenheit.
31. The Btu rating of a heating or refrigerating machine is the number of Btus the machine can produce or remove in a time of one hour.
32. The ton rating of a refrigerating machine is the number of tons of ice the machine can produce in 24 hours by removing latent heat (144 Btus per pound) from water already cooled to 32°F.

CHAPTER REVIEW QUESTIONS

Determine whether the following questions are true or false.

18–1. One horsepower equals 550 watts per second.
18–2. One horsepower equal 33,000 foot-pounds per minute.
18–3. Torque is rotational power.
18–4. Horsepower is a combination of torque and the speed of shaft rotation.
18–5. An electric stator consists of a motor and a rotor.
18–6. A single-phase induction motor needs a revolving stator field to start but does not need one to run.

18–7. A three-phase induction motor needs a revolving stator field both to start and to run.
18–8. A commutator repositions the rotor field by switching current into different sections of the armature.
18–9. The brushes are connected to the ends of the armature.
18–10. All motors use windings to create the stator field.
18–11. Induction motors create rotor current by transformer action, with the field winding acting as a primary winding and the armature acting as a secondary winding.
18–12. Synchronous speed is a multiple of 60 Hz.
18–13. In calculating ac power, the mechanical work done by a motor is treated like resistance.
18–14. The ton rating reflects the amount of latent heat the refrigeration unit can remove in a given time.

Select the response that best answers the question or completes the statement.

18–15. A device is mounted on a rotor so that each end of the armature always contacts the same brush as the shaft rotates. This device is
   a. a commutator.
   b. a slip ring.
   c. a squirrel cage.
   d. a stator.
18–16. A device is mounted on a rotor so that each end of the armature contacts first one brush, then the other brush, as the shaft rotates. This device is
   a. a commutator.
   b. a slip ring.
   c. a squirrel cage.
   d. a shaded pole.
18–17. The minimum number of magnetic fields necessary for a motor to operate is
   a. one.
   b. two.
   c. three.
   d. six.
18–18. Current may be created in a rotor by
   a. a stator or slip rings.
   b. a commutator or induction.
   c. a shaded pole or capacitor.
   d. a permanent magnet or electromagnet.
18–19. A squirrel-cage motor is so called because of the design of the
   a. stator.
   b. rotor.
   c. brushes.
   d. slip rings.
18–20. One Btu is the energy necessary
   a. to produce one ton of ice in a 24-hour period.
   b. to change the weight of water by one degree per pound.
   c. to change the temperature of water by one degree per pound.
   d. to change the energy of water by 746 watts per pound.
18–21. The Btu rating of a heating or refrigerating machine is
   a. the number of Btus the machine can produce or remove in a time of one hour.
   b. the number of Btus the machine can produce or remove in a time of one day.
   c. the pounds of water the machine can turn to ice in one hour.
   d. the pounds of water the machine can turn to ice in one day.

Solve the following problems.

18–22. A 65-pound child hangs from a 4-foot tree branch. How much torque is applied to the tree?

18–23. A pump has a 277-V single-phase motor that produces 6.3 hp. What is the motor's current?

18–24. A 120-V single-phase motor draws 9.6 A and runs at 1,100 rpm. What is the motor's torque?

18–25. A motor has an apparent power of 3.4 kVA and a true power of 2.8 kW. What is the reactive power value?

18–26. A single-phase motor draws 16 A at 240 V and converts 85% of this into mechanical work and heat. What is the phase angle?

18–27. What power factor is necessary for a 240-V single-phase motor to draw 13 A and produce 3.7 shaft horsepower?

18–28. A 240-V single-phase motor draws 17 A. How much current would the motor draw if it were a three-phase motor?

18–29. How many amps would a 240-V 15,000-Btu air conditioner draw?

GLOSSARY
------

apparent power   Total voltage multiplied by total current; measured in VA or kVA.

armature   The winding on the rotor.

Btu   British thermal unit; the energy necessary to change the temperature of 1 pound of water by 1° Fahrenheit.

brush   A block of soft carbon that makes electrical contact inside a motor by rubbing against the commutator.

commutator   A brass band on a rotor, split into segments connected to the ends of the armature windings.

field coil   The winding on the stator.

horsepower   A unit of mechanical power equal to 550 foot-pounds per second, or 746 watts.

power factor   The ratio between true and apparent power.

reactive power   The result of back voltage (CEMF); the reason for power factor; measured in VAR or kVAR.

rotor   The inner part of an electric motor; the part that turns.

rpm   Revolutions per minute; the unit of rotor speed.

slip rings   Two separate brass rings on the rotor connected to the ends of the armature.

stator   The outer part of an electric motor; the part that does not turn.

ton   A rating of refrigeration equipment: the number of tons of ice that can be produced in 24 hours by removing latent heat from water chilled to 32°.

torque   Rotational energy or work.

true power   The power actually dissipated in mechanical work and heat: apparent power times the power factor; measured in watts or kW.

ANSWERS TO CHAPTER REVIEW QUESTIONS

CHAPTER 1

1–1 false
1–3 true
1–5 true
1–7 false
1–9 true
1–11 true
1–13 c. electrical forces
1–15 b. molecules
1–17 b. negative charge
1–19 c. equal numbers of protons and electrons
1–21 b. leave the shell as free electrons
1–23 a. positive charge
1–25 b. electrons
1–27 d. all of the above
1–29 b. space around a charged object
1–31 d. polarity
1–33 b. electromagnetic energy that moves through space
1–35 b. number of waves per second arriving in the beam
1–37 b. heat

CHAPTER 2

2–1 true
2–3 false
2–5 false
2–7 true
2–9 false
2–11 true
2–13 true
2–15 true
2–17 true
2–19 b. newton-meter
2–21 a. voltage
2–23 b. emf
2–25 b. free electrons
2–27 c. ions and free electrons
2–29 b. ampere
2–31 d. resistance
2–33 d. watt
2–35 e. nilo is not a prefix
2–37 b. 285×10^6
2–39 140 ft-lb

2–41 677.5 J
2–43 108,000 J
2–45 4.5 Ω
2–47 25 A
2–49 3 hp; 2,238 W
2–51 3.46×10^{10}
2–53 7.34×10^{-8}
2–55 37,600,000
2–57 23,900,000,000
2–59 a. 37,500 V
　　　 b. 7,500,000,000 W
　　　 c. 0.0223 A
　　　 d. 0.000015 V

CHAPTER 3

3–1 false
3–3 true
3–5 false
3–7 false
3–9 true
3–11 b. 60 Hz
3–13 b. is the correct symbol
3–15 b. voltage rating
3–17 b. power rating
3–19 b. linear and tapered
3–21 a. DIP
3–23 c. limit
3–25 c. overload
3–27 a. short
3–29 a. potentiometer
　　　 b. fuse
　　　 c. dc voltage source
　　　 d. resistor
　　　 e. switch
　　　 f. rheostat
　　　 g. ac voltage source
　　　 h. lamp
3–31 a. 2,100 Ω　1,900 Ω
　　　 b. 616,000 Ω　　　504,000 Ω
　　　 c. 18,000 Ω 12,000 Ω
3–33 a. yes. $|2,850\ \Omega - 2,700\ \Omega| \div 2,700\ \Omega \times 100 =$ 5.5% < 10%
　　　 b. yes. $|31,800\ \Omega - 33,000\ \Omega| \div 33,000\ \Omega \times 100 =$ 3.6% < 5%

c. no. $|470\text{ k}\Omega - 453\text{ k}\Omega| \div 470\text{ k}\Omega \times 100 = 3.6\% > 2\%$

d. no. $|12\text{ }\Omega - 11.28\text{ }\Omega| \div 12\text{ }\Omega \times 100 = 6\% > 5\%$

e. no. $|5.1\text{ M}\Omega - 6.25\text{ M}\Omega| \div 5.1\text{ M}\Omega \times 100 = 22.5\% > 20\%$

f. yes. $|3,600\text{ }\Omega - 3,660\text{ }\Omega| \div 3,600\text{ }\Omega \times 100 = 1.7\% < 2\%$

3–35 Use a low-voltage source and/or a current-limiting resistor in series with the equipment.

CHAPTER 4

4–1 true
4–3 false
4–5 true
4–7 true
4–9 false
4–11 false
4–13 true
4–15 true
4–17 true
4–19 false
4–21 false
4–23 false
4–25 false
4–27 false
4–29 **d.** current is inversely proportional to resistance
4–31 **c.** jack
4–33 **a.** 250 V dc
    b. 6 kΩ
    c. 25 μA dc
    d. 0.5 mA dc
    e. 125 mV dc
    f. 125 V ac
4–35 **b.** infinity
4–37 **c.** 3 electrical points
4–39 3 A
4–41 1.5 A
4–43 120 V
4–45 1,520 W
4–47 560 W
4–49 120 V
4–51 20-V dc range
4–53 112.1 MW
4–55 317.25 MW
4–57 1 part in 19,999
4–59 200-mA range

CHAPTER 5

5–1 true
5–3 false
5–5 false
5–7 true
5–9 false
5–11 false
5–13 true

5–15 false
5–17 true
5–19 **a.** yes
    b. no
    c. no
    d. no
    e. no
    f. yes
    g. yes
    h. no
5–21 **b.** resistance
5–23 **a.** current flows through them
5–25 **c.** the metal frame of the machine
5–27 **a.** a short
5–29 **b.** the same
5–31 **c.** either zero or full source voltage
5–33 22 Ω
5–35 9.7 Ω
5–37 46 V
5–39 24 V
5–41 3 V
5–43 4 V
5–45 **a.** +56.36 V
    b. −48.18 V
5–47 31.9 Ω
5–49 R_2 is shorted

CHAPTER 6

6–1 false
6–3 true
6–5 false
6–7 false
6–9 **a.** R_2 and R_3
    c. R_2 and R_4
6–11 **d.** Kirchhoff's Current Law
6–13 7.6 Ω
6–15 12.5 Ω, 6.25 Ω
6–17 72 W, 96 W, 57.6 W, 225.6 W
6–19 22 Ω
6–21 $R_1 = 20\text{ }\Omega$, $R_2 = 10\text{ }\Omega$, $R_3 = 20\text{ }\Omega$

CHAPTER 7

7–1 true
7–3 false
7–5 true
7–7 true
7–9 true
7–11 true
7–13 false
7–15 true
7–17 false
7–19 **b.** all current paths can be determined by examining the connections of the components

7–21 **c.** decrease total equivalent resistance, and thereby increase total current and total power

7–23 **c.** a short between the point and the source, or an open between the point and ground

7–25 **a.** is lower than the resistance of the component alone

7–27 **c.** multiplied

7–29 **b.** $R_6 = R_2 \parallel R_3$

7–31 $V_1 = 2.69$ V, $I_1 = 134.615$ mA, $V_2 = 2.31$ V, $I_2 = 76.923$ mA, $V_3 = 2.31$ V, $I_3 = 57.692$ mA

7–33 270.53 Ω

7–35 V_1 will remain at 2.2974 V if R_5 is removed. R_5 is grounded on both sides and has no effect on the circuit.

7–37 **a.** 7.5 kΩ
b. 75 kΩ
c. 250 kΩ
d. 750 kΩ

7–39 If V_A and V_B are equal, the drop between them is zero. If this is caused by a short, then R_2 is shorted.

7–41 If V_A equals V_B because of an open, then R_4 is open. The circuit consists of R_1 and R_3. No current flows in R_2, and the voltage at A is present at B through the continuity of R_3. Proof: 18 V × 360 Ω ÷ (100 Ω + 360 Ω) = 14.09 V

CHAPTER 8

8–1 false
8–3 true
8–5 false
8–7 false
8–9 true
8–11 true
8–13 **c.** output
8–15 **c.** Viewed from the output terminal, any practical voltage source is equivalent to an ideal voltage source in series with an internal resistance.
8–17 **a.** output voltage changes with load current
8–19 **d.** five
8–21 4 Ω
8–23 37.8 Ω
8–25 $V_L = 3.47$ V $\quad I_L = 385.6$ mA
8–27 Loop 1: 20 V − 7.7907 V − 12.0535 V − 0.1557 V = 0
Loop 2: 24 V − 11.8477 V − 12.0535 V − 0.0987 V = 0
Loop 3: 20 V − 7.7907 V + 11.8477 V − 24 V + 0.0987 V − 0.1557 V = 0

CHAPTER 9

9–1 true
9–3 false
9–5 true
9–7 false
9–9 **c.** an automobile
9–11 **b.** a flashlight
9–13 **c.** the dielectric strength of the conductor is not a factor
9–15 50 A

9–17 500 nA
9–19 625 kV

CHAPTER 10

10–1 false
10–3 true
10–5 true
10–7 false
10–9 false
10–11 true
10–13 **a.** a bar magnet
10–15 150 At
10–17 1.43 T

CHAPTER 11

11–1 false
11–3 false
11–5 false
11–7 false
11–9 true
11–11 **c.** direction of relative motion is the chief factor in the polarity of induced voltage
11–13 **a.** period
11–15 **a.** peak
11–17 **c.** average
11–19 **b.** and **d.**: period times the speed of light, speed of light divided by frequency
11–21 $V_{peak} = 1.6$ V $\quad V_{rms} = 1.13$ V
11–23 $V_{max} = 31.33$ V $\quad V_{min} = 12.07$ V
11–25 $T = 34$ μs $\quad f = 29.41$ kHz
11–27 85 V
11–29 $V_A = 12.87$ V $\quad V_B = 21.37$ V
11–31 300 MHz 60 MHz

CHAPTER 12

12–1 false
12–3 false
12–5 false
12–7 false
12–9 true
12–11 **c.** the sum of the inductance values
12–13 **b.** self-induced voltage
12–15 **d.** transient time
12–17 **c.** tau
12–19 **d.** the ratio of reactance to resistance
12–21 4.55 mH
12–23 1.626 ms
12–25 1 τ: 52.8 mA, 2.4 ms 2 τ: 67.2 mA, 4.8 ms
3 τ: 72.0 mA, 7.2 ms 4 τ: 73.2 mA, 9.6 ms
5 τ 75.0 mA, 12 ms
12–27 83.95 mA

CHAPTER 13

13–1 false
13–3 false
13–5 true
13–7 false
13–9 false
13–11 true
13–13 false
13–15 true
13–17 true
13–19 false
13–21 false
13–23 false
13–25 false
13–27 true
13–29 true
13–31 false
13–33 **b.** prevents charge from passing between the plates
13–35 **a.** an uncharged object will become charged when a charged object is brought near
13–37 **b.** plate area
13–39 **c.** in series
13–41 **c.** 32 μF
13–43 **c.** effective plate area
13–45 **b.** ability of the atoms to withstand the dislocating forces
13–47 **c.** connecting another capacitor of equal value in series
13–49 **b.** a dc open and an ac short
13–51 **d.** phase shifting
13–53 **d.** arc suppression
13–55 576 μC
13–57 750 μF
13–59 7.44 μF
13–61 58 μF
13–63 24.6 mA

CHAPTER 14

14–1 **d.** both inductance and capacitance produce a 90° phase shift between voltage and current
14–3 **b.** phase
14–5 **d.** the adjacent
14–7 **a.** the hypotenuse
14–9 **d.** a solution of the Pythagorean Theorem
14–11 **a.** the sine function
14–13 **a.** series RL circuit
14–15 **c.** series RLC circuit
14–17 **d.** parallel RL circuit
14–19 **f.** parallel RLC circuit
14–21 **b.** parallel RL circuit
14–23 **c.** series RC circuit
14–25 **a.** series RL or series RC circuit
14–27 **d.** parallel RL or series RC circuit
14–29 21.95 units
14–31 111.74 units
14–33 $\cos \theta = 0.69336$ $\theta = 46.10°$

14–35 15.25 units
14–37 142.99 units
14–39 25.65 units
14–41 $X_L = 1{,}319\ \Omega$ $Z = 1{,}367\ \Omega$ $I = 4.54$ mA
      $V_R = 1.63$ V $V_L = 5.99$ V $\theta = 74.73°$
14–43 $X_L = 5{,}812\ \Omega$ $I_L = 0.71$ mA $I_R = 2.05$ mA
      $I_T = 2.17$ mA $Z = 1{,}889\ \Omega$ $\theta = 18.99°$
14–45 $X_L = 349\ \Omega$ $X_C = 130\ \Omega$ $X_T = 219\ \Omega$
      $Z = 297\ \Omega$ $I = 14.14$ mA $V_L = 4.94$ V
      $V_C = 1.84$ V $V_R = 2.83$ V $\theta = 47.60°$
      $f_r = 22.62$ kHz

CHAPTER 15

15–1 false
15–3 true
15–5 true
15–7 true
15–9 true
15–11 true
15–13 true
15–15 **b.** 1 kHz
15–17 **b.** 5 kHz
15–19 **d.** attenuated
15–21 **c.** Q times the source voltage
15–23 **c.** low impedance at its resonant frequency
15–25 **b.** band-pass filter
15–27 **d.** band-reject filter
15–29 **c.** high selectivity
15–31 −1.25 dB
15–33 149.6 kHz
15–35 $f_c = 34.9$ kHz V_{out} @3.5 kHz = 428 mV
      V_{out} @350 kHz = 4.28 V
15–37 $f_c = 9.46$ kHz V_{out} @ 950 Hz = 189 mV
      V_{out} @ 95 kHz = 1.89 V
15–39 36.2 kΩ

CHAPTER 16

16–1 false
16–3 **b.** the number of repetitions per second of the pulse pattern
16–5 **b.** sharpens the waveform of a pulse
16–7 **a.** a differentiator
16–9 $T = 11.78$ μs   PRR = 84.89 kHz
      Duty cycle = 63.33%
16–11 slight distortion: the duration of the pulse is much longer than the transient time of the integrator

CHAPTER 17

17–1 false
17–3 true
17–5 false

17–7 true

17–9 true

17–11 false

17–13 b. secondary winding

17–15 a. primary winding

17–17 c. the number of primary turns divided by the number of secondary turns

17–19 a. the input impedance of the load equals the output impedance of the source

17–21 d. ac allows the use of transformers to transmit large amounts of power at low current levels

17–23 b. between circuit stages in a radio receiver

17–25 a. instantaneous polarity

17–27 c. uses one winding as both its primary and secondary

17–29 1:38.1

17–31 277.5 kV

17–33 $V_{pri} = 3.6$ V $\quad I_{pri} = 100$ mA
$P = 360$ mW $\quad Z_{pri} = 36$ Ω

17–35 $P_{app} = 18.56$ kVA

CHAPTER 18

18–1 false

18–3 false

18–5 false

18–7 true

18–9 false

18–11 true

18–13 true

18–15 b. slip ring

18–17 b. two

18–19 b. rotor

18–21 a. the number of Btus the machine can produce or remove in a time of one hour

18–23 16.97 A

18–25 1.93 kVAR

18–27 27.8°

18–29 18.75 A

INDEX

A

Absolute permittivity, 354
See Alternating current
Active state of a pulse, 485
Adjacent side of a right triangle, 391
Adjustable resistor, 67
Algebraic sum, 120
Alkaline cells, 228
Alloys, 264
Alnico, 264
Alternating current (ac):
   in a capacitor, 373–75
   defined, 57, 288
   inductive opposition to, 334–40
   open, 340
   power, 548–59
   short, 378
   single-phase ac electric motors, 545
   three-phase ac electric motors,
      546–47
Alternating voltage:
   alternators, 58, 286–88
   bands, 305–6
   frequency, 300–301, 302–3
   graph of, 288–89
   instantaneous and noninstantaneous
      values, 299
   multimeters used to measure, 295
   oscillators and, 290
   oscilloscopes and, 291–92, 295–97
   peak-to-peak values, 297–99
   peak voltage, 293
   period, 301–3
   phase, 306–14
   rms (root-mean-square) voltage, 293
   sine wave, 288–92
   sources, 290
   true power, 299
   voltage values, 292–99
   wavelength, 304–5
Alternators, defined, 58, 286–88
American Wire Gauge (AWG) system,
      233, 234, 235
Ammeters:
   clamping, 108–9
   symbol for, 104
Ampacity, 230–31
Ampere, 32
Ampere-hour rating of cells, 225
Amplification, 449
Analog multimeters, 96, 97–99, 188

B

Back voltage, 327
Balanced bridge, 210
Band-pass filter, 446
Band-pass filter, resonant:
   applications of, 477
   parallel, 476
   series, 475–76
Band-reject filter, 447
Band-reject filter, resonant:
   applications of, 477
   parallel, 477
   series, 476–77
Bands:
   center frequency, 442
   defined, 442
   frequency, 305–6, 441–44
   half-power points, 443
Bandwidth, 442
Bar magnets, 264
Battery (ies):
   automobile, 225–26
   defined, 58, 224
Bell, Alexander Graham, 448
Bias voltage, 378
Blown fuses, 73, **74**
Bode plot, 451
Boosting the signal, 461
Branch, 150
Breakdown voltage, 356–57
Bridges, 210–12
British thermal unit (Btu), 559
Brushes, 540
Buss bars, 244

Apparent power, 548–49
Arc function, 396–98
Arc suppression, capacitors and, 380
Armature, 540
Atomic structure, 2
Atoms
   electromagnetic, 18
   excited, 18
Attenuation, 449
Audio frequency (AF), 301
Automotive circuits, parallel circuits and,
      170
Autotransformers, 526–28
Axial leads, 358

C

Cable:
   coaxial, 236, 238
   defined, 236
   ribbon, 238
   shielded, 236, 238
   twin-lead, 238
   twisted-pair, 236
Calculators, use of scientific, 88, 155
      398–99
Capacitance:
   charge and discharge cycle, 369–71
   defined, 346
   description of amount of charge that is
      stored, 348–49
   determining, 355
   effect of, on current, 369–77
   measuring, 380–81
   in parallel *RC* circuits, 427
   physical formula for, 354–55
   in series *RC* circuits, 409
   unit of, 350–52
Capacitive reactance, 375–77
Capacitors:
   ac current in, 373–75
   applications for, 377–80
   basic, 347–48
   color code, 356
   construction features, 358, 360
   dc working (breakdown) voltage,
      356–57
   dielectric constant, 353–54
   dielectric material, 360–62
   dielectric thickness, 353
   electrolytic, 362–63
   leakage resistance, 357–58
   in parallel, 367–69
   physical characteristics of, 352–54
   plate area, 352
   safety precautions when working with,
      383
   schematic symbols for, 363
   in series, 364–67
   testing and troubleshooting, 380–83
   tolerance, 356
   types of, 358–63
Capacitor-start motor, 546
Carbon-composition resistor, 66
Carbon-zinc cells, 226–27
Caustic soda cells, 229

Cells
    alkaline, 228
    ampere-hour rating of, 225
    carbon-zinc, 226–27
    caustic soda, 229
    defined, 57–58, 224
    fuel, 229
    lead-acid, 225–26
    lithium, 228
    mercury, 228
    nickel-cadmium, 228
    nickel-iron, 228
    nuclear, 230
    primary, 225
    reserve, 229
    secondary, 225
    solar/photovoltaic, 229
    wet versus dry, 225
    zinc chloride, 227
CEMF (counter-electromotive force), 327
Center frequency, 442
Characteristic impedance, 240–41
Charge. *See* Electrical charge
Charge carriers, 8
Charged particles, 4
Chassis ground, 134–35
Christmas tree lights, as series circuit, 143
Circuit breakers
    role of, 244–45
    as series circuit, 142
Circuits. *See* Electrical circuits; under
        type of
Circular mil, 232
Clamping ammeters, 108–9
Coefficient of coupling, 507
Commercial power distribution, 518
Commutator, 540–41
Compound dc motor, 544, 545
Compounds, 3
Conductance, 155
Conductance formula, 155
Conductors, 8:
    ohmmeters used for testing, 106–7
Conservation of charge, 349
Contactors, 72, 269–70
Continuity, 106
Core materials, coupling and, 507
Corona discharges, 14
Cosine function, 392–96
Coulomb, 11
Counter-electromotive force (CEMF), 327
Coupling:
    capacitors and, 378
    coefficient of, 507
Cross-sectional area, resistance and, 35
Crystalline structure, 31
Current:
    alternating, 57, 286–315
    control device, 60
    defined, 29
    direct, 57
    energizing, 516–17
    formula for, 32, 110

formula wheel for calculating, 89
in gases, 31
Kirchhoff's Current Law, 151, 216
leakage, 242
in liquids, 31
measuring, 104–5
in parallel circuits, 150–54
in parallel *RC* circuits, 423–25
in parallel *RL* circuits, 417–18
path, 58–59
in series circuits, 117–19
in solids, 30–31
symbol for, 29
total reactive current in parallel *RLC*
        circuits, 428–31
unit for, 32
Cutoff frequency, 443–44, 454
Cutting the signal, 461
Cycle, sine wave, 289–90

D

Damped oscillation, 463
dc. *See* Direct current
Decade, 447
Decibel, 448
    levels based on input and output
        voltages, 449–50
Decoupling, capacitors and, 378
Degrees, in cycle time, 289
Delta connection, 555
Demagnetization, 265
Dielectric:
    constant, 353–54
    defined, 347
    material, 360–62
    strength, 242–43, 357
    thickness, 353
Differentiation:
    applications of, 502
    Fourier Series and, 501
Differentiation, *RC*:
    electrical analysis of the waveform,
        497–98
    relationship between pulse period,
        circuit transient time, and, 498
    visual analysis of the waveform,
        498–500
Differentiation, *RL*:
    electrical analysis of the waveform, 494
    relationship between pulse period,
        circuit transient time, and, 495
    visual analysis of the waveform, 495–96
Differentiator:
    defined, 494
    RC 496
    relationship between integrator
        waveforms and, 501
    RL, 494
Digital multimeters (DMMs), 96, 99–100,
        188, 356
DIP (double in-line package) switches, 70
Direct current (dc), 57:
    electric motors, 544–45

inductive opposition to, 327–33
open, 378
power supplies, 60–61
short, 340
working voltage, 356–57
Divisions, analog multimeter, 97–99
Domains, 260
Duty cycle, 485–86

E

Earth ground, 135
Edison, Thomas, 228, 518
Efficiency, 43
Electrical charge:
    charged and uncharged particles, 4
    induced, 12
    law of charges, 4
    in objects, 8–13
    quantities and units, 271–77
    static, 11, 12–13
    symbol for, 11
    transfer of, 11–12
    unit of, 11
Electrical points, 101
Electrical potential, 28
Electrical shocks, 35
Electrical symbols and diagrams, 74–76
Electric circuits:
    See Parallel circuits; Series circuits;
        Series-parallel circuits
    components of, 56
    current control device, 60
    current path, 58–59
    equivalent, 154
    load, 59
    overcurrent protection device, 60
    terminal pairs and viewing of, 196
    voltage source, 57–58
Electric field:
    defined, 13
    intensity, 14–15
    lines of force, 13–14
Electric motors:
    dc, 544–45
    horsepower, 536
    principles of, 536–44
    relationship between horsepower,
        13
    torque, and rpm, 538–40
    requirements of, 542–44
    revolutions per minute (rpm), 537
    single-phase ac, 545
    single-phase induction/squirrel-cage,
        541, 545–46
    terminology, 540–41
    three-phase ac, 546–47
    torque, 536–37
Electrolyte, 31
Electrolytic capacitors, 362–63
Electromagnetic:
    energy, 16–19
    progagation of waves, 279–81

radiation, 17
waves and atoms, 18
Electromagnetic interference (EMI), 328
Electromagnetism:
See also Magnetism
defined, 266
devices, 268–71
differences between permanent magnets and electromagnets, 266–67
relays, 269–70
solenoids, 268–69
Electromotive force, 28
Electron beam, 8
Electronic color code, 62, **62**
Electrons:
charged, 4
defined, 2
excited, 18
free, 6
spin of, 256–57
Electrostatic induction, 348–49
Elements:
defined, 3
partial list of, **3**
Energizing current, 516–17
Energy:
conversion, 27
defined, 27
joules, 27
symbol for, 27
Energy storage, capacitors and, 379–80
Engineering notation, 48
EPROM, 19
Equivalent circuits, 154, 198
use of Thevenin, 202–6
Eutectic alloy, 239
Excited atoms/electrons, 18
Exponential curves, 331–32
Exponents:
key on scientific calculators, 88
negative, 45
positive, 44–45
zero, 44

F

Farad, 350
Ferrite, 322
Ferromagnetic materials, 257
Field coil, 540
Field intensity, 14–15, 276–77
Film-type resistor, 67
Filters:
See also Pass filters, *RC* and *RL*
band-pass, 446
band-reject, 447
Bode plot, 451
decibel, 448–50
high-pass, 446
logarithmic scales, 447
logarithms, 447
low-pass, 446
purpose of, 445–46
rolloff, 450–51

Flux:
defined, 238
density, 275–76
in magnetism, 272–73
symbol for, 272
unit of, 273
Flux lines, 259
rules of magnetic, 262
Foot-pound:
defined, 26
formula for converting to newton-meters, 27
Formula wheel, 88–89
Fourier Series, 441, 501
Franklin, Benjamin, 348
Free electrons, 6
Frequency
See also Pass filters, *RC* and *RL*
ac wave, 300–301, 302–3
audio, 301
bands, 305–6, 441–44
center, 442
cutoff, 443–44, 454
electromagnetic energy wave, 17
filters, 378, 445–52
fundamental, 440–41
harmonic, 440
in parallel *RC* circuits, 426
in parallel *RL* circuits, 422
in parallel *RLC* circuits, 431
radio, 301
resonant, 462–77
resonant, in parallel *RLC* circuits, 431
resonant, in series *RLC* circuits, 414–16
-response curves, 443–44
in series *RC* circuits, 408
in series *RL* circuits, 404
Frequency filters, capacitors as, 378
Fuel cells, 229
Function switches, 97
Fundamental frequency, 440–41
Fuses:
description of, 72–74
ohmmeters used for testing, 107
as series circuit, 142

G

Gain, 449
Gases, current in, 31
Ground-fault circuit interrupter (GFCI), 247, 249
Grounding wire, 246–47
Ground-referenced voltage readings, 131–35
effects of opens and shorts on, 184–87

H

Half-power points, 443
Harmonic frequency, 440
Harmonics, 440–41
Henry, 322

High-pass filter, 446
series *RC* circuit, 460–61
series *RL* circuit, 456–58
Holes, 6
Horsepower, 42
relationship between torque, rpm and, 538–40
relationship between watts and, 536
Horseshoe magnets, 264
Hot wire, 245
Hydrogen, 2–3
Hypotenuse of a right triangle, 391

I

Ideal device/circuit, 197
Ideal voltage source, 57, 197–98
Impedance, 206
characteristic, 240–41
matching, 207, 514–16
in parallel *RC* circuits, 425–26
in parallel *RL* circuits, 418–22
in parallel *RLC* circuits, 431
reflected, 513–14
of resonant frequency in series *RLC* circuits, 416
in series *RC* circuits, 406–7
in series *RL* circuits, 400–402
in series *RLC* circuits, 411
Induced charge, 12
Induced voltage, 277–81
Inductance:
defined, 320–21
mutual, 506–8
opposition to ac current, 334–40
opposition to dc current, 327–33
in parallel *RL* circuits, 422
in series *RL* circuits, 404
symbol for, 321
Inductive kick, 327
Inductive reactance, 334–36
Inductors:
defined, 322
in parallel, 325–26
physical characteristics of, 323–24
in series, 324–25
types of, 322–23
unit of, 322
Infinity, 100, 107
Infrared waves, 18–19
Input, 196
Instantaneous polarity, 525–26
Instantaneous values, 299
Insulation testers, 109
Insulators, 8, 242–43
Integration:
applications of, 502
Fourier Series and, 501
Integration, *RC*
electrical analysis of the waveform, 492
relationship between pulse period, circuit transient time, and, 492
visual analysis of the waveform, 492–93

Integration, *RL*
electrical analysis of the waveform, 489
relationship between pulse period,
circuit transient time, and, 489
visual analysis of the waveform, 489–91
Integrator:
defined, 488
RC, 491
relationship between differentiator
waveforms and, 501
RL, 488
Internal resistance
defined, 198–99
determining the value of, 199–201
International numerical prefixes
defined, 44
engineering notation, 48
scientific notation, 44–47
Intrinsic polarity, 125
Ionizing radiation, 19
Ions:
defined, 6
negative, 6
positive, 6
Isolation feature of transformers, 519

J

Jacks, 97
Joules, 27

K

Kilo-circular-mil (kcmil) system,
233–35
Kilowatt-hour meters, 109
Kilowatt-hours, 40–41
formulas for, 41
Kirchhoff's Current Law, 151, 216
Kirchhoff's Voltage Law, 120, 216–17,
369–71
Knife switch, 69

L

Lagging wave, 309–11
Lasers, principle of, 18
Latent heat, 560
Law of charges, 4
Lead-acid cells, 225–26
Leading wave, 309–11
Leakage current, 242
Leakage resistance, capacitor, 357–58
Length, resistance and, 36
Lightning, 14–15, 31
Limit switches, 71–72
Line losses, 58
Liquids, current in, 31
Lithium cells, 228
Load, 59
Logarithmic curves, 331–32
Logarithmic scales, 447
Logarithms, 447

Low-pass filter, 446
series *RC* circuit, 458–60
series *RL* circuit, 455–56

M

Magnetic:
attraction, 262
domains, 260
fields, 258–59
flux lines, 259
force, 256
horseshoe, 264
materials, 257–58
moment, 256–57
polarity and physical forces, 263–64
poles, 260–61
repulsion, 262–63
retentivity, 264
rules of magnetic flux, 262
waves, progagation of, 279–81
Magnetism, 16
See also Electromagnetism
demagnetization, 265
field intensity, 276–77
flux density, 275–76
flux in, 272–73
induced voltage, 277–81
permeability in, 277
quantities and units, 271–77
reluctance in, 273–74
residual, 264–65
saturation, 265
Magnetomotive force (mmf),
273
Magnets:
bar, 264
development of, 260
differences between permanent magnets
and electromagnets, 266–67
poles, 260–61
retentivity, 264
shapes of, 264
Magnet wire, 236
Magnification factor, 466
Maximum Power Transfer Theorem,
206–7, 514
Maximum value, 64
Maxwell, James, 258
MCM.　*See* Kilo-circular-mil (kcmil)
system
Measuring current, 104–5
Measuring resistance, 105–7
Measuring voltage, 100–103
Meggers, 109
Mercury cells, 228
Mercury switches, 72
Meter loading, effects of, 187–91
Meter movement, 96
Meters:
See also Multimeters; Ohmmeters
clamping ammeters, 108–9
insulation testers/meggers, 109

kilowatt-hour, 109
panel, 107–8
wattmeters, 109
Micro, symbol for, 48
Micro switches, 71
Minimum value, 64
Modulation, 280
Molecules, 3
Multimeters, 95
analog, 96, 97–99, 188
defined, 96
digital, 96, 99–100, 188
parts of, 97
setting up, 97
used to measure ac voltage, 295
Multiple-tap transformers, 522–23
Multiple-winding transformers, 521–22
Mutual inductance, 506–8

N

Napier, John, 331
National Electrical Code (NEC), 230
National Fire Protection Association, 230
Negative charge, 4
Negative ions, 6
Negative voltage readings, 101
Networks, 209–10
Neutral charge, 4
Neutral wire, 245
Neutrons:
defined, 2
uncharged, 4
Newton-meter:
defined, 27
formula for converting foot-pounds to,
27
Nickel-cadmium cells, 228
Nickel-iron cells, 228
Noise, 238, 290
electromagnetic interference (EMI), 328
radio frequency interference (RFI), 328
Nominal value, 64
Nuclear cells, 230
Nucleus of an atom, 2

O

Objects, charged, 8–13
Off-scale readings:
analog, 99
digital, 100
Ohm, 34
Ohm-centimeters, 37
Ohmic value, 62–63
Ohmmeters:
symbol for, 105
testing capacitors with, 382–83
testing conductors with, 106–7
testing fuses with, 107
testing switches and relays with, 107

Ohm's Law:
    calculating, 82–83
    formula for, 34, 82
    for magnetism, 274
    for total values, 117
    value of internal resistance and,
        199–200
    voltage drops and, 120
Opens:
    ac, 340
    dc, 378
    defined, 141
    effects on ground-referenced voltage
        readings, 184–87
    in parallel circuits, 167
    in series circuits, 141
    in series-parallel circuits, 182–87
Opposite side of a right triangle, 391
Oscillation, 462–63
Oscillators, 290, 477
Oscilloscopes, 291–92, 295–97
Output, 196
Overcurrent protection device, 60
Overload, blown fuses and, 73

P

Panel meters, 107–8
Parallel circuits:
    capacitors in, 367–69
    conductance formula, 155
    current flow direction in, 208
    current in, 150–54
    defined, 150
    formula for equal branches, 157
    inductors in, 325–26
    multiple voltage sources in, 162–63
    power in, 160–62
    practical examples of, 169–70
    product-over-sum formula, 156
    resistance in, 154–58
    RIPE table in, 163–66
    rule for parallel equivalent resistance,
        158
    shorts and opens in, 166–69
    voltage in, 158–60
Parallel Circuit Visual Recognition Rule,
    150
Parallel Current Rule, 151
Parallel Power Rule, 160–61
Parallel *RC* circuits:
    capacitance and, 427
    current in, 423–25
    frequency and, 426
    impedance and phase-angle formulas
        for, 425–26
    resistance and, 427
Parallel Resistance Rule, 154–55
Parallel *RL* circuits:
    current in, 417–18
    frequency and, 422
    impedance and phase-angle formulas
        for, 420–22
    impedance in, 418–22

inductance and, 422
    resistance and, 422–23
Parallel *RLC* circuits, 427
    frequency and, 431
    impedance in, 431
    total reactive current in, 428–31
Parallel Voltage Rule, 158
Particles:
    charged and uncharged, 4
    defined, 2
Pass filters, *RC* and *RL*, 452–54
    applications of, 461–62
    cutoff frequency, 454
    high-pass filter (series *RC* circuit),
        460–61
    high-pass filter (series *RL* circuit),
        456–58
    low-pass filter (series *RC* circuit),
        458–60
    low-pass filter (series *RL* circuit),
        455–56
Peak-to-peak values, 297–99
Peak voltage, 293
Period, 301–3
    of a pulse, 485
    relationship between phase angle and,
        311–14
Permeability in magnetism, 277
Phase:
    angle, 307–9
    defined, 306–7
    dot notation, 525–26
    effect of transformer loading on phase
        angle, 516–17
    leading and lag, 309–11
    relationship between period and,
        311–14
Phase shifting, capacitors and, 379
Phase splitting, 523–25
Phasor diagrams, 337, 390
Phasors, 337
Phasor sum, 390
Photovoltaic cells, 229
Plates, 347, 352–54
Plugs and receptacles, 249–50
Polarity:
    defined, 9–10
    instantaneous, 525–26
    intrinsic, 125
    in series circuits, 125–26
Polarized plugs, 249–50
Poles, switch, 70
Positive charge, 4
Positive ions, 6
Potentiometers, 67
Power:
    ac power cycle and problems, 550–55
    apparent, 548–49
    commercial distribution, 518
    defined, 38
    efficiency, 43
    factor, 555
    formula for, 38, 109

horsepower, 42
    kilowatt-hours, 40–41
    loss in transmission lines, 241, 518
    of ten, 47, **47**
    in parallel circuits, 160–62
    primary and secondary, 512–13
    as a product of voltage and current, 32,
        40, 85–87
    reactive, 549–50
    in series circuits, 123–24
    three-phase, 555–58
    triangle, 548, 554
    true, 299, 549
    unit of, 38
    watts, 38–39
Power ratings:
    British thermal unit (Btu), 559
    resistor, 66
    ton, 560
Power supplies:
    dc, 60–61
    in electronic equipment, 518–19
Power supply filters, capacitors as, 377–78
Practical device/circuity, 197
Practical voltage source, 57, 197–98
Primary cells, 225
Primary power, 512–13
Primary winding, 506–7
Prime movers, 537
Probes, 97
Product-over-sum formula, 156
Proportionality, voltage drops and, 122
Protons
    charged, 4
    defined, 2
Pulses:
    active state of, 485
    average voltage of, 486–88
    defined, 484
    duty cycle, 485–86
    period of, 485
    repetition rate (PRR), 485
    resting state of, 485
    used to test circuits, 500–502
Push-button switches, 70–71
Pythagorean Theorem, 391–92

Q

Quality (Q) factor of a coil, 338–39
    magnification factor, 466
    selectivity and, 472
Quantum mechanics, 33

R

Radial leads, 358
Radical, 89
Radio frequency (RF), 301
Radio frequency interference (RFI),
    328
Range switches, 97
Reactive power, 549–50
Receptacles, 249–50
Reflected impedance, 513–14

Relays:
  electromagnetism and, 269–70
  ohmmeters used for testing, 107
Reluctance in magnetism, 273–74
Repulsion-start induction motor, 545
Reserve cells, 229
Residential wiring, 245–49
Residual magnetism, 264–65
Resistance:
  defined, 33
  electrical shocks and, 35
  factors of, 35–37
  formula for, 110
  formula wheel for calculating, 89
  measuring, 105–7
  in parallel circuits, 154–58
  in parallel RC circuits, 427
  in parallel RL circuits, 422–23
  in series circuits, 116–17
  in series RC circuits, 409
  in series RL circuits, 404
  unit of, 34
Resistivity:
  of metals, **37**
  symbol for, 37
  unit of, 37
Resistors:
  defined, 61
  electronic color code, 62, **62**
  high-precision, 65
  less than 10 ohms, 65
  ohmic value, 62–63
  power rating, 66
  standard values, 65
  types of, 66–67
  value and tolerances, 64–65
Resolution, digital multimeter, 99–100
Resonance, defined, 463
Resonant band-pass filter:
  applications of, 477
  parallel, 476
  series, 475–76
Resonant band-reject filter:
  applications of, 477
  parallel, 477
  series, 476–77
Resonant frequency, 462–77
  defined, 463
  energy storage in a resonant circuit, 464–67
  parallel equivalent resistance, 469–71
  in parallel RLC circuits, 431
  reasons for differences in low and high impedance, 467–69
  relationship between Q, bandwidth and, 472–75
  selectivity, 471–72
  in series RLC circuits and, 414–16
  transfer of reactive energy, 463–64
Resting state of a pulse, 485
Retentivity, 264
Revolutions per minute (rpm), 537, 538
Rheostats, 67

Right triangle, 391
Ringing, 468
RIPE table:
  in parallel circuits, 163–66
  in series circuits, 128–31
  in series-parallel circuits, 177–82
Rms (root-mean-square) voltage, 293
Rocker switches, 70
Rolloff, 450–51
Rotary switches, 71
Rotor, 540
Rule of Eight, 6

S

Salient poles, 546
Sampling, digital multimeter, 100
Saturation, magnetic, 265
Scales, analog multimeter, 97–99
Schematic diagrams, 74–76
Scientific notation, 44–47
  converting from standard notation to, 45–46
  converting to standard notation from, 46–47
  negative exponent, 45
  positive exponent, 44–45
  power of ten, 47, **47**
  purpose of, 44
  zero exponent, 44
Secondary cells, 225
Secondary power, 512–13
Secondary winding, 506–7
Selectivity, 471–72
Semiconductors, 31
Series ac motor, 545
Series aiding, 126
Series circuits:
  capacitors in, 364–67
  current flow direction in, 208
  current in, 117–19
  defined, 116
  ground-referenced voltage readings, 131–35
  inductors in, 324–25
  multiple voltage sources in, 126–27
  polarity in, 125–26
  power in, 123–24
  practical examples of, 142–43
  resistance in, 116–17
  RIPE table in, 128–31
  shorts and opens in, 136–42
  voltage in, 119–23
Series Circuit Visual Recognition Rule, 116
Series Current Rule, 117–18
Series dc motor, 544–45
Series opposing, 126
Series-parallel circuits:
  defined, 176
  loading effects of meters, 187–91
  RIPE table in, 177–82
  shorts and opens in, 182–87
  simplification of, 176–77

Series Power Rule, 123
Series RC circuits, 405
  See also Pass filters, RC and RL
  capacitance and, 409
  frequency and, 408
  impedance in, 406–7
  resistance and, 409
  voltage drops in, 407–8
Series Resistance Rule, 116
Series RL circuits:
  See also Pass filters, RC and RL
  ELI the ICE mnemonic, 399–400
  frequency and, 404
  impedance in, 400–402
  inductance and, 404
  resistance and, 404
  theta (phase angle), 403–4
  voltage drops in, 402
Series RLC circuits:
  impedance in, 411
  resonant frequency and, 414–16
  theta positive and negative, 411–13
  total reactance and, 410–11
  voltage drops in, 413–14
Series Voltage Rule, 120
Shaded-pole motor, 546
Shells, 2
Short(s):
  ac, 378
  blown fuses and, 73
  dc, 340
  defined, 136
  effects on ground-referenced voltage readings, 184–87
  in parallel circuits, 166–67
  in series circuits, 136–41
  in series-parallel circuits, 182–87
  voltmeters used to find, 138
Shunt dc motor, 544, 545
Siemen, 155
Sine function, 290, 392–96
Sine wave:
  cycle, 289–90
  graph of, 288–92
  leading and lag, 309–11
  peak-to-peak values, 297–99
  peak voltage, 293
  period, 301–3
  phase, 306–14
Single-phase ac electric motors, 545
Single-phase induction motors, 541, 545–46
  capacitors and, 379
Slide switches, 70
Slip rings, 540, 541
Solar cells, 229
Soldering, 238–40
Solenoids, 268–69
Solids, current in, 30–31
Speed of light, 18
Spike, voltage, 327
Split-phase motor, 546
Square and square root tip, 89

Squirrel-cage induction motor, 541, 545–46
    capacitors and, 379
Static charge/electricity:
    benefits of, 12
    description of, 11
    hazards of, 13
Stator, 540
Step-down transformers, 511
Step-up transformers, 511
Superconductors, 33–34
Superposition Theorem, 213–17
Surge, blown fuses and, 73
Surge suppressors, 251, 328
Switches:
    description of, 69–72
    ohmmeters used for testing, 107
    as series circuit, 142
    voltmeters used for checking, 142
Symbols and diagrams, electrical, 74–76
Synchronization, 309
Synchronous motor, 547
Synchronous speed, 544, 547

T

Tangent function, 392–96
Tank circuit, 469
Temperature, resistance and, 36–37
Temperature coefficient, 37
Terminal pairs, 196
Tesla, Nikolai, 518
Test leads, 97
Thermal runaway, 37
Thermionic emission, 8
Theta, 391
    in series *RL* circuits, 403–4
    in series *RLC* circuits, 411–13
Thevenin's Theorem, 201–7
    for calculating bridge voltage and current, 211–12
Thevenin voltage, 202
Three-phase ac motors, 546–47
Three-phase power, 555–58
Throws, switch, 70
Time constant, 332, 371–73
Time-delay fuse, 74
Timing, capacitors and, 378
Toggle switches, 70
Tolerances, 64–65, **64**
    capacitor, 356
Ton rating, 560
Torque, 536–37
    relationship between horsepower, rpm and, 538–40
Transformer(s):
    applications for, 518–21
    auto-, 526–28
    basic characteristics, 508–17
    effect of loading on phase angle, 516–17
    function of, 509
    impedance matching, 514–16
    instantaneous polarity, 525–26

isolation feature of, 519
    multiple-tap, 522–23
    multiple-winding, 521–22
    phase splitting, 523–25
    primary and secondary power, 512–13
    reflected impedance, 513–14
    step-down, 511
    step-up, 511
    troubleshooting, 529–30
    tuned coupling, 519–21
    turns ratio, 509–10
    variations, 521–29
    voltage ratio, 510–11
Transients, 327–30
Transient time, 328, 373
Transition, wave, 303
Trigonometry:
    arc function, 396–98
    defined, 391
    functions, 392–96
    Pythagorean Theorem, 391–92
    right triangle, 391
    using a scientific calculator in, 398–99
True power, 299, 549
Tune circuit, 464
Tuned coupling transformers, 519–21
Turns ratio, 509–10

U

Ultraviolet light, 19
Uncharged particles, 4

V

Vacuum-tube-volkt-milliammeter (VTVM), 188
Valence shell, 5–6
Value, resistor, 64–65
Volt, 28
Voltage:
    average, 295
    back, 327
    breakdown, 356–57
    divider formula, 122
    dividers, 142
    electrical potential and, 28
    electromotive force, 28
    formula for, 29, 110
    formula wheel for calculating, 89
    induced, 277–81
    instantaneous values, 299
    Kirchhoff's Voltage Law, 120, 216–17, 369–71
    measuring, 100–103
    multiple voltage sources in parallel circuits, 162–63
    multiple voltage sources in series circuits, 126–27
    in parallel circuits, 158–60
    peak, 293
    ratio, 510–11

regulation and use of Thevenin's Theorem, 206
    regulator, 57, 206
    rms (root-mean-square) voltage, 293
    in series circuits, 119–23
    source, 57–58
    spike, 327
    transients, 327–30
    unit of, 28
Voltage drop:
    defined, 120
    Ohm's Law and, 120
    proportionality and, 122
    in series *RC* circuits, 407–8
    in series *RL* circuits, 402
    in series *RLC* circuits, 413–14
Voltage readings, ground-referenced, 131–35
Voltmeters
    symbol for, 101
    used to check capacitors, 382
    used to check switches, 142
    used to find shorts, 138
Volt-ohm-milliameters (VOMs), 96, 188

W

Wattmeters, 109
Watts, 38–39
    formula for relationship between horsepower and, 42
    formula for relationship between kilowatts and, 41
    relationship between horsepower and, 536
    symbol for, 38
Watt's Law:
    calculating, 84–85
    formula for, 40, 84
    power as a product of voltage and current and, 40, 85–87
Waveforms, 440–41
    See also Differentiation; Integration
Wavelength, 17, 304–5
    bands, 305–6
Waves
    See also Sine wave
    electromagnetic, 18
    infrared, 18–19
    ultraviolet, 19
Weber, 273
Westinghouse, George, 518
Wheatstone bridge, 210
Winding, primary and secondary, 506–7
Wire:
    ampacity, 230–31
    circular mil, 232
    ground-fault circuit interrupter (GFCI), 247, 249
    grounding, 246–47
    hot, 245
    magnet, 236
    neutral, 245
    solid versus stranded, 236

Wire gauge systems, 233–35
Wire-wound resistor, 66
Wiring:
    ground-fault circuit interrupter (GFCI),
        247, 249
    parallel circuits and interior, 169–70
    plugs and receptacles, 249–50

    residential, 245–49
    surge suppressors, 251
Work:
    defined, 26
    foot-pound, 26
    formula for, 26
    newton-meter, 27

    symbol for, 27, 38
Wye connection, 555–56

Z

Zinc chloride cells, 227